T0137331

Springer Tracts in Advanced Robotics 117

Editors

Prof. Bruno Siciliano
Dipartimento di Ingegneria Elettrica
e Tecnologie dell'Informazione
Università degli Studi di Napoli
Federico II
Via Claudio 21, 80125 Napoli
Italy
E-mail: siciliano@unina.it

Prof. Oussama Khatib
Artificial Intelligence Laboratory
Department of Computer Science
Stanford University
Stanford, CA 94305-9010
USA
E-mail: khatib@cs.stanford.edu

More information about this series at http://www.springer.com/series/5208

Jean-Paul Laumond · Nicolas Mansard
Jean-Bernard Lasserre
Editors

Geometric and Numerical Foundations of Movements

Springer

Editors
Jean-Paul Laumond
LAAS-CNRS
Toulouse
France

Jean-Bernard Lasserre
LAAS-CNRS
Toulouse
France

Nicolas Mansard
LAAS-CNRS
Toulouse
France

ISSN 1610-7438 ISSN 1610-742X (electronic)
Springer Tracts in Advanced Robotics
ISBN 978-3-319-84680-4 ISBN 978-3-319-51547-2 (eBook)
DOI 10.1007/978-3-319-51547-2

Printed on acid-free paper

This Springer imprint is published by Springer Nature
The registered company is Springer International Publishing AG
The registered company address is: Gewerbestrasse 11, 6330 Cham, Switzerland

Series Foreword

Robotics is undergoing a major transformation in scope and dimension. From a largely dominant industrial focus, robotics is rapidly expanding into human environments and vigorously engaged in its new challenges. Interacting with, assisting, serving, and exploring with humans, the emerging robots will increasingly touch people and their lives.

Beyond its impact on physical robots, the body of knowledge robotics has produced is revealing a much wider range of applications reaching across diverse research areas and scientific disciplines: biomechanics, haptics, neurosciences, virtual simulation, animation, surgery, and sensor networks among others. In return, the challenges of the new emerging areas are proving an abundant source of stimulation and insights for the field of robotics. It is indeed at the intersection of disciplines that the most striking advances happen.

The *Springer Tracts in Advanced Robotics (STAR)* is devoted to bringing to the research community the latest advances in the robotics field on the basis of their significance and quality. Through a wide and timely dissemination of critical research developments in robotics, our objective with this series is to promote more exchanges and collaborations among the researchers in the community and contribute to further advancements in this rapidly growing field.

The volume by J.P. Laumond, N. Mansard and J. Lasserre provides a broad edited collection on movement analysis, which is the outcome of the workshop "Geometric and Numerical Foundations of Movements" held at LAAS-CNRS in Toulouse in November 2015. Following a tutorial presentation of the problem, the contents are effectively organized into four main sections: geometry, action and movement; numerical analysis and optimization; foundation of human movement; robot motion generation. The unique feature of the volume stands in its inherent multidisciplinary character spanning robotics, control theory, neurosciences and mathematics. New optimization techniques are presented, based on recent results from real algebraic geometry, which shed new light for advancements on motion research.

Rich of results by the most active teams in the field, this volume constitutes a very fine addition to STAR!

Naples, Italy
November 2016

Bruno Siciliano
STAR Editor

Preface

This book aims at gathering roboticists, control theorists, neuroscientists, and mathematicians, in order to promote a multidisciplinary research on movement analysis. It follows the workshop "Geometric and Numerical Foundations of Movements" held at LAAS-CNRS in Toulouse in November 2015.[1] Its objective is to lay the foundations for a mutual understanding that is essential for synergetic development in motion research. In particular, the book promotes applications to robotics—and control in general—of new optimization techniques based on recent results from real algebraic geometry.

Starting from a robotics perspective, the generation of goal oriented motion for robots obeys classically to a two-step paradigm. The first step is the planning, where the typical problem is to find a geometric path that allows the robot to reach the desired configuration starting from the current position while ensuring obstacle avoidance and enforcing the satisfaction of kinematic constraints. Motion planning lays its grounding on the decidability properties of this classic geometrical problem. Moreover, the traditional approaches that are used to find solutions rely on the global probabilistic certainty of the convergence of path construction stochastically sampled in the configuration space. The second step of motion generation is the control, where the robot has to perform the planned motion while ensuring the respect of dynamical constraints. Motion control seeks primarily for local controllability or at least the stability of the motion. The basic instances of these problems have long been tackled using local state-space control. However, the typical nonlinearity of the dynamics, together with the non-controllability of its linearization, leads more and more solutions to resort to model preview control. These methods allow to predict the outcome of a control strategy in a future horizon and to improve it accordingly, usually by using numerical optimizations which take

[1]The workshop took place in the framework of the Anthropomorphic Motion Factory launched by the European project ERC-ADG 340050 Actanthrope (2014–2018) devoted to exploring the computational foundations of anthropomorphic action. The workshop was also supported by the European project ERC-ADV 666981 Taming (2015–2019) and the French ANR project Entracte (2014–2017).

into account the safety constraints and efficiency intents. However, since few years, the improvement of computational capabilities and numerical algorithms allows more and more to deal with complex dynamical systems and for longer horizons. This allows these approaches to untighten the local nature of their applications and progressively start wider explorations of their reachable space. This evolution brings us to the question of the rising overlap between planning and control. Today, most planning problems would take too much time to be solved online with numerical approaches. Does that imply that the generation of motion will theoretically never be free of the necessity of a prior planning? Or on the contrary, is planning only a numerical issue?

All these questions are also addressed in Life Sciences. Indeed, movement is a fundamental characteristic of living systems. How roboticists may benefit from neurophysiologist know-how and vice versa? System modelling is one way to gather both communities.

While actions operate in a physical space, motions begin in a motor control space. For robots and living beings, the link between actions expressed in the physical space and motions originated in the motor space, turns to geometry in general and, in particular, to linear algebra. Geometric control theory and numerical analysis highlight two complementary perspectives on optimal human and humanoid motion. Among all possible motions performing a given action, optimization algorithms tend to choose the best motion according to a given performance criterion. Optimal motions then appear as plausible action signatures.

How to express actions in terms of motions? How to face the computational complexity of bridging the 3D physical space with the high-dimension control space? How to reveal movement synergies? How to account for the underactuation of the locomotion? What optimality criterion underlies a given action? All these questions open challenging issues to direct and inverse optimal control, with recent developments in polynomial optimization and real algebraic geometry.

The multidisciplinary perspective on movement analysis is reflected in the book by its table of content. After a specific chapter introducing the rational above, the chapters are gathered within four main parts addressing respectively mathematics (Part Geometry, Action and Movement), applied mathematics (Part Numerical Analysis and Optimization), life science (Part Foundations of Human Movement), and robotics (Part Robot Motion Generation).

Editing a book with a multidisciplinary perspective is not an easy task. We thank all the authors for their effort in making their own research field accessible to others and all the reviewers who helped us in reaching this objective.

Toulouse, France
August 2016

Jean-Paul Laumond
Nicolas Mansard
Jean-Bernard Lasserre

Contents

Part III Foundation of Human Movement

Part IV Robot Motion Generation

Robot Motion Planning and Control: Is It More than a Technological Problem?

Mehdi Benallegue, Jean-Paul Laumond and Nicolas Mansard

Abstract The generation of motion for robots obeys classically to a two-step paradigm. The first step is the planning, where the typical problem is to find a geometric path that allows the robot to reach the desired configuration starting from the current position while ensuring obstacle avoidance and enforcing the satisfaction of kinematic constraints. Motion planning lays its grounding on the decidability properties of this classic geometrical problem. Moreover, the traditional approaches that are used to find solutions rely on the global probabilistic certainty of the convergence of path construction stochastically sampled in the configuration-space. The second step of motion generation is the control, where the robot has to perform the planned motion while ensuring the respect of dynamical constraints. Motion control seeks primarily for local controllability or at least the stability of the motion. The basic instances of this problems have long been tackled using local state-space control. However, the typical nonlinearity of the dynamics, together with the non controllability of its linearization, lead more and more solutions to resort to model predictive control. These methods make it possible to predict the outcome of a control strategy in a future horizon and to improve it accordingly, commonly by using numerical optimizations which take into account the safety constraints and efficiency intents. However, since few years, the improvement of computational capabilities and numerical algorithms allows more and more to deal with complex dynamical systems and for longer horizons. This allows then these approaches to untighten the local nature of their applications and progressively start wider explorations of their reachable space. This evolution brings us to the question of the rising overlap between planning and control. Today, most planning problems would take too much time to be solved online with numerical approaches. Does that imply that the generation of

M. Benallegue (✉)
National Institute of Advances Industrial Science and Technology (AIST), Tsukuba, Japan
e-mail: mehdi@benallegue.com

J.-P. Laumond · N. Mansard
LAAS-CNRS, Université de Toulouse, Toulouse, France
e-mail: jpl@laas.fr

N. Mansard
e-mail: nmansard@laas.fr

J.-P. Laumond et al. (eds.), *Geometric and Numerical Foundations of Movements*, Springer Tracts in Advanced Robotics 117, DOI 10.1007/978-3-319-51547-2_1

motion will theoretically never be free of the necessity of a prior planning? Or on the contrary, is planning only a numerical issue?

1 The Classic Paradigm

The first prominent book dedicated to motion generation in robotics was subtitled "Planning and Control" [1]. Under this perspective, this volume theorized a pattern for the production of robot action which is widely established until today. Indeed, later reference books such as "Robot motion planning" [2], "Modelling and control of robot manipulators" [3] and the unescapable "Handbook of robotics" [4], crystallized this scheme and the subsequent dichotomy and scheduling of two major branches of robotics. This scheme can be synthesized into two successive and poorly connected steps. At first the motion planning, which occurs usually offline and allows to find a geometric path avoiding obstacles and kinematic constraints and achieving the required task. After that, the second step controls the robot to actually follow the path and generate the motion in the physical world, guaranteeing at the same time its dynamical feasibility, stability and robustness. This second step generally occurs online, and may resort to alter the geometric path provided by planning step, for example through dynamical filtering.

1.1 Geometric Search and Numerical Control

Although these two steps coexist often in the same motion generation frameworks, they are rooted in different soils and resulted in a deeply historical separation. For instance, planning techniques are mostly algorithmic answers provided by computational geometry to the issue of point-to-point path finding, related to the archetypal "Piano mover's problem" [2]. Indeed, the problem of path planning of a robot in geometrically constrained environment (obstacles, joint limits, auto-collisions) boils down to finding a path for a point in a commonly higher dimensional manifold called configuration space [5]. Furthermore, sufficient conditions were provided to ensure the decidability of point-connection queries in this manifold [6]. However, no general effective solution was produced through these theoretical yet powerful guarantees [7]. Therefore, computer science community had recourse to the weaker certainty provided by stochastic search methods. These methods, derived from Monte-Carlo techniques, are able to find a path between two points lying in a connected component in a finite time. This gave birth to the well known probabilistic algorithms based on sampling in configuration-space: the Probabilistic Road Map (PRM) [8] and Rapidly-exploring Random Trees (RRT) [9]. Afterwards, these paths are commonly improved to minimize a criteria such as length or energy using either probabilistic or deterministic approaches [10, 11].

Subsequently, the classic paradigm moves this path to controllers in order to ensure the achievement of the actual motion. Robot motion control has naturally emerged from the classic control theory through the control of manipulator robots for manufacturing [12]. It is mainly about achieving the desired path while providing several guarantees. The most important of these guarantees is feasibility, meaning that the generated trajectory is possible regarding laws of physics and robot limitations. This commonly requires to augment the path representation in the configuration manifold with a curve in its tangent bundle representing the time-derivative of the trajectory. This augmented space is called phase space or more commonly state space. But sometimes the workspace of the controller includes second order time-derivatives of the configuration if not higher and the space is then named 'control-space' for which not only all possible states of the system are represented, but also all possible controls. Other guarantees are mandatory for the successful achievement of the motion, and include stability and robustness, which allow the achievement in the task disregarding perturbations to the system or errors in the model of the robot dynamics. This constrained and dynamical system makes the generation of trajectories far from reach of purely geometrical methods. Moreover, the dimensional multiplication of the search manifold, compared to the configuration space, explodes the required time and space for stochastic sampling to inconceivable amounts, even for the simplest systems. Therefore the three guarantees, dynamical feasibility, stability and robustness are commonly considered as exclusively delegated to the control, and classic motion planning does not deal with these dimensions. Control, on its side either considers local representations of this space around the paths provided by planning, or more and more often resorts to numerical techniques such as constrained optimization to generate reliable, stable and hopefully efficient trajectories.

Despite the broad general definitions provided above, control is a name given to various kinds of processes. Often, several of them run concurrently on a robot for a same motion. They are usually organized into abstraction layers or levels. The lowest layer deals with the physical world, such as controlling motor currents, facing sometimes complex dynamics but reduced to only atomic parts of the motion and using relatively simple solutions. It runs the fastest control loops, and particularly involves sensor feedback. The highest layers have more global knowledge about the performed motion, action or task, runs complex control algorithms but usually receives lower amounts of inputs from embedded sensors. Every layer uses a model of the layer underneath and constitutes the interface with the upper one.

1.2 *Control as a Layered Scheme: The Example of the Biped Locomotion*

A good example is the generation of walking motion for humanoid robots. This example not only shows the layered nature of control, but also describes how complex tasks are tackled today in robotics. Indeed, humanoid robots have no actuation generating translations and rotations of the body. These can only be produced by using external forces, specifically contacts. The joint actuation allows usually to generate contact forces in infinite ways. However not all of them are able to guarantee the safety of the robot since the dynamics is intrinsically unstable and limited by feasibility constraints.

To tackle these problems, there is a classic scheme involving layers of control. The highest level is the so-called Walking Pattern Generator (WPG). It is the controller which considers a reference in translation and orientation. To guarantee the long term stability, it is almost mandatory for the humanoid to consider the outcome of the control in a sufficiently long future horizon. If the robot is always able to safely stop in the future, the generated motion in the present is also safe. However, this process is usually time-consuming and it is applied only in a simplified version of the dynamics. One famous and broadly used simplification is the *linear inverted pendulum model* also called *cart-table model*, which considers that the robot is a point mass and constrains the center of mass to stay on a horizontal plane [13]. A layer below is the inverse kinematics which is an open-loop process allowing to generate convenient joint trajectories to achieve the reference position of the center of mass. The next level is the stabilizer, it ensures that the desired contact forces are correctly respected, together with the reference posture, but it deals also with short term balance issues such as perturbations recovery. It usually modifies the inverse kinematics outcome. Finally the motor micro controllers constitute the lowest level of control, ensuring in a high-frequency closed-loop control that the desired joint position, velocity and/or torques are correctly tracked.

The example above shows different processes referred to as control. It is interesting to note that the highest level control, the WPG, generates a prediction of the motion on a future horizon in which the upcoming trajectory is *planned*: the proximity of model predictive control with planning is an important object of discussion presented in the next section.

On the other hand, the goal provided to the WPG can be produced either by a controller in the case of visual servoing for example [14], but it is more often the outcome of an authentic planning algorithm as defined earlier. And here we hit one of the limitations of the definitions given above. All along our description of layered control, the planning appears naturally as the actual highest layer of the control, having a global view on the motion to generate and relying on lower layers to achieve the actual trajectory. This aspect, among others constitutes the core of rising semantic collisions between planning and control, sometimes because of what could be considered as misnomers, but more and more often due to an increasing proximity leading to the beginning of a true overlap.

We have exploited in this section the example of legged locomotion. However, similar layered paradigms arise in various types of robot. For example, motion generation on quadcopters use some flatness properties to plan a guide path among obstacles, which is then refined using trajectory optimization then tracked by a complete controller. When embedding manipulation effectors on the quadcopter, an additional layer based on inverse kinematics is also added [15].

2 The Expanding Overlap

2.1 *Is There Any Fundamental Difference Between Planning and Control?*

The haziness between the definitions is maintained by the absence of clear nomenclature stating the classification between planning and control. For example, sometimes open-loop control is referred to as planning, sometimes as motion control. But despite this definition uncertainty, a consensus emerges to state that planning is more than sampling paths in the configuration space, which would be easy to distinguish from control.

Among the aspects blurring the lines lies the increasing ability of planning techniques to compute new solutions fast enough to consider interactive, if not real-time, performances. This is often referred to as motion replanning or reactive planning. Thanks to new techniques, together with faster computers, the sensors feedback could be immediately taken into account and new plans could be computed in a more and more reasonable time.

One more important aspect of overlap is the introduction of velocities, accelerations and forces in trajectory planning, giving what is called *kinodynamic* motion planning [16]. With this paradigm, the dynamic feasibility constraints could be directly taken into account and anticipated for lower lever controller. It had also the benefit to allow a wider range of motions which would not be possible within purely geometric constraints.

In the kinodynamic planning methods, even when the planning approaches use road-map sampling, the path itself is often a result of an optimization scheme, and here is the most important way planning treads on control's toes [17]. Optimization techniques were introduced in planning to enable the planner to generate relevant and efficient trajectories. These methods are also usually called *optimal control*.

Optimal trajectories have been studied since centuries ago. First, early mechanics has stated that a body always move in a way that optimizes a certain cost, giving formulations ranging from Maupertuis's to Hamilton's principles. Then, in the 1900's, the properties of these optimal trajectories were extensively studied, especially for the case of actuated systems, in order to be able to generate efficient motions. Two major principles characterize the optimality of a control system.

One prominent necessary condition for the optimality of a controlled trajectory is the *Pontryagin's maximum principle* (PMP) which gives a local property that has to be valid each instant. The application of this principle allows the generation of trajectories optimizing a given criterion for robots such as redundant manipulators. One other important solution to generate optimal trajectories is the *dynamic programming*, generally using the Hamilton–Jacobi–Bellman (HJB) equation. This gives necessary and sufficient condition of the optimality of a trajectory, but its theoretical consequence allows to build the control minimizing the criterion in all the state space. This powerful approach is usually very computationally costly for nonlinear systems. Therefore, to be applicable, it requires to be approximated, in terms of dynamics or control space.

Despite their similarities, these two principles reveal a fundamental gap, which might be seen as the major clustering between planning and control. Pontryagin's principle only considers the properties of the optimal path itself, without considering the dynamical behavior in its neighborhood. We then obtain one unique path to the goal, that must be carefully tracked as the principle does not reveal how to behave aside the optimal trajectory. In the majority of the cases, if supposing some good properties of the dynamics and the cost function, this is only a technological issue, and as soon as the replanning is fast enough one can achieve closed-loop performances with satisfying performances. However such fast replanning lacks theoretical guarantees of stability, especially for heavily constrained systems. Therefore these approaches are unsafe from a control-theory point-of-view. On the other hand, dynamic programming predicts the dynamics at least in the space around the optimal trajectory. A local controller is then provided when the state is deviated from the optimal trajectory. In general this enables the controller to check the stability of the optimal solution, at least locally.

This consideration may constitute a boundary between pure geometrical planning approaches (even when dealing with dynamic feasibility) and actual motion control, which could be part of planning methods. Combining motion planning and motion control in a single motion generation scheme remains a challenging issue. The elastic strip framework introduced in [18] tends to enable the execution of a previously planned motion in a dynamic environment. Control is here dominated by real-time execution constraints. This general scheme applies also to motion planning for nonholonomic systems [19] as well as to feedback-based motion planning [20] for autonomous mobile manipulation.

To summarize in few words, planning is about reasoning on the global trajectory, without systematically considering the stability of the underlying control scheme, but can become control, if it is able to recompute a trajectory online and if it guarantees the stability of the system regarding perturbations and uncertainties in the state-space.

2.2 Model Predictive Control

Up to today, most controllers, including those for complex robots still generate motion according to instantaneous feedback linearization (e.g. inverse kinematics, inverse dynamics), and heavily rely on geometric paths provided by classic planners. But as seen earlier, the planning [21–23] and control [24, 25] have already started overlapping around the corpus of fast numerical optimization approaches of trajectories on the base of a model of the dynamics. Thanks to these approaches, the controller can see further in the future, but unlike geometric planning, it is generally too costly to optimize over the entire trajectory. The trajectory covers a limited time-horizon, predicting the outcome of a given control, and it has to be recomputed at each instant to update the state on the basis of sensors feedback and to keep predicting far enough in the future. This approach is designated as Model Predictive Control [1] (MPC).

Predictive control is a theoretically well-posed methodology [26]. For instance, it has been shown that if a system is stabilizable, then there exists a finite horizon for which the optimal control is stable. In addition, in most situation, it provides efficient and relevant motions if not optimal regarding a cost function.

Bellman's principle generates intrinsically a control for the whole state space, taking implicitly into account its constraints and geometry. When controlling a system on the base of this principle, we have to explore the state-space. If the dynamics is linear, this is possible thanks to Riccati's equation. If the control space is discretized, standard dynamic programming requires the full exploration of the resulting graph. Otherwise, in the general case of nonlinear non-discretized system, it is generally possible through local modifications of the predicted trajectory. This provides the model predictive control with the ability to produce its own plan at best and to modify it according to the state and the constraints at least. The controller takes the role of the planner, and this with increasing efficiency and independence.

However, the success of this approach is built upon a fundamental shift of paradigm, where the globality of the dynamic principle is traded against local properties, that are well expressed by the naming "differential dynamic programming". In most of the case, working on local deformation of a candidate trajectory would bring the search in a local minimum, and it is in general not possible to characterize the global optimality of the resulting trajectory, but only its local optimality. Mathematically, the trade indeed corresponds to exchanging the curse of dimensionality of dynamic programming against the nonconvexity (hence local property) of a corresponding numerical program. Reformulating the trajectory optimization as a convex numerical program is typically a very nontrivial task. We will also discuss in the second part of this book rising alternatives to solve numerical programs with global algorithms based on positivity certificate. However such alternatives are yet limited to simple systems (low-dimension and sparse).

Aside of the problem of local optimality, one of the main challenges for application of *differential* dynamic programming to complex robots is its computational complexity [27]: a large nonlinear optimization problem needs to be solved online at control frequency. Yet most current optimization solvers result in long optimization times keeping them out of the reach of real-time uses, even on the most powerful processing units. For example, whole body motion generation for humanoid robots implies solver computation times about minutes of computation, while milliseconds of reaction times are necessary on the real robot.

In brief, MPC is theoretically able to discover complex movements, while the recent progresses in computational complexity and numerical mathematics makes it now relevant in many robotic applications. Therefore, one may ask the following question: is the difficulty to generate motions for complex robot due to intrinsic properties lying in the space geometry and the structure of the optimization problem?

---

[1] Sometimes it is called also model *preview* control, but this latter designation seems restrained to linear systems.

Or it is only a technological issue that can be solved with the emergence of new optimization techniques implemented on more powerful computers?

3 A Technological Problem?

3.1 The Room for Improvement

The simplest possible solution for the problem is to wait for better processors and optimization solutions which will allow faster computations. However, the generation of complex motions for a humanoid robot could take hours if not days. Therefore, there is a need of at least four orders of magnitude in the reduction of computation time. And with the end of Moore's law [28], the perspective of such a purely technological revolution is rather uncertain.

In fact, the complexity of nonlinear optimizers comes from two theoretical aspects: the nominal cost of one iteration and the number of iterations [29]. The first aspect depends on the size of the optimization problem (number of variables and constraints). However, recent progress in both numerical mathematics and computer power makes it now possible to perform an iteration in less than a fraction of a second, even for complex robots like humanoids [25, 30]. The second aspect depends on the convexity of the numerical problems. If the problem is convex, tens of iterations are sufficient to converge to the global solution. However, most problems in robotics are not convex, and therefore require thousands of iterations to converge to a (possibly local) minimum [21, 22, 31].

Due to these aspects, the computational cost of a trajectory optimization is variable, not only from a problem to another, but also depending on the initialization of the optimizer. For instance, there are many optimal control problems in humanoid robotics which guarantee that if the initial state is close to the minimum, the cost function will likely be locally convex and the trajectory will be optimized with few iterations, in a time short enough for real-time use. Therefore, one important scope for progress lies in this issue: is it possible to guess in advance the search area of any instance of this optimization problem?

Answering this question does likely neither call on a new planning approach nor designing a higher level of control. That is because the answer is not a reference for the controller but only a guess of a relevant starting value for the optimizer. This could be considered as a different branch of motion generation.

This novel point-of-view needs a conceptual evolution of model predictive control, which is more than a technological issue, but tends to show that the motion generation is not intrinsically condemned to perpetuate the classic paradigm. Furthermore, other solutions may raise and solve the issue in other different ways such as the promising polynomial optimization. This may finally establish that a simple technological improvement would solve this issue.

3.2 *What Are the Good Questions to Ask?*

All along this text, we seem to show that control is intended to occupy the entire scene of motion generation in a near future. And the reader may deduce that planning methods are globally inferior to control in all aspects. However, this exclusive view would deprive us from the results of one of the most outstanding research areas in robotics. Indeed, there are still cases for which planning performs better than control alone, such as the generation of contacts which pose difficult problems to control because of the associated non-smooth dynamics. In addition, there will always be motion generation problems which are better solved with geometric reasoning than with numerical optimization, such as trajectory generation for nonholonomic mobile robots.

Finally, imagine replacing robots by humans in the context of this discussion. If we dare to throw an eye on the living world, we would see that sometimes, despite the motor and control performances of humans, they still take the time to devise and plan their next actions. Among all the possible motions to grasp an object, what is the underlying principle that makes the selection of a particular motion? How do humans organize their behaviors to reach a given objective? Where does the reasoning take place? What are the relative contributions of voluntary actions computed in frontal cortex, to reflexive actions computed by spinal reflexes? How and why are different actions computed by different mechanisms? What are the musculoskeletal synergies simplifying the control of complex motions? Such questions lie at the core of the research in Computational Neuroscience and Biomechanics.

Exploring all these questions in a pluridisciplinary perspective constitutes the core objective of the current book.

References

1. M. Brady, *Robot Motion: Planning and Control* (MIT press, Cambridge, 1982)
2. J.-C. Latombe, *Robot Motion Planning* (Kluwer Academic Publishers, Boston, 1991)
3. L. Sciavicco, B. Siciliano, *Modelling and Control of Robot Manipulators* (Springer, London, 2001)
4. B. Siciliano, O. Khatib, *Springer Handbook of Robotics* (Springer, Berlin, 2008)
5. T. Lozano-Perez, Spatial planning: a configuration space approach. IEEE Trans. Comput. **100**(2), 108–120 (1983)
6. J.T. Schwartz, M. Sharir, On the "piano movers" problem. ii. general techniques for computing topological properties of real algebraic manifolds. Adv. Appl. Math. **4**(3), 298–351 (1983)
7. J.E. Hopcroft, J.T. Schwartz, M. Sharir, *Planning, Geometry, and Complexity of Robot Motion* (Ablex Publishing Corporation, New Jersey, 1987)
8. L.E. Kavraki, P. Švestka, J.-C. Latombe, M.H. Overmars, Probabilistic roadmaps for path planning in high-dimensional configuration spaces. IEEE Trans. Robot. Autom. **12**(4), 566–580 (1996)
9. S.M. Lavalle, J.J. Kuffner, Jr, Rapidly-exploring random trees: progress and prospects, in *Algorithmic and Computational Robotics: New Directions* (2000), pp. 293–308
10. J. Pan, L. Zhang, D. Manocha, Collision-free and smooth trajectory computation in cluttered environments. Int. J. Robot. Res. **31**(10), 1155–1175 (2012)

11. S. Sekhavat, P. Svestka, J.-P. Laumond, M.H. Overmars, Multilevel path planning for nonholonomic robots using semiholonomic subsystems. Int. J. Robot. Res. **17**(8), 840–857 (1998)
12. R.P. Paul, *Robot Manipulators: Mathematics, Programming, and Control: The Computer Control of Robot Manipulators* (MIT Press, Cambridge, 1981)
13. S. Kajita, F. Kanehiro, K. Kaneko, K. Fujiwara, K. Harada, K. Yokoi, H. Hirukawa, Biped walking pattern generation by using preview control of zero-moment point, in *Proceedings of the 2003 IEEE International Conference on Robotics and Automation, ICRA'03*, vol. 2 (IEEE, 2003), pp. 1620–1626
14. C. Dune, A. Herdt, O. Stasse, P-B. Wieber, K. Yokoi, E. Yoshida, Cancelling the sway motion of dynamic walking in visual servoing, in *2010 IEEE/RSJ International Conference on Intelligent Robots and Systems (IROS)* (IEEE, 2010) pp. 3175–3180
15. A. Boeuf, J. Cortes, R. Alami, T. Siméon, Planning agile motions for quadrotors in constrained environments, in *IEEE/RSJ International Conference on Intelligent Robots and Systems (IROS)* (2014)
16. B. Donald, P. Xavier, J. Canny, J. Reif, Kinodynamic motion planning. J. ACM (JACM) **40**(5), 1048–1066 (1993)
17. R. Tedrake, I. Manchester, M. Tobenkin, J. Roberts, Lqr-trees: feedback motion planning via sums-of-squares verification. Int. J. Robot. Res. **29**(8), 1038–1052 (2010)
18. O. Brock, O. Khatib, Elastic strips: a framework for motion generation in human environments. Int. J. Robot. Res. **21**(12), 1031–1052 (2002)
19. H. Jaouni, M. Khatib, J.-P. Laumond, Elastic bands for nonholonomic car-like robots: algorithms and combinatorial issues, in *Algorithmic Foundations of Robotics on Robotics: The Algorithmic Perspective, WAFR '98* (A. K. Peters, Ltd, Natick, 1998), pp. 69–80
20. Y. Yang, O. Brock, Elastic roadmaps–motion generation for autonomous mobile manipulation. Auton. Robot. **28**(1), 113–130 (2010)
21. I. Mordatch, E. Todorov, Z. Popović, Discovery of complex behaviors through contact-invariant optimization. ACM Trans. Graph. (TOG) **31**(4), 43 (2012)
22. N. Ratliff, M. Zucker, J. Andrew Bagnell, S. Srinivasa, Chomp: gradient optimization techniques for efficient motion planning, in *IEEE International Conference on Robotics and Automation, 2009. ICRA'09* (IEEE, 2009), pp. 489–494
23. M. Toussaint, H. Ritter, O. Brock, The optimization route to robotics. KI-Künstliche Intell. **29**(4), 379–388 (2015)
24. J. Koenemann, A. Del Prete, Y. Tassa, E. Todorov, O. Stasse, M. Bennewitz, N. Mansard, Whole-body model-predictive control applied to the hrp-2 humanoid, in *2015 IEEE/RSJ International Conference on Intelligent Robots and Systems (IROS)* (IEEE, 2015), pp. 3346–3351
25. Y. Tassa, T. Erez, E. Todorov, Synthesis and stabilization of complex behaviors through online trajectory optimization. in *2012 IEEE/RSJ International Conference on Intelligent Robots and Systems (IROS)* (IEEE, 2012), pp. 4906–4913
26. L. Grüne, J. Pannek, *Nonlinear Model Predictive Control* (Springer, London, 2011)
27. G. Schultz, K. Mombaur, Modeling and optimal control of human-like running. IEEE/ASME Trans. Mechatron. **15**(5), 783–792 (2010)
28. M. Mitchell Waldrop, On nature website: the chips are down for moore's law, February 2016. http://www.nature.com/news/the-chips-are-down-for-moore-s-law-1.19338
29. A. Forsgren, *On Warm Starts for Interior Methods* (Springer, Heidelberg, 2005)
30. P. Hämäläinen, J. Rajamäki, C. Karen Liu, Online control of simulated humanoids using particle belief propagation. ACM Trans. Graph. (TOG) **34**(4), 81 (2015)
31. J. Schulman, Y. Duan, J. Ho, A. Lee, I. Awwal, H. Bradlow, J. Pan, S. Patil, K. Goldberg, P. Abbeel, Motion planning with sequential convex optimization and convex collision checking. Int. J. Robot. Res. **33**(9), 1251–1270 (2014)

Part I
Geometry, Action and Movement

Several Geometries for Movements Generations

Daniel Bennequin and Alain Berthoz

Abstract In previous works we reanalyzed the kinematics of hand movements and locomotion, and suggested that several geometries are used conjointly by the brain for according the shape and the duration along trajectories; this was done in collaboration with Tamar Flash and her collaborators [10, 64, 67], and with Quang-Cuong Pham [79]. The variety of geometries which were implied in this process, were associated to sub-groups of the affine group of a plane: full affine, equi-affine and Euclidean. Other studies have shown how the above geometries constrain the production of the movements [92], or began to use the affine geometry in Robotics [80]. In this article, we propose to use a new variety of geometries which extends the preceding series in another direction, to cover wider contexts and more complex movements, like prehension, initiation of walking, locomotion, navigation, imagined motion. The new spaces adapted to those geometries have no points; they come from topos theory, which is an extension of set theory replacing sets by fields and graphs of dynamics. Any given topos generates a variety of different geometries, which can be mixed as in the preceding studies. Such geometries take into account efforts, forces and dynamics; they do not neglect them aside as does traditional geometry. In this preliminary report we indicate the simplest characteristics of spaces which underly the above examples. The hypothesis is also that these spaces are implemented in different, although overlapping, central nervous system networks in the brain, corresponding to the different action spaces mentioned above. Here, as for the known classical geometries, the most concrete suggestion concerns the timing of movement: we predict that different components of the controlled system are using *different intrinsic time courses*, and that the mapping between these different internal durations is an important part of the dynamic under geometrical control. This reminds us of a well

D. Bennequin (✉)
Institut de Mathématiques de Jussieu, Université Paris VII, Paris, France
e-mail: bennequin@math.univ-paris-diderot.fr

A. Berthoz
Collège-de-France, Paris, France
e-mail: alain.berthoz@college-de-france.fr

J.-P. Laumond et al. (eds.), *Geometric and Numerical Foundations of Movements*, Springer Tracts in Advanced Robotics 117, DOI 10.1007/978-3-319-51547-2_2

known psychological observation, for instance that time in imagination does not flow as ordinary clocks time, but this also suggests that reaching an object with the hand has its own time, or that equilibrium control in walking works within a specific time, which is different from the walking trajectory displacement time.

1 Introduction. Geometry, from Spaces to Transformations

Geometries as Transformations Groups; Multiple Geometries.

Geometry appeared first as a science of space (or planes) and simple objects (or figures). Accumulation of mathematical knowledge about measuring, constructing, cutting and pasting has begun five thousands of years ago, in Sumer and Babylonia, cf. Eleanor Robson [95], and in Egypt, cf. Annette Imhausen [51]. A formally perfect theory, based on axioms and demonstrations, was invented in Greece, cf. Euclide, Archimedes, Apollonius. But with centuries of practice and reflection, Geometry became gradually a science of *transformations*. A strong emphasis in this direction was proposed by arabian and persian mathematicians of the IX-th to XII-th centuries, in particular Abd Al-Jalīl al-Sijzī (Al-Sijistani), Abu Sahl Al Quhrī, Ibn Al-Haytham (Alhazeen), cf. [94]. A clear formulation of this evolution was proposed at the end of XIXth century, by F. Klein in his Erlangen program [53], when he told that *the essence of a Geometry is contained in a group and a family of subgroups*, cf. Appendix 1.

For our theory, we shall retain first, the idea of *multiple geometries*, and second the Galois idea of a *spectrum of conjugated subgroups in a group*, giving a new notion of what is a general space in a geometry.

The notion of parallelism was explicitly separated from the notion of distance in geometry by Euler in the XVIIIth century, under the name of affine geometry; then an independent treatment of points, lines and planes was at the core of projective geometry, as it was foreseen by Kepler an Desargues in the first part of the XVIIth century, when they introduced the idea of points at infinity. In addition a theory of curved forms and their qualitative relations, anticipated by Pascal and Leibniz, was offered in the middle of XIXth century by the rising science of Topology, mostly due to Riemann. This was also the time of the discovery of several geometries which violate the axiom of parallels of Euclide, in particular the hyperbolic geometry of Bolyai, Lobatchevski and Gauss, but also the elliptic geometry on the projective space. We therefore have to admit the multiplicity of geometries. These geometries are essentially characterized by the type of change of reference frames allowed by them, which belong to a given group of transformations. Moreover, and crucial to our proposal, the points which constitute a space, are characterized by the sub-groups of transformations that let them unchanged. (Cf. Appendix 1.)

The Origin of Spaces; Indifference Spectrum.

Finally Helmholtz and Poincaré [86–89, 104] invited us to question the nature of space. Their essential conclusion is that the only way for an animal to organize itself for acting in space, is to incorporate what is due to its action and what is due to an external change in the world. For this purpose, the animal (or any living organism) has to perform an active internal comparison between the sensory effects which result of voluntary self motions and the ones resulting from modifications of the world outside. To internalize this difference is a necessity for survival.

A change of apparent visually perceived form of an object can be due either to the movement of the object in space or to my own movements around it, this defines a special set of "ambiguous" transformations (because they deal with this dual potential interpretation); they form a group G. Every group can be interpreted as defining a particular structure of *ambiguity*. However, the form of the thick frontier between the inner and the outer world is not the group itself, it is a certain faithful representation of G by permutations on a set (ensemble). In our case, with Euclidean displacements, Poincaré showed that this set (ensemble) is the collection of certain sub-groups (of the form gHg^{-1}, cf. Appendix 1), that are transformations having no effect on particular end sensors, like the end of fingers or the retinal fovea (cf. [28]); those sub-groups form a structure of *indifference*. All this framework was already present in the seminal work of Galois about the ambiguity on the solutions of an algebraic equation [38]. Forgetting the internal structure of the sub-groups, and considering each one as a "point", induces a set, named a quotient set, on which G acts transitively; by definition, this is a *geometrized space*. (Cf. the up cited Poincaré books, and [8, 28].)

From this point of view, spatial knowledge is equivalent to the organisation of the command of motions, and geometrical rules describe a form of interaction with the world and with agents acting in the world.

From an experimental point of view, D. Philipona, K. O'Reagan, J.-P. Nadal [82] have succeeded to implement the approach of Poincaré on a virtual robot, to recover the dimension 6 of the group of isometries in $3D$ space.

The consideration of variable curvatures induced a revolution in Geometry: starting with plane and space curves and with surfaces in the space (Monge, Gauss), the study of curvature was extended to manifolds of every dimension by Riemann. Here the infinitesimal reference was Euclidean, but after Klein and Sophus Lie, it became evident that all kinds of Klein geometries associated to a differentiable Lie group provide an extended notion of curvature. The complete theory was developed by Elie Cartan in the first part of XXth century, and was named Cartan geometry, cf. [96]. This considerably extended the range of geometrized spaces.

However several new directions appeared in Geometry in the second part of XXth century; for instance, coming from Topology, the geometrical study of the dynamical fields on manifolds and their deformations (Whitney, Thom, Milnor, Smale et al.), cf. [101, 102], or coming from Algebraic Geometry and Arithmetics, the development of categories and topos (Eilenberg, Mac Lane, Grothendieck, Verdier et al.), cf. [42, 62]. As the name "functor" for natural maps between categories is not very appealing for a non-mathematician, we prefer in this text to use the name "field" in its place, which gives a better intuition. This is justified by the following example: the simplest physical field is a scalar depending on the place, for instance the temperature $T(x, y, z)$, where x, y, z denote Cartesian coordinates of the points in the usual space, and it is also a functor from the category $0 \to 1$, with the two objects 0 and 1, and one arrow between them, to the category of sets (mathematical word for ensembles). In the same manner, any vector field $V(x, y, z)$ is such a functor. Note that we can replace the simple oriented graph $0 \to 1$ by any oriented graph Γ, and get in this way a representation of interacting fields. We will see in what follows how this categorical framework permits to enrich our description of geometries adapted to complex movements.

We can summarize the above evolution of geometry as follows:
a geometry is made by a certain set of transformations; in the traditional point of view, this set constitutes a group of transformations of a space; in the new extended point of view, it is a field of natural transformations of a field of spaces into itself.

The Relation with Biology and Neuroscience.

Several biologists and psychologists have suggested that the inner representation of space is associated to movement production; many of them have insisted on the importance of group theory, and geometrical invariance. They explained that groups organize perception and action together. In particular, the experiments and the theories of J.J. and E.J. Gibson [39, 97] deserve to be cited. J. Piaget reported that the psychological evolution of children follows an ordered sequence of different geometries, first topological, then projective, then affine, and finally Euclidean [83, 84].

In the domain of vision, we must mention the works of J. Koenderinck and A. van Dorn [54], about the role of affine geometry in visual motion perception, and F. Wolf and his collaborators, who attributed a decisive role to the group of displacements in the visual plane for the organization of cortical maps of $V1$ [106]. In addition, J. Koenderinck used all kinds of possible groups for planar geometries arising in the perception and the analysis of images [55]. Considerations of Differential geometry and Lie groups theory were also used in the context of visual neuroscience by J. Petitot, P. Chossat and O. Faugeras, D. Barbieri, G. Citti and A. Sarti.

As reminded by R. Llinas in his book, *I of the vortex: From neurons to self* [61], the structure of vertebrates brains appeared in schematic form in the larvae of the ascidian, just before the vertebrates (5.10^8 years ago): the tunicate larva has one eye, one otolith, a chord and several muscles to control movements and to perceive space. In particular, the elements for the Euclidian group were already present. Thus, the origin of our brain's structure and dynamics is motor control, in the wide sense, to orient itself, to move in water, to navigate and decide where it will be the best to stop.

As claimed by A. Pellionisz and R. Llinas [74, 75], the brain is a geometric machine, because there is the need to transform sensory information coded co-variantly in sensor space into the contra-variant space of the effectors. In particular, they had attributed to the cerebellum this task of transformation between covariant and contra-variant coding.

Following their suggestion detailed analysis of the "eigen vectors" or the six vestibular organs, six eye muscles, thirty two neck muscles revealed interesting invariance in their organization. See [18], and the work of Barry Peterson on neck muscles geometry and its correspondence with vestibular neuron geometrical coding, cf. for instance [77, 78].

Hypothesis: Different Geometries for Different Spaces of Actions.

A. Berthoz [14–16] describe many aspects of the fundamental link between brain, movement and decision. In particular he proposed that several geometries are necessary for guiding several networks controlling actions in different spaces. Neuropsychological observations of pathological behaviors following brain lesions have revealed that different neural networks are involved in action in different spaces. (See reviews in [16, 17, 43, 73].) It has been proposed that at least five spaces are subserved by at least four different mechanisms and networks:
(1) ***Body space***, which is reconstructed in a "body schema" in networks located in the temporo-parietal junction, as first shown in epileptic patients by the neurologist Wilder Penfield in Canada, who identified this brain region as responsible for "awareness of body schema and spatial relationships" [76]. It is known that this schema takes into account all the mechanical and dynamic properties of the real physical body, and it has been also proposed that the temporo-parietal junction contains an "internal model" of gravity, cf. [57, 66].
(2) ***Near action and prehension space***, which is equivalent to the space at which we can reach things with the extended hand. In this space the geometries have to include forces and dynamic properties of the objects that one manipulates or obstacles that we may encounter. Simplifying laws of movement are at work to control gestures (see above and [10]). Actions can be made in ego-centric reference frame or in object centered reference frame or, if another person is involved, in hetero-centric reference frame.
(3) ***Far action space***, that is the space that we reach with a short locomotor trajectory (typically a room). In this space it has been shown that optimizing principles induce stereotyped trajectories. Both ego and allocentric reference frames can be used as well as heterocentric ones. Evidence shows that the neural networks involved in this space are not the same as those for near action space (cf. [85, 105]).
(4) ***Environmental navigation space***, that cannot be explored by a short walk. Typically a city or a park that requires an allocentric cartographic coding to be able to navigate and find new paths. Cf. [71]. (5) In addition to this modularity recent studies have identified multiple reference frames and different neural structures for "egocentric" (referred to an observer own body viewpoint), "allocentric" (map like, independent of an observer view point), or even "heterocentric" (taking an other person as a reference) ([6, 12, 24, 37, 58]. This diversity of reference frame has given rise to a number of terminologies (like first or third person perspective etc.).

Our hypothesis is that evolution has applied a principle of modularity and designed different networks for actions and perception in these different spaces because each had different requirements and therefore different "geometries".

In the present paper we show, in addition to the already published combination of Euclidian, affine and equi-affine geometries mentioned above, how a variety of different geometries is useful (and even necessary) to understand various aspects of motor control and sensory-motor interaction with the world. We explain how these geometries intervene for adaptation of neuronal dynamics by virtual systems of "homological nature", and how the movements durations reflect geometrical invariants and coordinate choices. Moreover, we suggest that new types of generalized geometries without points, are necessary for guiding the neural networks underlying complex actions, movements preparation and execution.

2 Geometries for Motions Timing

2.1 Euclidean and Galilean Brains Structures

It is amazing to see how precisely the geometrical principles of Physics are reflected in the organization and dynamics of the visual and vestibular system for controlling posture, locomotion, active vision and equilibrium in highly dynamic conditions. In particular, the vestibular end sensors of vertebrates, the semi-circular canals and the otoliths which record heads rotations and translations. (See a recent review in [40].) Even at the first level of transduction, in the hair cells, there exists a coherent recording of linear acceleration and rotational velocity, or at higher order, linear jerk and rotational acceleration. We have described recently the remarkable geometrical organisation of the otolithic maculae which allow this transducer through the creation of a "virtual dynamic line" to detect $3D$ acceleration very rapidly and efficiently [27]; and we have shown that a peculiar geometry of the semi-circular canals ampullae optimizes the distribution of forces for the detection or rotational forces [65].

All this is compatible with the natural analysis of a Galilean group. From principles of the Theory of Relativity, linear acceleration and gravitation are a priori non separable; however, after two neuronal relays, in the cerebellum, gravitation and acceleration information are both accessible. (Cf. [3, 107]). With vision (and/or hearing), we get the ten dimensions of the complete *Galilean group* R the rotation (3), V a uniform speed (3), T a spatial translation (3) and τ a time translation (1). Cf. [9, 41].

In addition, vestibular, visual and proprioceptive information flows are able to produce in the hippocampal formation a variety of geometrical neurons for navigation in the Euclidian plane. This is performed by the system of *place cells*, *head direction cells*, *grid cells*, *frame cells*, *boundary cells* etc. Cf. [1]. A variety of frames for navigation can be obtained with this diversity of modes of coding.

Note that this network involves many structures from other regions of the brain, for instance in the Thalamus [52], and it exchanges information with other neocortical areas, for instance prefrontal or parietal cortex [13, 21], and even cerebellum [22, 50].

2.2 *Affine Evidence and Multiplicity of Geometries*

The brain uses other geometries than Euclidean. For instance Flash and Handzel [31, 32], Pollick and Shapiro [90] have remarked that the 2/3 law [56] which gives a non-linear relationship between tangential velocity and curvature during a natural movement, can be interpreted in terms of affine geometry. The starting point of the 2/3 power law was an old observation [19], that when drawing, or writing, the end effector moves slower in the more curved parts of the trajectory; the more precise law tells that the linear velocity $V(t)$ is proportional to $R(t)^{1/3}$, i.e. the radius of curvature of the trajectory at time t elevated to the exponent $1/3$ (which makes $-2/3$ for the angular velocity, and gives its name to the law). Handzel and Flash, Pollick and Shapiro [31, 90] re-expressed this law as follows: the time course along a path corresponds to the unique equi-affine invariant way of parameterizations. The equi-affine group is the subgroup of affine transformations that preserves the area.

For hand movements in space, Maoz et al. [64] have shown that torsion with exponent $1/6$ comes into the play, and this also can be interpreted by the equi-affine invariant parameterization. Concerning the shape of trajectories, Polyakov et al. [91, 92] studying scribbling in monkeys, have shown a dominance of arcs of parabola, which are the curves with the highest dimension of affine symmetries.

A remarkable finding was that the same kind of law also holds for human locomotion, but with an exponent smaller than $1/3$ and depending upon the form of the trajectory. References [47, 103] The idea that similar laws subserve the generation of a trajectory for a similar gesture executed by different effectors was known in Physiology under the name of "the principle of motor equivalence". In accordance, it was shown recently that in the motor system a large distributed population encode handwriting movements in scale independent manner [44].

Studies on locomotion have also suggested that general optimizing principles probably involving non trivial euclidian geometries subserve the formation of locomotor trajectories [4, 48, 81].

A systematic exploration of kinematics of drawing and walking [10] showed that different geometries (Euclidian, equi-affine, affine) are used together for generating the time course of a trajectory, depending on its local shape. This incorporates the observation of Binet and Courtier that isochrony guides successive productions of point to point motion with the hand. And this is compatible with the necessary interplay between vision and motion, because, for instance, when we have to walk on a circle what we see on the ground is an ellipse. Geometry appears to serve the action/perception coupling. (For the comparison between perception and production with respect to the geometric parameterizations see [60].)

The combination of the different geometries during a movement was modeled by logarithmic combinations of invariant abscissae. This allowed to revisit the time course of velocities from the point of view of shifts between different geometries. As a result, we have compared the duration in drawing and walking, and shown that the main difference between them is a larger impact of Euclidean geometry for walking and a larger impact of pure affine deformation for drawing.

The fact that affine geometry underlies hand movements production was established from a statistical analysis of a large set of scribbling [79]. More precisely, if two paths segments can be transformed accurately one into each other by an affine transformation, then the timing on these segments are accurately matched by the affine transportation, once it is normalized by total time. Also these finding have been recently applied to robotics.[1]

Of course, we do not assert that geometry alone is responsible of movement planing and generation; geometry must conjugate its role with other principles, like min jerk, min variance, min time or min energy (cf. [67]).

3 Geometries for Adaptation

3.1 *Homological Spaces and Galois Operations*

Below we propose some suggestions to explain how the brain creates several different geometries and why these geometries allow a great flexibility i.e a capacity of adaptation to the variety of conditions in which action has to be made.

Specificity.

According to Poincaré, Euclidean geometry has its origin in the overlap of information between the inner world and the outer world, during movements and explorations. As reminded in the first sections, this overlap produces an ambiguity which possesses a structure, described by a convenient group and a convenient space. Every geometry used by the brain offers a *specific process* to overcome the complexity of the interaction with the external world, it guides the choice of pertinent aspects in an excessively rich set of interaction, and it allows to plan actions in various spaces. This is compatible with the with the concept of modularity and simplexity proposed by Berthoz [16]. Our hypothesis is that during evolution, living organisms have created in the brain new neuronal structures adapted to different action spaces, extending the range and abilities of interactions with the world. These new structures are organized according to specific geometries.

Let it be clear that we do not suppose that geometries are organised as such in the brain, or as described by the abstract concepts of mathematics like those below.

[1]Recently, Q.-C. Pham and Y. Nakamura developed a new trajectory deformation algorithm based on affine transformations. Reference [80]. The idea is to apply a set of predefined affine transformations to a set of trajectory segments, to avoid unexpected obstacles or to achieve a new objective goal. They also conjugate this idea with optimization algorithms for better accuracy, respecting C^1 continuity, keeping fixed final configuration and avoiding joint limits. The method was tested on a virtual planar three-links manipulator, and compared to polynomial interpolations; the main result is a considerable gain in *computation time* for equal accuracy. This can also be efficiently applied for minimizing curvature's changes in $3D$ point to point deformation. The method was applied to rapid motion transfer from humans to robots, with better performance in kinematics than polynomial interpolation methods.

But these geometries correspond to specific processing characteristics of certain cells and networks, defined by particular sets of transformations which occur within these networks and which constrain and modulate the interactions between several brain areas, dedicated to a given type of functions. A simple example is given by Affine geometry, present in movement preparation and execution, as in visual or somatosensory perception; there is apparently no center dedicated to this geometry, to the contrary it seems operating indirectly in some dispersed co-variant states of neuronal networks, premotor and cerebellar areas or in higher visual areas for instance.

Adaptation.

We suggest that adaptation is the main goal of most of the non Euclidean geometries that are used by the brain. In Biology, adaptation is an ubiquitous and essential property of sensory and motor processing, allowing the living systems to sense and anticipate what is changing in the world [14]. An example is the decline in the frequency of firing of a neuron in response to constantly applied environmental conditions, or more generally, any change in the relationship between stimulus and response that is induced by the level of stimulus [59]. For surviving, acting and perceiving efficiently, every living entity must dispose of functional systems that adapt themselves continuously to the changing environment and to internal modifications.

A general hypothesis of *homology for adaptation* [8] can be expressed as follows: for every decision organ, or every function (that are dynamics transforming an input in output, or predicting an output from prior experience), denoted by X, which depends on parameters, or modules, denoted by U (corresponding to the mathematical notion of *unfolding*, that is a deployment), there exists necessarily a secondary organ, or functional system, I, which "builds" the schema (plan) of the functional dynamics in X, and at the same time "shapes" the modules in U (gives them a structure), for guiding the adaptation of X. Up to this point I could be the support of any "internal model" or "supervisor" as it was already suggested in the neurosciences or robotics or artificial intelligence literatures; for instance, a notion of observer was proposed by J.-J. Slotine and W. Lohmiller [99, 100], with interesting applications to dynamical systems that satisfy a property of contraction. However this would not generate a *new geometry*; thus the most important point here is that I comes naturally equipped with a set of *virtual transformations*, forming a group G, which characterizes the Galoisian nature of homology. This group corresponds to a *scheme* of deformations of the modules in U. The definition of a group (or more generally a groupoid, cf. Appendix 2), i.e. operational associativity and reversibility, corresponds to the main requirements for an adaptation scheme. In the simplest examples, where X describes the competitions between minima of a potential function, the space U is made by significant minimal deformations coefficients of the potential, the space I introduces imaginary linear combinations of vanished minima, that do not correspond to realizable situations, and the geometry G on I is an extension of the Galois group of a generic numerical equation. The correspondence between I and U is induced by the map from the roots (real or imaginary) to the coefficients, it is not one to one, thus the induced *real deformations* in U are only the shadows of the geometrical operations of G.

Consequently, the natural homological operations in I create a new world, including the various plans for adaptation. This is a new formulation, leading to a control. In other terms, I is at the same time, able to "understand" when the dynamics of X needs to be adapted, and able to "drive" the required adequate deformation of the parameters of U, by working in an imaginary space. This represents a higher level of control than a simple direct coupling between the dynamics of X and U.

This process involves therefore a true *ternary structure*, with three equals actors, X, U and I, where I expresses the convenient geometry G. This structure can be generalized in the framework of localized categories, to get new geometries without points, corresponding to a field of groups. An interesting example is given by the co-homology of information topos, where X is a category of observables, U a set of probability laws, and I the localized mutual information quantities of Shannon, see [7, 8].

Let us insist on an important point: the geometry G on I introduces a particular structure of ambiguity between the brains states and the external world. As this point is surprising and certainly difficult to accept for lectors who are not familiar with homology or Galois theory, let us take the example of Affine geometry in movement production. In this case, there is no compensation between a transformation in neuronal states induced by a body motion and a transformation in neuronal states induced by a change in the world, but there is a compensation between a transformation in neuronal states induced by an internal dynamics and a virtual coordinates change on the world. For instance, preparing a hand writing "up to dilatation" expresses an internal ambiguity with respect to the achieved motion. A dilatation of the dimensions of the ambient space has no experimental support, it is only a virtual change of the world, that can be reflected by a transformation of the dynamics in a particular structure of the brain and compensated by the dynamics of another structure. (Cf. [44] for an experimental support of this hypothesis in the motor system.)

Properties of Homology.

Thus our hypothesis is that geometries are properties of spaces I which are of homological nature with respect to a neuronal dynamics X, they define the relevant characteristics of X, they structure these characteristics, and define virtual operations on them (see [8, 20]). In the same manner that Poincaré associated a group and a geometrized space to the ambiguity structure of rigid motions, a geometry, associated to a group G and a spectrum of sub-groups (gHg^{-1}), is implemented in I to make standard and flexible the deformations for adaptation in the space U which represents the control of the changes in X (the unfolding of the dynamics). In general, the relationship between I and U is ambiguous, reflecting the ambiguity between I and the real world, but in most known cases, they are related by a locally one-to-one correspondence, in particular they have the same dimension.

We suggest that concretely, X, U and I are probably implemented in interacting neuronal networks, frequently organized in brain areas, or interacting nuclei. For instance, an interesting brain structure for containing homological areas is the thalamus. Even if he used other words and concepts, D. Mumford [68, 69] described the thalamus as a kind of black-board were transiently complex messages are schematized. This is compatible with another important role of the thalamus, which is to transmit information between brains areas, in particular neo-cortical ones [29], and to systematically send traces of this information to the motor system [98].

A paradigmatic example of homological space I, with an affine geometry on it, is for color in LGN (cf. [8]). Of course other regions are possibly of this nature, for instance the Entorhinal cortex EC, in relation to CA3, CA1 and Subiculum in the Hippocampal formation, and several thalamic nuclei (cf. [1, 52]), may play this role for navigation and more generally for memory.

An homological network I is also a dynamical function, thus it can itself be transformed by a higher degree homology for adaptation. This can generate a cascade of homological folds: $H(H(...))...)$. ... In fact, there exist in a developed brain many ternary structures, supported by different but overlapping material structures, that are interacting cellular networks. Several ternary structures can share a given geometry, or several different geometries on different ternary structures can interact coherently. We saw this result in our analysis of duration of curved trajectories [10].

Therefore our general hypothesis is that ***(1) the brain is a creator of ambiguities (on the model of the initial ambiguity between inner and external world, but comparing now internal modulations with external coordinates changes), (2) the associated geometries*** G ***are virtual operations on ideal homology spaces*** I, ***that are engaged in ternary structures*** X, U, I, ***for guiding control and adaptation***.

In reality, during the Evolution, it is likely that the three components and the corresponding geometry evolved together.

3.2 Canonical Times and Moving Frames

A geometry G on a space I can offer a repertoire of virtual movements and plans of actions. In the present text, we focus on trajectories of body systems and their generalizations, and we replace I by a geometrized space E that represents the considered body system. This space E is identified with an homogeneous quotient G/H of G by a sub-group H (cf. Appendix). Elie Cartan (1937) summarized in a beautiful concrete method one century of research about the geometric invariants of curves in E; this is the *moving frames method*. (Cf. [10, 23, 35, 72].) In [10, 32], this method was used to define a series of natural timing on trajectories for drawing and walking, and compare it with experiments.

The Cartan's method is inspired by the Galois theory [38], and consists in establishing a natural bijection from a product of spaces issued from the groups G and H to the manifold of infinitesimal elements of every order of curves in E. On the side of G, this gives a sequence of homogeneous spaces G_n/G_{n+1}, where G_n is a decreasing sequence of groups, starting with $G_0 = G$ and $G_1 = H$, plus numerical values for geometrical invariants until the order n (i.e. quantities which do not depend on the frames and depend only on the derivatives of the trajectory up to the order n), and on the side of curves, this corresponds to the sequence of derivatives (jets) of order n modulo derivatives (jets) of order $n-1$, that define the *infinitesimal curve* up to the order n. The decreasing sequence of sub-groups i.e. $\{e\} = G_N \subset G_{N-1} \subset \cdots \subset G_1 \subset G_0 = G$ ends at the order N, which coincides with the dimension of H.

For the concrete algorithm of the moving frame method, see the above references. [Briefly, for mathematicians readers, it consists to fix a frame F_0 and to choose order after order the frames $g_n F_0$ attached to the given curve in such a manner that the differential equations which describes their displacement along Γ involve the less possible number of free parameters. It follows that a given curve Γ determines step by step, starting with $n = 0$ a canonical sequence of subgroups $H_n = g_n G_n g_n^{-1}$. The element g_n being defined modulo multiplication to the right by a sub-group G_n (ambiguity of order n), the frame at this order is only partially defined. At the order N the frame becomes fixed, then the following orders, larger than N, give new invariant quantities for Γ, that correspond in general to higher order derivatives of a finite set of curvatures.]

Up to a simple ambiguity, like t changes $at + b$, the canonical parametrization emerges as the only one were the moving frames equations keep their form.

Classical examples are the Serret–Frenet frame in Euclidean geometry, parameterized by arc-length, giving rise to usual Euclidean curvature and torsion. Less classical examples are the equi-affine frame, corresponding to the osculating rational helix, in the $1/3$ parametrization, giving the equi-affine curvature, and the more general affine frame, cf. [10]. The order N is given by the dimension of H; in the case of planar geometry, this gives 1 for Euclidean, 3 for equi-affine, 4 for affine, 6 for projective.

The final canonical moving frame equation can be seen as an optimal form of sparseness for transportation description. The decreasing sequence of sub-groups can be intuitively understood as a manner to break progressively the ambiguity on the space surrounding a curve, by choosing canonically a system of coordinates, attached to our trajectory. Thus, we see that in this context, of a trajectory in a geometrized space, information is formulated with groups theory, as in other contexts it is formulated with probability theory.

For preparation of action (or its reverse face, that is perception, as proposed by Rodolfo Llinas, [61]), the above geometrical description represents an economy of planning (e.g. the affine group invariance for seeing or imagining, to prepare walking); then, step by step, from anticipation to execution, a sequence of sub-groups breaks the indetermination.

Timing along trajectories and canonical coordinates on geometrized spaces arise from the group structure of the geometry, by an analysis of the ambiguity about coordinates in space for stabilizing position, velocity, acceleration, jerk, and so on.

However new concepts are needed to address complex organizations in different spaces, like, as mentioned in the introduction: action on *body space*, or reaching in *near action and prehension space*, or *locomotor out of reach space*, or *environmental space*. Composite motions can require more than two geometrized spaces. We would like to consider here what are the neural underlying “spaces” in these cases.

4 Geometrical Spaces for Movements

4.1 Topos and New Geometries Without Points

We have previously, as described above, found evidence of combinations of several geometrical systems for the production and control of point's trajectories, their rhythms and timing of duration. In the case of drawing and writing, the considered effector is the end of a finger, or a pointer, and in the case of walking locomotion, the point which is considered in place of the effector is the projection on the floor of a convenient point in the body (for instance the center of the head), thus we characterised these movements by *precisely defined points* in space or on the ground. However, most of the natural voluntary movements, or gestures, cannot be represented by the motion of only one point.

A first natural suggestion is to work in a large "parameter space" with many dimensions corresponding to the control of many points attached to the body, as it is frequently done in robotics. However, this method can hide a geometry, as for instance was first the case in astronomy by the description of planets motions by composition of circular motions with constant velocity; this method occluded for a long time the Kepler motion on conics. Thus we prefer to look at the possibility of *geometries without points*.

Such geometries and dynamics do exist, they are associated to Topos, (cf. Appendix 2). The idea of topos [42, 93] is to replace sets and points in these sets by networks of arrows with fixed topology, between sets, and we forget about the notion of points. For example a chain $\bullet \to \bullet \to \cdots \to \star$ of fixed length $n+1$, gives rise to all sequences of length $n+1$ of transformations. More generally, an oriented graph Γ gives rise to the articulated structures of several maps between spaces acting coherently as indicated by Γ; we proposed to designate generically these families of maps by the name of "fields". Thus a usual set is replaced by a field. And the usual transformations from a set to a set are now replaced by a collection of maps between the sets that are placed over the same vertex of Γ, which satisfy a condition of compatibility: that going on two different paths from a set to another one, either following the fields or following the transformations of fields, we get the same result. For instance in the above example of the chains, a generalized transformation from a field $A_0 \to A_1 \to \cdots \to A_n$ to a field $B_0 \to B_1 \to \cdots \to B_n$, consists in a collection of transformations $A_0 \to B_0, A_1 \to B_1, \ldots, A_n \to B_n$, such that for any index $k < n$, the two manners of going from A_k to B_{k+1} give the same map.

The theory of topos was applied to intuitionist logic. (William Lawvere et al., cf. [93] and the references inside.) Even the simplest example of the chain of length 2, that is $0 \to 1$, gives a topos (associated in fact to the site which is the dual graph $1 \to 0$, because presheaves are contra-variant functors). This topos is named the *topos of Shadoks* by Alain Prouté, that has a table of truth with more than two values, then violating the contradiction principle. The Boolean calculus of characteristics functions of sets is replaced by a 3-valued logic (Heyting algebras): true, false and

... uncertain. See the exact statement and its proof at the end of the second appendix below.

All the usual objects in mathematics can be reconsidered in this setting. This applies in particular to dynamical systems and their deformation. The construction of homology I can be extended for giving "homological fields", that are fields $I_0 \to I_1$ in the case of Shadoks. We can look again at the familiar geometries in the plane, affine, equi-affine and Euclidean, where each geometry corresponds to a group G and a subgroup H; in the new setting of Shadoks, G has to be replaced by a morphism of groups $G_0 \to G_1$ and H by has to be replaced by a restricted morphism $H_0 \to H_1$ from a subgroup $H_0 \subset G_0$ to a subgroup $H_1 \subset G_1$.

Then, assume we can develop a generalization of the algorithm of the Cartan's moving frame in this context, each group of the chain will contain a canonical sequence of sub-groups, order by order; then the usual sequence is replaced by a grid of morphisms.

Continuing this way, we obtain a new notion of moving frames, which is a mapping between sets of frames. But remind that each pair (G, H) determines a parametrization of trajectories that gives a virtual timing in accord with the geometry, thus two pairs (G_0, H_0) and $G_1, H_1)$ gives two notions of timing. As a consequence something very interesting happens with *time*: we discover that there is no reason for a unified timing, or unified rhythm. The canonical parametrization becomes a change between two parametrization times.

Of course the executed motion is realized in a coherent timing, our assertion applies to the working systems which prepare and execute the action; like in elaboration of a movie, each image and scene being worked in their own timings, and then glued together for projection.

Remark this is not in contradiction with the known relation between the timing of actual executions of action and the timing of the mental imagination of this action, cf. [26, 70], because, as said in [25], imagined and executed actions share, to some extent, the same central structures. On the other side, our assertion accords with the adaptive compression of timing in imagination and memory for navigation, cf. [5], because, in this case, there are probably two different topos geometries at work, one for locomotion and another one for navigation.

In fact, the notion of trajectory itself is now problematic: for example, in the chain, like the shadoks one, a real interval I of numbers for parameterizing the time is only needed at the end, over 1, but at the origin, over 0, we can get a manifold Y, not necessarily an interval, equipped with a map $f : Y \to I$. Thus the parametrization of a trajectory by the time is replaced in the case of a shadoks space $\pi : E_0 \to E_1$ by a field $F : Y \to E_0$ underlying the ordinary trajectory $\gamma : I \to\to E_1$, in the sense that $\pi(F(y)) = \gamma(f(y))$ for every y in Y. This leads to the notion of a non-linear time, which can have many dimensions, corresponding to a network in movement preparation and execution.

Thus we obtain a well established psychological fact: that ***inner times are manifolds***.

Applications to Different Action Spaces.

We shall now apply these concepts to the question of the different action spaces: *prehension, initiation of locomotion, near locomotion, environmental imagined navigation.* In these example the action relies on fields of forces of very different nature. It is natural to expect that the timing to keep an object, to stand up, to press the floor with a foot or to imagine and dream of a city are different from the timing of an ordinary clocks. Theses action-times should give an insight of the different geometries at works.

4.2 Examples

We will consider shortly several sorts of natural motions, and sketch a first preliminary essay to model them by topos geometries. Our first approximation will use only two levels, i.e. geometries in *Shad* (the topos of shadoks), which are associated to an homomorphism of groups $G_B \to G_A$, that respects subgroups $H_B \to H_A$ of G_B and G_A respectively. This will correspond to a division in two parts of the motor system, representing the body and a point of interest. In the same spirit, we could have used a longer chain, taking into account the torso, the shoulder, the hand position and finally the end effected move. Of course this should have been more realistic and efficient in applications. But in this section, we want to present the main idea without modelling experiments, and for that purpose, two levels seem sufficient.

A general fact will appear: that G_A is a group, in accord with usual geometry, but G_B is not a group, only a *groupoid* (the notion that extends groups in category theory, cf. appendix), corresponding to the fact that physical articulations of a body's part limit the iteration of its motions. However this makes no profound difference for the following discussion, and we neglect this fact, considering G_B and H_B as if they were ordinary groups.

4.2.1 Prehension

The convenient geometry has to be *polarized*, from the surface of body to the external world containing objects; this polar structure was also underlined by Tamar Flash. We model this geometry by a two levels sequence $G_B \to G_A$, where G_A is a geometry for the classical external space, that could be the Euclidean one or an affine extension, and a two levels sequence of subgroups $H_B \to H_A$. We consider for simplicity that the object of interest must be reached and moved by the hand. The compensation scheme of Poincaré implies that the group G_B describes the configurations frames that the body can attain with respect to the hand position, indifferently on the fact that the movement is voluntary or imposed from another subject. It describes the allowed equilibrium states of the arm. Stabilizing mechanisms of neuro-muscular activity at equilibrium of the arm should play an important role here, for allowing the compensation of transformations of G_B on neural activities. This can be seen as the set of postures of the arm. In this respect, we can cite the results of Tamar Flash on the stiffness field of the hand in multi-joints arm movements [30].

The map from G_B to G_A represents the induced hand movement. The sub-group H_B of G_B must go into H_A, thus it is a set of deformations of the arm with a fixed end. The choice of H_B defines a sort of redundancy. The quotient space $E_B = G_B/H_B$ defines the geometrical degrees of freedom of the arm, or postural schemes. The minimal choice for E_B is the linear space of vectors from the shoulder to the wrist, which is probably not sufficient for most neural control; a more interesting choice considers in addition the three articulations, at the shoulder, at the elbow and at the wrist. In the Euclidean framework, this gives three angles, but in affine geometry the space is more intricate.

If we want to take in account the eyes movements information to the reaching system, a better representation of the geometry should involve two groups, G_{B_1}, G_{B_2}, one for the arm, the other for the eyes, going to the rigid motions G_A, and two sub-groups H_{B_1} in G_{B_1} fixing the hand position, and H_{B_2} in G_{B_2} fixing the direction of the gaze.

What appears on this example is an interesting possibility to revisit the notion of spatial reference frames used in reaching (cf. [2, 46]. For instance the notion of center of frame is replaced by at least two centers, one in the object, with a possible virtual identification with the subject, and one in the articulation points of the arm, and/or the eyes center. The flexibility in defining a reference frame on the aye, the shoulder or even in arbitrary points has been documented in several papers from the groups of John Soechting, Francesco Lacquaniti, Paolo Viviani and Alain Berthoz. In fact, in the spirit of topos, what replaces the notion of a center is the graph which relates arm, eyes and objects, here the graph has two arrows from arm and eye respectively to the object; thus, in some sense, two new geometries are acting together, one for the eye and the object, the other one for the arm.

4.2.2 Initiation of Locomotion

The convenient geometry has also to be polarized, from the support of the body to to the Euclidean objective space, where locomotion is effectuated. Consequently we take again, as for prehension, for G_A and H_A an ordinary space geometry. This geometry can be Euclidean or affine, depending on the level of preparation in the brain. The second level ambiguity group G_B describes the configurations that the body can attain; it corresponds to the *posture* of the whole body. The arrow (morphism) from G_B to G_A is given by the movement of a rigid frame attached to the body, that can be the head, or the torso. We see that probably, several arrows shall be considered, corresponding to several new geometries, working together.

We observe again that G_B is not a complete group, it is only a groupoid, because the body has to keep contact with the floor, and cannot be deformed arbitrarily; in particular the map from G_B to G_A is not surjective. The group H_A is the set of rigid motion fixing the center of gravity of the chosen body frame. The constraint on the sub-group H_B is that it must go to H_A, and that $E_B = B = G_B/H_B$ defines sufficiently simple parameters to be controlled, like particular angles at the articulations.

If we take for H_B the subgroup of all postures giving a fixed body's center of gravity, we obtain for G_B/H_B a part of the ambient space, representing vectors of standing.However this is not the most interesting choice, we could impose more constraints on H_B, at articulations under control, for instance the neck, or head stabilization, this would give a larger space G_B/H_B, defining a more interesting *postural scheme*. This corresponds precisely to a choice of reduction of number of degrees of freedom (cf. [11]).

In addition, as in the case of reaching, we can take into account the visual information, that enlarges greatly the dimension of the adapted frames dynamics. Note also the possibility to enlarge the geometry by coupling to the topos geometry of another moving subject.

4.2.3 Locomotion

Here G_A and H_A respectively correspond to a choice of geometry in the horizontal plane. We must be careful to not confound this plane with the ground floor, which has a role in the second level group G_B, where resistance and stand up are taken into account. Here, the horizontal plane means the plane that is transversal to the vertical gravity direction. In G_A we can suppose that only the horizontal movement is planed.

The group G_B is again a definition of posture, but now far from equilibrium (at least in humans), and H_B determines the posture with respect to the vertical. A possible choice of H_B could be the sub-group which stabilizes a given vector attached to the body, for instance the axis of torso, or the vector joining the contact on the floor with the center of mass.

It would be interesting to enlarge the postural group G_B by including the transformations of the relief and the nature of the ground floor, modulating the feet positions and the control of the center of mass. Also obstacles and objects in the environment could also be integrated in the geometry through a variety of changing reference frames.

We constat that in each situation, prehension, initiation of walking, locomotion, a variety of spaces and topos geometries appear naturally, as we previously saw that several geometries were useful for the analysis of points trajectories in hand drawing and locomotory motion.

Note that in all the above examples, G_B and H_B have their origin in the usual forces fields and dynamical mechanisms that underly the motions of bodies and objects described by G_A and H_A, but they shall not be confounded with these forces and dynamics, they define operational schemes for them, that can be used for instance to overcome environmental changes for maintaining the success of an action, without much changes of its form.

Also, in all these examples, it is tempting to extend the geometry for including an external time into the geometry itself, taking for G_A the Galilean group in place of the Euclidean group. This will accord better with the fundamental role of the visuo-vestibular system for controlling motor coordination, locomotion, navigation, awareness of space, and social interactions.

5 Discussion

In this article, we began by reviewing the evolution of the mathematical point of views on Geometry, underlying the growing role of transformations and group theory. Then we developed the hypothesis that the brain's evolution used a principle of modularity and designed different networks for actions and perception in different spaces implementing different geometries for different requirements, and we added the hypothesis that the necessity of geometries is found in their role for shaping and guiding adaptation. Note that, when Andras pellionisz Rodolfo Llinas claimed that the *the brain is a geometric machine*, they had particularly in mind the fact that neurons with their particular form of dendrites and axon collaterals, and neural systems, in particular in the cerebellum, but also in other structures of the brain, are performing a geometrical work, but they also linked this work with the hidden geometrical functioning, which links actions and perception.

We had previously suggested that several geometries, Euclidean, affine and equi-affine, serve as a guide for humans hand drawing and locomotion. In particular the timing along the trajectories of significant points effectors was reanalyzed and well explained by a mixture of geometrically invariant parameterizations (cf. [10]). More generally, it appeared that preparation and execution of these movements can be better understood by using operations that are organized by those geometries (cf. [31, 32, 34, 64, 67, 79, 80, 92]).

However in more complex situations, the body motion cannot be described conveniently by one point, as an end effector, or even by a finite collection of points. In fact, this remark applies if we want to analyze more precisely the movements of the hand in drawing or the movements of the body in walking. In the present article we have addressed the problem to extend the geometrical approach to such movements. Our new suggestion is to use the extension of the traditional geometries in the context of *topos theory*.

This theory generalizes the theory of sets (i.e. ensembles), by replacing sets by diagrams of fields between sets (cf. the appendices), thus the notion of point disappear but the concept of geometry are maintained and enlarged.

The examples we had in mind were prehension, initiation of walking, locomotion, navigation, imagined motion. In all those cases A. Berthoz had suggested before that several geometries are necessary for guiding several networks in the brain controlling actions in different spaces (cf. [14–16]).

In the present article we have indicated preliminary suggestions of useful topos geometries for prehension, initiation of locomotion and locomotion. In each case, the geometry has a polarized structure, for prehension or initiation of locomotion the polarization goes from the body to the environment, for locomotion the polarization corresponds to the vertical direction opposed to the gravity vector, and in both cases, principles are emerging to define a workspace of postures.

The goal is to describe a field of spaces with scheme for preparing, controlling, executing and adapting movements. For that purpose we have connected the new suggestion to previous developments, in two directions: *homology for adaptation* (cf. [8]) and *time correspondence* (cf. [10, 32]). The first notion is the formation, for each neuronal function, of a secondary functional space that extracts and defines dynamical characteristics and admit ideal operations (Galoisian in nature) forming a geometry, and then applies these operations for guiding the necessary adaptation of the dynamics, at this level invariants are created. The second notion is a kind of Galois correspondence between parameterized manifolds in the geometrized space and frames in the geometry (Cartan moving frame). Then we suggested in the present study, that these two ingredients, homology for adaptation and time correspondence, must be extended in the context of topos geometries.

Topos were invented by Grothendieck and Verdier to unify homological algebra and co-homology theories. Thus, certainly, the co-homological spaces that are adapted to represent the working of chained ternary structures for adaptation of neuronal functions, would come from homological algebras of convenient modules over a ringed topos. We mention that a step in this direction was the definition of Information Homology in [7].

However much has to be done for giving theoretical hypotheses about the geometries underlying the different classes of movements, to be experimentally tested. One direction where we can look for that, would be given by concrete predictions of timing and duration for the body's segments, in relation to stiffness fields and EMG signals. An article in preparation will give more details on the hypothetical brain's networks that are involved in the new geometries.

These new geometries must be compatible with the notion of internal models of the world, as it is well expressed for instance by the Free Energy minimization principle [36], with the optimization of smoothness [33, 49] and with optimal control, including the minimal variance principle [45], but they give a different point of view, for complementing them.

With respect to the more usual applications of geometry and dynamics, as in optimal feedback control, that were used for the understanding of principles guiding voluntary motions in neurosciences or robotics, the main difference of our approach is the research of *invariance principles that are based on a fully developed geometry*.

These invariance principles based on geometry, are starting with physical forces and physical constraints, but in some sense they are integrating them, and become free from them, to get a scheme for preparation, control and execution of movements. For that, planning is elaborated in neuronal networks that are working as if there were ideal geometrized spaces, not as spaces mimicking the real external physical world, because these ideal spaces are structured for allowing the operations of the more convenient geometry for adaptation, and then, paradoxically, only at the end of the process, movements are realized in the physical space through the determination of configurations in time.

6 Appencices

6.1 Groups

Definitions: a *group* is a set G, with a privileged element e, in which a product law is defined, which associates an element gh to every pair of elements g, h of G, satisfying three axioms: (*i*) associativity $(gh)k = g(hk)$; (*ii*) neutrality $ge = eg = g$; (*iii*) the existence of an inverse for each element g, i.e. g^{-1} such that $gg^{-1} = g^{-1}g = e$. The principal example is given by the group $Aut(X)$ of permutations of a set X, that are all the manners to exchange the elements of X between themselves; the product law being the composition of successive permutations; the neutral element e being the identity, when nothing changes; the inverse of a permutation consisting to replace things in the former order. A *subgroup* H of G is a subset containing e which is preserved by products and inversions. Examples are given by subsets of transformations of a set X, containing the identity, and closed by composition and reversibility.

In general, we define an *action* (or *representation*) of G on a set X, as an operation law, $(g, x) \mapsto g.x$, satisfying the two axioms (*i*) $\forall x, e.x = x$ (*ii*) $\forall g, h, x, (gh).x = g.(h.x)$. An action gives a map from G to $Aut(X)$, which respects the neutral elements and the multiplication laws. More generally, such a map between two groups is named an *homomorphism*.

Conjugation by g in the group is the important operation which sends every element g' to the *conjugated element* $gg'g^{-1}$. The set of all conjugations in G is a subgroup $Int(G)$ of $Aut(G)$.

The group $Iso_+(E)$ of rigid isometric displacements of the ordinary Euclidean space E, infinitely extended around us, is a subgroup of the group of all permutations of the points of E. This group $Iso_+(E)$ is made by translations, rotations and twists, i.e. compositions of a rotation and a translation in the direction of the axis of rotation. The group $GA(E)$ of *affine transformations* of E contains $Iso_+(E)$, but its is richer, it is made by all permutations f for which there exists a linear transformation $\varphi = Tf$ of vectors of translations, such that, for any point P and any vector $\vec{v}$, the image of $P + \vec{v}$ by f is equal to $f(P) + \varphi(\vec{v})$; for instance all dilatation, stretching, squeezing are affine transformations. The number of dimensions of $Iso_+(E)$ is 6, and for $GA(E)$ it is 12. They are permutations which send parallel lines to parallel lines. Generally these transformations do not respect the distance, nor the volume. The subgroup which preserves the volume is named *equi-affine*.

Larger than $GA(E)$, we can consider the 3D projective group $PGL(\mathbb{P})$, acting on the projective space $\mathbb{P}$, which is obtained from the affine space E by adding a set of *points at infinity*. In fact, these ideal points are in one to one correspondence with the non-oriented directions of parallel lines in E; and a sequence of points x_n in E approaches a point at infinity corresponding to the direction d if it escapes to infinity in the direction of d. Projective transformations are not required to respect the parallelism relation, but they must send any straight line to a straight line, and respect the incidence properties; they depend on 15 dimensions.

Much larger is the *diffeomorphism group*, which preserves continuity and differential regularity of figures; it has an infinite number of dimensions.

More generally, Felix Klein suggested that a Geometry is associated to a group G and a subgroup H, in such a manner that the *points* that G transforms are the subsets of G of the form $xH = \{xh \in G|h \in H\}$ when x describes G; two elements x, y of G are said equivalent modulo H when $xH = yH$, and the set of equivalence classes is denoted by G/H and named a Klein space. However, if we adopt this definition, a special marked point appears in the space, which is H itself. To forget this special point, geometers introduced general *homogeneous spaces*, which are sets X where G acts in such a manner that, each time a point x is chosen in X, the stabilizer G_x of x (i.e. the subgroup of G formed by the elements g such that $g.x = x$), is conjugated to H; therefore X can be identified with a Klein space G/G_x.

However, for most of the useful geometries, it appeared that the set G/H, or the isomorphic homogeneous space X with its action of G, can be identified with the set of the subgroups gHg^{-1} that are conjugated to H; this is due to the fact that in these cases, the application of G onto $Int(G)$ is almost an embedding. Therefore, in such a case, the points of the given geometry identify with the subgroups fixing them.

For all the above examples of groups, there exist natural families of sub-groups that define convenient geometries, which are useful for Mathematics, Physics and Biology. For instance, the traditional Euclidean space, certainly the more useful of all for the ordinary life, is given by the set E of all sub-groups SO_P, where SO_P designates the set of rotations of any angles and any axis containing the point P. It can be easily seen that such a subgroup characterizes a point. Moreover, in this case, the group of translations acts transitively on E: for any pair of points (P, Q) the translation be the vector $\overrightarrow{PQ}$ gives a canonical choice of transformation g in $Iso_+(E)$ that conjugates SO_P to SO_Q, i.e. $SO_P = gSO_Qg^{-1}$. Therefore, once a point of origin is chosen, all other points are described by vectors.

Remark: it is necessary to distinguish a family of conjugated subgroups for characterizing a geometry, the whole group G is not sufficient by itself; this can be seen on a simple example: the group is $G = PSL_2(\mathbb{C})$ is made by the two by two matrices with complex coefficients, where g and λg are identified when λ is a nonzero complex number. If we consider in it, the the family of maximal compact sub-groups conjugated to unitary matrices $H = PU_2$, we get the setting for the hyperbolic space, i.e. the $3D$ simply connected complete Riemannian manifold with curvature -1. The pair G, H also appears to be the setting for the conformal geometry of the Euclidean plane, i.e. the science of circles, the geometry where distances are forgotten but angles are preserved. They are not really two different geometries, one is the natural boundary of the other. We see the pertinence of Klein's formulation. However, G gives also the setting for the complex projective line $\mathbb{P}^1(\mathbb{C})$; every projective transformation in dimension one over complex numbers is an homography. But the conjugated subgroups in this case have to be the stabilizers of lines through 0 in $\mathbb{C}^2$, or equivalently of points in $\mathbb{P}^1(\mathbb{C})$, which are the groups of an affine complex line, whose standard form is made by triangular matrices with one element of the diagonal equals to 1.

6.2 Categories and Topos

A *category* $\mathcal{C}$ is formed by a set of objects $\mathcal{C}_0$ and a family of sets of arrows $\mathcal{C}_1(a, b)$ between pairs of objects (a, b), such that, each time an arrow g goes from an object a to an object c and an arrow f goes from c to b, there exists an associated arrow $f \circ g$ from a to b, satisfying the two following axioms: (i) (associativity) for three consecutive arrows f, g, h we have $(f \circ g) \circ h = f \circ (g \circ h)$, ($ii$) (neutral elements) for each object c there is an identity arrow 1_c such that for each arrow g coming to c and each arrow f leaving c we have $g = 1_c \circ g$ and $f = f \circ 1_c$. An arrow from a to b in a category is also called a morphism from a to b, and the set of arrows from a to b is also denoted by $Mor(a, b)$ or $Mor_{\mathcal{C}}(a, b)$. A good reference is the book of Mac Lane [62].

A group G can be seen as a category with only one object e, and where any arrow g has an inverse g^{-1}, such that $1_e = gg^{-1} = g^{-1}g$.

Between two categories $\mathcal{C}$, $\mathcal{C}'$ there is also a good notion of morphism; it is a *functor* $T : \mathcal{C} \to \mathcal{C}'$, which associates to each object c an object $T(c)$ and to each arrow $f : a \to b$ an arrow $T(f) : T(a) \to T(b)$, sending the identity morphisms to the identity morphisms and respecting compositions. This kind of functor is named co-variant; there is also a notion of contra-variant functor which reverses the sense of arrows; that is $T^*(f)$ goes from $T^*(b)$ to $T^*(a)$. And between two functors $T : \mathcal{C} \to \mathcal{C}'$ and $S : \mathcal{C}' \to \mathcal{C}''$ of the same variance, there is also a nice notion of morphism: a *natural transformation* α, which associates to each object c of $\mathcal{C}$ of a morphism $\alpha(c) : T(c) \to S(c)$ in $\mathcal{C}'$, in such a way that for each morphism $f : a \to b$ in $\mathcal{C}$ we have the commutativity relation: $\alpha(b) \circ T(a) = S(b) \circ \alpha(a)$. As said by Mac Lane [63], "intuitively, a natural transformation is one which is defined in the same way or by the same formula for every object in the category in question."

The following example makes all these definitions quickly understandable; it corresponds to rectangular matrices of every finite dimensions over a field of numbers K; the set $\mathcal{M}_0$ is the set of natural integers $\mathbb{N}$, including 0, and for any pair (n, m) of natural integers the set $Mor_{\mathcal{M}}(n, m)$ is the set of $m \times n$ matrices with coefficients in K; when $n = m$ we have the identity matrix 1_n, and the composition in the category is given by the multiplication of matrices. An example of a much larger category $\mathcal{V}$ is given by the set of structures of finite dimensional K-vector spaces on a given large set, and by the family of linear applications between these spaces. An example of functor is the embedding ε of $\mathcal{M}$ in $\mathcal{V}$ which sends an integer n to K^n and a matrix M of size $m \times n$ to the operation of multiplication of M with a column vector in K^n. A less canonical functor β in the other direction, from $\mathcal{V}$ to $\mathcal{M}$, is obtained by choosing at the level of objects a basis on every vector space, and at the level of morphisms, the matrix which expresses a linear operator in the chosen basis at the source and the goal. Let us consider the two manner of composing these functors $\varepsilon \circ \beta$ and $\beta \circ \varepsilon$; a natural transformation A from the identity functor $Id_{\mathcal{V}}$ to $\varepsilon \circ \beta$ associates to V the map which sends each vector of V to its coordinates in K^n, and a natural transformation B from the identity functor $Id_{\mathcal{M}}$ to $\beta \circ \varepsilon$ associates to n the element 1_n. For any V, $A(V)$ is a bijection, and for any n, $B(n)$ is a bijection; it is said in such a case that ε and β defines an *equivalence of category*. If we think that the most important things are the properties of morphisms and not the properties of objects, the equivalence of two categories means that they contain essentially the same information; even if one of them contains much less objects, as this is the case for $\mathcal{M}$ with respect to $\mathcal{V}$.

Remark: in category theory it is frequent to encounter collections of objects that are not sets, for instance the collection of all the vector spaces, all the groups, or all the sets; the paradoxes found by Russell and Cantor, for instance, have shown that these examples cannot be sets; but the notion of collection is too vague to justify most aimed theorems, consequently the inventors (Eilenberg, Mac Lane, Grothendieck, ...) managed to constrain the objects and the arrows to be sets $\mathcal{C}_0$, $\mathcal{C}_1$. They tried to find a setting where most of the general constructions that could be expected for "collections" can be done, but also can be controlled. For instance the fact that there cannot exist a set of all sets is overturned by the introduction of a category Set_U, whose objects are the sets belonging to a given set U, which is named the *universe* and which verifies convenient properties. The same thing can be made for groups $Grps_U$ or K-vector spaces $Vect_U^K$. The axioms for a universe are the following ones: (1) if $C \in U$ and $a \in C$ then $a \in U$, (2) if C and D are elements of U, the set with two elements C, D belongs to U, and the set with one element (C, D) (the pair (C, D), which exists in every set theory), and also the product $C \times D$ (made by the pairs (c, d) for $c \in C$ and $d \in D$) belongs to U; (3) the set $\mathcal{P}(C)$ of parts of an element C of U belongs to U, as does the union $\bigcup C$ of all the elements x of the elements a of C; (4) natural integers are included in U; (5) if $C \in U$ and $D \subset U$ (i.e. $d \in D$ implies $d \in U$) and if there exists a surjection f from C onto D, it follows that $D \in U$.

(Again see [62].)

A problem is that the set of integers $\mathbb{N}$ is the only known universe, in fact the existence of other universes was proved to be undecidable in the usual axiomatics of set theory (as Zermelo–Frankel theory for instance). Real numbers are so useful that Grothendieck and Verdier have suggested to add to set theory the following axiom: for every set X there exists a universe U such that $X \in U$. There is no proof that it is safe, i.e. without contradiction, because this is already the case for ZF theory.

By definition, a category is a U-category when all the sets $Mor(a, b)$ are elements of U, a category is inside U when $\mathcal{C}_0$ and $\mathcal{C}_1$ are parts of U, a category is said to be U-small when $\mathcal{C}_0$ and $\mathcal{C}_1$ are elements of U.

In general, when $\mathcal{C}_0$ and $\mathcal{C}_1$ are only unspecified "collections", Mac Lane prefers to speak of a meta-category, and he reserves the name of category to the case where $\mathcal{C}_0$ and $\mathcal{C}_1$ are sets. In what follows, we assume that a universe U is chosen, but we do not mention it.

An important sort of category generalizes the notion of group; it is named a *groupoid*. The special axiom specifies that every arrow has an inverse; in particular for every a, the set $Mor(a, a)$ is a group.

When a is an object of a category $\mathcal{C}$, the category $\mathcal{C}|a$ (a fiber over a) has for objects the pairs (b, f) where b is an object of $\mathcal{C}$ and where $f \in Mor(b, a)$, and for set of morphisms between (b, f) and (c, g) the subset of the $h \in Mor(b, c)$ which satisfy $f = g \circ h$. A *refinement* (more frequently named a *sieve*) R of the object a is a sub-category of $\mathcal{C}|a$, that contains $(c, f \circ g)$ each time $g \in Mor(c, b)$ and (b, f) belongs to R.

A *topology* on a category $\mathcal{C}$, in the sense of Grothendieck, is made by the association of a set $J(a)$ of *refinements* to every object a, which satisfies the following axioms: (i) for every $f : b \to a$, if $R \in J(a)$, then $f^*R \in J(b)$, where f^*R is made by the $g : c \to b$ such that $f \circ g$ belongs to R; (ii) for any a and $R \in J(a)$, the refinement R' of a belongs to $J(a)$ if and only if, for any $b \to a$ in R, the category f^*R' belongs to $J(b)$. A *site* is a category equipped with a topology. In the examples below, most of them being in a finite setting, we take the *discrete topology*, where, by definition, for any a, the set $J(a)$ has only one element, which is the category $\mathcal{C}|a$ itself. All the site we will consider are U-small.

An usual topological space X gives a topology in this sense: the objects of the site are the open sets and its morphisms are the inclusions between them, and for every open set U, an element of $J(U)$ is an open covering of U, taken with all the finer coverings.

A *presheaf* on a category $\mathcal{C}$ is a contra-variant functor from $\mathcal{C}$ to the category *Set* of the sets in a given universe U. In the category *Set*, there is a notion of infinite product, and for any presheaf F on $\mathcal{C}$, and any refinement R of a in $\mathcal{C}$, if $F_R : R \to Set$ denotes the restriction, we can consider the projective limit $limF_R$, which is the subset of the product of all the $F(b)$ for $b \in R$, made by the families (x_b); $b \in R$, such that $f : c \to b$, implies $x_c = F(f)(x_b)$. A *sheaf* on a site $\mathcal{C}$ is a presheaf $F : \mathcal{C} \to Set$, such that, for any a and any $R \in J(a)$, the natural map from $F(a)$ to $limF_R$ is a bijection. In the discrete case, every presheaf is a sheaf.

Definition: A *topos* is a category isomorphic to the category of sheaves over a site.

The category *Set* is the simplest topos, associated to the site with one element. An interesting generalization is given by the category of G-sets, where G is a group, a G-set being a set on which G acts to the left, and a G-morphism being an equivariant map.

Most of the constructions that are possible in *Set*, are also possible in every topos. For instance projective and injective limits exist. In particular *Cartesian squares* do exist; they are defined as follows: if $f : a \to c$ and $g : b \to c$, there exists a (non-necessarily unique) object d, equipped with two morphisms $h : d \to a$, $k : d \to b$, satisfying $f \circ h = g \circ k$, such that for any pairs of morphisms $u : x \to a$, $v : x \to c$ satisfying $f \circ u = g \circ v$, there exists a unique morphism $w : x \to d$ which satisfies $u = h \circ w$ and $v = k \circ w$. The object d is denoted by $a \times_c b$.

In every topos, there exists a final object, i.e. an object 1 such that any for other object F the set $Mor(F, 1)$ has one and only one element. This object is interpreted as a singleton.

There exists also an initial object $\emptyset$ which is the empty functor, having a unique morphism to every object in the topos.

Starting with the category having two elements and one arrow between them, a very interesting generalization of set theory occurs, cf. Prouté [93], which corresponds to an intuitionist point of view, where a property can be true, false, or uncertain in various manners. We will come back soon to this example, named the topos of Shadoks.

First we have to generalize the notion of parts of an object in any category, where the notion of point is absent: an arrow $f : b \to a$ is said *monic* (or injective, or a monomorphism) if, for any pairs of arrows $g, h : c \to b$, the equality $f \circ g = f \circ h$ implies $g = h$. Two monics $f : b \to a$ and $f' : b' \to a$ are said equivalent, when there exist arrows (necessarily monics) g and h such that $f = f' \circ g$ and $f' = f \circ h$. By definition and equivalence class of monics going to a is a *subobject* of a in $\mathcal{C}$.

A special Cartesian square occurs when we consider a morphism $f : a \to c$ and an injective morphism $g : b \to c$; the particularity comes from the fact that $k : d \to b$ is determined by g; in this case, $d = a \times_c b$ is named the *pull-back* of the sub-object X defined by g and is written $f^{-1}(X)$; the morphism $h : d \to a$ is a monic. Its universal property is that a morphism $u : x \to a$ can be factorized by a morphism to d (as $u = h \circ w$) if and only if the morphism $f \circ u$ can be factorized by a morphism to b (as $f \circ u = g \circ v$). Therefore in a topos, even if points do not exist, every morphism $f : a \to c$ induces a map f^{-1} from the sub-objects of c to the sub-objects of a.

The relation with intuitionist logic comes from the fact that every set *Sub*(X), of the sub-objects of an object X, possesses a natural structure of Heyting pre-algebra (cf. [93]). This is the natural origin of *contextual logic*.

An important property of a topos $\mathcal{F}$ is the existence of a classifying object for sub-objects, i.e. an object Ω marked with a special sub-object T, which is given by a morphism from 1 to Ω, such that for any monic $f : G \to F$ there exists a unique morphism $\chi_f : F \to \Omega$ satisfying $T \circ 1_G = \chi_f \circ f$, i.e. G is the pull-back of T by χ_f. Moreover the correspondence between sub-objects f and morphisms χ_f is natural, in the sense that sub-objects of sub-objects go to composition of characteristic morphisms χ, and so on. The sub-objects of Ω give the different values of truth of the logic associated to the topos; for instance T is true, $\emptyset$ is false.

As we said, a simple and surprising example of topos is given by the category *Shad* made by the presheaves over the category with two objects A, B and three arrows, which are the neutral elements 1_A, 1_B and only one more element $\alpha : A \to B$; it is named the topos of Shadoks by Alain Prouté [93]. An object of this topos can be seen as a pair of sets X, Y (X for the shadoks set, Y for their eggs) respectively associated to A and B, and a map f from Y to X, associated to α, like the map which associates to each egg its unique parent. The singleton 1 of *Shad* is given by a point 1 for A and a point 1 for B. But there is a natural embedding of *Set* in *Shad*, made by the shadoks without eggs, which corresponds to the case where Y is empty. In particular, the singleton of *Set* is an intermediary object between the empty element of *Shad* and the singleton of *Shad*; it is denoted by the symbol $1/2$.

A morphism in *Shad* is a pair of maps $G : Y \to Y'$, $F : X \to X'$ making a commutative diagram, i.e. $f' \circ G = F \circ f$. This morphism is monic (resp. epic) if and only if the two maps are injective (resp; surjective). A sub-object of $f : Y \to X$ can be represented by a subset X' of X and a subset Y' of Y (necessarily empty if X'

is empty) such that $f(Y') \subset X'$. If a map $(F, G) : (Z, W) \to (X, Y)$ is given, the pull-back of (X', Y') is simply given by $(F^{-1}(X'), G^{-1}(Y'))$.

Let us consider the sheaf Ω defined by $\Omega(A) = \{0, 1\}$, $\Omega(B) = \{0, 1/2, 1\}$, and $\Omega(\alpha)$ sending 0 to 0 and 1 and 1/2 to 1. We write T for the sub-object of source the shadok singleton 1 (one bird and one egg), which sends the egg to 1 in $\Omega(B)$ and the bird to 1 in $\Omega(A)$. Now, given the sub-object $S = (X', Y')$ of $f : Y \to X$, we define χ_S by sending X' to 1 and $X \backslash X'$ to 0 (no choice here if we want to recover the Boolean characteristic function of set theory), and by sending Y' to 1 (the only possibility if we want that $S = \chi_S^{-1}(T)$), the elements of $Y \backslash Y'$ that are sent in X' by f to 1/2 (the only possibility if we want that χ_S is a morphism), and the rest of the elements of Y to 0 (also the only possibility if we want that χ_S is a morphism). This proves that Ω and T are respectively the classifying space of sub-objects and its universal element.

We also see, by comparison with the case of sets, that the sub-object obtained by sending $1(A)$ to 0 in $\Omega(A)$ and $1(B)$ to 0 in $\Omega(B)$ can be interpreted as describing the failure of S, a strict complementary, but the third possible sub-object of source 1, which is obtained by sending $1(A)$ to 1 in $\Omega(B)$ and $1(B)$ to 1/2 in $\Omega(A)$, gives neither the failure of S neither its success, in some sense, if the egg is considered as the future of the bird, the temporary success of S at the level X becomes a failure at the level of Y, but it could be better to tell that an incertitude is maintained here. This clearly gives a logic with more possibilities than true or false, something like undecidable.

A group object in *Shad* is an homomorphism of ordinary groups $\varphi : G_B \to G_A$, a subgroup is a pair of respective sub-groups H_A, H_B such that $\varphi(H_B) \subset H_A$.

For spaces, behind the usual objective space G_A/H_A or the set of groups conjugated to H_A, several situations are distinguished: the elements of the complementary subset of H_B in G_B that go to H_A, and the elements that do not.

In [7] it is shown that the Shannon information quantities have their origin in the first co-homology of a module associated to probability laws for a canonical sheaf over the site of random variables for observation of a system. Relations with Galois theory was also suggested in this article, and relation with geometry has to be developed. It would be nice to go one step further and connect this information co-homology or a derived more homotopical theory to the structures that are needed in ternary structures for adaptation.

References

1. P. Andersen, R. Morris, D. Amaral, T. Bliss, J. O'Keefe, *The Hippocampus Book* (Oxford University Press, Oxford, 2006)
2. R.A. Andersen, Visual and eye movement function of the posterior parietal cortex. Annu. Rev. Neurosci. **12**, 377–403 (1989)
3. D.E. Angelaki, A.G. Shaikh, A.M. Green, Neurons compute internal models of the physical laws of motion. Nature **430**(6999), 560–564 (2004)

4. G. Arechavaleta, J.-P. Laumond, H. Hicheur, D. Le Bihan, A. Berthoz, The nonholonomic nature of human locomotion: a modeling study, in *Biomedical Robotics and Biomechatronics* (2006), pp. 158–163
5. A.E. Arnold, G. Iaria, A.D. Ekstrom, Mental simulation of routes during navigation involves adaptive temporal compression. Cognition **25**(157), 14–23 (2016)
6. J. Barra, L. Laou, J.-B. Poline, D. Le Bihan, A. Berthoz, Does an oblique/slanted perspective during virtual navigation engage both egocentric and allocentric brain strategies? PLoS One **7**(11), e49537 (2012)
7. P. Baudot, D. Bennequin, The homological nature of entropy. Entropy **17**(5), 3253–3318 (2015)
8. D. Bennequin, Remarks on invariance in the primary visual systems of mammals, in *Neuromathematics of Vision*, ed. by G. Citti, A. Sarti (Springer, Berlin, 2014), pp. 243–333
9. D. Bennequin, A. Berthoz, Non-linear Galilean vestibular receptive fields. Conf. Proc. IEEE Eng. Med. Biol. Soc. **2011**, 2273–2276 (2011)
10. D. Bennequin, R. Fuchs, A. Berthoz, T. Flash, Movement timing and invariance arise from several geometries. PLoS Comput. Biol. **5**(7), e1000426 (2009)
11. N.A. Bernstein, *The Co-ordination and Regulation of Movements* (Pergamon Press, New York, 1967)
12. A. Berthoz, Reference frames for the perception and control of movement, in *Brain and Space*, ed. by J. Paillard (Oxford University Press, Oxford, 1991), pp. 81–111
13. A. Berthoz, Parietal and hippocampal contribution to topokinetic and topographic memory. Philos. Trans. R. Soc. **352**, 1437–1448 (1997)
14. A. Berthoz, *Le sens du mouvement* (Odile Jacob, 1997)
15. A. Berthoz, *La décision* (Odile Jacob, 2003)
16. A. Berthoz, *Simplexity. How to Deal with a Complex World* (Yale University Press, New Haven, 2011)
17. A. Berthoz, *La vicariance* (Odile Jacob, 2013). transl. *Vicariousness* (Harvard University Press, 2016)
18. A. Berthoz, W. Graf, P.P. Vidal, *The Head-Neck Sensory Motor System* (Oxford University Press, Oxford, 1992)
19. A. Binet, J. Courtier, Sur la vitesse des mouvements graphiques. Rev. Philos. **25**, 664–671 (1893)
20. M. Bompard-Porte, D. Bennequin, *Pulsions et politique. Suivi de Le non-être homologique* (Editions L'Harmattan, Paris, 1998)
21. N.E. Burgess, K.J. Jeffery, J.E. O'Keefe, *The Hippocampal and Parietal Foudations of Spatial Cognition* (Oxford University Press, Oxford, 1999)
22. E. Burguiere, A. Arleo, M. reza Hojjati, Y. Elgersma, C.I. De Zeeuw, A. Berthoz, L. Rondi-Reig, Spatial navigation impairment in mice lacking cerebellar LTD: a motor adaptation deficit? Nat. Neurosci. **8**(10), 1292–1294 (2005)
23. E. Cartan, *La théorie des groupes finis et continus et la géométrie différentielle traitées par la méthode du repère mobile* (Gauthier-Villars, 1937)
24. G. Committeri, G. Galati, A.-L. Paradis, L. Pizzamiglio, A. Berthoz, D. Le Bihan, Reference frames for spatial cognition: different brain areas are involved in viewer-, object- and landmark-centered judgments about object location. J. Cogn. Neurosci. **16**(9), 1517–1535 (2004)
25. J. Decety, Do imagined and executed actions share the same neural substrate? Brain Res. Cogn. Brain Res. **3**(2), 87–93 (1996)
26. J. Decety, M. Jeannerod, C. Prablanc, The timing of mentally represented actions. Behav. Brain Res. **34**(1–2), 35–42 (1989)
27. M. Dimiccoli, B. Girard, A. Berthoz, D. Bennequin, Striola magica. A functional explanation of otolith geometry. J. Comput. Neurosci. **35**(2), 125–154 (2013)
28. J. Droulez, D. Bennequin, Perception des symétries et invariances perceptives, in *Symétries, symétries et asymétries du vivant*, ed. by M. Siksou (Lavoisier, 2005), pp. 155–171

29. G.M. Edelman, *The Remembered Present: A Biological Theory of Consciousness* (Basic Books, New York, 1989)
30. T. Flash, The control of hand equilibriul trajectories in multi-joint arm movements. Biol. Cybern. **57**, 257–274 (1987)
31. T. Flash, A. Handzel, Affine differential geometry of human arm trajectories. Abstr. Soc. Neurosci. **22**, 1635 (1996)
32. T. Flash, A. Handzel, Affine differential geometry analysis of human arm movements. Biol. Cybern. **96**(6), 577–601 (2007)
33. T. Flash, N. Hogan, The coordination of arm movements: an experimentally confirmed mathematical model. J. Neurosci. **5**(7), 1688–1703 (1985)
34. T. Flash, Y. Meirovitch, A. Barliya, Models of human movement: trajectory planning and inverse kinematics studies. Robot. Auton. Syst. **61**(4), 330–339 (2013)
35. O. Faugeras, Cartan's moving frame method and its application to the geometry and evolution of curves in the Euclidean, affine and projective planes. Preprint Inria (1994)
36. K. Friston, The free-energy principle: a unified brain theory? Nat. Rev. Neurosci. **11**(2), 127–138 (2010)
37. G. Galati, G. Pelle, A. Berthoz, G. Committeri, Multiple reference frames used by the human brain for spatial perception and memory. Exp. Brain Res. **206**(2), 109–120 (2010)
38. É. Galois, *Œuvres mathématiques* (Gauthier-Villars et fils, 1897)
39. J.J. Gibson, E.J. Gibson, Continuous perspective transformations and the perception of rigid motion. J. Exp. Psychol. **54**(2), 129–138 (1957)
40. J.M. Goldberg, V.J. Wilson, K.E. Cullen, *The Vestibular System: A Sixth Sense* (Oxford University Press, Oxford, 2012)
41. W. Graf, F. Klam, The vestibular system: functional and comparative anatomy, evolution and development. Comptes Rendus Palevol **5**(3–4), 637–655 (2006)
42. A. Grothendieck, J.-L. Verdier, *Théorie des topos (sga 4, exposés i-vi)*, Springer Lecture Notes in Mathematics, pp. 269–270
43. O.J. Grüsser, T. Landis (eds.), *Visual Agnosias and Other Disturbances of Visual Perception and Cognition* (Macmillan, London, 1991)
44. N.K. Harpaz, T. Flash, I. Dinstein, Scale-invariant movement encoding in the human motor system. Neuron **81**, 452–462 (2014)
45. C.M. Harris, D.M. Wolpert, Signal-dependent noise determines motor planning. Nature **394**(6695), 780–784 (1998)
46. B. Heider, A. Karnik, N. Ramalingam, R.M. Siegel, Neural representation during visually guided reaching in macaque posterior parietal cotex. J. Neurophysiol. **104**, 3494–3509 (2010)
47. H. Hicheur, S. Vieilledent, M.J. Richardson, T. Flash, A. Berthoz, Velocity and curvature in human locomotion along complex curved paths: a comparison with hand movements. Exp. Brain Res. **162**(2), 145–154 (2005)
48. H. Hicheur, Q.-C. Pham, G. Arechavaleta, J.-P. Laumond, A. Berthoz, The formation of trajectories during goal-oriented locomotion in humans. I. A stereotyped behaviour. Eur. J. Neurosci. **26**(8), 2376–2390 (2007)
49. N. Hogan, An organizing principle for a class of voluntary movements. J. Neurosci. **4**(11), 2745–2754 (1984)
50. K. Igloi, C.F. Doeller, A. Berthoz, L. Rondi-Reig, N. Burgess, Lateralized human hippocampal activity predicts navigation based on sequence or place memory. Proc. Natl. Acad. Sci. USA **107**(32), 14466–14471 (2010)
51. A. Imhausen, *Mathematics in Ancient Egypt. A Contextual History* (Princeton University Press, Princeton, 2016)
52. M.M. Jankowski, J. Passecker, M.N. Islam, S. Vann, J.T. Erichsen, J.P. Aggleton, S.M. O'Mara, Evidence for spatially-responsive neurons in the rostral thalamus. Front. Behav. Neurosci. **9**, 256 (2015)
53. F. Klein, *Vergleichende Btrachtungen über neurer geometrische Forschungen*. Reedition H. Wussing, Das Erlanger Programm, 1974 (Deichert, 1872)
54. J.J. Koenderink, A.J. van Doorn, Facts on optic flow. Biol. Cybern. **56**(4), 247–254 (1987)

55. J.J. Koenderink, A.J. van Doorn, *Pictorial Space* (MIT Press, Cambridge, 2003)
56. F. Lacquaniti, C. Terzuelo, P. Viviani, The law relating the kinematic and figural aspects of drawing movements. Acta Psychol. **54**(1), 115–130 (1983)
57. F. Lacquaniti, G. Bosco, S. Gravano, I. Indovina, B. La Scaleia, V. Maffei, M. Zago, Gravity in the brain as a reference for space and time perception. Multisens. Res. **28**(5–6), 397–426 (2015)
58. S. Lambrey, C. Doeller, A. Berthoz, N. Burgess, Imagining being somewhere else: neural basis of changing perspective in space. Cereb. Cortex **22**, 166–174 (2011)
59. S.B. Laughlin, The role of sensory adaptation in the retina. J. Exp. Biol. **146**(1), 39–62 (1989)
60. N. Levit-Binnun, E. Schechtman, T. Flash, On the similarities between the perception and production of elliptical trajectories. Exp. Brain Res. **172**, 533–555 (2006)
61. R.R. Llinás, *I of the Vortex: From Neurons to Self* (MIT Press, Cambridge, 2002)
62. S.M. Lane, *Categories for the Working Mathematician*, vol. 5 (Springer, New York, 1998)
63. S.M. Lane, *Homology* (Springer, New York, 2012)
64. U. Maoz, A. Berthoz, T. Flash, Complex unconstrained three-dimensional hand movement and constant equi-affine speed. J. Neurophysiol. **101**(2), 1002–1015 (2009)
65. P. Marianelli, A. Bethoz, D. Bennequin, Crista egregia: a geometrical model of the crista ampullaris, a sensory surface that detects head rotations. Biol. Cybern. **109**(1), 5–32 (2015)
66. J. Mc Intyre, M. Zago, A. Berthoz, F. Lacquaniti, Does the brain model Newton's laws? Nat. Neurosci. **4**, 693–694 (2001)
67. Y. Meirovitch, D. Bennequin, T. Flash, Geometrical invariance and smoothness maximization for task-space movement generation. Preprint (2016)
68. D. Mumford, On the computational architecture of the neocortex. Biol. Cybern. **65**(2), 135–145 (1991)
69. D. Mumford, On the computational architecture of the neocortex. Biol. Cybern. **65**(3), 241–251 (1992)
70. J. Munzert, Temporal accuracy of mentally simulated transport movements. Percept. Mot. Skills **94**(1), 307–318 (2002)
71. F. Nemmi, M. Boccia, L. Piccardi, G. Galati, C. Guariglia, Segregation of neural circuits involved in spatial learning in reaching and navigational space. Neuropsychologia **51**(8), 1561–1570 (2013)
72. P.J. Olver, A survey of moving frames, *Computer Algebra and Geometric Algebra with Applications* (Springer, Berlin, 2005), pp. 105–138
73. J. Paillard (ed.), *Brain and Space* (Oxford University Press, New York, 1991)
74. A. Pellionisz, R. Llinás, Tensorial approach to the geometry of brain function: cerebellar coordination via a metric tensor. Neuroscience **5**(7), 1125–1136 (1980)
75. A. Pellionisz, R. Llinás, Tensor network theory of the metaorganization of functional geometries in the central nervous system. Neuroscience **16**(2), 245–273 (1985)
76. W. Penfield, E. Boldrey, Somatic, motor and sensory represntation in the cerebral cortex of man as studied by electirc stimulation. Bran **60**, 389–443 (1937)
77. B. Peterson, Current approaches and future directions to understanding control of head movement. Prog. Brain Res. **143**, 369–381 (2004)
78. B. Peterson, J. Baker, E. Keshner, Multidimensional analysis of head stabilization-progress and prospects, *The Head-Neck Sensory Motor System* (Oxford University Press, Oxford, 2012)
79. Q.-C. Pham, D. Bennequin, Affine invariance of human hand movements: a direct test (2012). arXiv:1209.1467
80. Q.-C. Pham, Y. Nakamura, A new trajectory deformation algorithm based on affine transformations. IEEE Trans. Robot. **31**(4), 1054–1063 (2015)
81. Q.-C. Pham, H. Hicheur, G. Arechavaleta, J.-P. Laumond, A. Berthoz, The formation of trajectories during goal-oriented locomotion in humans. II. A maximum smoothness model. Eur. J. Neurosci. **26**(8), 2391–2403 (2007)
82. D. Philipona, K. O'Regan, J.-P. Nadal, Is there something out there? Inferring space from sensorimotor dependencies. Neural Comput. **15**(9), 2029–2049 (2003)

83. J. Piaget, B. Inhelder, *La représentation de l'espace chez l'enfant* (Presses Universitaires de France, 1948)
84. J. Piaget, B. Inhelder, A. Szeminska, *La géométrie spontanée de l'enfant* (Presses Universitaires de France, 1948)
85. L. Piccardi, F. Bianchini, R. Nori, A. Marano, F. Iachini, L. Lasala, C. Guariglia, Spatial location and pathway memory compared in the reaching vs. walking domains. Neurosci. Lett. **30**(566), 226-230 (2014)
86. H. Poincaré, *La Science et l'hypothèse* (Flammarion, 1902)
87. H. Poincaré, *La Valeur de la science*. Bibliothèque de philosophie scientifique (Flammarion, 1905)
88. H. Poincaré, *Science et méthode*. Bibliothèque de philosophie scientifique (Flammarion, 1908)
89. H. Poincaré, *Dernières Pensées*. Bibliothèque de philosophie scientifique (Flammarion, 1913)
90. F.E. Pollick, G. Sapiro, Constant affine velocity predicts the 1/3 power law of planar motion perception and generation. Vis. Res. **37**(3), 347–353 (1997)
91. F. Polyakov, R. Drori, Y. Ben-Shaull, M. Abeles, T. Flash, A compact representation of drawing movements with sequences of parabolic primitives. PLoS Comput. Biol. **5**(7), e1000427 (2009)
92. F. Polyakov, E. Stark, R. Drori, M. Abeles, T. Flash, Parabolic movement primitives and cortical states: merging optimality with geometric invariance. Biol. Cybern. **100**(2), 159–184 (2009)
93. A. Prouté, Introduction à la logique catégorique. Cours à Paris 7. Preprint (2016)
94. R. Rashed, *The Development of Arabic Mathematics: Between Arithmetic and Algebra*, vol. 156 (Springer Science and Business Media, Berlin, 2013)
95. E. Robson, *Mathematics in Ancient Iraq: A Social History* (Princeton University Press, Princeton, 2008)
96. R.W. Sharpe, *Differential Geometry: Cartan's Generalization of Klein's Erlangen Program* (Springer, New York, 1997)
97. R.E. Shaw, M. McIntyre, W.M. Mace, The role of symmetry in event perception, *Perception: Essays in Honor of James J. Gibson* (Cornell University Press, Ithaca, 1974)
98. S.M. Sherman, R.W. Guillery, *Exploring the Thalamus and Its Role in Cortical Function*, 2nd edn. (MIT Press, Cambridge, 2006)
99. J.J.E. Slotine, Modular stability tools for distributed computation and control. Int. J. Adapt. Control Signal Process. **17**(6), 397–416 (2002)
100. J.J. Slotine, W. Lohmiller, Modularity, evolution, and the binding problem: a view from stability theory. Neural Netw. **14**(2), 137–145 (2001)
101. R. Thom, Stabilité structurelle et morphogenèse: essai d'une théorie générale des modèles. Interédition (1977)
102. R. Thom, Esquisse d'une sémiophysique. physique aristotéliciene et théorie des catastrophes (1988)
103. S. Vieilledent, Y. Kerlirzin, S. Dalbera, A. Berthoz, Relationship between velocity and curvature of a human locomotor tralectory. Neurosci. Lett. **305**(1), 65–69 (2001)
104. H. von Helmholtz, The origin and meaning of geometrical axioms. Mind **3**, 301–321 (1876)
105. P.H. Weiss, J.C. Marshall, G. Wunderlich, L. Tellmann, P.W. Halligan, H.-J. Freund, K. Zilles, G.R. Fink, Neural consequences of acting in near versus far space: a physiological basis for clinical dissociations. Brain **123**(12), 2531–2541 (2000)
106. F. Wolf, T. Geisel, Universality in visual cortical pattern formation. J. Physiol.-Paris **97**(2), 253–264 (2003)
107. T.A. Yakusheva, A.G. Shaikh, A.M. Green, P.M. Blazquez, J.D. Dickman, D.E. Angelaki, Purkinje cells in posterior cerebellar vermis encode motion in an inertial reference frame. Neuron **54**(6), 973–985 (2007)

On the Duration of Human Movement: From Self-paced to Slow/Fast Reaches up to Fitts's Law

Frédéric Jean and Bastien Berret

Abstract In this chapter, we present a mathematical theory of human movement vigor. At the core of the theory is the concept of the cost of time. According to it, natural movement cannot be too slow because the passage of time entails a cost which makes slow moves undesirable. Within this framework, an inverse methodology is available to reliably and robustly characterize how the brain penalizes time from experimental motion data. Yet, a general theory of human movement pace should not only account for the self-selected speed but should also include situations where slow or fast speed instructions are given by an experimenter or required by a task. In particular, the limit case of a "maximal speed" instruction is linked to Fitts's law, i.e. the speed/accuracy trade-off. This chapter first summarizes the cost of time theory and the procedure used for its accurate identification. Then, the case of slow/fast movements is investigated but changing the duration of goal-directed movements can be done in various ways in this framework. Here we show that only one strategy seems plausible to account for both slow/fast and self-paced reaching movements. By relying upon a free-time optimal control formulation of the motor planning problem, this chapter provides a comprehensive treatment of the linear-quadratic case for single degree of freedom arm movements but the principles are easily extendable to multijoint and/or artificial systems.

1 Introduction

Everyday actions are usually performed at a pace that people would commonly qualify of "comfortable", which is neither too fast nor too slow. Movement duration or average speed are inherent characteristics of biological and artificial sensorimotor

---

F. Jean (✉)
Unité de Mathématiques Appliquées, ENSTA ParisTech, Université Paris-Saclay, 91120 Palaiseau, France
e-mail: frederic.jean@ensta-paristech.fr

B. Berret
CIAMS, Univ. Paris-Sud, Université Paris-Saclay, 91405 Orsay, France
e-mail: bastien.berret@u-psud.fr

J.-P. Laumond et al. (eds.), *Geometric and Numerical Foundations of Movements*, Springer Tracts in Advanced Robotics 117, DOI 10.1007/978-3-319-51547-2_3

control, a process that takes place both in space and time. Understanding the underpinnings of movement pace formation is of crucial importance not only in motor neuroscience (as many disorders lead to bradykinesia, [3, 36]) but also in fields where humans are brought to interact with artificial systems, such as humanoid robotics, robot-assisted rehabilitation, neuroprosthetics or computer animation. The presence of temporal discrepancies may considerably affect the way humans perceive and collaborate with such entities. More generally, to improve the human-likeness of artificial sensorimotor systems, high-level computational principles leading to appropriate movement pace must be developed. In human motor control, most research efforts on the topic have been turned toward specific paradigms such as the speed/accuracy trade-off [23, 24, 49] where movements are assumed to be performed as fast as possible for a given level of accuracy (see [15, 41], for reviews). This empirical observation has been formalized as Fitts's law [17] and successfully implemented in human-computer interaction to model movement time [34]. An interesting observation is that any system assuming an exponential decay of the distance left to the center of the target will trivially yield Fitts's law [10, 12]. Actually, robotic studies often exploit this property to drive reliably a robot to some desired spatial target in an adjustable amount of time (e.g. [35]). This is typically achieved by tuning a parameter that the modeler must set by hand. A similar tuning of parameters is required to vary movement time when using PID controllers and even more involved feedback schemes (e.g. [40]). Therefore, task duration is often hard coded by fixing a desired movement time at the planning stage or merely results from the application of a (possibly finely tuned) feedback gain at the execution stage. The approach undertaken in this chapter lies in-between.

A recent hypothesis advanced the idea that the duration of biological movement could be driven by a "cost of time" [44, 46, Chap. 11]. In this view, slow movements are undesirable because the passage of time incurs a cost: it is "better" to achieve a task soon than later. This would be a property of the neural controller for reasons that may relate to the functioning of the reward system (i.e. temporal discounting of reward [44, 47]) via the cortico-basal ganglia loop. Movement vigor may indeed originate from the basal ganglia [45, 54] and its interaction with cortical areas encoding movement speed [11, 28]. In [4], we developed an inverse approach allowing to automatically infer, from experimental data, what would be the cost of time for reaching movements. The time cost then proved to allow elaborating and predicting the duration of upcoming reaching movements of various amplitudes and directions performed at a self-selected speed: motion time was thus an emergent property of the motor preparation stage. Here, we further analyze how this framework can embrace task instructions such as "move slow" or "move fast". We also give an account of Fitts's law in this context. This work was conducted within the optimal control framework and, more precisely, the free-time optimal control formalism. Optimal control theory relies upon the choice of cost functions that define what is optimal behavior for a given system [50, 51]. One great feature of optimal control is the high-level of abstraction that it enables, allowing to easily port findings from biological to artificial systems and vice-versa. For our purpose, we shall distinguish between subjective and objective cost functions throughout the chapter. An objective cost function is specified or imposed by the task itself. Typical examples are the specification of a target location (e.g. endpoint error) or a reference trajectory to track (e.g. draw an ellipse). In contrast, a subjective cost function is specified by the sensorimotor system itself and crucially serves to resolve the remaining degrees of freedom that are left free by the (redundant) task. It may measure energy expenditure, effort, jerk or any other quantity such as the cost of time which is at the core of the present work.

This chapter is organized as follows. First, we briefly review how the cost of time can be characterized unequivocally from real data in the proposed framework. We then analyze quite extensively the linear quadratic case and explain how the theory can account for speed changes resulting from explicit constraints given by an experimenter such as Fitts's like instructions. Throughout the chapter, we give a theoretical treatment of the problem together with illustrations in the context of a single degree of freedom arm performing reaching movements in a horizontal plane. The concepts are however easily transferable to more complex systems and tasks.

2 Theory and Results

2.1 *Theory of the Cost of Time*

The present theory is derived within the framework of optimal control (OC) theory, which assumes that the signature of human movement is optimality (with respect to a certain cost function) [50]. It implicitly supposes that the trajectories triggered by the central nervous system can be accounted for by a certain infinitesimal cost $h(\mathbf{x}, \mathbf{u}, t)$, which depends on the system state $\mathbf{x} \in \mathbb{R}^n$, the motor command $\mathbf{u} \in \mathbb{R}^m$ and the time $t \in I \subset \mathbb{R}$, respectively. In a sense, biological trajectories would adhere to a principle of least action where the "action" would be the time integral of h. In seminal studies assuming this framework [18, 38, 55], the time window of integration was set a priori by the modeler: movement time was simply fixed in accordance with experimental measurements. However, since movement time or average speed are motor decision variables, then a free-time formulation of the problem should rather be used [30, 42]. In this way, the duration of movement would emerge implicitly from the optimality of behavior, as already proposed by [25] who assumed to penalize the total motion duration itself. In the same vein, at the core of the present theory aiming to account for the vigor of movement is the idea of the "cost of time" (CoT) [4, 44]. The theory assumes that h can be separated into a term that penalizes time only, $g(t)$ (the infinitesimal CoT), plus a term that depends on the state/control variables, $l(\mathbf{x}, \mathbf{u})$, which allows to shape the trajectories followed by the system. Thus, if $h(\mathbf{x}, \mathbf{u}, t) = g(t) + l(\mathbf{x}, \mathbf{u})$, a mathematical analysis shows that it is actually possible to compute the value $g(t)$ by resolving an OC problem in fixed time t with known initial/final states (denoted by $\mathbf{x}^0$ and $\mathbf{x}^f$ respectively), given a system dynamics $\frac{d\mathbf{x}}{dt} = \dot{\mathbf{x}} = \mathbf{f}(\mathbf{x}, \mathbf{u})$ and a trajectory cost $l(\mathbf{x}, \mathbf{u})$. We briefly recall how this is achieved but the reader is referred to [4] for more details.

Given an input $\mathbf{u}(\cdot)$ defined on an interval $[0, t_\mathbf{u}]$, we denote by $\mathbf{x}_\mathbf{u}(\cdot)$ the trajectory of $\dot{\mathbf{x}}(t) = \mathbf{f}\big(\mathbf{x}(t), \mathbf{u}(t)\big)$ satisfying $\mathbf{x}_\mathbf{u}(t_\mathbf{u}) = \mathbf{x}^f$. As explained above, we consider the following cost function:

$$C(\mathbf{u}, t_\mathbf{u}) = \int_0^{t_\mathbf{u}} \Big(g(t) + l\big(\mathbf{x}_\mathbf{u}(t), \mathbf{u}(t)\big)\Big)\, dt, \tag{1}$$

where the functions g and l are non-negative. The function l has been the subject of extensive investigations in motor control (e.g. [7, 18, 55]) and may capture both subjective (related to an individual's decision) and objective (task related) goals. The trajectory cost $l(\mathbf{x}, \mathbf{u})$ is assumed to be known or identifiable (in fixed time OC formulations). The function g is the infinitesimal (i.e. instantaneous) CoT we can identify and whose antiderivative is the actual CoT, $G(t) = \int_0^t g(s)ds$ (we assume $G(0) = 0$ for simplicity).

We consider the following *free-time* OC problems:

Given an initial state $\mathbf{x}^0$, *minimize the cost* $C(\mathbf{u}, t_\mathbf{u})$ *among all inputs* $\mathbf{u}(\cdot)$ *and all times* $t_\mathbf{u}$ *such that* $\mathbf{x}_\mathbf{u}(0) = \mathbf{x}^0$ *and* $\mathbf{x}_\mathbf{u}(t_\mathbf{u}) = \mathbf{x}^f$ *(by definition of* $\mathbf{x}_\mathbf{u}$*).*

We will assume the existence of minimal solutions $\mathbf{u}(\cdot)$ with a finite time $t_\mathbf{u}$, which may be guaranteed under some technical conditions on the dynamics and on the cost [31].

Next, let $V_{\mathbf{x}^f}(t, \mathbf{x}^0)$ be the value function[1] of the OC problem joining $\mathbf{x}^0$ to $\mathbf{x}^f$ in fixed-time t, that is

$$V_{\mathbf{x}^f}(t, \mathbf{x}^0) = \inf \int_0^t l\big(\mathbf{x}_\mathbf{u}(s), \mathbf{u}(s)\big)ds, \tag{2}$$

where the infimum is taken among all inputs $\mathbf{u}(\cdot)$ such that $\mathbf{x}_\mathbf{u}(0) = \mathbf{x}^0$ and $\mathbf{x}_\mathbf{u}(t) = \mathbf{x}^f$. It is the optimal cost of a motion in time t between $\mathbf{x}^0$ and $\mathbf{x}^f$.

Then the movement time τ, that is the time $t_\mathbf{u}$ of an optimal solution $\mathbf{u}(\cdot)$ of the free-time OC problem, satisfies

$$\tau \in \operatorname{argmin}_{t \geq 0} \left(\int_0^t g(s)ds + V_{\mathbf{x}^f}(t, \mathbf{x}^0)\right), \tag{3}$$

and, assuming that $V_{\mathbf{x}^f}$ is differentiable with respect to t, we get:

$$g(\tau) = -\frac{\partial V_{\mathbf{x}^f}}{\partial t}(\tau, \mathbf{x}^0). \tag{4}$$

It is well-known from the Hamilton–Jacobi–Bellman theory that $\frac{\partial V_{\mathbf{x}^f}}{\partial t}(\tau, \mathbf{x}^0) = \mathscr{H}_0^\star\big(\mathbf{x}(\tau), \mathbf{p}(\tau)\big)$, with $\mathscr{H}_0^\star(\mathbf{x}, \mathbf{p}) = \max_\mathbf{v} \mathscr{H}_0(\mathbf{x}, \mathbf{p}, \mathbf{v})$ where $\mathscr{H}_0 = \mathbf{p}^\top \mathbf{f}(\mathbf{x}, \mathbf{v}) +$

[1] Note that we did not use the standard way to define the value function: for a movement duration equal to t, this is usually $\tilde{V}_{\mathbf{x}^f}(w, \mathbf{x}^0(w)) = \inf \int_w^t l\big(\mathbf{x}_\mathbf{u}(s), \mathbf{u}(s)\big)ds$. Here we set $V_{\mathbf{x}^f}(t - w, \mathbf{x}^0(w)) = \tilde{V}_{\mathbf{x}^f}(w, \mathbf{x}^0(w))$, hence $\frac{\partial V_{\mathbf{x}^f}}{\partial t} = -\frac{\partial \tilde{V}_{\mathbf{x}^f}}{\partial t}$.

$l(\mathbf{x}, \mathbf{v})$ is the Hamiltonian associated with the fixed-time OC problem,[2] $\mathbf{x}(t)$ is an optimal solution, and $\mathbf{p}(t) \in \mathbb{R}^n$ is the *co-state vector* [42]. Since it is obvious that the corresponding optimal control $\mathbf{u}(\cdot)$ is also a minimal solution of the OC problem in *fixed time* τ we then have $\mathscr{H}_0^\star(\mathbf{x}(\tau), \mathbf{p}(\tau)) = \mathscr{H}_0(\mathbf{x}(\tau), \mathbf{p}(\tau), \mathbf{u}(\tau))$, we get in this way $g(\tau) = -\mathscr{H}_0(\mathbf{x}(\tau), \mathbf{p}(\tau), \mathbf{u}(\tau))$.

Interestingly, the above analysis shows that the derivation extends to stochastic settings [48, 51]. In particular in the linear quadratic Gaussian (LQG) case, the infinitesimal CoT can be easily computed because the value function has a parametric form whose parameters can be evaluated via the resolution of decoupled ordinary differential equations [29].

In summary, it suffices to solve a stochastic or deterministic OC problem in fixed time t to recover the value of $g(t)$. This will be exemplified in the linear quadratic (LQ) case in the next section, before the problem of tuning movement time (around the optimal one) will be addressed.

To test the above methodology, we asked subjects to perform 1-dof arm movements in the horizontal plane. These reaching movements were of different amplitudes and, for each amplitude, the duration was estimated from motion capture data. In Fig. 1, we depict the main results. Overall, an affine relationship between movement extent and time can be drawn from the experimental data. When identifying $g(t)$ for several movement times t, one can characterize the shape of g on the interval of actual movement durations. This can be done either using the regression line and single data points. For the depicted subject, movement times varied between about 600 ms (for an amplitude of 5°) to about 1400 ms (for an amplitude of 95°). Therefore, we were able to identify the CoT in a robust and reliable manner on the interval 600–1400 ms when using the affine amplitude-duration relationship. Outside of this interval, extrapolation was required. However, it must be noticed that the shape of g on the range of empirical movement times was sufficient to conclude that the CoT was neither linear nor purely convex or concave. Actually, its shape tended to be sigmoidal. The present shapes were obtained when assuming the torque change [55] as trajectory cost l. Assuming the angle jerk [18] as trajectory cost would not change the sigmoidal shape. It is visible that when identified from single trials, the cost of time $g(t)$ appears to be quite noisy (gray dots in Fig. 1). This might be due to the discrepancy between planning (the frame of our model) and execution where all the musculoskeletal properties of the arm and sensorimotor noise do perturb the planned trajectory and the movement time (see also Sect. 3.3). This may also be linked to the way the brain actually finds optimal strategies and to the shape of the total cost: even though it has a "U" shape with respect to motion duration, a relative flatness around the optimal time may induce variability in duration even at the planning stage. For a more thorough analysis of the identification process of $g(t)$ with additional assessments, the reader is referred to [4]. The biomechanical model of the arm is described in Sect. "Model for arm reaching movements". Finally, it must also be noted that the

[2] We assume here that there are no abnormal extremals (an hypothesis which is satisfied in particular by controllable linear systems). As a consequence, it is not necessary to put a Lagrange multiplier in front of l in $\mathscr{H}_0$.

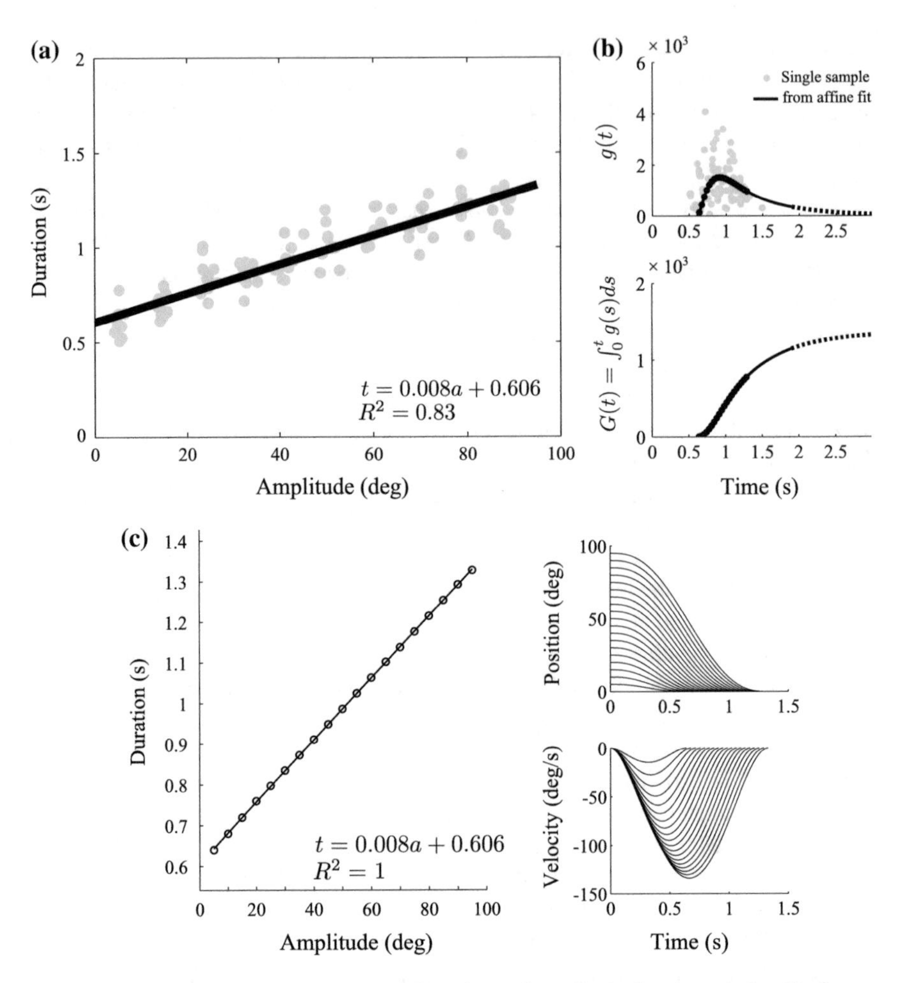

Fig. 1 Movements at spontaneous speed. **a** Experimental amplitude-duration relationship for one individual. **b** Infinitesimal and integral time costs. Values of the infinitesimal CoT were recovered either in single trials (*gray dots*) or by making use of the affine fit of the amplitude-duration relationship presented in the first panel. *Dotted lines* are extrapolated values. **c** OC simulations in free-time using the CoT $G(t)$. The duration movement is exactly recovered for every amplitude, and the corresponding trajectories are displayed (standard bell-shaped velocity profiles for such 1-dof movements in the horizontal plane)

free-time optimal control model predicts smooth and bell-shaped velocity profiles, which agrees with classical observations for such planar arm movements.

At this point, it is useful to make some comments about the present modeling. First of all, we chose to rely upon the optimal control formalism but alternative approaches have been proposed to account for the duration of motion. Among them, an interesting approach is based on invariance principles in affine geometry [2], which proved to explain well both isochrony and isogony laws during drawing movements (e.g.

ellipses). However the link between both approaches is not obvious and it is likely that considering the same kind of cyclical curve-drawing movements in our framework could help to understand how each approach relates to the other. Interestingly, a recent study introducing the concept of "drive", which is mathematically equivalent to assuming a constant value of $g(t)$ here, analyzed such curve-drawing movements and was able to replicate the two-third power law and the overall size and speed scaling of biological movement from this simple principle. An interpretation in our framework could be that the asymptotic value of the $g(t)$ is not zero, but the confirmation of such an hypothesis requires tasks with larger durations in order to sample $g(t)$ on a larger interval. Another central assumption in our approach is the additive separability between the cost of time $g(t)$ and the trajectory cost $l(\mathbf{x}, \mathbf{u})$ that does not depend on time. This is a strong assumption which has nonetheless the advantage of leading to a model that extends previous classical optimal control models developed in fixed time with time-independent trajectory costs (i.e. the optimal fixed-time trajectories coincide with the free-time trajectories of the same duration). This additive separability is also in the spirit of a first-order approximation of a general cost $h(\mathbf{x}, \mathbf{u}, t)$ and can thus be viewed as a simplification of a more general problem. The reader is referred to [4, p. 1057] for other comments about this hypothesis.

2.2 *Linear Quadratic Models*

2.2.1 General Settings and Solutions

Let us focus on deterministic LQ models for 1 degree-of-freedom (dof) motions. This framework is relevant to model simple arm reaching movements. The state of such systems can be described by $\mathbf{x} = (\theta, \ldots, \theta^{(n-1)}) \in \mathbb{R}^n$ and then the dynamics has the form

$$\theta^{(n)} + c_{n-1}\theta^{(n-1)} + \cdots + c_0\theta = u, \tag{5}$$

which is a single-input linear system $\dot{\mathbf{x}} = A\mathbf{x} + Bu$, $u \in \mathbb{R}$. Typically $n = 2$ or 3 for dynamical models of the arm (see below). The single-input LQ case is also interesting from a theoretical point of view as strong results of well-posedness of the inverse problem exist. In particular (see [4]), the uniqueness and robustness to perturbations of experimental data can be proven for $g(t)$. Indeed, the underlying quadratic cost can be identified unequivocally [39] from the empirical and presumably optimal trajectories and in a continuous way (roughly speaking, the mapping between the optimal trajectories and the quadratic cost is continuous, so that small changes in experimental trajectories result in small changes in the quadratic cost). This remarkable theoretical result motivates a deeper investigation of the LQ scenario.

A quadratic cost for the system given in Eq. (5) is a function $\alpha u^2 + \mathbf{x}^T Q\mathbf{x} + 2\mathbf{x}^T Su$, with $\alpha > 0$, which is a positive semidefinite quadratic form in $(\mathbf{x}, u)$. Up to a normalization we can assume $\alpha = 1$ and then consider a cost of the form $l(\mathbf{x}, u) = u^2 + \mathbf{x}^T Q\mathbf{x} + 2\mathbf{x}^T Su$.

The associated OC problem in fixed time $\tau > 0$ is the following: *given terminal conditions* $\mathbf{x}^0, \mathbf{x}^f \in \mathbb{R}^n$, *minimize the cost*

$$C_\tau(u) = \int_0^\tau \big(u(t)^2 + \mathbf{x}_u(t)^T Q\mathbf{x}_u(t) + 2\mathbf{x}_u(t)^T Su(t)\big)dt,$$

among all controls u *such that the solution* $\mathbf{x}_u$ *of* $\dot{\mathbf{x}} = A\mathbf{x} + Bu, \mathbf{x}_u(0) = \mathbf{x}^0$, *satisfies* $\mathbf{x}_u(\tau) = \mathbf{x}^f$.

Initial and final points $\mathbf{x}^0$ and $\mathbf{x}^f$ are always chosen as equilibrium states of the system, that is, $\mathbf{x}^0 = (\theta^0, 0, \ldots, 0)$ and $\mathbf{x}^f = (\theta^f, 0, \ldots, 0)$. We also make the technical assumption that the pair $(A, Q^{1/2})$ is observable (this assumption is necessary for Eq. (6) below to hold). Ferrante et al. [16] showed that the optimal trajectory $\mathbf{x}_u$ of this problem is given by

$$\mathbf{x}_u(t) = e^{tA_+}\mathbf{p}_1 + e^{tA_-}\mathbf{p}_2, \tag{6}$$

where the vectors $\mathbf{p}_1, \mathbf{p}_2 \in \mathbb{R}^n$ are the unique solution of

$$\begin{cases} \mathbf{x}^0 &= \mathbf{p}_1 + \mathbf{p}_2, \\ \mathbf{x}^f &= e^{\tau A_+}\mathbf{p}_1 + e^{\tau A_-}\mathbf{p}_2. \end{cases} \tag{7}$$

The matrices A_-, A_+ are respectively anti-stable and stable (the eigenvalues of A_+ are actually the opposite of the ones of A_-). These matrices are determined through a Riccati equation and do not depend on $\mathbf{x}^0$, $\mathbf{x}^f$, and τ, but only on the parameters (A, B) of the dynamic and (Q, S) of the cost.

Remark 1 When $\mathbf{x}^f = 0$, the vectors $\mathbf{p}_1$ and $\mathbf{p}_2$ depend linearly on $\mathbf{x}^0$ and so do $\mathbf{x}_u(t)$ for every t. This is also true for $u(t)$ since its expression has a form similar to Eq. 6 (see [16]).

Remark 2 Note that the corresponding OC problem in infinite time is the following: *given an initial condition* $\mathbf{x}^0 \in \mathbb{R}^n$, *minimize the cost*

$$C_\infty(u) = \int_0^\infty \big(u(t)^2 + \mathbf{x}_u(t)^T Q\mathbf{x}_u(t) + 2\mathbf{x}_u(t)^T Su(t)\big)dt,$$

among all controls u, *where* $\mathbf{x}_u$ *is the solution of* $\dot{\mathbf{x}} = A\mathbf{x} + Bu$, $\mathbf{x}_u(0) = \mathbf{x}^0$. The solution of this problem is given again by Eq. (6), with the same matrices A_-, A_+, but with parameters $\mathbf{p}_1 = \mathbf{x}^0$ and $\mathbf{p}_2 = 0$.

2.2.2 Computation of the Infinitesimal CoT

Up to a translation in θ (position variable), we can always assume $\mathbf{x}^f = 0$. We then choose a family of initial conditions $\mathbf{x}^0(a) = (a, 0, \ldots, 0)$, parameterized by the movement extent $a > 0$ (i.e. the amplitude of the motion). For every amplitude $a > 0$ we denote by $t^*(a)$ the duration (which can be estimated experimentally) of the motion between $\mathbf{x}^0(a)$ and $\mathbf{x}^f$, and by $u^a(\cdot)$ a control minimizing the integral cost $C_{t^*(a)}(u)$ in fixed time $t^*(a)$ between $\mathbf{x}_u(0) = \mathbf{x}^0(a)$ and $\mathbf{x}_u(t^*(a)) = \mathbf{x}^f = 0$. By standard computations (see [4]) we obtain $\frac{\partial V_{\mathbf{x}^f}}{\partial t}(t^*(a), \mathbf{x}^0(a)) = -u^a(t^*(a))^2$, and so from Eq. 4,

$$g\big(t^*(a)\big) = u^a(t^*(a))^2. \tag{8}$$

Moreover, the value $u^a(t^*(a))$ can be seen to depend linearly on $\mathbf{x}_u(0)$ in the LQ case (see Remark 1), and so it depends linearly on a since $\mathbf{x}_u(0) = \mathbf{x}^0(a) = a\mathbf{x}^0(1)$. In other words, $u^a(t^*(a)) = a\varphi(t^*(a))$, where the function $\varphi(\cdot)$ is defined as follows: *for every* $\tau > 0$, $\varphi(\tau)$ *is the value* $u^1(\tau)$ *of the control minimizing the integral cost* $C_\tau(u) = \int_0^\tau (u^2 + \mathbf{x}^T Q\mathbf{x} + 2\mathbf{x}^T Su)dt$ *in fixed time* τ *between* $\mathbf{x}_u(0) = \mathbf{x}^0(1)$ *and* $\mathbf{x}_u(\tau) = \mathbf{0}$. Note that $\varphi(\cdot)$ is a *universal function of time* that depends only on the system dynamics (A, B) and the trajectory cost and not on the specific behavior of an individual. This universal function of time can be computed explicitly thanks to the equations given in [16]. We finally obtain $g\big(t^*(a)\big) = \varphi\big(t^*(a)\big)^2 a^2$.

Empirical observations show that the time $t^*(a)$ is typically an increasing function of the amplitude, so that its inverse $a^*(t)$ exists. We can then determine the function $g(\cdot)$ by $g(t) = \varphi(t)^2 a^*(t)^2$. In particular, if it appears from experiments that the function t^* is approximately affine of the form $t^*(a) = \alpha a + \beta$, then the infinitesimal CoT can be written $g(t) = \varphi(t)^2(\frac{1}{\alpha}t - \frac{\beta}{\alpha})^2$. Hence, it suffices to compute $\varphi(t)$, which can be done explicitly, to recover the actual infinitesimal CoT from the experimental

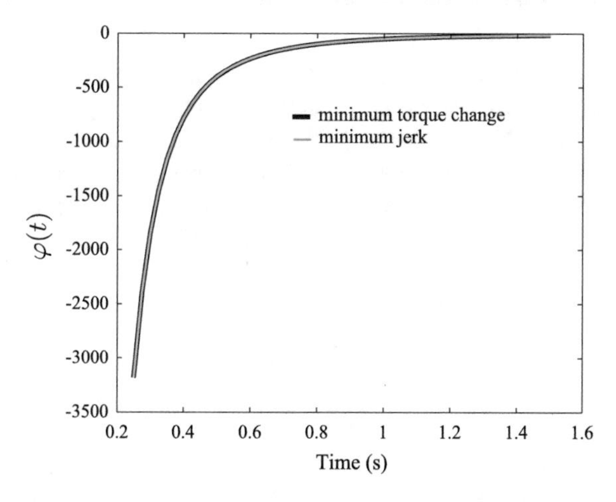

Fig. 2 The universal function φ for the minimum torque change and minimum jerk optimality criteria, for a 1-dof arm moving in the horizontal plane. The function was found to be quasi identical for these two costs. From this function, the infinitesimal CoT can be recovered as $g(t) = \varphi(t)^2 a(t)^2$ where $a(t)$ is the amplitude corresponding to a movement in time t (which can be determined experimentally)

duration/amplitude mapping. For illustration, the function φ is plotted in the Fig. 2 for the two main trajectory costs considered here, namely the angle jerk and torque change optimality criteria.

3 Accounting for Other Motion Speeds

A general theory of human movement vigor should also be able to account for movement times departing from the self-chosen ones. It is clear that motion duration can vary in function of the task, in particular verbal instructions given by an experimenter such as go quickly/slowly to the target. How can the CoT theory take account of this variation? Remind that, from Eq. (3), the duration of a motion satisfies

$$\tau \in \operatorname{argmin}_{t\geq 0}\left(\int_0^t g(s)ds + \inf_{\mathbf{u}} \int_0^t l\big(\mathbf{x_u}(s), \mathbf{u}(s)\big)ds\right). \tag{9}$$

Hence variations of the motion duration can be explained by changes either of $g(t)$ or of $l(\mathbf{x}, \mathbf{u})$. The first question is: is it possible to explain changes of motion duration by playing on the CoT $g(t)$?

3.1 The Sole Modification of the Cost of Time Cannot Explain Slower/Faster Movements

Let us assume first that the cost of the trajectory is independent of the task and hence that changes in motion duration only result from changes of the CoT. Typically, it is clear from Eq. (9) that any increase of the values of $G(t) = \int_0^t g(s)ds$ implies a decrease of the duration τ. The effect of instructions such as "go quickly to the target" could then simply correspond to an increase of the CoT $G(t)$. Conversely, instructions such as "go slowly to the target" could produce a decrease of the CoT, which implies in turn an increase of the motion duration. Let us examine the consequence of this hypothesis for the model described in Sect. 2.2, that is, in the context of linear quadratic (LQ) models. In this case, motions are always solutions of a LQ optimal control problem in fixed finite time, with always the same cost $l(\mathbf{x}, u) = u^2 + \mathbf{x}^T Q\mathbf{x} + 2\mathbf{x}^T Su$ but with a time τ that depends on the term $G(t)$.

Rescaling of g.

An intuitive idea would be to rescale g by multiplying it by some positive parameter κ. A simple investigation however proves that such an approach is falsified by experimental findings (see for instance [9, 58]). Indeed, such a rescaling would induce a new amplitude $\sqrt{\kappa}a^*(t)$ for a movement in time t (we use the notations of Sect. 2.2.2, i.e. $\kappa g(t) = \varphi(t)^2\big(\sqrt{\kappa}a^*(t)\big)^2$). Therefore, the rescaled CoT would yield the affine amplitude-duration relationship $t^*(a) = \alpha \frac{a}{\sqrt{\kappa}} + \beta$. It can be concluded that just rescaling the CoT does not allow to change both the slope and the intercept of the amplitude/duration relationship. Hence, since both the intercept and slope are found to change experimentally when the instructed speed is varied, this observation cannot be attributed to a global rescaling of the CoT g. Typically, when a subject is asked to move faster, not only α is reduced significantly but also β.

Arbitrary change of g.

We now consider that g can be changed both in shape and magnitude. Consider first the case of an overall decrease of the CoT, which produces a longer duration τ, i.e. slower movements. The asymptotic analysis of Appendix "Asymptotic Study for Small/Large Time and Fixed Cost" shows that, for a large duration τ, the solution in time τ of the LQ problem associated with the cost $l(\mathbf{x}, u)$ looks alike the solution of the same LQ problem in infinite time (see Lemma 4). The latter solutions have an exponential decay to the final state and, moreover, a single peak of velocity whose magnitude is independent of the time τ. These characteristics are not compatible with what is known of slow reaching movements where velocity traces are gradually more multipeaked [27, 56], which moreover seems to be a preplanned property not simply due to sensory feedback processing [14].

Consider now the case of an increase of the CoT, which induces a shorter duration τ and hence faster movements. The asymptotic analysis when $\tau \to 0$ shows that, for small enough durations, the solutions of the LQ problem are almost identical, up to a change of time-parameterization, and are of polynomial form (see Lemma 5). More precisely, the theory would predict that for faster and faster movements the velocity profiles are dilatations of each other and have a symmetric shape (see Remark 6). Such a strict scaling law of symmetric speed profiles is falsified by experimental observations. Indeed, it was shown that movements become more asymmetric as speed increases, where the relative duration of the deceleration phase increases during extremely fast reaches (from ~50% of total motion duration for rapid reaches to ~70% for maximally fast reaches) [32, 33]. This is moreover incompatible with the exponential decay of the distance left to the target observed in Fitts's like studies.

In summary, the sole modification of the CoT cannot explain slower/faster movements in the LQ framework. Of course we could consider different models than the linear-quadratic ones. Indeed, in the latter models we make several hypotheses: first, the evolution of the state is given by linear differential equations; second, the state and the control are unbounded; third, we restrict ourselves to the class of costs function which are quadratic function of both state and control. The first assumption is not questionable as soon as we do not finely model the dynamics of muscles or do not consider multijoint systems, which is consistent at the present level of investigation. The third one seems to be reasonable since the class of quadratic costs is sufficient to reproduce accurately simple arm motions at least [18, 26, 38]. Moreover, the conclusions above should be very similar for a slightly larger class of costs functions (for instance a class including the absolute work as in [5]), even if the asymptotic study would be much more difficult in that case. The most critical hypothesis actually is the second one. Indeed it is evident that the state and the control, being physical quantities, are bounded. In a LQ model, this fact is taken into account implicitly since high values are penalized in the cost. This approach is valid as long as the values of the state and the control in the optimal solutions do not exceed the bounds. This condition is not easy to check since most of the bounds are not really known, it is however clearly satisfied when the duration of the motion is not too small. Another approach would be to take into account explicitly these bounds. In that case, it exists a minimum time to go from one given state to another one. When the CoT $G(t)$ increases (for instance because of instructions such as "go quickly to the target"), the time τ of the motion converges to the minimum time and, under standard convexity hypotheses on the cost [19], the optimal solutions converge to the minimum time solutions. This scenario is not really plausible for different reasons. First, minimum time solutions present, in general time-intervals, saturations of the control's bounds. Since such characteristic saturations have never been observed in fast motions for quantities such as velocity, acceleration and jerk, the control would necessarily be a higher-order quantity. Even if they existed, such saturations would hardly be compatible with trajectories satisfying Fitts's law (exponential decay of the end-effector position to the goal). Conceivably, saturation may occur at the level of motoneurons activity but experimental data of surface electromyography (EMG), the main non-invasive approach to estimate the overall activity of motor units, indicate that EMG activity is relatively far from maximal during rapid reaching [1]. Moreover, no plateau is visible on any sensible time window and the so-called triphasic pattern, with well-distinguished EMG bursts, is known to govern ballistic movements [8, 22]. Secondly, the hypothesis of minimum time trajectories has already been studied in [49] and contradicted in [58]. Intuitively, the reason is that humans do not always move as fast as possible for a given level of accuracy: in most daily activities, we could move faster without degrading task performance. At last, one may mention that even when instructed to move as fast as possible, the actual maximal speed of a subject is not attained. It has been proven that subjects can move faster without altering accuracy when explicitly asked to co-contract muscles, an energy consuming strategy [37].

In summary, we presented strong arguments supporting that the sole modification of the CoT cannot be put forward to explain neither slower nor faster movements. Then, it seems necessary to assume that task-induced changes of motion duration are due to changes of the cost of the trajectory, i.e. $l(\mathbf{x}, u)$.

3.2 *From Self-paced Motion to Slower/Faster Movements*

How do changes of the cost $l(\mathbf{x}, u)$ affect the duration? Consider the example of the linear quadratic models. For our purpose, we propose to interpret the total infinitesimal cost

$$g(t) + l(x, u) = g(t) + u^2 + \mathbf{x}^T Q\mathbf{x} \quad [+ \text{ possibly mixed terms } \mathbf{x}^T Su].$$

by distinguishing three types of terms:

- $g(t)$ is the cost of the time, it penalizes slow motion by accumulating infinitesimal values during the passage of time;
- u^2 is a subjective cost that evaluates the "effort" associated with a movement. It can reflect mechanical energy expenditure, amount of joint torques, smoothness etc., depending on the modeling; we can also include the mixed terms in the subjective cost, and possibly some part $\mathbf{x}^T Q' \mathbf{x}$ of the quadratic terms in $\mathbf{x}$. In essence, the subjective part of the trajectory cost reflects an individual's motor decision (often useful to resolve all residual task redundancy).
- $\mathbf{x}^T Q \mathbf{x} = (\mathbf{x} - \mathbf{x}^f)^T Q (\mathbf{x} - \mathbf{x}^f)$ is an objective cost, also part of the trajectory cost. Here, it penalizes the fact of being away from the goal $\mathbf{x}^f$ (recall that $\mathbf{x}^f = 0$ here without loss of generality) and can be modulated by the requirements of the task. It is objective in the sense that it is directly related to the task's demand.

Hence we postulate that a change in the description of the task (e.g. go quickly/slowly to the target) will affect only the objective cost, not the two other ones. Let us explain how it could work. To simplify, we assume that the matrix Q is diagonal (i.e. cost function with separate variables), $\mathbf{x}^T Q \mathbf{x} = r\theta^2 + s_1\dot{\theta}^2 + \cdots + s_{n-1}(\theta^{(n-1)})^2$. Since the term $r\theta^2 = r(\theta - \theta^f)^2$ penalizes the fact of being away from the goal, the instruction "go quickly to the goal" translates as "increase r". In the same way, since the term $s_1\dot{\theta}^2$ penalizes high velocities, the instruction "go slower" translates as "increase s_1" (it can also increase the other parameters s_i). And it appears actually that the duration of a trajectory minimizing the cost

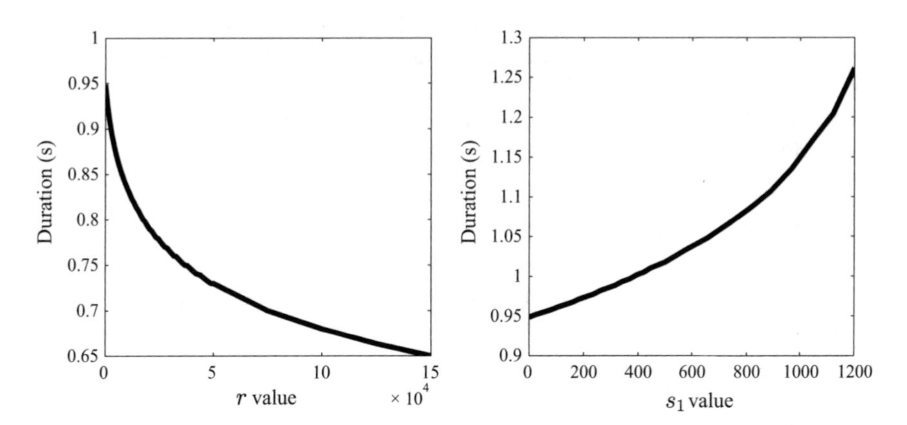

Fig. 3 Slower and faster movements for the 1-dof planar arm movement under consideration. Effects of varying r and s_1 on the motion duration. These graphs were drawn from the CoT of the individual presented in Fig. 1 and amplitude was set to $a = 45°$. Increasing the positional weight r induces an decrease of movement duration, as predicted by the free-time optimal solutions. The opposite effect can be observed when increasing the velocity weight s_1. Modifying the objective trajectory cost is a sensible way to tune movement duration around the individual's self-selected one

$$\int_0^{t_u} \left(g(t) + u^2 + r\theta^2 + s_1\dot{\theta}^2 + \cdots + s_{n-1}(\theta^{(n-1)})^2\right)dt, \tag{10}$$

decreases with r and increases with s_1, as expected. This fact was checked through numerical simulations and the dependence of motion duration on r and s_1 is reported in Fig. 3. Tuning the objective cost provides a means to modify the motion duration around the reference value corresponding to a self-selected movement pace. Verbal instructions such as "produce a quick movement" or "produce a slow movement" can thus be accounted for in this way. It should be noted that increasing r or s_1 breaks the affine relationship between movement amplitude and time. If linearity is preserved for relatively small enough values of r and s_1, the correlation coefficients nevertheless go down as these weights increase. The next section actually shows that there is a gradual distortion of the amplitude/time relationship such that Fitts's law is actually recovered for very large values of r.

3.3 Towards Fitts's Law and the Speed/Accuracy Trade-Off

To take into account accuracy constraints, we propose to consider goal-directed movements such as arm pointing as the superposition of an open-loop motion (the planned trajectory) and of a feedback process whose role is to provide on-line corrections and in particular to stabilize the hand around the target [13, 52]. We then distinguish two different motion times:

- the *planning time*, denoted by τ_p, which is the duration of the planned trajectory and can be determined by solving a free time OC problem involving the CoT as described previously;
- the *execution time*, denoted by τ_s, which is the actual duration of the motion; it may differ from the planning time because of the feedback process.

In general planning and execution times may differ for two reasons: the presence of perturbations and the fact that the point aimed at differs from the actual stopping point. The former situation occurs because of the presence of sensorimotor noise in the nervous system and the latter may occur in the case of accuracy requirements. For example, if the target has a width w, and if the instruction is "as fast and as accurate as possible", the subject will conceivably aim at a point inside or near the center of the target to ensure target achievement (see [53]), whereas the motion can actually be stopped once the trajectory meets the target via the activation of terminal feedback processes.

Again, we will consider a 1-dof LQ model as in Sect. 2.2. Let $\mathbf{x}^0$ be the starting point, $\mathbf{x}^f = 0$ be the center of a target of width w and $\mathbf{x}(\cdot)$ be the planned trajectory between these points. On the one hand, the planning time τ_p satisfies $\mathbf{x}(\tau_p) = 0$. In other words, the end-effector attains the center of the target exactly in time τ_p (no perturbations are assumed here). On the other hand, the movement will be stopped as soon as $\theta(t) \leq w/2$, i.e. the stopping time τ_s satisfies approximately $\theta(\tau_s) = w/2$.

Consider for instance the case of as fast as possible and as accurate as possible movements, which is the standard scenario behind Fitts's law [17]. As explained in the previous section, the instruction "move fast" corresponds to a cost with a very large coefficient r (see Eq. (10)). It can be shown (see Appendix "Asymptotic Study for Fixed Time") that in that case the planning time τ_p is rather small and the planned trajectory is of the form

$$\theta(t) \approx c\theta^0 e^{-\alpha_r t}, \qquad \text{for } t/\tau_p \text{ large enough,}$$

where c, α_r are positive constants with $\alpha_r \tau_p$ large (i.e. $\alpha_r \tau_p \to \infty$ as $r \to \infty$).

The stopping time is determined by the constraint $\theta(\tau_s) = w/2$. Therefore it satisfies

$$\tau_s \approx \frac{1}{\alpha_r} \log c + \frac{1}{\alpha_r} \log(2\theta^0/w),$$

which is of the same form than the original formulation of Fitts's law, that is, $t = \tilde{\alpha} \log_2(2a/w) + \tilde{\beta}$ with $t = \tau_s$ and $a = \theta^0$ [17]. Hence Fitts's law can be accounted for by our theory, although developed in a deterministic context, without explicitly assuming a linear feedback control law that would lead to an exponential decay of the distance left to the target as done in [12] or [43].

Remark 3 Note that the distinction between planning and stopping movement times allows one to recover Fitts's law as soon as the planned trajectory decreases exponentially. It is in particular the case in all models with infinite horizon and quadratic costs (either deterministic, i.e. LQR, or stochastic, i.e. LQG), and more generally in all linear models with a proportional feedback $u = K\mathbf{x}$, even though in those cases there would be no planned movement time. Hence, this is mainly the shape of the trajectory which explains Fitts's law in such models: one does not increase motion duration specifically because of a higher accuracy demand but rather motion duration increases as a consequence of the exponential decrease of the distance left to the target during maximally fast reachings. It is likely that the planned motion duration could also be increased on purpose if modeling signal-dependent noise and adding a terminal error term in a stochastic context [23] but we did not consider stochastic formulations of the present deterministic free-time OC problems.

To illustrate the convergence to Fitts's law, we performed simulations for the same 1-dof arm model with $r = 10^8$ and $s_1 = 0.05r$ in Eq. (10). The results are depicted in Fig. 4 where the switch from affine to logarithmic relationships between amplitude and duration is illustrated. In accordance with experimental findings, velocity profiles also become more asymmetrical in the sense that the relative duration of deceleration drastically increases for maximally fast reaches [32, 33]. These graphs also explain why Fitts's law does not hold for self-paced movements but is mainly a limit case, which agrees with experimental observations (see [58]).

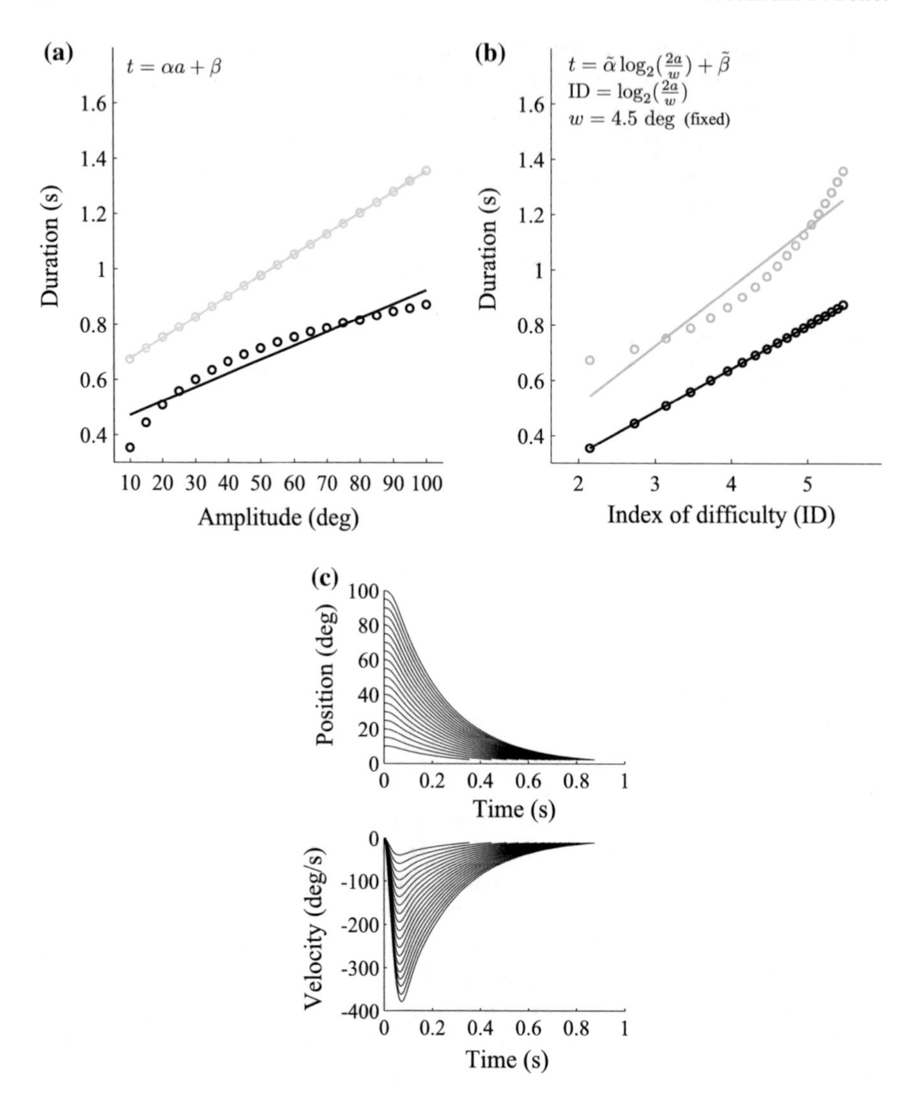

Fig. 4 The case of Fitts's law for 1-dof planar arm pointing movements. **a** Relationship between amplitude and duration. In *gray*, the original amplitude/duration of the example individual is recalled (i.e. $r = s_1 = 0$ for self-paced movements). When r becomes very large (here $r = 10^8$ and $s_1 = 0.05r$), movements become faster and the amplitude/duration relationship departs from its linear shape (*black traces*, where a logarithmic profile is visible). **b** Relationship between index of difficulty ($\mathrm{ID} = \log_2(\frac{2a}{w})$) and duration. In *gray*, for self-paced movements ($r = s_1 = 0$). In *black*, for Fitts's instructions ($r = 10^8$ and $s_1 = 0.05r$). Fitts's law is recovered very accurately in the latter case in contrast to self-paced motions where a convex, instead of linear, trend is observed. **c** Position and velocity profiles corresponding to Fitts's law simulations in the free time OC formalism. The exponential decrease of the distance *left* to the target is visible for the position variable and the asymmetry of speed profiles can be compared to those of Fig. 1c

4 Conclusion

In this chapter, we have presented a theoretical view of the computational principles that may underlie the control of movement vigor within the central nervous system. We tackled the issue of how reach duration can be adjusted to speed instructions in this framework. At the core of the theory is the hypothesis of the existence of a "cost of time" [44]. It assumes that the passage of time has a cost per se, which explains why our movements are not slower. Using an inverse optimal control approach, we showed that this hypothetical time cost can be reliably identified from experimental data of movement extent and duration and without resorting to any parametric adjustment [4]. Yet, the cost of time aims at explaining the spontaneous/natural movement vigor, i.e. self-chosen motion pace. When explicitly asked to move slower or faster, we argued that humans do not seem to modify the cost of time itself but rather an objective trajectory cost reflecting the specific task constraints. Whereas the time cost and the subjective trajectory cost seem to be relatively invariant (at least on a short time scale), we provided evidence that the introduction of an objective trajectory cost is crucial to capture speed instructions given by an experimenter. In particular, Fitts's law is recovered in our framework as a limit case.

Acknowledgements This work is supported by a public grant overseen by the French National Research Agency (ANR) as part of the "Investissement d' Avenir" program, through the "iCODE Institute project" funded by the IDEX Paris-Saclay, ANR-11-IDEX-0003-02.

Appendix: Technical Details

Asymptotic Study

We describe in this asymptotic studies the behavior of the solutions of the linear quadratic model introduced in Sect. 2.2 when some parameters of the problem go to zero or infinity.

Asymptotic Study for Small/Large Time and Fixed Cost

Let us study the behavior of the optimal solutions when the final time τ varies, the quadratic cost $l(\mathbf{x}, u) = u^2 + \mathbf{x}^T Q\mathbf{x} + 2\mathbf{x}^T Su$ and the terminal conditions $\mathbf{x}^0 = (\theta^0, 0, \ldots, 0)$, $\mathbf{x}^f = 0$ being fixed. For every $\tau > 0$ we denote by $\mathbf{x}_\tau(t) = \left(\theta_\tau(t), \ldots, \theta_\tau^{(n-1)}(t)\right)$, $t \in [0, \tau]$, the solution of the free-time OC problem in fixed time τ whose expression is given by Eq. (6). Consider first the case of large times, that is the case where $\tau \to \infty$. Remind that in this case $e^{\tau A_+}$ and $e^{-\tau A_-}$ tend to zero.

Lemma 4 *When $\tau \to \infty$, there holds*

$$\mathbf{x}_\tau(t) = e^{tA_+}\mathbf{x}^0 + O(\|e^{\tau A_+}\mathbf{x}^0\|).$$

As a consequence, there exists constants $c, \alpha > 0$, and $\varepsilon \in (0, 1)$, such that

$$\theta_\tau(t) = c\theta^0 e^{-\alpha t} + O(e^{-\alpha\tau}), \qquad \textit{for any } t \in [\varepsilon\tau, \tau].$$

We thus recover a somewhat intuitive result: the solution of a LQ problem in fixed time converges to the solution of the same LQ problem with infinite horizon when the time goes to infinity (see Remark 2).

Proof We deduce directly from the conditions of Eq. (7) the values of $\mathbf{p}_1 = \mathbf{p}_1(\tau)$ and $\mathbf{p}_2 = \mathbf{p}_2(\tau)$ in function of $\mathbf{x}^0$. By putting these values into Eq. (6), we obtain

$$\mathbf{x}_\tau(t) = e^{tA_+}\left(I - e^{-\tau A_-}e^{\tau A_+}\right)^{-1}\mathbf{x}^{\mathbf{0}} - e^{(t-\tau)A_-}e^{\tau A_+}\mathbf{x}^{\mathbf{0}},$$

which is of the form $e^{tA_+}\mathbf{x}^0 + O(\|e^{\tau A_+}\mathbf{x}^0\|)$ since A_+ is stable and A_- is anti-stable. Now, $e^{tA_+}\mathbf{x}^0$ is a function of t which can be written as a sum of decreasing exponential terms. Denoting by $e^{-\alpha t}$ the less decreasing term in this sum, it appears that all other exponential terms in $e^{tA_+}\mathbf{x}^0$ are negligible in front of $e^{-\alpha\tau}$ for t/τ not too small and we obtain the formula for θ_τ (note that in general $\alpha = \min\{-\Re(\lambda) \,:\, \lambda \ eigenvalue\ of\ A_+\}$). □

Consider now the case of small times, i.e. the case where $\tau \to 0$. In that case we can prove the following result.

Lemma 5 *Let $p(s)$ be the polynomial function of degree $2n-1$ defined by $\big(p(0), p'(0), \ldots, p^{(n-1)}(0)\big) = \mathbf{x}^0$ and $\big(p(1), p'(1), \ldots, p^{(n-1)}(1)\big) = \mathbf{x}^f$. Then*

$$\theta_\tau(t) = p\left(\frac{t}{\tau}\right) + O(\tau).$$

As a consequence, $\theta_\tau(t) \approx p(\frac{t}{\tau})$ for small times τ: a change of the final time induces approximately a temporal rescaling of the solutions.

Remark 6 Note that since the terminal conditions are equilibriums, the polynomial $p(\cdot)$ satisfies $\dot{p}(t) = \dot{p}(1-t)$, which implies that the velocity profiles of θ_τ have an almost symmetric shape for small times τ. Indeed, the polynomial function $\widetilde{p}(t) = \theta^0 - p(1-t)$ satisfies the same conditions at $t = 0$ and $t = 1$ as $p(t)$, which implies by unicity of the solution that $\widetilde{p}(t) = p(t)$, and so the conclusion.

Proof Let us start with a preliminary remark on the optimal solution θ_τ. On one hand, it follows from Eq. (6) that $\theta_\tau(t)$ is an analytic function (i.e. it is equal to its Taylor series) which depends linearly on the vectors $\mathbf{p}_1 = \mathbf{p}_1(\tau)$ and $\mathbf{p}_2 = \mathbf{p}_2(\tau)$. Hence, all derivatives of θ_τ at 0 depend linearly on the pair $(\mathbf{p}_1, \mathbf{p}_2)$. On the other hand, due to the particular properties of the matrices A_-, A_+ (see [16, Lemma 1]), there is a one-to-one correspondence between $(\mathbf{p}_1, \mathbf{p}_2)$ and the $2n$ first derivatives of θ_τ at 0, i.e. by $\theta_\tau^{(k)}(0),\ 0 \leq k \leq 2n-1$. As a consequence, all derivatives of θ_τ at 0 depend linearly on the $2n$ first ones: for every integer k there exists a constant C_k such that, for any τ, $|\theta_\tau^{(k)}(0)| \leq C_k \Theta_\tau$, where

$$\Theta_\tau = \max\left\{|\theta_\tau^{(k)}(0)|,\ 0 \leq k \leq 2n-1\right\}.$$

Set $\phi_\tau(t) = \theta_\tau(t) - p(\frac{t}{\tau})$. We have to prove that $\phi_\tau(t) = O(\tau)$. The above remark and the fact that $\left(p(0), \ldots, p^{(n-1)}(0)\right) = \left(\theta_\tau(0), \ldots, \theta_\tau^{(n-1)}(0)\right) = (\theta^0, 0, \ldots, 0)$ imply that the Taylor expansion of ϕ_τ has the form,

$$\phi_\tau(t) = \sum_{k=n}^{2n-1} \frac{t^k}{k!}\left(\theta_\tau^{(k)}(0) - \frac{p^{(k)}(0)}{\tau^k}\right) + \Theta_\tau O(t^{2n}), \tag{11}$$

where all $O(\cdot)$ are uniform with respect to τ. By definition of $p(\cdot)$ we have also $\phi_\tau^{(j)}(\tau) = 0$ for $j = 0, \ldots, n-1$, and from Eq. (11) we get

$$\sum_{k=n}^{2n-1} \frac{1}{(k-j)!}\left(\tau^k \theta_\tau^{(k)}(0) - p^{(k)}(0)\right) = \Theta_\tau O(\tau^{2n}), \qquad j = 0, \ldots, n-1.$$

It follows that, for $k = n, \ldots, 2n-1$ there holds $\tau^k \theta_\tau^{(k)}(0) - p^{(k)}(0) = \Theta_\tau O(\tau^{2n})$, and thus from the definition of Θ_τ we obtain that $\Theta_\tau \tau^{2n} = O(\tau)$. This and Eq. (11) give $\phi_\tau(t) = O(\tau)$, which proves the lemma. □

Asymptotic Study for Fixed Time

Let us try to understand now how the optimal solutions behave when some coefficients in the cost function are modified. We fix an initial state $\mathbf{x}^0 = (\theta^0, 0, \ldots, 0)$, a final one $\mathbf{x}^f = 0$, and an infinitesimal CoT $g(t)$. We consider a family of costs $l_r(\mathbf{x}, u)$ depending on a parameter r of the form

$$l_r(\mathbf{x}, u) = u^2 + r\theta^2 + \mathbf{x}^T Q_0 \mathbf{x} + 2\mathbf{x}^T S u,$$

that is, with a matrix $Q(r) = Q_0 + r e_1 e_1^T$ ($e_1 = (1, 0, \ldots, 0)$ denotes the first vector of the canonical basis of $\mathbb{R}^n$). We want to study the behavior when r tends to ∞ of the optimal solutions of the following free-time OC problem: *minimize the cost*

$$C_r(u, t_u) = \int_0^{t_u} \left(g(t) + l_r\left(\mathbf{x}_u(t), u(t)\right)\right)\, dt,$$

among all inputs $u(\cdot)$ *and all times* t_u *such that* $\mathbf{x}_u(0) = \mathbf{x}^0$ *and* $\mathbf{x}_u(t_u) = \mathbf{x}^f$. As we have seen previously, the time $\tau = \tau(r)$ is determined by Eq. (9) and the optimal solutions are the one of the OC problem $\min \int_0^\tau l_r(\mathbf{x}, u)$ in fixed time τ.

Lemma 7 *For every* $r > 0$ *we denote by* $\mathbf{x}_r(t) = \left(\theta_r(t), \ldots, \theta_r{}^{(n-1)}(t)\right)$, $t \in [0, \tau(r)]$, *the solution of the free-time OC problem associated with* C_r. *Assume that the infinitesimal cost of time* $g(\cdot)$ *is a bounded function. Then there exists constants* $c, \alpha > 0$, *and* $\varepsilon \in (0, 1)$, *such that, when* $r \to \infty$, *we have* $r^{1/2n}\tau(r) \to \infty$ *and*

$$\theta_r(t) = c\theta^0 e^{-\alpha r^{1/2n} t} + O\left(e^{-\alpha r^{1/2n}\tau(r)}\right), \qquad \textit{for any } t \in [\varepsilon\tau(r), \tau(r)].$$

Note that the boundedness assumption on g is very natural and seems to be verified experimentally since we obtain functions $g(t)$ that are decreasing for large t.

Proof To simplify the study, we give only the proof in the case where the matrices Q_0 and S are zero, and the dynamics (Eq. (5)) is of the form $\theta^{(n)} = u$. The proof of the complete result can be obtained by showing that this case actually gives the highest order terms with respect to r. With the preceding hypothesis, $\theta_r(t)$ is the solution of the OC problem in fixed time $\tau = \tau(r)$ associated with the infinitesimal cost $u^2 + r\theta^2$, or equivalently with $\frac{1}{r}u^2 + \theta^2$. Set $\widetilde{\theta_r}(t) = \theta_r(tr^{-1/2n})$. Then $\widetilde{\theta_r}(t)$ is the solution of the OC problem in fixed time $r^{1/2n}\tau$ associated with the infinitesimal cost $u^2 + \theta^2$. In the latter problem, nothing depends on r except the duration $r^{1/2n}\tau$. It results from the analysis of Sect. 2.2.2 that there exists a universal function of time $\varphi(\cdot)$ such that $\widetilde{u_r}(r^{1/2n}\tau) = \theta^0\varphi(r^{1/2n}\tau)$. Since we have $u_r(t) = r^{1/2}\widetilde{u_r}(r^{1/2n}t)$, we obtain

$$u_r(\tau) = r^{1/2}\theta^0\varphi(r^{1/2n}\tau).$$

Now remember (see Eq. (8)) that the time τ must satisfy $g(\tau) = (u_r(\tau))^2$, which gives $g(\tau) = r\left(\theta^0\varphi(r^{1/2n}\tau)\right)^2$. Assume by contradiction that the quantity $r^{1/2n}\tau(r)$ is bounded as $r \to \infty$. Then $\varphi(r^{1/2n}\tau)$ is bounded away from zero (φ is positive and continuous on $(0, +\infty)$, and converges to $+\infty$ as $t \to 0$, see Fig. 2), and therefore $g(\tau(r)) \to \infty$ as $r \to \infty$, which contradicts the boundedness of g. Thus we get $r^{1/2n}\tau(r) \to \infty$.

Since $\widetilde{\theta_r}(t)$ is the solution of an OC problem in fixed time with a very large time $r^{1/2n}\tau(r)$, it results from Lemma 4 that $\widetilde{\theta_r}(t) = c\theta^0 e^{-\alpha t} + O\left(e^{-\alpha r^{1/2n}\tau(r)}\right)$ for t larger than $\varepsilon r^{1/2n}\tau(r)$ for some $\varepsilon \in (0, 1)$. The conclusion follows from $\theta_r(t) = \widetilde{\theta_r}(tr^{1/2n})$. □

Model for Arm Reaching Movements

Single degree-of-freedom (dof) limb. For a 1-dof arm moving in the horizontal plane, the basic model used throughout the study was already described in numerous other studies (e.g. [4, 20, 21, 26, 49]) and is as follows:

$$\begin{cases} I\ddot{\theta} & = \tau - b\dot{\theta} \\ \dot{\tau} & = u \end{cases} \tag{12}$$

where is θ the shoulder joint angle, τ is the muscle torque, b is the friction coefficient ($b = 0.87$ here), I is the moment of inertia of the arm with respect to the shoulder joint (value estimated based upon Winter's table for each participant; [57]) and u is the single control variable.

For the trajectory cost we typically considered canonical quadratic costs of the form $l(\mathbf{x}, u) = u^2 + \mathbf{x}^T Q\mathbf{x} + 2\mathbf{x}^T Su$, where $\mathbf{x} = (\theta, \dot{\theta}, \ddot{\theta}) \in \mathbb{R}^3$ denotes the system state. The two most famous examples are the minimum torque change corresponding to $l(\mathbf{x}, u) = u^2$ [55] and the minimum jerk corresponding to $l(\mathbf{x}, u) = \dddot{\theta}^2$ [18]. Other costs, possibly composite, may account for such planar movements in fixed time but such an investigation is out of the scope of the present chapter (but see [4–7, 19] for studies related to the trajectory cost identification).

References

1. L.B. Bagesteiro, R.L. Sainburg, Handedness: dominant arm advantages in control of limb dynamics. J. Neurophysiol. **88**(5), 2408–2421 (2002). doi:10.1152/jn.00901.2001
2. D. Bennequin, R. Fuchs, A. Berthoz, T. Flash, Movement timing and invariance arise from several geometries. PLoS Comput. Biol. **5**(7), e1000,426 (2009). doi:10.1371/journal.pcbi.1000426
3. A. Berardelli, J.C. Rothwell, P.D. Thompson, M. Hallett, Pathophysiology of bradykinesia in parkinson's disease. Brain **124**(Pt 11), 2131–2146 (2001)
4. B. Berret, F. Jean, Why don't we move slower? the value of time in the neural control of action. J. Neurosci. **36**(4), 1056–1070 (2016). doi:10.1523/JNEUROSCI.1921-15.2016
5. B. Berret, C. Darlot, F. Jean, T. Pozzo, C. Papaxanthis, J.P. Gauthier, The inactivation principle: mathematical solutions minimizing the absolute work and biological implications for the planning of arm movements. PLoS Comput. Biol. **4**(10), e1000,194 (2008). doi:10.1371/journal.pcbi.1000194
6. B. Berret, J.P. Gauthier, C. Papaxanthis, How humans control arm movements. Proc. Steklov Inst. Math. **261**, 44–58 (2008)
7. B. Berret, E. Chiovetto, F. Nori, T. Pozzo T, Evidence for composite cost functions in arm movement planning: an inverse optimal control approach. PLoS Comput. Biol. **7**(10), e1002,183 (2011). doi:10.1371/journal.pcbi.1002183
8. J.M.M. Brown, W. Gilleard, Transition from slow to ballistic movement: development of triphasic electromyogram patterns. Eur. J. Appl. Physiol. Occup. Physiol. **63**(5), 381–386 (1991). doi:10.1007/BF00364466
9. S.H. Brown, H. Hefter, M. Mertens, H.J. Freund, Disturbances in human arm movement trajectory due to mild cerebellar dysfunction. J. Neurol. Neurosurg. Psychiatry **53**(4), 306–313 (1990)
10. S. Card, T. Moran, A. Newell, *The Psychology of Human-computer Interaction* (L. Erlbaum Associates, Hillsdale, 1983)
11. M.M. Churchland, G. Santhanam, K.V. Shenoy, Preparatory activity in premotor and motor cortex reflects the speed of the upcoming reach. J. Neurophysiol. **96**(6), 3130–3146 (2006). doi:10.1152/jn.00307.2006
12. E.M. Connelly, A control model: an alternative interpretation of fitts' law. Proc. Hum. Factors Ergon. Soc. Annu. Meet. **28**(7), 625–628 (1984). doi:10.1177/154193128402800722

13. M. Desmurget, S. Grafton, Forward modeling allows feedback control for fast reaching movements. Trends Cogn. Sci. **4**(11), 423–431 (2000)
14. J.A. Doeringer, N. Hogan, Intermittency in preplanned elbow movements persists in the absence of visual feedback. J. Neurophysiol. **80**(4), 1787–1799 (1998)
15. D. Elliott, W.F. Helsen, R. Chua, A century later: Woodworth's (1899) two-component model of goal-directed aiming. Psychol. Bull. **127**(3), 342–357 (2001)
16. A. Ferrante, G. Marro, L. Ntogramatzidis, A parametrization of the solutions of the finite-horizon lq problem with general cost and boundary conditions. Automatica **41**, 1359–1366 (2005)
17. P.M. Fitts, The information capacity of the human motor system in controlling the amplitude of movement. J. Exp. Psychol. **47**(6), 381–391 (1954)
18. T. Flash, N. Hogan, The coordination of arm movements: an experimentally confirmed mathematical model. J. Neurosci. **5**(7), 1688–1703 (1985)
19. J.P. Gauthier, B. Berret, F. Jean, A biomechanical inactivation principle. Proc. Steklov Inst. Math. **268**, 93–116 (2010)
20. J. Gaveau, B. Berret, L. Demougeot, L. Fadiga, T. Pozzo, C. Papaxanthis, Energy-related optimal control accounts for gravitational load: comparing shoulder, elbow, and wrist rotations. J. Neurophysiol. **111**(1), 4–16 (2014). doi:10.1152/jn.01029.2012
21. R. Gentili, V. Cahouet, C. Papaxanthis, Motor planning of arm movements is direction-dependent in the gravity field. Neuroscience **145**(1), 20–32 (2007). doi:10.1016/j.neuroscience.2006.11.035
22. M. Hallett, C.D. Marsden, Ballistic flexion movements of the human thumb. J. Physiol. **294**, 33–50 (1979)
23. C.M. Harris, D.M. Wolpert, Signal-dependent noise determines motor planning. Nature **394**(6695), 780–784 (1998). doi:10.1038/29528
24. C.M. Harris, D.M. Wolpert, The main sequence of saccades optimizes speed-accuracy trade-off. Biol. Cybern. **95**(1), 21–29 (2006). doi:10.1007/s00422-006-0064-x
25. B. Hoff, A model of duration in normal and perturbed reaching movement. Biol. Cybern. **71**, 481–488 (1994)
26. N. Hogan, An organizing principle for a class of voluntary movements. J. Neurosci. **4**(11), 2745–2754 (1984)
27. C. Isenberg, B. Conrad, Kinematic properties of slow arm movements in parkinson's disease. J. Neurol. **241**(5), 323–330 (1994)
28. M.T. Johnson, J.D. Coltz, T.J. Ebner, Encoding of target direction and speed during visual instruction and arm tracking in dorsal premotor and primary motor cortical neurons. Eur. J. Neurosci. **11**(12), 4433–4445 (1999)
29. H.J. Kappen, Optimal control theory and the linear bellman equation, in *Bayesian Time Series Models*, ed. by D. Barber, A.T. Cemgil, S. Chiappa (Cambridge University Press, Cambridge, 2011), pp. 363–387. http://dx.doi.org/10.1017/CBO9780511984679.018. Cambridge Books Online
30. D.E. Kirk, *Optimal Control Theory: An Introduction* (Prentice-Hall, New Jersey, 1970)
31. E.B. Lee, L. Markus, *Foundations of Optimal Control Theory* (Wiley, New York, 1967)
32. C. MacKenzie, T. Iberall, *The Grasping Hand*, Advances in Psychology (North-Holland, London, 1994)
33. C.L. MacKenzie, R.G. Marteniuk, C. Dugas, D. Liske, B. Eickmeier, Three-dimensional movement trajectories in fitts' task: implications for control. Q. J. Exp. Psychol. Sect. A **39**(4), 629–647 (1987). doi:10.1080/14640748708401806
34. I.S. MacKenzie, Fitts' law as a research and design tool in human-computer interaction. Hum.-Comput. Interact. **7**(1), 91–139 (1992)
35. N. Mansard, O. Stasse, P. Evrard, A. Kheddar, A versatile generalized inverted kinematics implementation for collaborative working humanoid robots: the Stack of Tasks, in *ICAR'09: International Conference on Advanced Robotics* (Munich, Germany, 2009), pp 1–6. http://hal-lirmm.ccsd.cnrs.fr/lirmm-00796736

36. P. Mazzoni, A. Hristova, J.W. Krakauer, Why don't we move faster? Parkinson's disease, movement vigor, and implicit motivation. J. Neurosci. **27**(27), 7105–7116 (2007). doi:10.1523/JNEUROSCI.0264-07.2007
37. O. Missenard, L. Fernandez, Moving faster while preserving accuracy. Neuroscience **197**, 233–241 (2011). doi:10.1016/j.neuroscience.2011.09.020
38. W.L. Nelson, Physical principles for economies of skilled movements. Biol. Cybern. **46**(2), 135–147 (1983)
39. F. Nori, R. Frezza, Linear optimal control problems and quadratic cost functions estimation, in *12th Mediterranean Conference on Control and Automation, MED'04* (Kusadasi, Aydin, Turkey, 2004)
40. U. Pattacini, F. Nori, L. Natale, G. Metta, G. Sandini, An experimental evaluation of a novel minimum-jerk cartesian controller for humanoid robots, in *2010 IEEE/RSJ International Conference on Intelligent Robots and Systems (IROS)* (2010), pp 1668–1674. doi:10.1109/IROS.2010.5650851
41. R. Plamondon, A.M. Alimi, Speed/accuracy trade-offs in target-directed movements. Behav. Brain Sci. **20**, 279–303 (1997)
42. L.S. Pontryagin, V.G. Boltyanskii, R.V. Gamkrelidze, E.F. Mishchenko, *The Mathematical Theory of Optimal Processes* (Pergamon Press, New York, 1964)
43. N. Qian, Y. Jiang, Z.P. Jiang, P. Mazzoni, Movement duration, fitts's law, and an infinite-horizon optimal feedback control model for biological motor systems. Neural. Comput. **25**(3), 697–724 (2013)
44. R. Shadmehr, Control of movements and temporal discounting of reward. Curr. Opin. Neurobiol. **20**(6), 726–730 (2010). doi:10.1016/j.conb.2010.08.017
45. R. Shadmehr, J.W. Krakauer, A computational neuroanatomy for motor control. Exp. Brain Res. **185**(3), 359–381 (2008). doi:10.1007/s00221-008-1280-5
46. R. Shadmehr, S. Mussa-Ivaldi, *Biological Learning and Control* (MIT Press, Cambridge, 2012)
47. R. Shadmehr, J.J. Orban de Xivry, M. Xu-Wilson, T.Y. Shih, Temporal discounting of reward and the cost of time in motor control. J. Neurosci. **30**(31), 10,507–10,516 (2010). doi:10.1523/JNEUROSCI.1343-10.2010
48. R. Stengel, *Optimal Control and Estimation*, Dover books on advanced mathematics (Dover Publications, Mineola, 1986)
49. H. Tanaka, J.W. Krakauer, N. Qian, An optimization principle for determining movement duration. J. Neurophysiol. **95**(6), 3875–3886 (2006). doi:10.1152/jn.00751.2005
50. E. Todorov, Optimality principles in sensorimotor control. Nat. Neurosci. **7**(9), 907–915 (2004). doi:10.1038/nn1309
51. E. Todorov, in *Optimal control theory, Bayesian Brain: Probabilistic Approaches to Neural Coding*, ed. by K. Doya (2006), pp. 269–298
52. E. Todorov, M.I. Jordan, Optimal feedback control as a theory of motor coordination. Nat. Neurosci. **5**(11), 1226–1235 (2002). doi:10.1038/nn963
53. J. Trommershäuser, L.T. Maloney, M.S. Landy, Decision making, movement planning and statistical decision theory. Trends Cogn. Sci. **12**(8), 291–297 (2008). doi:10.1016/j.tics.2008.04.010
54. R.S. Turner, M. Desmurget, Basal ganglia contributions to motor control: a vigorous tutor. Curr. Opin. Neurobiol. **20**(6), 704–716 (2010). doi:10.1016/j.conb.2010.08.022
55. Y. Uno, M. Kawato, R. Suzuki, Formation and control of optimal trajectory in human multijoint arm movement minimum torque-change model. Biol. Cybern. **61**(2), 89–101 (1989)
56. R.P.R.D. van der Wel, D. Sternad, D.A. Rosenbaum, Moving the arm at different rates: slow movements are avoided. J. Mot. Behav. **42**(1), 29–36 (2010). doi:10.1080/00222890903267116
57. D. Winter, *Biomechanics and Motor Control of Human Movement* (Wiley, New York, 1990)
58. S.J. Young, J. Pratt, T. Chau, Target-directed movements at a comfortable pace: movement duration and fitts's law. J. Mot. Behav. **41**(4), 339–346 (2009). doi:10.3200/JMBR.41.4.339-346

Geometric and Numerical Aspects of Redundancy

Pierre-Brice Wieber, Adrien Escande, Dimitar Dimitrov and Alexander Sherikov

Abstract If some resources of a robot are redundant with respect to a given objective, they can be used to address other, additional objectives. Since the amount of resources required to realize a given objective can vary, depending on the situation, this gives rise to a limited form of decision making, when assigning resources to different objectives according to the situation. Such decision making emerges in case of conflicts between objectives, and these conflicts appear to be situations of linear dependency and, ultimately, singularity of the solutions. Using an elementary model of a mobile manipulator robot with two degrees of freedom, we show how standard resolution schemes behave unexpectedly and inefficiently in such situations. We propose then as a remedy to introduce carefully tuned artificial conflicts, in the form of a trust region.

1 Preamble on Redundancy in Robotics

According to the Oxford Dictionary of English, redundancy is the state of being not or no longer needed or useful. If some resources of a robot appear to be not needed or useful to realize a given objective, a common idea is to make use of them to address another, additional objective. This gives rise to so-called *redundancy resolution schemes* [1]. In the typical, iterative procedure, the robot is assigned to:

P.-B. Wieber (✉) · D. Dimitrov · A. Sherikov
INRIA Grenoble Rhône-Alpes, Montbonnot-Saint-Martin, France
e-mail: Pierre-Brice.Wieber@inria.fr

D. Dimitrov
e-mail: mail@drdv.net

A. Sherikov
e-mail: Alexander.Sherikov@inria.fr

A. Escande
CNRS-AIST Joint Robotics Laboratory UMI3218/RL, Tsukuba, Japan
e-mail: adrien.escande@gmail.com

J.-P. Laumond et al. (eds.), *Geometric and Numerical Foundations of Movements*,
Springer Tracts in Advanced Robotics 117, DOI 10.1007/978-3-319-51547-2_4

1. Realize a first objective. If some resources appear to be redundant, they can be used to additionally…
2. Realize a second objective. If some resources still appear to be redundant, they can be used to additionally…
3. Realize a third objective, *etc.*

1.1 Kinematic Redundancy

Historically, the primary target of redundancy resolution schemes has been the kinematics of robots, where the resources considered are basically degrees of freedom, and the objectives are standard kinematic tasks (reaching, pointing) [1]. Recent developments in this field include the capacity to consider tasks expressed either as equality or inequality constraints [2]. This allows handling kinematic tasks such as avoiding, or staying within a region. As a typical example, proposed in [2], a humanoid robot is assigned with the following objectives:

1. Maintain balance. If possible, additionally…
2. Avoid collisions. If possible, additionally…
3. Reach an object. If possible, additionally…
4. Keep this object within sight.

A key property of inequality constraints is that they can be active or not, depending on the situation. As a result, the amount of resources required to realize an objective can vary. In the example given above, depending on the position of obstacles, the robot may have enough remaining resources to reach the target object or not, and keep it within sight or not (see Fig. 1). This gives rise to a limited form of decision making.

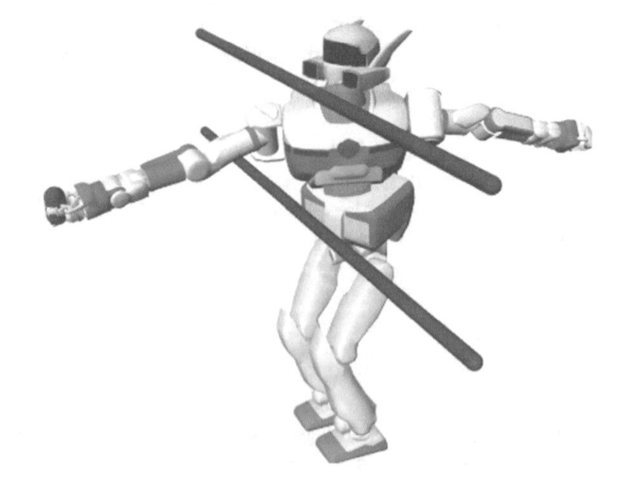

Fig. 1 As a typical example, proposed in [2], a humanoid robot is assigned with the following objectives:
1. Maintain balance;
2. Avoid collisions; 3. Reach an object; 4. Keep this object within sight

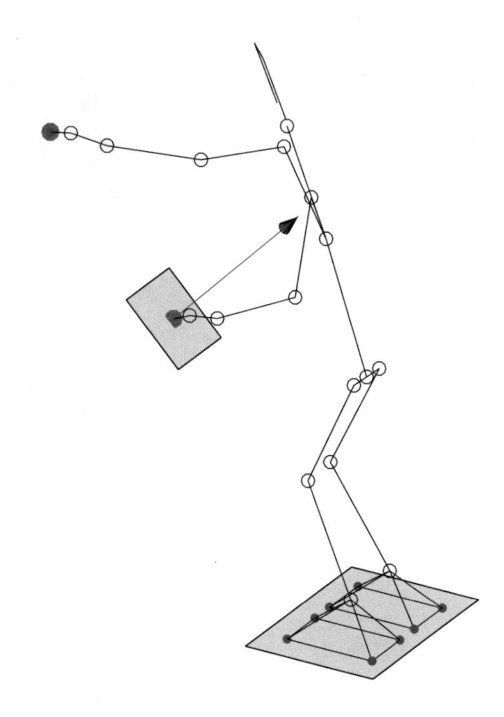

Fig. 2 In this example, a humanoid robot avoids putting weight with its left hand on a potentially risky surface, unless absolutely necessary to reach a target object with the right hand

1.2 Force Redundancy

Forces are another resource than can classically become redundant in a robot. This can be actuator forces, or contact forces with the environment. As an example, proposed in [3], a humanoid robot is assigned with the following objectives:

1. Maintain balance. If possible, additionally…
2. Reach an object. If possible, additionally…
3. Use only specified contact forces.

In this case, the redundancy resolution scheme decides whether to use only the specified contact forces or not, depending on the position of the target object. This is used to avoid putting weight on a potentially risky surface, unless absolutely necessary (see Fig. 2).

1.3 Time Redundancy

Another resource which is less often discussed is time. Typical circumstances are that some objectives should be realized either:

- as much as possible (it is equally important to realize the objectives now and later), as actually investigated in [3] to make decisions on balanced contact phases,

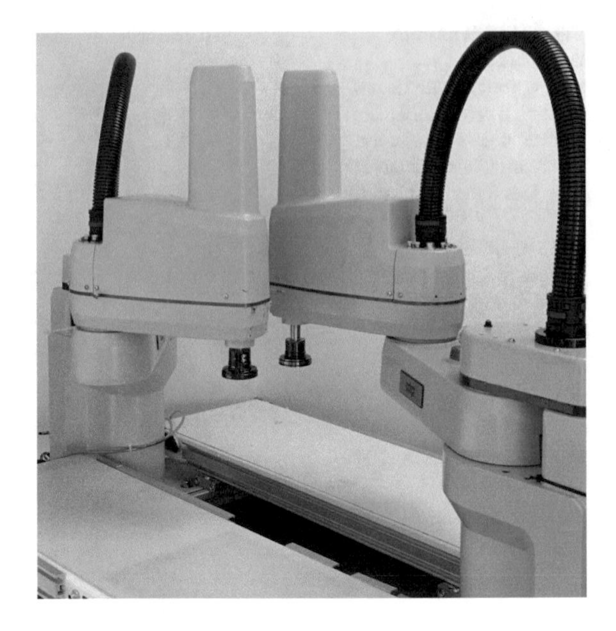

Fig. 3 In this example, two industrial manipulator robots are assigned with the following objectives:
1. Avoid collisions;
2. Do not accelerate too rapidly; 3. Reach their targets as soon as possible

- as long as possible (they should be realized now, and if possible, they should be realized later as well), as investigated in [4] to generate safe motions in uncontrolled environments,
- as soon as possible (they should be realized at some point in the future, and if possible, they should be realized earlier as well), as used in the example below to generate time optimal motions.

As an example, proposed in [5], two industrial manipulator robots (Fig. 3) are assigned with the following objectives:

1. Avoid collisions. If possible, additionally…
2. Do not accelerate too rapidly. If possible, additionally…
3. Reach their targets as soon as possible.

This results in generating time-optimal reaching motions online, under acceleration constraints and, more importantly, collision avoidance.

1.4 A General and Fundamental Issue

Beyond kinematics, forces, and time, the idea of making use of redundant resources to address additional objectives is very general. Even Isaac Asimov's famous Three Laws of Robotics can be reformulated to emphasize an underlying aspect of redundancy of resources:

1. A robot may not injure a human being or, through inaction, allow a human being to come to harm. If possible, additionally…
2. A robot must obey the orders given to it by human beings. If possible, additionally…
3. A robot must protect its own existence.

These three laws exemplify our final goal as roboticists, which is to have robots able to tackle complex situations, where different objectives have to be addressed concurrently. These objectives may conflict, and the robot has to make decisions accordingly. These decisions may involve important safety issues such as 'maintain balance', 'avoid collisions', 'do not injure a human being'. Our observation here is that redundancy resolution schemes are all about that: resolving conflicts between objectives, making decisions accordingly, and enforcing in the end the safety of the robot and of its environment. This is undeniably a fundamental issue in robotics, and the main motivation for the following analysis.

2 The Case with Two Objectives

Suppose that a robot is assigned with two objectives, that can be expressed as two functions $f(r)$ and $g(r)$ that should be equal to 0. If there is a conflict and these functions can't be equal to 0, they should at least be as close as possible to 0. Traditional approaches are to weight the (squared) norm of these objective functions, or to prioritize them.

2.1 *Weighting*

Weighting would lead to solving the following, unconstrained nonlinear program:

$$\underset{r}{\text{minimize}} \ \frac{1}{2}\|f(r)\|^2 + \frac{\omega}{2}\|g(r)\|^2, \tag{1}$$

with some given positive weight ω. In this case, the first order necessary condition for optimality is that

$$f(r)^T \frac{\partial f(r)}{\partial r} + \omega\, g(r)^T \frac{\partial g(r)}{\partial r} = 0. \tag{2}$$

If there is no conflict between these objectives, a solution with both $f(r) = 0$ and $g(r) = 0$ can be realized, and the necessary optimality condition above is trivially satisfied. However, if there is a conflict, the solutions are with $f(r) \neq 0$ or $g(r) \neq 0$ or both, and the necessary optimality condition above reveals that the rows of the Jacobian matrices are linearly dependent.

2.2 *Prioritizing*

Prioritizing would lead to imposing that $g(r) = 0$ (if possible), and solving the following, constrained nonlinear program:

$$\underset{r}{\text{minimize}} \ \frac{1}{2}\|f(r)\|^2 \tag{3}$$

$$\text{such that } g(r) = 0. \tag{4}$$

In this case, the first order necessary condition for optimality is that

$$f(r)^T \frac{\partial f(r)}{\partial r} + \lambda^T \frac{\partial g(r)}{\partial r} = 0 \tag{5}$$

with some Lagrange multipier λ to be determined.

Once again, if there is no conflict between these objectives, a solution with $f(r) = 0$ can be realized, and the necessary optimality condition above is trivially satisfied (with $\lambda = 0$). However, if there is a conflict, the solutions are with $f(r) \neq 0$ (and potentially $\lambda \neq 0$), and the necessary optimality condition above reveals that the rows of the Jacobian matrices are again linearly dependent.

2.3 *Conflicts and Linear Dependency*

It appears that situations of conflict are actually situations of linear dependency. This shouldn't come as a surprise, as this linear dependency is nothing more than the mathematical expression of the fact that objectives would like to draw from the same set of resources of the robot (expressed here in the variable r). We're going to see that this linear dependency can be the source of significant problems.

3 An Elementary Mobile Manipulator Robot

Consider an elementary mobile manipulator robot with two degrees of freedom: able to translate along the x axis, and equipped with an arm of unit length attached to a rotary joint with an angle θ, as depicted on Fig. 4, so $r = (x, \theta)$. Suppose that it is assigned with the following two objectives:

1. Reach a target with its end effector, at a coordinate equal to 2,

$$g = 2 - (x + \cos\theta) \to 0. \tag{6}$$

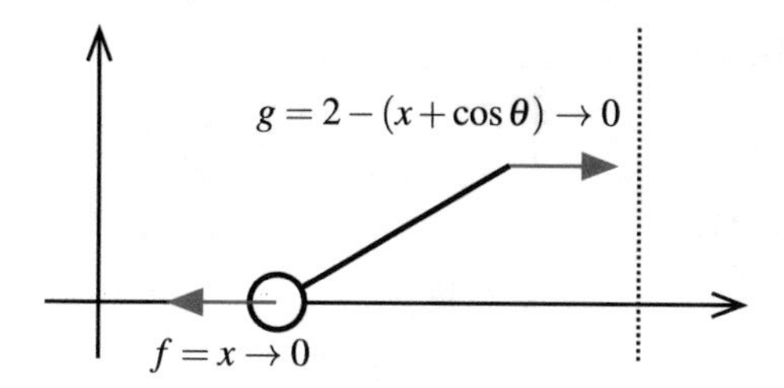

Fig. 4 An elementary mobile manipulator robot, able to translate along the x axis, and equipped with one rotary joint of angle θ, is assigned with the following objectives: 1. Reach an object with its end effector, $g = 2 - (x + \cos\theta) \to 0$; 2. Stay at the origin, $f = x \to 0$

2. If possible, additionally, stay at the origin,

$$f = x \to 0. \tag{7}$$

3.1 *Unreachability in the Objectives Space*

In the space defined by the two functions (f, g), the common objective is to be as close as possible to the origin $(0, 0)$ (see Fig. 5). We can observe however that this point is not reachable since, following the definitions (6) and (7) of these functions, we obviously always have

$$f + g \geq 1. \tag{8}$$

This unreachability means that these two functions can't be equal to 0 at the same time: there is a conflict between these objectives.

3.2 *Solutions Lie at Singularities*

We have seen that situations of conflict are situations of linear dependency. In this example, depending on the choice of resolution scheme, weighted (1) or prioritized (3) and (4), solutions will lie at different points on the boundary of the reachable space, where $f + g = 1$ (see Fig. 5). We can see from the definitions (6) and (7) that on this boundary, $\theta = 2k\pi$. Our mobile manipulator looks then as in Fig. 6, and the Jacobian matrix of the two objective functions

$$\begin{bmatrix} df \\ dg \end{bmatrix} = \begin{bmatrix} 1 & 0 \\ -1 & \sin(\theta) \end{bmatrix} \begin{bmatrix} dx \\ d\theta \end{bmatrix} \tag{9}$$

becomes singular.

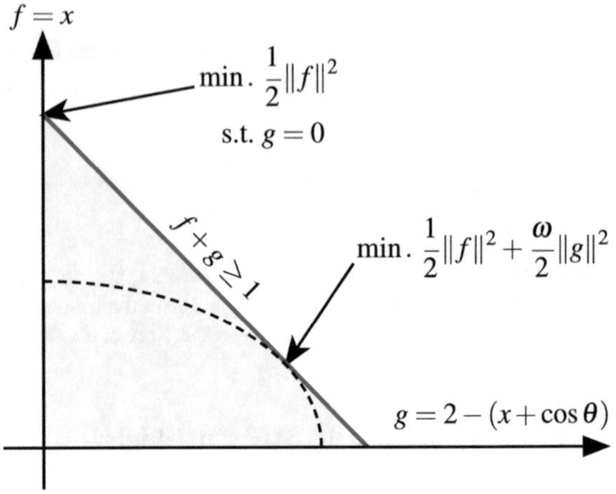

Fig. 5 In the space defined by the two functions (f, g), the common objective is to be as close as possible to the origin (0, 0). However, this point is not reachable since, following the definitions of these functions, we obviously always have $f + g \geq 1$. Depending on the choice of resolution scheme, solutions will lie at different points on the boundary, where $f + g = 1$

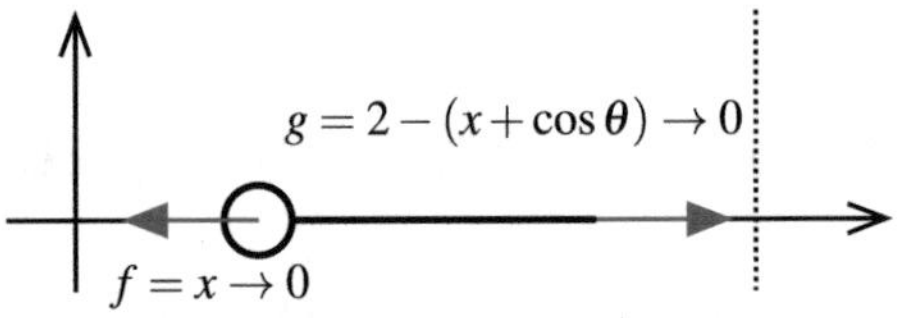

Fig. 6 When $f + g = 1$, $\theta = 2k\pi$, and the Jacobian matrix of the two objective functions becomes singular

3.3 A Common Outcome of Redundancy Resolution Schemes

Most examples provided in the literature on redundancy resolution schemes typically consider conflicts with objective functions $\|f(r)\|^2$ of the form $\|r - r_p\|^2$, where r_p is some preferred value for the resources r [6–8]. In those cases, the Jacobian matrix $\frac{\partial f(r)}{\partial r}$ is the identity matrix, which is always trivially linearly dependent with any Jacobian matrix $\frac{\partial g(r)}{\partial r}$. It follows that in those cases, linear dependency does not materialize only at the optima: there are no particular losses of matrix rank at the solutions, they do not particularly lie at singularities.

But such objective functions involve by construction all the resources of the robot, so there are no redundant resources left afterwards for the redundancy resolution scheme to proceed with. Objective functions which do not involve all the resources of the robot are naturally much more frequent. However, few examples of conflicts in such situations have appeared in the literature, and mostly in the case of humanoid robots [2, 9–11], probably because such robots can be particularly redundant. In these cases, linear dependency may materialize only at the optima, so there is a loss of matrix rank there: solutions lie at singularities, as above. This has already been acknowledged in [12] to be a common occurrence in redundancy resolution schemes. This singularity at the solutions is what will make standard approaches behave inefficiently.

4 Resolution Process

The usual resolution method for Nonlinear Least Squares Programs is to perform a sequence of Gauss–Newton steps, where nonlinear functions are replaced with their first-order linear approximations [13]. In the case of the nonlinear programs introduced earlier for the weighted and prioritized approaches, this leads to Quadratic Programs (QPs)

$$\underset{dr}{\text{minimize}}\ \frac{1}{2}\left\| f(r) + \frac{\partial f(r)}{\partial r} dr \right\|^2 + \frac{\omega}{2}\left\| g(r) + \frac{\partial g(r)}{\partial r} dr \right\|^2 \tag{10}$$

and

$$\underset{dr}{\text{minimize}}\ \frac{1}{2}\left\| f(r) + \frac{\partial f(r)}{\partial r} dr \right\|^2 \tag{11}$$

$$\text{such that } g(r) + \frac{\partial g(r)}{\partial r} dr = 0. \tag{12}$$

In robotics, this would correspond to a Closed-Loop Inverse Kinematics (CLIK) method, or to a task-space velocity control (but similar results are actually observed for task-space acceleration or torque control).

4.1 On the Boundary of the Reachable Space

When the state of our robot is on the boundary of the reachable space, when $f + g = 1$ and $\theta = 2k\pi$, we have seen that the rows of the Jacobian matrices $\frac{\partial f(r)}{\partial r}$ and $\frac{\partial g(r)}{\partial r}$ are linearly dependent, and the robot is in a singularity (as in Fig. 6). In this case, the solutions to the QPs above are such that

$$dx = \frac{\omega}{1+\omega} - x \tag{13}$$

for the weighted approach, and

$$dx = 1 - x \tag{14}$$

for the prioritized approach. Considering a least-norm solution, as usual in this case, we would also have $d\theta = 0$. The two numerical schemes appear therefore to converge to their respective solutions (on the boundary of the reachable space): $x = \frac{\omega}{1+\omega}$ and $\theta = 2k\pi$ for the weighted approach, $x = 1$ and $\theta = 2k\pi$ for the prioritized approach.

4.2 Away from the Boundary

When the state of our robot is away from the boundary of the reachable space, when $f + g > 1$, rows of the Jacobian matrices are linearly *independent*, so the linearized objectives

$$f(r) + \frac{\partial f(r)}{\partial r} dr = 0 \tag{15}$$

and

$$g(r) + \frac{\partial g(r)}{\partial r} dr = 0 \tag{16}$$

do not conflict and can be realized altogether. In this case, both numerical schemes lead the robot in the same direction dr, such that

$$df = \frac{\partial f(r)}{\partial r} dr = -f \tag{17}$$

and

$$dg = \frac{\partial g(r)}{\partial r} dr = -g. \tag{18}$$

This direction points towards the origin $(0, 0)$ of the objectives space, towards the boundary of the reachable space (Fig. 5).

4.3 Not Going in the Right Direction

We have seen that both the weighted and the prioritized numerical schemes converge to their respective solutions when on the boundary of the reachable space, and when away from this boundary, they first move towards it. It appears however that they don't go really in the right direction in that case. Indeed, following definition (7), equation (17) means

$$dx = -x, \tag{19}$$

so the mobile base of the robot (with coordinate x) is actually moving towards the origin $x = 0$, instead of any of the desired solutions $x = \frac{\omega}{1+\omega}$ or $x = 1$.

This is because the two linearized objectives do not conflict when away from the boundary, so the resolution schemes can temporarily aim at satisfying both, and that means having x go towards 0 in the first place. This behavior can be grasped visually on Fig. 4, where the two linearized objectives can be represented as two red arrows: the mobile base will move in the direction of the origin until the robot looks as on Fig. 6, when the state of the robot arrives on the boundary of the reachable space, and the two linearized objectives finally conflict. This conflict is then resolved satisfactorily, and the robot is eventually led to the desired solution.

The robot must therefore reach first a situation of conflict between the linearized objectives, *i.e.*, a singularity, before going in the right direction. And in case the numerical scheme is not made robust to singularities, it might get stuck oscillating around this singularity, never reach it, and never start moving in the right direction. This exact behavior was clearly observed in [14]. The usual approach to singularity robustness is to introduce damping (Tikhonov regularization) [15], which has been observed to work properly in the case of conflicting objectives [10], although it has been acknowledged to be difficult to tune by hand, as the problem is actually very sensitive [2]. This is one of the issues addressed in the method proposed next, a key feature of which is automatic tuning of parameters.

5 Introducing Artificial Conflicts

Having the robot go in a wrong direction, and waiting until it reaches a singularity before it starts going in the right direction, introduces an unpredictable delay before convergence begins, what is inefficient and undesirable. This happens because the desired solutions lie at singularities, and to provide a meaningful approximation, the first-order linear models have to be singular in a similar way. Otherwise, these models basically point in a wrong direction. The idea then is to introduce carefully tuned artificial conflicts, to interfere with these models and have them point in a better direction.

5.1 A Trust-Region Method

A standard remedy to rank-deficiency issues in Nonlinear Least Squares Programs is to resort to the Levenberg–Marquardt method, which can be seen as combining Gauss–Newton steps with a trust-region method [13]. For the prioritized approach, the QP (11) and (12) can be modified in the following way:

$$\underset{dr}{\text{minimize}} \ \frac{1}{2}\left\| f(r) + \frac{\partial f(r)}{\partial r} dr \right\|^2 \tag{20}$$

$$\text{such that } g(r) + \frac{\partial g(r)}{\partial r} dr = 0, \tag{21}$$

$$\|dr\|_\infty \leq \Delta, \tag{22}$$

introducing a bound Δ on the norm of the step dr, using here an L^∞-norm to obtain a standard QP formulation.

The crucial consequence of introducing this bound is that it doesn't only affect the size of dr, it also affects its direction, by interfering with the minimization of the objective in (20). This is unlike a line search method, which keeps the direction constant and adapts only the size of dr (that would correspond in robotics to simply varying control gains, such as in task scaling [16]).

If the bound Δ is chosen large enough, it will not interfere, and the solution of the QP will be unaffected. But if chosen small enough, it will conflict with the minimization of the objective in (20), producing a direction dr that fully reflects the objective in the constraint (21), much less the objective to minimize in (20), as if these two objectives were themselves conflicting.

Introducing such an artificial conflict is appropriate only if the two objectives do conflict in the end. The problem is that there is usually no way to know in advance if this is the case or not. The bound Δ must therefore be carefully adapted online, based on some heuristics. It has been proposed in [14] to adapt this bound, depending on the occurrence of oscillations, due to singularity. In that case however, it is only when approaching singularity that the artificial conflict would be introduced, so the problem of going in a wrong direction in the first place would remain.

We propose here to follow a standard trust-region method, where this bound is adapted, depending on how much the nonlinear objective functions differ from their first-order linear approximations throughout the Gauss–Newton iterations [13]. As a result, artificial conflict can be introduced early in the process, taking care automatically that it does not interfere with convergence.

5.2 Adaptive Damping

Following a theorem due to Moré and Sorenson [13], trust-region methods can also be considered as a form of adaptive damping. For the weighted approach, the QP (10) can be modified in the following way:

$$\underset{dr}{\text{minimize}}\ \frac{1}{2}\left\|f(r)+\frac{\partial f(r)}{\partial r}dr\right\|^2+\frac{\omega}{2}\left\|g(r)+\frac{\partial g(r)}{\partial r}dr\right\|^2+\frac{\Lambda}{2}\|dr\|^2 \tag{23}$$

with a regularization coefficient Λ that has to be carefully chosen and adapted.

If $\Lambda = 0$, the QP and its solution are unaffected. But if $\Lambda > 0$, the regularization term $\|dr\|^2$ naturally conflicts with both objectives. And if Λ is large enough with respect to min. $\{1, \omega\}$, this regularization term will interfere significantly with the objective with lower weight, producing a direction dr that reflects relatively much more the objective with higher weight, as if these two objectives were themselves conflicting.

As before, introducing such an artificial conflict is appropriate only if the two objectives do conflict in the end, and there is usually no way to know in advance if this is the case or not. The regularization weight Λ must therefore be adapted online as well, based on some heuristics. As before, we propose to follow a standard trust-region method, where this weight is adapted, depending on how much the nonlinear objective functions differ from their first-order linear approximations throughout the Gauss–Newton iterations [13].

Adjusting the damping coefficient is usually proposed in robotics with the goal to interfere as little as possible with the linearized objectives, only when they become close to singular [1]. The situation here is opposite, since the goal is to interfere significantly with the linearized objectives, and especially in situations where they are far from singular.

5.3 Implementation Aspects

Concerning the weighted approach, the QP (10) could also be modified to include a bound (22) instead of a regularization, but that would transform the original, unconstrained Least Squares problem into a slightly more complex, constrained one. This is the reason why the regularized formulation (23) is usually favored.

Concerning the prioritized approach, the QP (11) and (12) could also be modified to include a regularization instead of a bound, as in [2, 10]. But solutions to the non-regularized prioritized problem can be obtained very efficiently, with the help of specific matrix factorizations, and lexicographic active set methods when handling inequality constraints [9, 17], which do not apply to the regularized case. The difference in computation time can be more than 30-fold – a good reason to favor the bounded formulation (20) and (22). An advantageous aspect of regularization however is that it precludes ill-conditioning when approaching singularities, and ill-conditioning can prevent the proper termination of the active set methods used to handle inequality constraints, by inducing a cycling in their iterations. Which option should be favored in the end is still an open question.

Finally, traditional trust-region methods are designed for unconstrained, single-objective optimization problems [13]. They apply naturally to the regularized Least Squares problem (23) that appears in the weighted approach. But for the constrained, multi-objective problems that appear in the prioritized approach, we have to resort to more recent and experimental, multidimensional filter methods [18].

6 Numerical Results

We are going to observe now more precisely how the numerical schemes discussed above behave with the problem (6) and (7). A set of initial guesses for the solutions is chosen randomly, each assigned a unique color, and subsequent iterations are plotted in the objectives space, defined by the two functions (f, g), using the same layout as in Fig. 5. The distance to the solutions is also provided as a function of the number of iterations in order to visualize convergence speed.

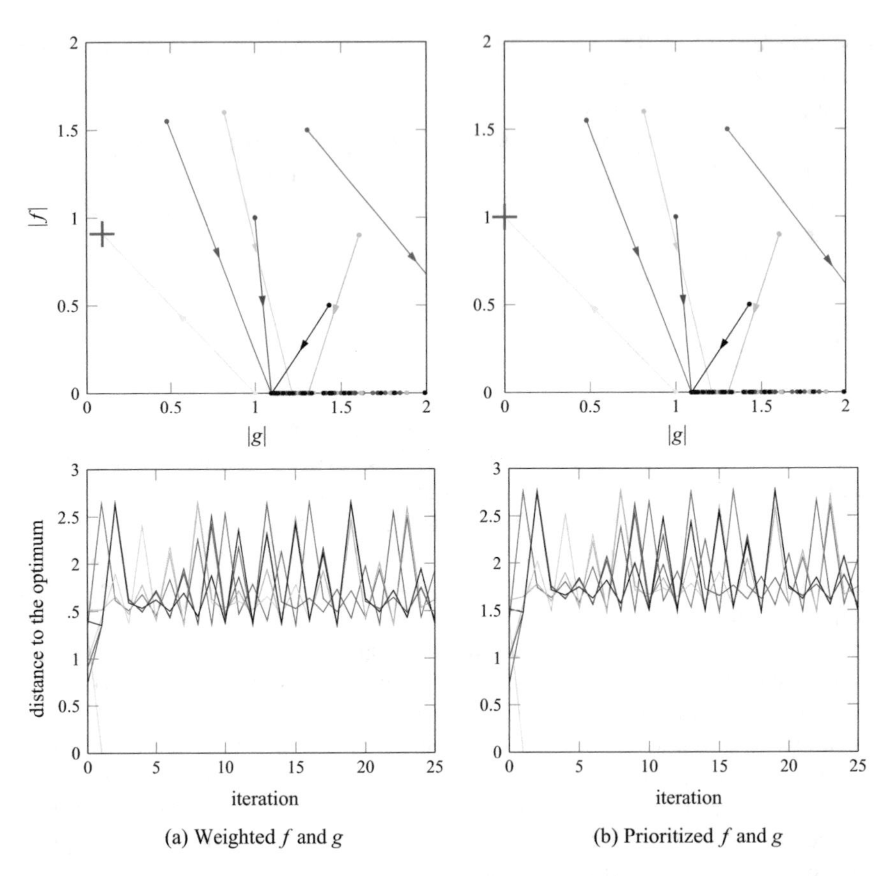

Fig. 7 Sequences of Gauss–Newton steps, as introduced in the QP (10) for the weighted approach (on the *left*, with a weight $\omega = 10$), and in the QP (11) and (12) for the prioritized approach (on the *right*), do not converge to the corresponding optima $(\frac{1}{11}, \frac{10}{11})$ and $(0, 1)$, represented by *red crosses* on the (f, g) plots above, except when initialized exactly on the boundary $f + g = 1$. The curves below show the corresponding evolutions of the distance to the optima with each iteration, which does not decrease at all in general

We can see in Fig. 7 that sequences of Gauss–Newton steps, as introduced in the QP (10) for the weighted approach (on the left, with a weight $\omega = 10$), and in the QP (11) and (12) for the prioritized approach (on the right), do not converge to the corresponding optima $(\frac{1}{11}, \frac{10}{11})$ and $(0, 1)$, represented by red crosses, except when initialized exactly on the boundary $f + g = 1$. Note how iterations are identical between the weighted approach (on the left) and the prioritized approach (on the right) when $f + g > 1$, as discussed earlier in Sect. 4.2.

We can see in Fig. 8 that simply dividing the length of the steps by 10, as would occur with a gain of 0.1 in a Closed Loop Inverse Kinematics scheme, naturally slows down each and every iteration, but does not lead to any improvement in convergence.

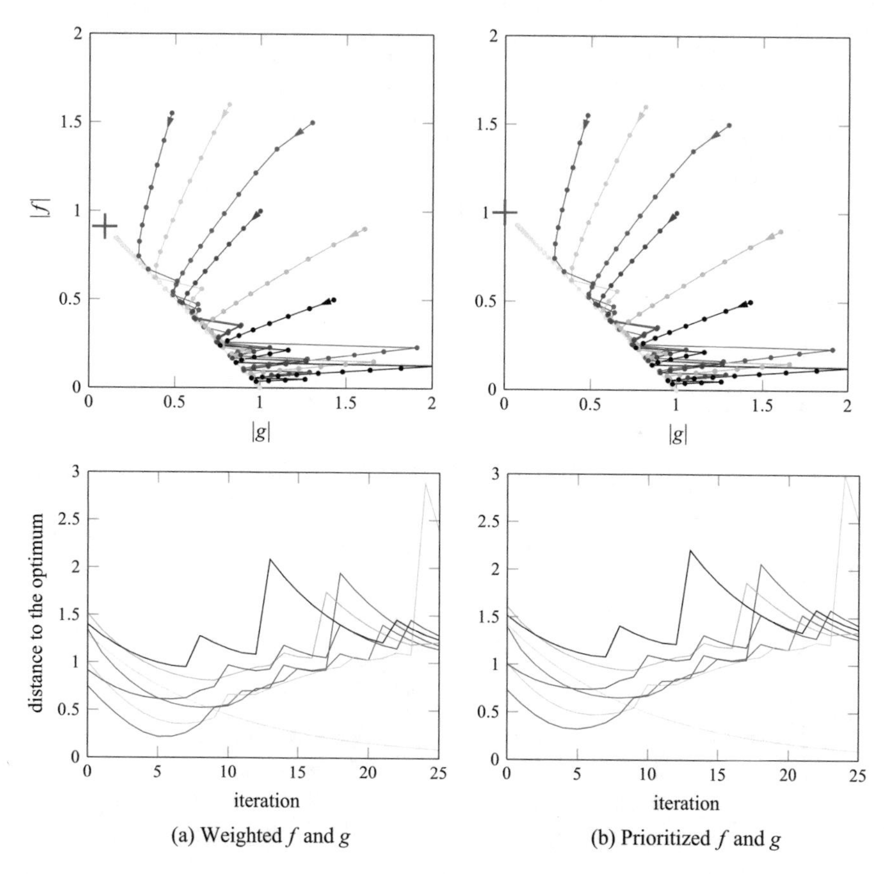

(a) Weighted f and g

(b) Prioritized f and g

Fig. 8 Simply dividing the length of the steps by 10, as would occur with a gain of 0.1 in a Closed Loop Inverse Kinematics scheme, naturally slows down each and every iteration, but does not lead to any improvement in convergence

We can see in Fig. 9 that for the weighted approach, introducing a regularization as in the QP (23), with a coefficient as high as $\Lambda = 0.4$, still does not lead to any better convergence. For the prioritized approach, introducing a bound as in the QP (20) and (21), with a constant $\Delta = 0.5$, leads to convergence in a few occasions, but not systematically. We can see in Fig. 10 that it is only with a regularization coefficient as high as $\Lambda = 0.55$ and a bound as small as $\Delta = 0.1$ that convergence begins to appear more reliably.

It appears that the regularization coefficient Λ and the bound Δ must be tuned carefully in order to reach convergence. This can be done automatically and efficiently with a simple trust-region method, as discussed in [13]. The corresponding behavior can be observed in Fig. 11. Of course, convergence speed may not always be as good as with a finely hand tuned coefficient or bound, but this does not account for the time required to hand tune these problem-specific parameters in the first place.

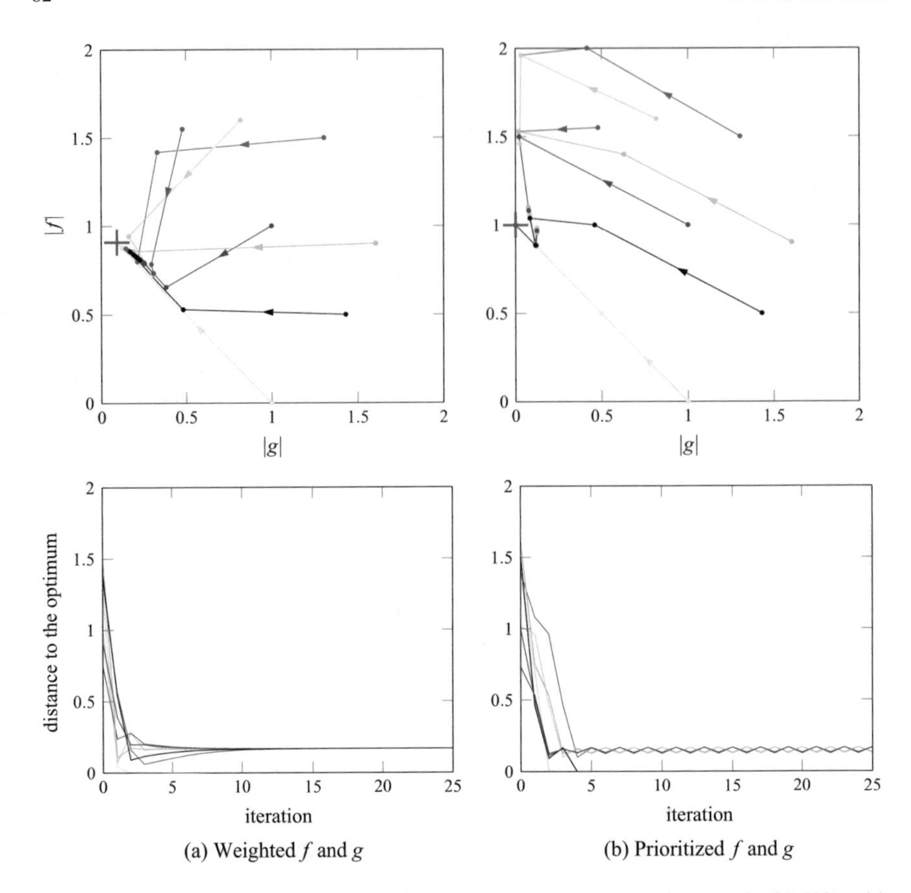

Fig. 9 For the weighted approach (on the *left*), introducing a regularization as in the QP (23), with a coefficient as high as $\Lambda = 0.4$, still does not lead to any better convergence. For the prioritized approach (on the *right*), introducing a bound as in the QP (20) and (21), with a constant $\Delta = 0.5$, leads to convergence in a few occasions, but not systematically

7 Conclusion

We have seen that redundancy resolution schemes can result in a limited form of decision making, when attributing the resources of a robot to its different objectives. This is because the amount of resources required to realize a given objective can vary, due particularly to the introduction of inequality constraints, that can be active or not, depending on the situation. Such decision making emerges in case of conflicts between objectives, and we have seen that these conflicts are directly related to situations of linear dependency.

The problem is that such linear dependency often results in singular solutions. In that case, standard resolution processes appear to go in the wrong direction, causing an unpredictable delay in convergence, because they rely on linearized objectives,

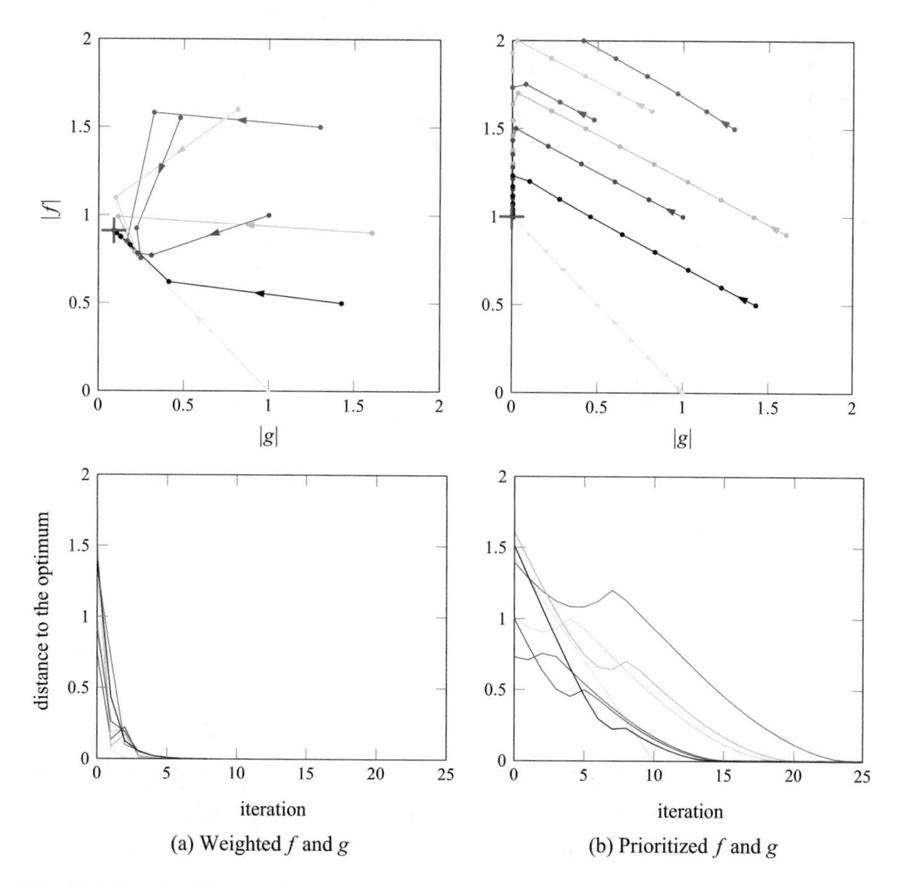

Fig. 10 It is only with a regularization coefficient as high as $\Lambda = 0.55$ (on the *left*) and a bound as small as $\Delta = 0.1$ (on the *right*) that convergence begins to appear more reliably

which can be inappropriate. It is striking to observe that the resolution processes behave nearly identically for the weighted and for the prioritized approach: both face exactly the same numerical difficulty here.

We propose to use trust-region methods as a remedy. They can be seen as a way to introduce artificial conflicts, automatically and carefully tuned to drive the resolution processes towards the desired solution. This has been demonstrated numerically with a simple system. More complex situations, with a greater number of objectives, need to be addressed now for a more thorough validation of the proposed method.

In conclusion, no claims can be made yet as to having fully solved the problem uncovered in this chapter. Further study of how humans handle such situations could prove enlightening in this regard, and may provide inspiration for more refined solutions.

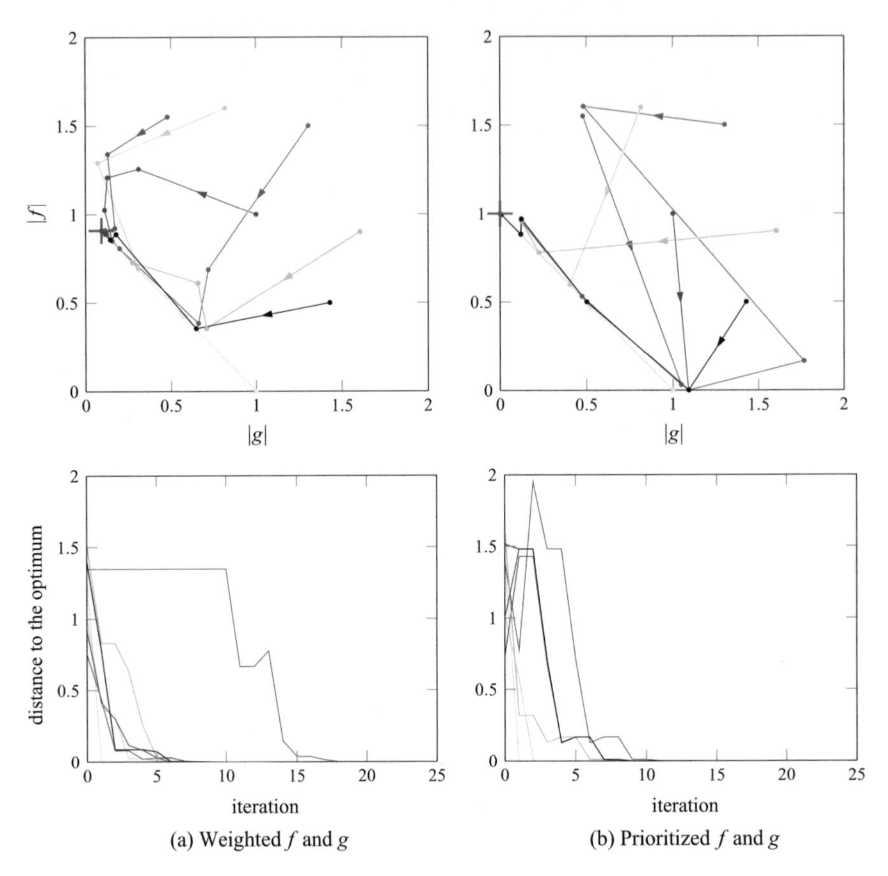

(a) Weighted f and g (b) Prioritized f and g

Fig. 11 It appears that the regularization coefficient Λ and the bound Δ must be chosen and adapted carefully in order to obtain convergence. This can be done automatically and efficiently with a simple trust-region method, as discussed in [13]

References

1. S. Chiaverini, G. Oriolo, I.D. Walker, Kinematically redundant manipulators, in *Springer Handbook of Robotics*, ed. By B. Siciliano, O. Khatib, Ch. 11 (Springer, Berlin, Heidelberg, 2008), pp. 245–265
2. O. Kanoun, F. Lamiraux, P.-B. Wieber, Kinematic control of redundant manipulators: generalizing the task-priority framework to inequality task. IEEE Trans. Robot. **27**(4), 785–792 (2011)
3. A. Sherikov, D. Dimitrov, P.-B. Wieber, Balancing a humanoid robot with a prioritized contact force distribution, in *IEEE-RAS International Conference on Humanoid Robots* (2015), pp. 223–228
4. N. Bohórquez, A. Sherikov, D. Dimitrov, P.-B. Wieber, Safe navigation strategies for a biped robot walking in a crowd, in *IEEE-RAS International Conference on Humanoid Robots* (2016)
5. S. Al Homsi, A. Sherikov, D. Dimitrov, P.-B. Wieber, A hierarchical approach to minimum-time control of industrial robots, in *IEEE/RSJ International Conference on Robotics and Automation* (2016)

6. B. Faverjon, P. Tournassoud, A local based approach for path planning of manipulators with a high number of degrees of freedom, in *IEEE/RSJ International Conference on Robotics and Automation* (1987), pp. 1152–1159
7. Y. Nakamura, H. Hanafusa, T. Yoshikawa, Task-priority based redundancy control of robot manipulators. Int. J. Robot. Res. **6**(2), 3–15 (1987)
8. B. Siciliano, J.J.E. Slotine, A general framework for managing multiple tasks in highly redundant robotic systems, in *International Conference on Advanced Robotics* (1991), pp. 1211–1216
9. A. Escande, N. Mansard, P.-B. Wieber, Hierarchical quadratic programming: fast online humanoid-robot motion generation. Int. J. Robot. Res. **33**(7), 1006–1028 (2014)
10. A. Del Prete, F. Nori, G. Metta, L. Natale, Prioritized motionforce control of constrained fully-actuated robots: "task space inverse dynamics". Robot. Auton. Syst. **63**(1), 150–157 (2015)
11. A.A. Maciejewski, C.A. Klein, Task-priority based redundancy control of robot manipulators. Int. J. Robot. Res. **4**(3), 109–116 (1985)
12. T. Sugihara, Robust solution of prioritized inverse kinematics based on Hestenes-Powell multiplier method, in *IEEE/RSJ International Conference on Intelligent Robots and System* (2014), pp. 510–515
13. J. Nocedal, S. Wright, *Numerical optimization*, 2nd edn. (Springer, 2006)
14. F.M. Bianchi, G. Gualandi, Handling and prioritizing inequality constraints in redundancy resolution optimization-based approach with task priority, Master's thesis, Sapienza University of Rome, 2011
15. Y. Nakamura, H. Hanafusa, Inverse kinematic solutions with singularity robustness for robot manipulator control. ASME J. Dyn. Syst. Meas. Control **108**(3), 163–171 (1986)
16. F. Flacco, A. De Luca, O. Khatib, Control of redundant robots under hard joint constraints: saturation in the null space. IEEE Trans. Robot. **31**(3), 637–654 (2015)
17. D. Dimitrov, A. Sherikov, P.-B. Wieber, *Efficient resolution of potentially conflicting linear constraints in robotics* (2015)
18. N. Gould, S. Leyffer, P. Toint, A multidimensional filter algorithm for nonlinear equations and nonlinear least squares. SIAM J. Optim. **15**(1), 17–38 (2004)

Part II
Numerical Analyzis and Optimization

Some Recent Directions in Algebraic Methods for Optimization and Lyapunov Analysis

Amir Ali Ahmadi and Pablo A. Parrilo

Abstract Exciting recent developments at the interface of optimization and control have shown that several fundamental problems in dynamics and control, such as stability, collision avoidance, robust performance, and controller synthesis can be addressed by a synergy of classical tools from *Lyapunov theory* and modern computational techniques from *algebraic optimization*. In this chapter, we give a brief overview of our recent research efforts (with various coauthors) to (i) enhance the scalability of the algorithms in this field, and (ii) understand their worst case performance guarantees as well as fundamental limitations. The topics covered include the concepts of "dsos and sdsos optimization", path-complete and non-monotonic Lyapunov functions, and some lower bounds and complexity results for Lyapunov analysis of polynomial vector fields and hybrid systems. In each case, our relevant papers are tersely surveyed and the challenges/opportunities that lie ahead are stated.

1 Algebraic Methods in Optimization and Control

In recent years, a fundamental and exciting interplay between *convex optimization* and *algorithmic algebra* has allowed for the solution or approximation of a large class of nonlinear and nonconvex problems in optimization and control once thought to be

This chapter is a revised and expanded version of the conference paper in [13], which was presented as a tutorial talk at the 53rd annual Conference on Decision and Control.

Amir Ali Ahmadi—His research is partially supported by an NSF CAREER Award, an AFOSR Young Investigator Program Award, and a Google Research Award.

A.A. Ahmadi (✉)
Department of Operations Research and Financial Engineering,
Princeton University, Princeton, NJ, USA
e-mail: a_a_a@princeton.edu
URL: http://aaa.princeton.edu/

P.A. Parrilo
Laboratory for Information and Decision Systems and The Department of Electrical Engineering and Computer Science, MIT, Cambridge, MA, USA
e-mail: parrilo@mit.edu
URL: http://www.mit.edu/parrilo/

J.-P. Laumond et al. (eds.), *Geometric and Numerical Foundations of Movements*, Springer Tracts in Advanced Robotics 117, DOI 10.1007/978-3-319-51547-2_5

out of reach. The success of this area stems from two facts: (i) Numerous fundamental problems in optimization and control (among several other disciplines in applied and computational mathematics) are *semialgebraic*; i.e., they involve optimization over sets defined by a finite number of possibly quantified *polynomial inequalities*. (ii) Semialgebraic problems can be reformulated as optimization problems over the set of *nonnegative polynomials*. This makes them amenable to a rich set of algebraic tools which lend themselves to *semidefinite programming*—a subclass of convex optimization problems for which global solution methods are available.

Application areas within optimization and computational mathematics that have been impacted by advances in algebraic techniques are numerous: approximation algorithms for NP-hard combinatorial problems [31], equilibrium analysis of continuous games [57], robust and stochastic optimization [23], statistics and machine learning [50], software verification [63], filter design [65], quantum computation [29], and automated theorem proving [32], are only a few examples on a long list.

In dynamics and control, algebraic methods and in particular the so-called area of *"sum of squares (sos) optimization"* [2, 26, 33, 41, 55] have rejuvenated Lyapunov theory, giving the hope or the outlook of a paradigm shift from classical linear control to a principled framework for design of nonlinear (polynomial) controllers that are provably safer, more agile, and more robust. As a concrete example, Fig. 1

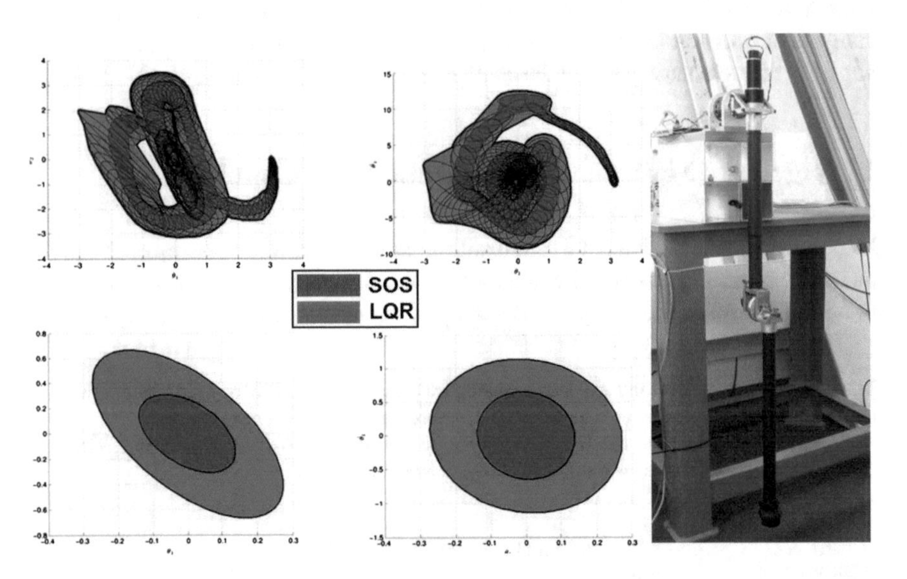

Fig. 1 From [51] (with Majumdar and Tedrake): The "swing-up and balance" task via sum of squares optimization for an underactuated and severely torque limited double pendulum (the "Acrobot"). *Top* projections of basins of attraction around a nominal swing-up trajectory designed by linear quadratic regulator (LQR) techniques (*blue*) and by SOS techniques (*red*). *Bottom* projections of basins of attraction of the unstable equilibrium point in the upright position stabilized by a linear controller via LQR (*blue*), and a cubic controller via SOS (*red*). To our knowledge, this work constitutes the first hardware implementation and experimental validation of sum of squares techniques in robotics

demonstrates our recent work with Majumdar and Tedrake [51] in this area applied to the field of robotics. As the caption explains, sos techniques provide controllers with much larger margins of safety along planned trajectories and can directly reason about the *nonlinear* dynamics of the system under consideration. These are crucial assets for more challenging robotic tasks such as walking, running, and flying. Sum of squares methods have also recently made their way to actual industry flight control problems, e.g., to explain the falling leaf mode phenomenon of the F/A-18 Hornet aircraft [25, 64] or to design controllers for hypersonic aircraft [20].

2 Our Target Areas in Algebraic Optimization and Control

Despite the wonderful advances in algebraic techniques for optimization and their successful interplay with Lyapunov methods, there are still many fundamental challenges to overcome and unexplored pathways to pursue. In this chapter, we aim at highlighting two concrete areas in this direction:

Area 1—Struggle with scalability Scalability is arguably the single most outstanding challenge for algebraic methods, not just in control theory, but in all areas of computational mathematics where these techniques are being applied today. It is well known that the size of the semidefinite programs (SDPs) resulting from sum of squares techniques (although polynomial in the data size) grows quickly and this limits the scale of the problems that can be efficiently and reliably solved with available SDP solvers. This drawback deprives large-scale systems of the application of algebraic techniques and perhaps equally importantly shuts the door on the opportunities that lie ahead if we could use these tools for *real-time optimization.*

In nonlinear control, problems with scalability also manifest themselves in form of complexity of Lyapunov functions. It is common for "simple" (e.g., low degree) stable systems to necessitate "complicated" Lyapunov functions as stability certificates (e.g., polynomials of high degree). The more complex the Lyapunov function, the more variables its parametrization will have, and the larger the sum of squares programs that search for it will have to be. In view of this, it is of particular interest to derive conditions for stability that are less stringent than those of classical Lyapunov theory. A related challenge in this area is the lack of a unified and comparative theory for various classes of Lyapunov functions available in the literature (e.g., polytopic, piecewise quadratic, polynomial, etc.). These problems are more pronounced in the study of *uncertain* or *hybrid system*, which are of great practical relevance in engineering.

Area 2—Lack of rigorous guarantees While most works in the literature formulate hierarchies of optimization problems that—*if* feasible—guarantee desired properties of a control system of interest (e.g., stability or safety), relatively few establish "converse results", i.e., proofs that if certain policies meet design specifications, then a particular level in the optimization hierarchy is guaranteed to find a certificate as a feasible solution. This is in contrast to more discrete areas of optimization where tradeoffs between algorithmic efficiency and worst-case performance guarantees are often quite well-understood.

A study of performance guarantees for some particular class of algorithms (in our case, sum of squares algorithms) naturally borders the study of lower bounds, i.e., fundamental limits on the efficiency of *any* algorithm that provably solves a problem class of interest. Once again here, the state of affairs in this area of controls is not entirely satisfactory: there are numerous fundamental problems in the field that while believed to be "hard" in folklore, lack a rigorous complexity-theoretic lower bound. One can attribute this shortcoming to some extent to the nature of most problems in controls, which typically come from continuous mathematics and at times describe qualitative behavior of a system rather than quantitative ones (consider, e.g., asymptotic stability of a nonlinear vector field).

The remainder of this chapter presents a brief report on some recent progress we have made on these two target areas, as well as some challenges that lie ahead. This is meant neither as a comprehensive survey paper, as there are many great contributions by other authors that we do not cover, nor as a stand-alone paper, as for the most part only entry points to a collection of relevant papers will be provided. The interested reader can find further detail and a more comprehensive literature review in the references presented in each section.

2.1 Organization of the Chapter

The outline of the chapter is as follows. We start by a short section on basics of sum of squares optimization in the hope that our chapter becomes accessible to a broader audience. In Sect. 4, we describe some recent developments on the optimization side to provide more scalable alternatives to sum of squares programming. This is the framework of *"dsos and sdsos optimization"*, which is amenable to linear and second order cone programming as opposed to semidefinite programming. In Sect. 5, we describe some new contributions to Lyapunov theory that can improve the scalability of algorithms meant for verification of dynamical systems (either continuous or hybrid). These include techniques for replacing high-degree Lyapunov functions with multiple low-degree ones (Sect. 5.1), and a methodology for relaxing the "monotonic decrease" requirement of Lyapunov functions (Sect. 5.2). The beginning of Sect. 5 also includes a list of recent results on complexity of deciding stability and on success/limitations of algebraic methods for finding Lyapunov functions. Both Sects. 4 and 5 are ended with a list of open problems or opportunities for future research.

We note that the chapter by J.B. Lasserre in this edited volume is also closely related to some of what we present here with respect to sum of squares optimization and its applications to control theory. The theory of *moments*, which is dual to the theory of sum of squares polynomials, is not present in the current chapter. For a treatment of this topic and its applications to control and robotics, see e.g. [34, 44, 45, 54].

3 A Quick Introduction to SOS for the General Reader

At the core of most algebraic methods in optimization and control is the simple idea of optimizing over polynomials that take only *nonnegative* values, either globally or on certain regions of the Euclidean space. A multivariate polynomial $p(x) := p(x_1, \ldots, x_n)$ is said to be (globally) *nonnegative* if $p(x) \geq 0$ for all $x \in \mathbb{R}^n$. As an example, consider the task of deciding whether the following polynomial in 3 variables and degree 4 is nonnegative[1]:

$$\begin{aligned} p(x) = & \ x_1^4 - 6x_1^3x_2 + 2x_1^3x_3 + 6x_1^2x_3^2 + 9x_1^2x_2^2 \\ & -6x_1^2x_2x_3 - 14x_1x_2x_3^2 + 4x_1x_3^3 \\ & +5x_3^4 - 7x_2^2x_3^2 + 16x_2^4. \end{aligned} \tag{1}$$

This may seem like a daunting task (and indeed it is as testing for nonnegativity is NP-hard), but suppose we could "somehow" come up with a decomposition of the polynomial as a sum of squares:

$$\begin{aligned} p(x) = & \ (x_1^2 - 3x_1x_2 + x_1x_3 + 2x_3^2)^2 + (x_1x_3 - x_2x_3)^2 \\ & +(4x_2^2 - x_3^2)^2. \end{aligned} \tag{2}$$

Then, we have at our hands an *explicit algebraic certificate* of nonnegativity of $p(x)$, which can be easily checked (simply by multiplying the terms out). A polynomial p is said to be a *sum of squares* (sos), if it can be written as $p(x) = \sum q_i^2(x)$ for some polynomials q_i. Interestingly, the question of existence of an sos decomposition (i.e., the task of going from (1) to (2)) can be cast as a *semidefinite program* (SDP) and be solved, e.g., by interior point methods. This is because of the following well-known theorem (see, e.g., [55]).

Theorem 3.1 *A multivariate polynomial p in n variables and of degree $2d$ is a sum of squares if and only if there exists a positive semidefinite matrix Q (often called the Gram matrix) such that*

$$p(x) = z^T Q z, \tag{3}$$

where z is the vector of monomials of degree up to d

$$z = [1, x_1, x_2, \ldots, x_n, x_1x_2, \ldots, x_n^d].$$

Proof If (3) holds, then we can do a Cholesky factorization on the Gram matrix, $Q = V^T V$, and obtain the desired sos decomposition as

$$p(x) = z^T V^T V z = (Vz)^T (Vz) = ||Vz||^2.$$

[1]The familiar reader may safely skip this section. For a more comprehensive introductary exposition, see: https://blogs.princeton.edu/imabandit/guest-posts/.

Conversely, suppose p is sos:

$$p = \sum_i q_i^2(x),$$

then for some vectors of coefficients a_i, we must have

$$p = \sum_i (a_i^T z(x))^2 = \sum_i (z^T(x) a_i)(a_i^T z(x)) = z^T(x) (\sum_i a_i a_i^T) z(x),$$

so the positive semidefinite matrix $Q := \sum_i a_i a_i^T$ can be extracted. As a corollary of the proof, we see that the number of squares in our sos decomposition is exactly equal to the rank of the Gram matrix Q. □

Note that the feasible set defined by the constraints in (3) is the intersection of an affine subspace (arising from the equality constraints matching the coefficients of p with the entries of Q) with the cone of positive semidefinite matrices. This is precisely the semidefinite programming (SDP) problem. The size of the Gram matrix Q is $\binom{n+d}{d} \times \binom{n+d}{d}$, which for fixed d is polynomial in n. Depending on the structure of p, there are well-documented techniques for further reducing the size of the Gram matrix Q and the monomial vector z. We do not pursue this direction here but state as an example that if p is homogeneous of degree $2d$, then it suffices to place in the vector z only monomials of degree exactly d.

Example 3.1 Consider the task proving nonnegativity of the polynomial in (1). Since this is a form (i.e., a homogeneous polynomial), we take

$$z = (x_1^2, x_1 x_2, x_2^2, x_1 x_3, x_2 x_3, x_3^2)^T.$$

One feasible solution to the SDP in (3) is given by

$$Q = \begin{pmatrix} 1 & -3 & 0 & 1 & 0 & 2 \\ -3 & 9 & 0 & -3 & 0 & -6 \\ 0 & 0 & 16 & 0 & 0 & -4 \\ 1 & -3 & 0 & 2 & -1 & 2 \\ 0 & 0 & 0 & -1 & 1 & 0 \\ 2 & -6 & 4 & 2 & 0 & 5 \end{pmatrix}.$$

Upon a decomposition $Q = \sum_{i=1}^3 a_i^T a_i$, with $a_1 = (1, -3, 0, 1, 0, 2)^T$, $a_2 = (0, 0, 0, 1, -1, 0)^T$, $a_3 = (0, 0, 4, 0, 0, -1)^T$, one obtains the sos decomposition

$$p(x) = (x_1^2 - 3x_1 x_2 + x_1 x_3 + 2x_3^2)^2 + (x_1 x_3 - x_2 x_3)^2 + (4x_2^2 - x_3^2)^2.$$

This is exactly how the expression in (2) was obtained. We remark that the task of generating semidefinite programs from sum of squares constraints has been automated in a number of freely-available software packages such as YALMIP [49] and

SOSTOOLS [60]. The interested reader can find a short tutorial and some examples here: https://yalmip.github.io/tutorial/sumofsquaresprogramming/. △

The question of *when* nonnegative polynomials admit a decomposition as a sum of squares is one of the central questions of real algebraic geometry, dating back to the seminal work of Hilbert [35, 62], and an active area of research today. This question is commonly faced when one attempts to prove guarantees for performance of algebraic algorithms in optimization and control.

In short, sum of squares decomposition is a sufficient condition for polynomial nonnegativity. It has become quite popular because of three reasons: (i) the decomposition can be obtained by semidefinite programming, (ii) the proof of nonnegativity is in form of an *explicit certificate* and is easily verifiable, and (iii) there is strong empirical (and in some cases theoretical) evidence showing that in relatively low dimensions and degrees, "most" nonnegative polynomials are sums of squares.

But why do we care about polynomial nonnegativity to begin with? We briefly present two fundamental application areas next: the polynomial optimization problem, and Lyapunov analysis of control systems.

3.1 The Polynomial Optimization Problem

The polynomial optimization problem (POP) is currently a very active area of research in the optimization community. It is the following problem:

$$\begin{array}{ll} \text{minimize } p(x) \\ \text{subject to } x \in K := \{x \in \mathbb{R}^n \mid g_i(x) \geq 0, h_i(x) = 0\}, \end{array} \tag{4}$$

where p, g_i, and h_i are multivariate polynomials. The special case of problem (4) where the polynomials p, g_i, h_i all have degree one is of course *linear programming*, which can be solved very efficiently. When the degree is larger than one, POP contains as special case many important problems in operations research; e.g., all problems in the complexity class NP, such as MAXCUT, travelling salesman, computation of Nash equilibria, scheduling problems, etc.

A set defined by a finite number of polynomial inequalities (such as the set K in (4)) is called *basic semialgebraic*. By a straightforward reformulation of problem (4), we observe that if we could optimize over the set of polynomials, *nonnegative on a basic semialgebraic set*, then we could solve the POP problem to global optimality. To see this, note that the optimal value of problem (4) is equal to the optimal value of the following problem:

$$\begin{array}{ll} \text{maximize } \gamma \\ \text{subject to } p(x) - \gamma \geq 0, \ \forall x \in K. \end{array} \tag{5}$$

Here, we are trying to find the largest constant γ such that the polynomial $p(x) - \gamma$ is nonnegative on the set K; i.e., the largest lower bound on problem (4). For ease of exposition, we only explained how a sum of squares decomposition provides a sufficient condition for polynomial nonnegativity globally. But there are straightforward generalizations for giving sos certificates that ensure nonnegativity of a polynomial on a basic semialgebraic set; see, e.g., [44, 56]. All these generalizations are amenable to semidefinite programming and commonly used to tackle the polynomial optimization problem.

3.2 Lyapunov Analysis of Dynamical Systems

Numerous fundamental problems in nonlinear dynamics and control, such as stability, invariance, robustness, collision avoidance, controller synthesis, etc., can be turned by means of "Lyapunov theorems" into problems about finding special functions (the *Lyapunov functions*) that satisfy certain sign conditions. The task of constructing Lyapunov functions has traditionally been one of the most fundamental and challenging tasks in control. In recent years, however, advances in convex programming and in particular in the theory of semidefinite optimization have allowed for the search for Lyapuonv functions to become *fully automated*. Figure 2 summarizes the steps involved in this process.

As a simple example, if the task in the leftmost block of Fig. 2 is to establish global asymptotic stability of the origin for a polynomial differential equation $\dot{x} = f(x)$, with $f : \mathbb{R}^n \to \mathbb{R}^n$, $f(0) = 0$, then the Lyapunov inequalities that a radially unbounded Lyapunov function V would need to satisfy are [43]:

$$\begin{aligned} V(x) &> 0 && \forall x \neq 0 \\ \dot{V}(x) &= \langle \nabla V(x), f(x) \rangle < 0 && \forall x \neq 0. \end{aligned} \tag{6}$$

Here, $\dot{V}$ denotes the time derivative of V along the trajectories of $\dot{x} = f(x)$, $\nabla V(x)$ is the gradient vector of V, and $\langle ., . \rangle$ is the standard inner product in $\mathbb{R}^n$. If we parametrize V as an unknown polynomial function, then the Lyapunov inequalities in (6) become polynomial positivity conditions. The standard sos relaxation for these inequalities would then be:

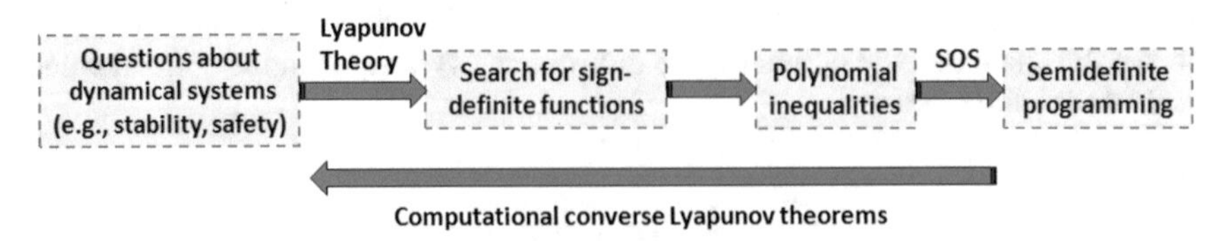

Fig. 2 The steps involved in Lyapunov analysis of dynamical systems via semidefinite programming. The need for "computational" converse Lyapunov theorems is discussed in Sect. 5

$$V \quad \text{sos} \quad \text{and} \quad -\dot{V} = -\langle \nabla V, f \rangle \quad \text{sos}. \tag{7}$$

The search for a polynomial function V satisfying these two sos constraints is a semidefinite program, which, if feasible, would imply[2] a solution to (6) and hence a proof of global asymptotic stability through Lyapunov's theorem. Let us spell out a similar application to formal verification of collision avoidance.

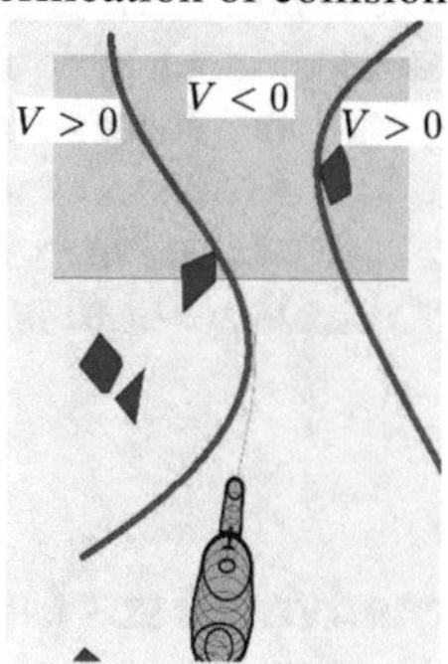

3.2.1 Barrier Certificates for Collision Avoidance

Whether on the ground or in the air, the fundamental requirement in motion planning for unmanned vehicles (UVs) in cluttered environments is collision avoidance. In recent years, the idea of "*barrier functions*" has been proposed for obtaining formal certificates of collision avoidance through sum of squares optimization [22, 58, 59]. The idea is simple and resembles Lyapunov theory: If we find a function $V : \mathbb{R}^n \to \mathbb{R}$ with the following properties,

$$V < 0 \quad \text{on the safe region,} \quad V > 0 \quad \text{on the obstacles,}$$

$$\dot{V} \leq 0 \quad \text{along all trajectories,}$$

then, trajectories starting from safe regions are certified to not collide with the obstacles. Once again, if we assume the dynamics of motion are given by some polynomial differential equation $\dot{x} = f(x)$ (or closely approximated as such via a Taylor expansion of high-enough degree), and further if we contain the obstacles and the region that requires safety verification with basic semialgebraic sets, then the search for a polynomial function V satisfying the inequalities above can be *fully automated* using sum of squares techniques. Indeed, suppose the safe region and an obstacle are respectively represented with a set of polynomial inequalities $S = \{x \in \mathbb{R}^n |\ h_i(x) \geq 0\}$, $O = \{x \in \mathbb{R}^n |\ g_j(x) \geq 0\}$. Then, we can use semidefinite programming to find the

[2]Here, we are assuming a strictly feasible solution to the SDP. Indeed a strictly feasible solution to (7) is required to get the strict inequalities in (6). Luckily, unless the SDP has an empty interior, a strictly feasible solution will automatically be returned by the interior point solver. See the discussion in [1, p. 41].

coefficients of a polynomial V and polynomials τ_i, σ_j satisfying

$$V - \sum_j \sigma_j g_j \quad \text{sos}, \quad -V - \sum_i \tau_i h_j \quad \text{sos}, \quad \text{and} \quad -\dot{V} = -\langle \nabla V, f \rangle \quad \text{sos}. \quad (8)$$

These algebraic identities indeed certify[3] the prescribed inequalities on V.

We remark that for this approach to be applicable to real-time motion planning, the underlying sum of squares programs need to be solved in real time as the obstacles enter the range of vision of the UV. At the moment, it is fair to say that sum of squares solvers are not quite capable of handling real-time applications. The same can be said about offline, but large-scale applications. Such limitations motivate the developments of our next section.

4 More Tractable Alternatives to sos Optimization [Area 1]

As explained in Sect. 3, a central question of relevance to applications of algorithmic algebra is to provide sufficient conditions for nonnegativity of polynomials, as working with nonnegativity constraints directly is in general intractable. The sum of squares (sos) condition achieves this goal and is amenable to semidefinite programming (SDP). Although this has proven to be a powerful approach, its application to many practical problems has been challenged by a simple bottleneck: *scalability*.

For a polynomial of degree $2d$ in n variables, the size of the semidefinite program that decides the sos decomposition is roughly n^d. Although this number is polynomial in n for fixed d, it can grow rather quickly even for low degree polynomials.

In addition to being large-scale, the resulting semidefinite programs are also often ill-conditioned and challenging to solve. In general, SDPs are among the most expensive convex relaxations and many practitioners try to avoid them when possible. In the field of integer programming for instance, the cutting-plane approaches used on industrial problems are almost exclusively based on linear programming (LP) or second order cone programming (SOCP). Even though semidefinite cuts are known to be stronger, they are typically too expensive to be used even at the root node of branch-and-bound techniques for integer programming. Because of this, many high-performance solvers, e.g., the CPLEX package of IBM [27], do not even provide an SDP solver and instead solely work with LP and SOCP relaxations. In the field of sum of squares optimization, however, a viable alternative to sos programming that can avoid SDP and take advantage of the existing mature and high-performance LP/SOCP solvers is lacking. This is precisely what we aim to achieve in this section.

Let $PSD_{n,d}$ and $SOS_{n,d}$ respectively denote the cone of nonnegative and sum of squares polynomials in n variables and degree d, with the obvious inclusion relation $SOS_{n,d} \subseteq PSD_{n,d}$. The basic idea is to approximate the cone $SOS_{n,d}$ from the

[3]Once again, strict feasibility of the constraints in (8) is required to rule out trivial solutions and lead to the strict inequalities that we would like to impose on V.

inside with new cones that are more tractable for optimization. Towards this goal, one may think of several natural sufficient conditions for a polynomial to be a sum of squares. For example, consider the following sets:

- The cone of polynomials that are sums of 4-th powers of polynomials: $\{p|\ p = \sum q_i^4\}$,
- The set of polynomials that are a sum of three squares of polynomials: $\{p|\ p = q_1^2 + q_2^2 + q_3^2\}$.

Even though both of these sets clearly reside inside the sos cone, they are not any easier to optimize over. In fact, they are much harder! Indeed, testing whether a (quartic) polynomial is a sum of 4-th powers is NP-hard [36] (as the cone of 4-th powers of linear forms is dual to the cone of nonnegative quartic forms [61]) and optimizing over polynomials that are sums of three squares is intractable (as this task even for quadratics subsumes the NP-hard problem of positive semidefinite matrix completion with a rank constraint [52]). These examples illustrate the rather obvious point that inclusion relationship in general has no implications in terms of complexity of optimization. Indeed, we would need to take some care in deciding what subset of $SOS_{n,d}$ we exactly choose to work with—on one hand, it has to comprise a "big enough" subset to be useful in practice; on the other hand, it should be computationally simpler for optimization.

4.1 *The Cone of r-dsos and r-sdsos Polynomials*

We now describe cones inside $SOS_{n,d}$ (and some incomparable with $SOS_{n,d}$ but still inside $PSD_{n,d}$) that are naturally motivated and that lend themselves to linear and second order cone programming. There are also several generalizations of these cones, including some that result in fixed-size (and "small") semidefinite programs. These can be found in [7] and are omitted from here.

Definition 4.1 (*Ahmadi, Majumdar, '13* [53])

- A polynomial p is *diagonally-dominant-sum-of-squares* (dsos) if it can be written as

$$p = \sum_i \alpha_i m_i^2 + \sum_{i,j} \beta_{ij}^+ (m_i + m_j)^2 + \beta_{ij}^- (m_i - m_j)^2,$$

for some monomials m_i, m_j and some constants $\alpha_i, \beta_{ij}^+, \beta_{ij}^- \geq 0$.

- A polynomial p is *scaled-diagonally-dominant-sum-of-squares* (sdsos) if it can be written as

$$p = \sum_i \alpha_i m_i^2 + \sum_{i,j} (\beta_i^+ m_i + \gamma_j^+ m_j)^2 + (\beta_i^- m_i - \gamma_j^- m_j)^2,$$

for some monomials m_i, m_j and some constants $\alpha_i, \beta_i^+, \gamma_j^+, \beta_i^-, \gamma_j^- \geq 0$.

- For a positive integer r, a polynomial p is *r-diagonally-dominant-sum-of-squares* (r-dsos) if $p \cdot \left(1 + \sum_i x_i^2\right)^r$ is dsos.
- For a positive integer r, a polynomial p is *r-scaled-diagonally-dominant-sum-of-squares* (r-sdsos) if $p \cdot \left(1 + \sum_i x_i^2\right)^r$ is sdsos.

We denote the set of polynomials in n variables and degree d that are dsos, sdos, r-dsos, and r-sdsos by $DSOS_{n,d}$, $SDSOS_{n,d}$, $rDSOS_{n,d}$, $rSDSOS_{n,d}$, respectively.

The following inclusion relations are straightforward:

$$DSOS_{n,d} \subseteq SDSOS_{n,d} \subseteq SOS_{n,d} \subseteq POS_{n,d},$$

$$rDSOS_{n,d} \subseteq rSDSOS_{n,d} \subseteq POS_{n,d}, \forall r.$$

Our terminology in Definition 4.1 comes from the following concepts in linear algebra.

Definition 4.2 A symmetric matrix A is *diagonally dominant* (dd) if $a_{ii} \geq \sum_{j \neq i} |a_{ij}|$ for all i. A symmetric matrix A is *scaled diagonally dominant* (sdd) if there exists an element-wise positive vector y such that:

$$a_{ii} y_i \geq \sum_{j \neq i} |a_{ij}| y_j, \forall i.$$

Equivalently, A is sdd if there exists a positive diagonal matrix D such that AD (or equivalently, DAD) is dd. We denote the set of $n \times n$ dd and sdd matrices with DD_n and SDD_n respectively.

Theorem 4.3 (Ahmadi, Majumdar,'13)

- *A polynomial p of degree $2d$ is dsos if and only if it admits a representation as $p(x) = z^T(x) Q z(x)$, where $z(x)$ is the standard monomial vector of degree d, and Q is a dd matrix.*
- *A polynomial p of degree $2d$ is sdsos if and only if it admits a representation as $p(x) = z^T(x) Q z(x)$, where $z(x)$ is the standard monomial vector of degree d, and Q is a sdd matrix.*

Theorem 4.4 (Ahmadi, Majumdar,'13) *For any nonnegative integer r, the set $rDSOS_{n,d}$ is polyhedral and the set $rSDSOS_{n,d}$ has a second order cone representation. For any fixed r and d, optimization over $rDSOS_{n,d}$ (resp. $rSDSOS_{n,d}$) can be done with linear programming (resp. second order cone programming), of size polynomial in n.*

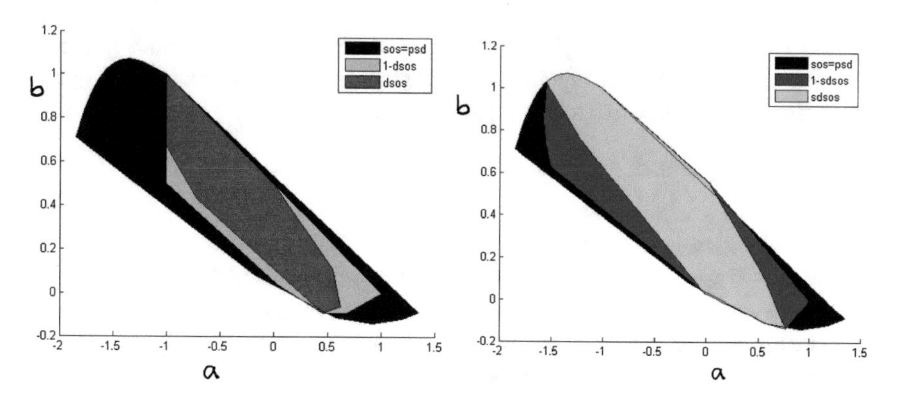

Fig. 3 A comparison of the different approximations to the set of nonnegative polynomials for a parametric family of bivariate quartics given in (9)

4.2 How Fast/powerful is the Dsos and Sdsos Methodology?

A good portion of our recent papers [6, 7, 53] is devoted to this question. We provide a sample of these results in this section.

As it is probably obvious, the purpose of the parameter r in Definition 4.1 is to have a knob for trading off speed with approximation quality. By increasing r, the hope is to obtain more accurate inner approximations to the set of nonnegative polynomials. The following example shows that even the linear programs obtained from $r = 1$ can outperform the semidefinite programs resulting from sum of squares.

Example 4.1 Consider the polynomial

$$p(x) = x_1^4 x_2^2 + x_2^4 x_3^2 + x_3^4 x_1^2 - 3x_1^2 x_2^2 x_3^2.$$

One can show that this polynomial is nonnegative but *not* a sum of squares [62]. However, we can give an LP-based nonnegativity certificate of this polynomial by showing that $p \in 1DSOS$. Hence, $1DSOS \nsubseteq SOS$.

Figure 3 considers a parametric family of bivariate quartic polynomials given by

$$p(x_1, x_2) = \frac{1}{2}x_1^4 + \frac{1}{2}x_2^4 + ax_1^3 x_2 + bx_1^2 x_2^2 + (1 - 2a - 4b)x_1 x_2^3 \tag{9}$$

and demonstrates the set of values for a and b for which the resulting polynomial is nonnegative, sos, sdsos, dsos, 1-sdsos, and 1-dsos. The fact that the sos restriction is exact here (i.e., recovers the set of nonnegative polynomials perfectly) is to be expected since it is easy to show that all nonnegative bivariate forms are sos. The dsos and sdsos inner approximations are doing reasonably well and indeed the 1-dsos and 1-sdsos versions improve the quality of approximation.

More generally, by employing appropriate results from real algebraic geometry, we can prove that some asymptotic guarantees that hold for sum of squares programming also hold for dsos and sdsos programming.

Theorem 4.5 (Ahmadi, Majumdar,'13)

- *Let p be an even form (i.e., a form where no variable is raised to an odd power). If $p(x) > 0$ for all $x \neq 0$, then there exists an integer r such that $p \in rDSOS$.*
- *Let p be any form. If $p(x) > 0$ for all $x \neq 0$, then there exists a form q such that q is dsos and pq is dsos. (Observe that this is a certificate of nonnegativity of p that can be found with linear programming.)*

On the practical side, we have preliminary evidence for major speed-ups with minor sacrifices in conservatism. Figure 5 shows our experiments for computing the region of attraction (ROA) for the upright equilibrium point of a stabilized inverted N-link pendulum with $2N$ states; see Fig. 4 for an illustration with $N = 6$ and [53] for experiments with other values of N. The same exact algorithm was run (details are in [53]), but polynomials involved in the optimization which were required to be sos, were instead required to be dsos and sdsos. Even the dsos program here is able to do a good job at stabilization. More impressively, the volume of the ROA of the sdsos program is 79% of that of the sos program. For this problem, the speed up of the dsos and sdsos algorithms over the sos algorithm is roughly a factor of 1400

Fig. 4 An illustration of the N-link inverted pendulum system (with N = 6)

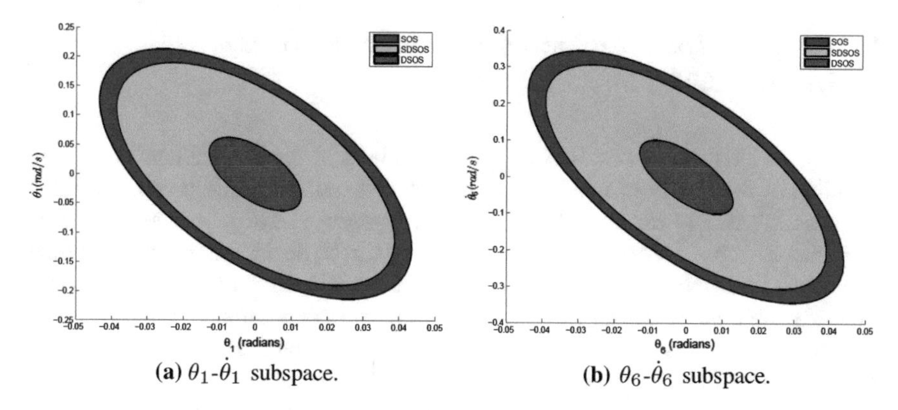

(a) θ_1-$\dot{\theta}_1$ subspace. **(b)** θ_6-$\dot{\theta}_6$ subspace.

Fig. 5 From [53] (with Majumdar and Tedrake): comparisons of projections of the ROAs computed for the 6-link pendulum system using DSOS, SDSOS and SOS programming, via LP, SOCP, and SDP respectively

(when SeDuMi is used to solve the SDP) and a factor of 90 (when Mosek is used to solve the SDP).

Perhaps more important than the ability to achieve speedups over the sos approach in small or medium sized problems is the opportunity to work in much bigger regimes where sos solvers have no chance of getting past even the first iteration of the interior point algorithm (at least with the current state of affairs). For example, in work with Majumdar and Tedrake [53], we use sdsos optimization to compute (in the order of minutes) a stabilizing controller and a region of attraction for an equilibrium point of a nonlinear model of the ATLAS robot (built by Boston Dynamics Inc. and used for the 2013 DARPA Robotics Challenge), which has 30 states and 14 control inputs. (See video made by Majumdar and Tedrake: https://www.youtube.com/watch?v=6jhCiuQVOaQ) Similarly, in [7], we have been able to solve dense polynomial optimization problems of degree 4 in 70 variables in a few minutes.

Opportunities for future research. We believe the most exciting opportunity for new contributions here is to reveal novel application areas in control, robotics, and polynomial optimization where problems have around 20–100 state variables and can benefit from tools for optimization over nonnegative polynomials. It would be interesting to see for which applications, and to what extent, our new dsos and sdsos optimization tools can fill the gap for sos optimization at this scale. To ease such investigations, a MATLAB package for dsos and sdsos optimization is soon to be released as part of the SPOTless toolbox.[4]

On the theoretical side, comparing worst-case approximation guarantees of dsos, sdsos, and sos approaches for particular classes of polynomial optimization problems (beyond our asymptotic results) remains a wide open area.

[4]https://github.com/spot-toolbox/spotless.

5 Computational Advances in Lyapunov Theory [Areas 1&2]

If we place the theory of dynamical systems under a computational lens, our understanding of the theory of nonlinear or hybrid systems is seen to be very primitive compared to that of linear systems. For linear systems, most properties of interest (e.g., stability, boundedness of trajectories, etc.) can be decided in polynomial time. Moreover, there are certificates for all of these properties in form of Lyapunov functions that are *quadratic*. Quadratic functions are tractable for optimization purposes. By contrast, there is no such theory for nonlinear systems. Even for the class of polynomial differential equations of degree two, we do not currently know whether there is a *finite time* (let alone polynomial time) algorithm that can decide stability. In fact, a well-known conjecture of Arnold from [19] states that there should not be such an algorithm. Likewise, the classical converse Lyapunov theorems that we have only guarantee existence of Lyapunov functions within very broad classes of functions (e.g. the class of continuously differentiable functions) that are a priori not amenable to computation. The situation for hybrid systems is similar, if not worse.

We have spent some of our recent research efforts [3, 9, 12, 14, 17] understanding the behavior of nonlinear (mainly polynomial) and hybrid (mainly switched linear) dynamical systems both in terms of computational complexity and existence of computationally friendly Lyapunov functions. In a nutshell, the goal has been to establish results along the "converse arrow" of Fig. 2 in Sect. 3. Some of our results are encouraging. For example, we have shown that under certain conditions, existence of a polynomial Lyapunov function for a polynomial differential equation implies existence of a Lyapunov function that can be found with sum of squares techniques and semidefinite programming [9, 12]. More recently, we have shown that stability of switched linear systems implies existence of an *sos-convex* Lyapunov functions [4]. These are Lyapunov functions that can be found with semidefinite programming and that have algebraic certificates of convexity [4, 11]. Unfortunately, however, we also have results that are very negative in nature:

Theorem 5.1 (Ahmadi, Krstic, Parrilo [14]) *The quadratic polynomial vector field,*

$$\begin{aligned}\dot{x} &= -x + xy \\ \dot{y} &= -y,\end{aligned} \tag{10}$$

is globally asymptotically stable but does not admit a polynomial Lyapunov function of any degree.

Theorem 5.2 (Ahmadi, Parrilo [12]) *For any positive integer d, there exist* homogeneous[5] *polynomial vector fields in 2 variables and degree 3 that are globally asymptotically stable but do not admit a polynomial Lyapunov function of degree $\leq d$.*

[5] A homogeneous polynomial vector field is one where all monomials have the same degree. Linear systems are an example.

Theorem 5.3 (Ahmadi, Jungers [5]) *Consider the switched linear system $x_{k+1} = A_i x_k$. For any positive integer d, there exist pairs of 2×2 matrices A_1, A_2 that are asymptotically stable under arbitrary switching but do not admit (i) a polynomial Lyapunov function of degree $\leq d$, or (ii) a polytopic Lyapunov function with $\leq d$ facets, or (iii) a piecewise quadratic Lyapunov function with $\leq d$ pieces. (This implies that there cannot be an upper bound on the size of the linear and semidefinite programs that search for such stability certificates.)*

Theorem 5.4 (Ahmadi [3]) *Unless P=NP, there cannot be a polynomial time (or even pseudo-polynomial time) algorithm for deciding whether the origin of a cubic polynomial differential equation is locally (or globally) asymptotically stable.*

Theorem 5.5 (Ahmadi, Majumdar, Tedrake [17]) *The hardness result of Theorem 5.4 extends to ten other fundamental properties of polynomial differential equations such as boundedness of trajectories, invariance of sets, stability in the sense of Lyapunov, collision avoidance, stabilizability by linear feedback, and others.*

These results show a sharp transition in complexity of Lyapunov functions when we move away from linear systems ever so slightly. Although one may think that such counterexamples are not representative of the general case, in fact it is quite common for simple nonlinear or hybrid dynamical systems to at least necessitate "complicated" (e.g., high degree) Lyapunov functions. In view of this, it is natural to ask whether we can replace the standard Lyapunov inequalities with new ones that are less stringent in their requirements but still imply stability. This would enlarge the class of valid stability certificates to include simpler functions and hence reduce the size of the optimization problems that try to construct these functions.

In this direction, we have developed two frameworks: *path-complete graph Lyapunov functions* (with Jungers and Roozbehani) [15, 18] and *non-monotonic Lyapunov functions* [1, 8]. The first approach is based on the idea of using *multiple* Lyapunov functions instead of one and brings in concepts from automata theory to establish how Lyapunov inequalities should be written among multiple Lyapunov functions. The second approach relaxes the classical requirement that Lyapunov functions should *monotonically* decrease along trajectories. We briefly describe these concepts next.

5.1 Lyapunov Inequalities and Transitions in Finite Automata [Areas 1&2]

Consider a finite set of matrices $\mathcal{A} := \{A_1, \ldots, A_m\}$. Our goal is to establish *global asymptotic stability under arbitrary switching* (GASUAS) of the difference inclusion system

$$x_{k+1} \in \text{co}\mathcal{A}\, x_k, \tag{11}$$

where co$\mathcal{A}$ here denotes the convex hull of the set $\mathcal{A}$. In other words, we would like to prove that no matter what the realization of our *uncertain and time-varying* linear system turns out to be at each time step, as long as it stays within co$\mathcal{A}$, then we have stability. Let $\rho(\mathcal{A})$ be the *joint spectral radius* (JSR) of the set of matrices $\mathcal{A}$:

$$\rho(\mathcal{A}) = \lim_{k\to\infty} \max_{\sigma\in\{1,\dots,m\}^k} \left\| A_{\sigma_k} \dots A_{\sigma_2} A_{\sigma_1} \right\|^{1/k}. \tag{12}$$

It is well-known that $\rho < 1$ if and only if system (11) is GASUAS.

Aside from stability of switched systems, computation of the JSR emerges in many areas of application such as computation of the capacity of codes, continuity of wavelet functions, convergence of consensus algorithms, trackability of graphs, and many others; see [42]. In [15, 18], we give SDP-based approximation algorithms for the JSR by applying Lyapunov analysis techniques to system (11). We show that considerable improvements in scalability are possible (especially for high dimensional systems) if instead of a common Lyapunov function of high degree for the set $\mathcal{A}$, we use *multiple* Lyapunov functions of low degree (quadratic ones). Motivated by this observation, the main challenge is to *understand which sets of inequalities among a finite set of Lyapunov functions imply stability*. We give a graph theoretic answer to this question by defining directed graphs whose nodes are Lyapunov functions and whose edges are labeled with matrices from the set of input matrices $\mathcal{A}$. Each edge of this graph defines a single Lyapunov inequality as depicted in Fig. 6a.

Definition 5.6 (*Ahmadi, Jungers, Parrilo, Roozbehani* [15]) Given a directed graph $G(N, E)$ whose edges are labeled with words (matrices) from the set $\mathcal{A}$, we say that the graph is *path-complete*, if for all finite words $A_{\sigma_k} \dots A_{\sigma_1}$ of any length k (i.e., for all words in $\mathcal{A}^*$), there is a directed path in the graph such that the labels on the edges of this path are the labels A_{σ_1} up to A_{σ_k}.

An example of a path-complete graph is given in Fig. 6b, with dozens more given in [18]. In the terminology of automata theory, path-complete graphs correspond precisely to finite automata whose language is the set $\mathcal{A}^*$ of all words

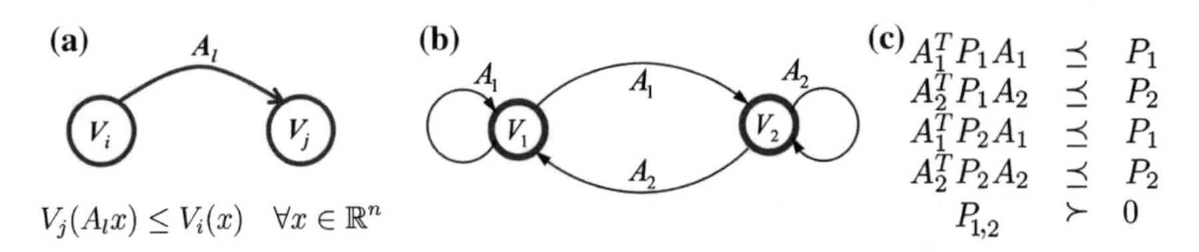

Fig. 6 Path-complete graph Lyapunov functions. **a** The nodes of the graph are Lyapunov functions and the directed edges, which are labeled with matrices from the set $\mathcal{A}$, represent Lyapunov inequalities. **b** An example of a path-complete graph on the alphabet $\{A_1, A_2\}$. This graph contains a directed path for every finite word. **c** The SDP associated with the graph in (**b**) when quadratic Lyapunov functions $V_{1,2}(x) = x^T P_{1,2} x$ are assigned to its nodes. This is an SDP in matrix variables P_1 and P_2 which if feasible implies $\rho(A_1, A_2) \leq 1$. We prove an approximation ratio of $1/\sqrt[4]{n}$ for this particular SDP

(i.e., matrix products) from the alphabet $\mathcal{A}$. There are well-known algorithms in automata theory (see e.g. [37, Chap. 4]) that can check whether the language accepted by an automaton is $\mathcal{A}^*$. Similar algorithms exist in the symbolic dynamics literature; see e.g. [48, Chap. 3]. Our interest in path-complete graphs stems from the following two theorems that relate this notion to Lyapunov stability.

Theorem 5.7 (Ahmadi, Jungers, Parrilo, Roozbehani [15]) *Consider* any *path-complete graph with edges labeled with matrices from the set $\mathcal{A}$. Define a set of Lyapunov inequalities, one per edge of the graph, following the rule in Fig. 6a. If Lyapunov functions are found, one per node, that satisfy this set of inequalities, then the switched system in (11) is GASUAS.*

Theorem 5.8 (Jungers, Ahmadi, Parrilo, Roozbehani [16]) *Consider any set of inequalities of the form $V_j(A_k x) \leq V_i(x)$ among a finite number of Lyapunov functions that imply GASUAS of system (11). Then the graph associated with these inequalities, drawn according to the rule in Fig. 6(a), is necessarily path-complete.*

These two theorems together give a characterization of *all stability proving Lyapunov inequalities*. Our result has unified several works in the literature, as we observed that many LMIs that appear in the literature [28, 30, 38–40, 46, 47] correspond to particular families of path-complete graphs. In addition, the framework has introduced several *new* ways of proving stability with new computational benefits. Finally, by relying on some results in convex geometry, we have been able to prove approximation guarantees (converse results) for the SDPs that search for Lyapunov functions on nodes of path-complete graphs. For example, the upper bound $\hat{\rho}$ that the SDP in Fig. 6c produces on the JSR satisfies

$$\frac{1}{\sqrt[4]{n}}\hat{\rho}(\mathcal{A}) \leq \rho(\mathcal{A}) \leq \hat{\rho}(\mathcal{A}).$$

5.2 *Non-monotonic Lyapunov Functions [Area 1]*

Our research on this topic is motivated by a very natural question: If all we need for the conclusion of Lyapunov's stability theorem to hold is for the value of the Lyapunov function to *eventually reach zero*, why should we require the Lyapunov function to decrease monotonically? Can we write down conditions that allow Lyapunov functions to increase occasionally, but still guarantee their convergence to zero in the limit? In [1, 8], we showed that this is indeed possible. The main idea is to invoke higher order derivatives of Lyapunov functions (or higher order differences in discrete time). Intuitively, whenever we allow $\dot{V} > 0$ (i.e., V increasing), we should make sure some higher order derivatives of V are negative, so the rate at which V increases decreases fast enough for V to be forced to decrease later in the future. An example of such an inequality for a continuous time dynamical system $\dot{x} = f(x)$ is [24]:

$$\tau_2 \dddot{V}(x) + \tau_1 \ddot{V}(x) + \dot{V}(x) < 0. \tag{13}$$

Here, τ_1 and τ_2 are nonnegative constants and by the first three derivatives of the Lyapunov function $V : \mathbb{R}^n \to \mathbb{R}$ in this expression, we mean

$$\dot{V}(x) = \langle \tfrac{\partial V(x)}{\partial x}, f(x) \rangle,$$

$$\ddot{V}(x) = \langle \tfrac{\partial \dot{V}(x)}{\partial x}, f(x) \rangle,$$

$$\dddot{V}(x) = \langle \tfrac{\partial \ddot{V}(x)}{\partial x}, f(x) \rangle.$$

In [1, 10], we establish a link between non-monotonic Lyapunov functions and standard ones, showing how the latter can be constructed from the former. The main advantage of non-monotonic Lyapunov functions over standard ones, however, is that they can often be much simpler in structure. As a simple example, consider the linear time-varying dynamical system

$$\dot{x}(t) = \begin{bmatrix} \cos(20t) - 0.2 & 1 \\ -1 & \cos(20t) - 0.2 \end{bmatrix} x(t). \tag{14}$$

Figure 7 shows a trajectory of this system on the left. By looking at this trajectory, it should be clear that a *time-invariant* standard Lyapunov function for this system either does not exist, or if it does, its structure should be extremely complicated. However, if one uses condition (13), the simple quadratic non-monotonic Lyapunov function $||x||^2$ provides a proof of stability. Indeed, if we let $V(x) = x^T x$, then we have

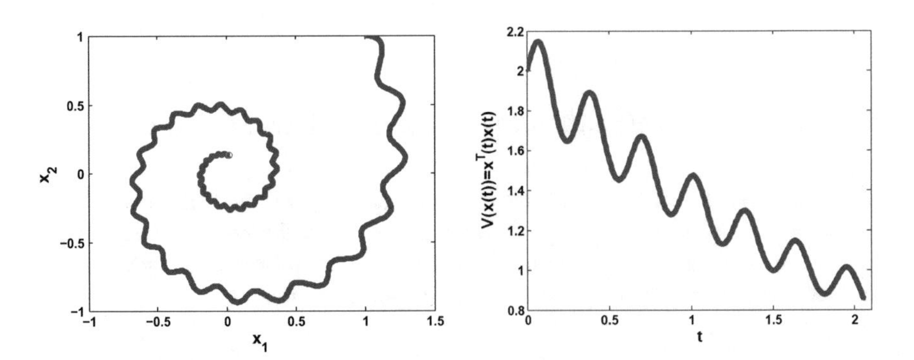

Fig. 7 *Non-monotonic Lyapunov functions* [1]. A typical trajectory of a linear time-varying dynamical system given in (14) (*left*). The value of a stability proving quadratic non-monotonic Lyapunov function along the trajectory (*right*). A classical time-independent Lyapunov function would have to be extremely complicated

$$\tau_2 \dddot{V}(x) + \tau_1 \ddot{V}(x) + \dot{V}(x) = x^T x \{ \\ \tau_2[-240\sin(20t)(\cos(20t) - 0.2) \\ +8(\cos(20t) - 0.2)^3 - 800\cos(20t)] \\ +\tau_1[-40\sin(20t) + 4(\cos(20t) - 0.2)^2] \\ +2[\cos(20t) - 0.2] \\ \}. \tag{15}$$

One can check that the expression in (15) can be made strictly negative for all x and all t for a range of positive values for τ_1 and τ_2; for example, for $\tau_1 = 0.0039$ and $\tau_2 = 0.0025$.

One disadvantage of a condition of the type (13) is that the expression is bilinear in the decision variables τ_1, τ_2 and the unknown coefficients of the polynomial V. Hence a joint search for these decision variables cannot be done via a convex program. We have shown in [10] however that one can replace condition (13) with other inequalities involving the first three derivatives which are at least as powerful, but also *convex* in the decision variables. This allows for sum of squares methods to become applicable for an automated search for non-monotonic Lyapunov functions. Once again, the concrete advantage of these Lyapunov functions over standard ones is the lower complexity associated with the parameterization of the candidate function. For optimization purposes, this directly translates to savings in the number of decision variables of the underlying sos programs; see, e.g., [10, Ex. 2.1].

Opportunities for future research. The body of work described in this section leaves several directions for future research:

- On the topic of complexity: What is the complexity of testing asymptotic stability of a polynomial vector field of degree 2? For degree 1, the problem can be solved in polynomial time; for degree 3, we have shown that the problem is strongly NP-hard [3, 12].
- On the topic of existence of polynomial Lyapunov functions: Is there a locally asymptotically stable polynomial vector field with rational coefficients that does not admit a local polynomial Lyapunov function? Our work in [14] presents an example with no *global* polynomial Lyapunov function. Bacciotti and Rosier [21, Prop. 5.2] present an independent example with no local polynomial Lyapunov function, but their vector field needs to have an *irrational* coefficient and the non-existence of polynomial Lyapunov functions for their example is not robust to arbitrarily small perturbations.
- On the topic of existence of sos Lyapunov functions: Does existence of a polynomial Lyapunov function for a polynomial vector field imply existence of an sos Lyapunov function (see [9] for a precise definition)? We have answered this question in the affirmative under a few assumptions [2, 9], but not in general.
- On the topic of path-complete graph Lyapunov functions: Characterize all Lyapunov inequalities among multiple Lyapunov functions that establish switched stability of a *nonlinear* difference inclusion. We know already that the situation is more delicate here than the characterization for the linear case presented in [18].

Indeed, we have shown [4] that path-complete graphs no longer guarantee stability and that *convexity* of Lyapunov functions plays a role in the nonlinear case.
- On the topic of non-monotonic Lyapunov functions: Characterize *all* Lyapunov inequalities involving a finite number of higher order derivatives that imply stability. Determine whether the search for Lyapunov functions satisfying these inequalities can be cast as a convex program.

Acknowledgements We are grateful to Anirudha Majumdar for his contributions to the work presented in Sect. 4 and to Russ Tedrake for the robotics applications and many insightful discussions.

References

1. A.A. Ahmadi, Non-monotonic Lyapunov functions for stability of nonlinear and switched systems: theory and computation. Master's Thesis, Massachusetts Institute of Technology, June 2008. http://aaa.lids.mit.edu/publications
2. A.A. Ahmadi, Algebraic relaxations and hardness results in polynomial optimization and Lyapunov analysis. Ph.D. Thesis, Massachusetts Institute of Technology, September 2011. http://aaa.lids.mit.edu/publications
3. A.A. Ahmadi, On the difficulty of deciding asymptotic stability of cubic homogeneous vector fields. in *Proceedings of the American Control Conference* (2012)
4. A.A. Ahmadi, R. Jungers, SOS-convex Lyapunov functions with applications to nonlinear switched systems. in *Proceedings of the IEEE Conference on Decision and Control* (2013)
5. A.A. Ahmadi, R. Jungers, On complexity of Lyapunov functions for switched linear systems. in *Proceedings of the 19th World Congress of the International Federation of Automatic Control* (2014)
6. A.A. Ahmadi, A. Majumdar, Some applications of polynomial optimization in operations research and real-time decision making (2014). Under review
7. A.A. Ahmadi, A. Majumdar, DSOS and SDSOS: more tractable alternatives to sum of squares and semidefinite optimization (in preparation, 2016)
8. A.A. Ahmadi, P.A. Parrilo, Non-monotonic Lyapunov functions for stability of discrete time nonlinear and switched systems. in *Proceedings of the 47th IEEE Conference on Decision and Control* (2008)
9. A.A. Ahmadi, P.A. Parrilo, Converse results on existence of sum of squares Lyapunov functions. in *Proceedings of the 50th IEEE Conference on Decision and Control* (2011)
10. A.A. Ahmadi, P.A. Parrilo, On higher order derivatives of Lyapunov functions. in *Proceedings of the 2011 American Control Conference* (2011)
11. A.A. Ahmadi, P.A. Parrilo, A complete characterization of the gap between convexity and sos-convexity. SIAM J. Optim. **23**(2), 811–833 (2013)
12. A.A. Ahmadi, P.A. Parrilo, Stability of polynomial differential equations: complexity and converse Lyapunov questions. Under review. Preprint available at http://arxiv.org/abs/1308.6833, 2013
13. A.A. Ahmadi, P.A. Parrilo, Towards scalable algorithms with formal guarantees for Lyapunov analysis of control systems via algebraic optimization. in *The Proceedings of the 53rd Annual Conference on Decision and Control* (2014), pp. 2272–2281
14. A.A. Ahmadi, M. Krstic, P.A. Parrilo, A globally asymptotically stable polynomial vector field with no polynomial Lyapunov function. in *Proceedings of the 50th IEEE Conference on Decision and Control* (2011)
15. A.A. Ahmadi, R. Jungers, P.A. Parrilo, M. Roozbehani, Analysis of the joint spectral radius via Lyapunov functions on path-complete graphs, *Hybrid Systems: Computation and Control 2011*, Lecture Notes in Computer Science (Springer, 2011)

16. A.A. Ahmadi, R.M. Jungers, P.A. Parrilo, M. Roozbehani, When is a set of LMIs a sufficient condition for stability? in *Proceedings of the IFAC Symposium on Robust Control Design* (2012)
17. A.A. Ahmadi, A. Majumdar, R. Tedrake, Complexity of ten decision problems in continuous time dynamical systems. in *Proceedings of the American Control Conference* (2013)
18. A.A. Ahmadi, R. Jungers, P.A. Parrilo, M. Roozbehani, Joint spectral radius and path-complete graph Lyapunov functions. SIAM J. Optim. Control **52**(1), 687 (2014)
19. V.I. Arnold. Problems of present day mathematics, XVII (Dynamical systems and differential equations). Proc. Symp. Pure Math., **28**(59) (1976)
20. A. Ataei-Esfahani, Q. Wang, Nonlinear control design of a hypersonic aircraft using sum-of-squares methods. in *Proceedings of the American Control Conference* (IEEE, 2007), pp. 5278–5283
21. A. Bacciotti, L. Rosier, *Liapunov Functions and Stability in Control Theory* (Springer, Heidelberg, 2005)
22. A.J. Barry, A. Majumdar, R. Tedrake, Safety verification of reactive controllers for UAV flight in cluttered environments using barrier certificates. in *Proceedings of the IEEE International Conference on Robotics and Automation* (IEEE, 2012), pp. 484–490
23. D. Bertsimas, D.A. Iancu, P.A. Parrilo, A hierarchy of near-optimal policies for multistage adaptive optimization. IEEE Trans. Autom. Control **56**(12), 2809–2824 (2011)
24. A.R. Butz, Higher order derivatives of Liapunov functions. IEEE Trans. Autom. Control **AC–14**, 111–112 (1969)
25. A. Chakraborty, P. Seiler, G.J. Balas, Susceptibility of F/A-18 flight controllers to the falling-leaf mode: nonlinear analysis. J. Guid. Control Dyn. **34**(1), 73–85 (2011)
26. G. Chesi, D. Henrion (eds.), Special issue on positive polynomials in control. IEEE Trans. Autom. Control **54**(5), 935 (2009)
27. CPLEX. V12. 2: Users manual for CPLEX. International Business Machines Corporation, **46**(53), 157 (2010)
28. J. Daafouz, J. Bernussou, Parameter dependent Lyapunov functions for discrete time systems with time varying parametric uncertainties. Syst. Control Lett. **43**(5), 355–359 (2001)
29. A.C. Doherty, P.A. Parrilo, F.M. Spedalieri, Distinguishing separable and entangled states. Phys. Rev. Lett. **88**(18), 187904 (2002)
30. R. Goebel, A.R. Teel, T. Hu, Z. Lin, Conjugate convex Lyapunov functions for dual linear differential inclusions. IEEE Trans. Autom. Control **51**(4), 661–666 (2006)
31. N. Gvozdenović, M. Laurent, Semidefinite bounds for the stability number of a graph via sums of squares of polynomials. Math. Program. **110**(1), 145–173 (2007)
32. J. Harrison, *Verifying nonlinear real formulas via sums of squares, Theorem Proving in Higher Order Logics* (Springer, Heidelberg, 2007), pp. 102–118
33. D. Henrion, A. Garulli (eds.), *Positive Polynomials in Control*, vol. 312 (Lecture Notes in Control and Information Sciences (Springer, Heidelberg, 2005)
34. D. Henrion, M. Korda, Convex computation of the region of attraction of polynomial control systems. IEEE Trans. Autom. Control **59**(2), 297–312 (2014)
35. D. Hilbert, Über die Darstellung Definiter Formen als Summe von Formenquadraten. Math. Ann. **32**, (1888)
36. C. Hillar, L.-H. Lim, Most tensor problems are NP-hard. arXiv preprint arXiv:0911.1393, 2009
37. J.E. Hopcroft, R. Motwani, J.D. Ullman, *Introduction to Automata Theory, Languages, and Computation* (Addison Wesley, Boston, 2001)
38. T. Hu, Z. Lin, Absolute stability analysis of discrete-time systems with composite quadratic Lyapunov functions. IEEE Trans. Autom. Control **50**(6), 781–797 (2005)
39. T. Hu, L. Ma, Z. Li, On several composite quadratic Lyapunov functions for switched systems. in *Proceedings of the 45th IEEE Conference on Decision and Control* (2006)
40. T. Hu, L. Ma, Z. Lin, Stabilization of switched systems via composite quadratic functions. IEEE Trans. Autom. Control **53**(11), 2571–2585 (2008)
41. Z. Jarvis-Wloszek, R. Feeley, W. Tan, K. Sun, A. Packard, Some controls applications of sum of squares programming. in *Proceedings of the 42th IEEE Conference on Decision and Control* (2003), pp. 4676–4681

42. R. Jungers, *The Joint Spectral Radius: Theory and Applications*, vol. 385 (Lecture Notes in Control and Information Sciences (Springer, Heidelberg, 2009)
43. H. Khalil, *Nonlinear Systems*, 3rd edn. (Prentice Hall, New Jersey, 2002)
44. J.B. Lasserre, Global optimization with polynomials and the problem of moments. SIAM J. Optim. **11**(3), 796–817 (2001)
45. J.B. Lasserre, D. Henrion, C. Prieur, E. Trélat, Nonlinear optimal control via occupation measures and LMI-relaxations. SIAM J. Control Optim. **47**(4), 1643–1666 (2008)
46. J.W. Lee, G.E. Dullerud, Uniform stabilization of discrete-time switched and Markovian jump linear systems. Automatica **42**(2), 205–218 (2006)
47. J.W. Lee, P.P. Khargonekar, Detectability and stabilizability of discrete-time switched linear systems. IEEE Trans. Autom. Control **54**(3), 424–437 (2009)
48. D. Lind, B. Marcus, *An Introduction to Symbolic Dynamics and Coding* (Cambridge University Press, Cambridge, 1995)
49. J. Löfberg, YALMIP: a toolbox for modeling and optimization in MATLAB. in *Proceedings of the CACSD Conference* (2004). http://control.ee.ethz.ch/~joloef/yalmip.php
50. A. Magnani, S. Lall, S. Boyd, Tractable fitting with convex polynomials via sum of squares. in *Proceedings of the 44th IEEE Conference on Decision and Control* (2005)
51. A. Majumdar, A.A. Ahmadi, R. Tedrake, Control design along trajectories with sums of squares programming. in *Proceedings of the IEEE International Conference on Robotics and Automation* (2013)
52. E. Marianna, M. Laurent, A. Varvitsiotis, *Complexity of the positive semidefinite matrix completion problem with a rank constraint, Discrete Geometry and Optimization* (Springer, Switzerland, 2013), pp. 105–120
53. A. Majumdar, A.A. Ahmadi, R. Tedrake, Control and verification of high-dimensional systems via dsos and sdsos optimization. in *Submitted to the 53rd IEEE Conference on Decision and Control* (2014)
54. A. Majumdar, R. Vasudevan, M.M. Tobenkin, R. Tedrake, Convex optimization of nonlinear feedback controllers via occupation measures. Int. J. Robot. Res. **33**(9), 1209 (2014)
55. P.A. Parrilo, Structured semidefinite programs and semialgebraic geometry methods in robustness and optimization. Ph.D. Thesis, California Institute of Technology, May 2000
56. P.A. Parrilo, Semidefinite programming relaxations for semialgebraic problems. Math. Program. **96**(2, Ser. B), 293–320 (2003)
57. P.A. Parrilo, Polynomial games and sum of squares optimization. in *Proceedings of the 45th IEEE Conference on Decision and Control* (2006)
58. S. Prajna, A. Jadbabaie, *Safety verification of hybrid systems using barrier certificates, Hybrid Systems: Computation and Control* (Springer, Heidelberg, 2004), pp. 477–492
59. S. Prajna, A. Jadbabaie, G.J. Pappas, A framework for worst-case and stochastic safety verification using barrier certificates. IEEE Trans. Autom. Control **52**(8), 1415–1428 (2007)
60. S. Prajna, A. Papachristodoulou, P.A. Parrilo, SOSTOOLS: sum of squares optimization toolbox for MATLAB, May 2002. http://www.cds.caltech.edu/sostools and http://www.mit.edu/~parrilo/sostools
61. B. Reznick, Uniform denominators in Hilbert's 17th problem. Math Z. **220**(1), 75–97 (1995)
62. B. Reznick, *Some concrete aspects of Hilbert's 17th problem, Contemporary Mathematics*, vol. 253 (American Mathematical Society, Rhode Island, 2000), pp. 251–272
63. M. Roozbehani, Optimization of Lyapunov invariants in analysis and implementation of safety-critical software systems. Ph.D. Thesis, Massachusetts Institute of Technology, 2008
64. P. Seiler, G.J. Balas, A.K. Packard, *Assessment of aircraft flight controllers using nonlinear robustness analysis techniques, Optimization Based Clearance of Flight Control Laws* (Springer, Heidelberg, 2012), pp. 369–397
65. R. Tae, B. Dumitrescu, L. Vandenberghe, Multidimensional FIR filter design via trigonometric sum-of-squares optimization. IEEE J. Sel. Top. Signal Process. **1**(4), 641–650 (2007)

Positivity Certificates in Optimal Control

Edouard Pauwels, Didier Henrion and Jean-Bernard Lasserre

Abstract We propose a tutorial on relaxations and weak formulations of optimal control with their semidefinite approximations. We present this approach solely through the prism of positivity certificates which we consider to be the most accessible for a broad audience, in particular in the engineering and robotics communities. This simple concept allows us to express very concisely powerful approximation certificates in control. The relevance of this technique is illustrated on three applications: region of attraction approximation, direct optimal control and inverse optimal control, for which it constitutes a common denominator. In a first step, we highlight the core mechanisms underpinning the application of positivity in control and how they appear in the different control applications. This relies on simple mathematical concepts and gives a unified treatment of the applications considered. This presentation is based on the combination and simplification of published materials. In a second step, we describe briefly relations with broader literature, in particular, occupation measures and Hamilton–Jacobi–Bellman equation which are important elements of the global picture. We describe the Sum-Of-Squares (SOS) semidefinite hierarchy in the semialgebraic case and briefly mention its convergence properties. Numerical experiments on a classical example in robotics, namely the nonholonomic vehicle, illustrate the concepts presented in the text for the three applications considered.

E. Pauwels
IRIT, Université de Toulouse, Toulouse, France
e-mail: edouard.pauwels@irit.fr

D. Henrion · J.-B. Lasserre (✉)
LAAS-CNRS, Université de Toulouse, CNRS, Toulouse, France
e-mail: lasserre@laas.fr

D. Henrion
Faculty of Electrical Engineering, Czech Technical University in Prague,
Technická 2, CZ-16626, Prague, Czech Republic
e-mail: henrion@laas.fr

J.-P. Laumond et al. (eds.), *Geometric and Numerical Foundations of Movements*,
Springer Tracts in Advanced Robotics 117, DOI 10.1007/978-3-319-51547-2_6

1 Introduction

1.1 Context

In the context of understanding and reproducing human movements and more generally in motion control, there has recently been a growing interest in using optimal control to model and account for the complexity of underlying processes [1, 7, 10, 20, 23, 24, 30, 32]. The question of the validity of this approach is still open and the interface between optimal control and human locomotion is an active field of research.

The so-called weak formulation of the optimal control problem has a long history in the control community [10, 16, 35], see also [9, Part III] for a detailed historical perspective. This approach comes with a rich convex duality structure [34], one side of which involves functional non negativity constraints. This type of constraint constitutes the focus of this chapter.

In general, functional positivity constraints are not tractable computationally. Advances in semialgebraic geometry on the representation of positive polynomials [29] have allowed us to construct provably *convergent hierarchies of sums-of-squares approximations* to this kind of intractable constraint when the problem at hand only involves polynomials [17, 18]. Based on semidefinite programming [33], these approximations provide a new perspective on infinite dimensional linear programs and functional non negativity constraints, along with tractable numerical approximations. Application of these hierarchies in control lead to the design of new methods to address control problems with global-optimality guaranties [6, 13, 15, 19, 25].

1.2 Content

This chapter is a tutorial which focuses on the application of infinite dimensional conic programming to control problems. This constitutes a very relevant tool for the human locomotion and humanoid robotics research communities. Indeed the sums-of-squares (SOS) hierarchy provides a systematic numerical scheme to solve related practical control problems. We will illustrate the power of this approach by focusing on three such particular problems, namely:

- region of attraction approximation.
- direct optimal control.
- inverse optimal control.

The infinite dimensional linear programming approach combined with its associated SOS hierarchy of approximations has been applied to these problems in [14, 19, 26, 27]. All in all, the content of this chapter is not new and is merely based on existing materials from the control and sums-of-squares approximations literature.

The purpose of the chapter is to reveal and highlight a few simple mechanisms and ideas that constitute a common denominator of all these applications.

Being concerned with accessibility to a broad audience, we deliberately hide important aspects of the approach. In particular, we focus on functional positivity constraints (one side of a coin in this approach) because we think that this is the most accessible way to present a general intuition regarding the weak formulation of optimal control problems. Another reason is that this simple notion of positivity allows us to provide very strong sub-optimality certificate stemming from elementary mathematics. Other facets of the same problem (the other side of the coin described in the dual of the infinite-dimensional linear program), including conic duality and details about the weak formulation of control problems on *occupation measures*, are only briefly mentioned in a second step with very few details. Indeed, this material is often perceived as more technical and less accessible from a mathematical point of view. Although we do not emphasize much the moment relaxation approximation and its relation with occupation measures, it would provide a more complete picture to speak about the *moment-SOS hierarchy* (instead of the SOS-hierarchy) because each semidefinite program of the SOS-hierarchy of approximations of functional positivity constraints has a dual (also a semidefinite program) which deals with "moments" of occupation measures. We mention this point only briefly and invite the reader interested in more details about these aspects to consult the existing literature.

1.3 Organization of the Chapter

The optimal control problem and its value function are introduced in Sect. 2. In Sect. 3, we introduce functional positivity constraints which involve surrogate value functions. We discuss implications of these types of constraints in the context of optimal control, and in particular we describe how they relate to the approximation of the value function. This constitutes a general and flexible core result that is useful in the control applications that we consider. Section 4 illustrates the concept in several control problems dealing with (i) the approximation of region of attraction, (ii) optimal control and (iii), inverse optimal control. Finally, Sect. 5 discusses connections with the optimal control literature and additional aspects of the approach that we do not describe explicitly. We also briefly describe how the sums-of-squares hierarchy of approximations can be implemented and discuss convergence issues.

2 Optimal Control and Value Function

2.1 Notations and Preliminaries

If A is a subset of $\mathbb{R}^n$, $\mathscr{C}(A)$ denotes the space of continuous functions from A to $\mathbb{R}$ while $\mathscr{C}^1(A)$ denotes the space of continuously differentiable functions from

A to $\mathbb{R}$. Let $X \subseteq \mathbb{R}^{d_X}$ and $U \subseteq \mathbb{R}^{d_U}$ denote respectively the state and control spaces, both supposed to be compact. The system dynamics are given by a continuously differentiable vector field $f \in \mathscr{C}^1(X \times U)^{d_X}$. Terminal state constraints are represented by a given compact set $X_T \subseteq X$.

Given all the above ingredients one may define *admissible trajectories* in the context of optimal control. We will use the following definition.

Definition 1 (*Admissible trajectories*) Consider an initial time $t_0 \in [0, 1]$ and a pair of functions (x, u) from $[t_0, 1]$ to $\mathbb{R}^{d_X}$ and $\mathbb{R}^{d_U}$ respectively. This pair constitutes an admissible trajectory if it has the following properties:

- u is a measurable function from $[t_0, 1]$ to U.
- For any $t \in [t_0, 1]$, $x(t) = x_0 + \int_{t_0}^{t} f(x(s), u(s))ds$.
- $x(1) \in X_T$.

Given $x_0 \in X$, denote by traj_{t_0,x_0} the set of such admissible trajectories starting at time t_0 with $x(t_0) = x_0$. Note that the second property implies that x is differentiable almost everywhere as a function of t, with $\dot{x}(t) = f(x(t), u(t))$ for almost all $t \in [t_0, 1]$.

The class of admissible trajectories constitutes the decision variables of an optimal control problem.

2.2 Optimal Control and Value Function

An optimal control problem consists of minimizing a functional over the set of admissible trajectories. The functional has a specific integral form involving a continuous Lagrangian $l \in \mathscr{C}(X \times U)$ and a continuous terminal cost $l_T \in \mathscr{C}(X_T)$. Given an initial time $t_0 \in [0, 1]$ and a starting point $x_0 \in X$, consider the infimum value:

$$\begin{aligned} v^*(t_0, x_0) := \inf \ & \int_{t_0}^{1} l(x(t), u(t))dt + l_T(x(1)) \\ \text{s.t. } & (x, u) \in \mathrm{traj}_{t_0,x_0} \end{aligned} \tag{OCP}$$

of the functional over all admissible trajectories. It is a well defined value that only depends on t_0 and x_0 and $v^* : [0, 1] \times X \to \mathbb{R} \cup \{+\infty\}$ is called the *value function* associated with the optimal control problem.

Note that the constraints in (OCP) ensure that we only consider admissible trajectories starting from x_0 at t_0, and therefore if traj_{t_0,x_0} is empty then $v^*(t_0, x_0) = +\infty$.

3 Bounds on the Value Function

The value function introduced in (OCP) can be a very complicated object. The existence of minimizing sequences, the question of the infimum being attained and the regularity of v^* are all quite delicate issues. In this section we show that functional positivity constraints that are expressible in a simple form lead to powerful approximation results. In addition, and remarkably, a striking feature of these results is that their proof arguments are elementary. We now focus on the description of these constraints while their origin and connection with control theory are postponed to Sect. 5.

3.1 Global Lower Bounds

We let "$\cdot$" denote the dot product between two vectors of the same size. For a given function $v \in \mathscr{C}^1([0,1] \times X)$, consider the following positivity conditions:

$$\begin{aligned} l(x,u) + \frac{\partial v}{\partial t}(t,x) + \frac{\partial v}{\partial x}(t,x) \cdot f(x,u) \geq 0 \quad & \forall (x,u,t) \in X \times U \times [0,1] \\ l_T(x) - v(T,x) \geq 0 \quad & \forall x \in X_T. \end{aligned} \tag{1}$$

Note that these conditions are indeed functional positivity constraints since both of them must hold for all arguments in certain sets. How to ensure or approximate such conditions in practical situations is discussed in Sect. 5.3. We focus for the moment on the consequences of condition (1) in terms of control, the following proposition being an elementary, yet powerful example.

Proposition 1 (Global lower bound on the value function) *If $v \in \mathscr{C}^1([0,1] \times X)$ satisfies condition (1) then $v(t_0, x_0) \leq v^*(t_0, x_0)$ for any $x_0 \in X$ and $t_0 \in [0,1]$.*

Proof Fix $x_0 \in X$ and t_0 and consider the set traj_{t_0,x_0} of admissible trajectories starting at x_0 at time t_0 as described in Definition 1. If this set is empty then $v^*(t_0, x_0) = +\infty$. Since v is continuous on a compact set, it is bounded and hence finite at (t_0, x_0) which ensures that $v(x_0, t_0) \leq v^*(x_0, t_0)$. If traj_{t_0,x_0} is not empty, consider an arbitrary but fixed admissible trajectory $(x, u)\colon [t_0, 1] \to X \times U$ which satisfies all the requirements of Definition 1 with $x(t_0) = x_0$. Combining admissibility with the first condition in (1) yields:

$$\begin{aligned} l(x(t),u(t)) + \frac{\partial}{\partial t}[v(t,x(t))] &= l(x(t),u(t)) + \frac{\partial v}{\partial t}(t,x(t)) + \frac{\partial v}{\partial x}(t,x(t)) \cdot \dot{x}(t) \\ &= l(x(t),u(t)) + \frac{\partial v}{\partial t}(t,x(t)) + \frac{\partial v}{\partial x}(t,x(t)) \cdot f(x(t),u(t)) \\ &\geq 0, \quad \text{for almost all } t \in [t_0, 1]. \end{aligned}$$

Integrating between t_0 and 1, and using non negativity of the first term, we obtain

$$\int_{t_0}^{1} l(x(t), u(t))\, dt + v(1, x(1)) - v(t_0, x_0) \geq 0.$$

Combining with the second condition in (1) yields

$$v(t_0, x_0) \leq \int_{t_0}^{1} l(x(t), u(t))\, dt + l_T(x(1)).$$

Since (x, t) was arbitrary among all admissible trajectories, this inequality is still valid if we take the infimum in the right hand side, which coincides with the definition of v^* in (OCP), and the proof is complete. □

Proposition 1 provides a sufficient condition to obtain global lower bounds on the value function v^*. A remarkable property of this condition is that it *does not depend explicitly on* v^*. In particular, condition (1) does not depend explicitly on regularity properties of v^* or on the existence of optimal trajectories in (OCP). Furthermore, they are expressed in a relatively compact form and the proof arguments are elementary.

3.2 Local Upper Bounds

We now turn to upper bounds on the value function v^* of problem (OCP). First, observe that if the set of admissible trajectories is empty in (OCP) then $v^*(t_0, x_0) = +\infty$. Hence upper bounding v^* using a continuous function only makes sense when the set of admissible trajectories is not empty. Therefore such upper bounds depend on admissible trajectories and only hold in a certain "local sense". In particular, global upper bounds do not exist in general, whence the local characteristic for the type of bounds derived in this section. We introduce the following notation

Definition 2 (*Domain of the value function*) Denote by $V \subset [0, 1] \times X$ the domain of v^*, that is, the subset of $[0, 1] \times X$ on which v^* takes finite values,

$$V := \left\{(t_0, x_0) \in [0, 1] \times X : \text{traj}_{t_0, x_0} \neq \emptyset\right\}$$

Consider a fixed pair $(t_0, x_0) \in V$ and a given fixed admissible trajectory $(x, u) \in \text{traj}_{t_0, x_0}$, starting at x_0 at time t_0. For a given $\varepsilon \geq 0$, the following conditions are a counterpart to the positivity condition in (1).

$$l(x(t), u(t)) + \frac{\partial v}{\partial t}(t, x(t)) + \frac{\partial v}{\partial x}(t, x(t)) \cdot f(x(t), u(t)) \leq \frac{\varepsilon}{2}, \quad \text{for almost all } t \in [t_0, 1]$$
$$l_T(x(T)) - v(1, x(1)) \leq \frac{\varepsilon}{2}. \tag{2}$$

They can be used to obtain the following upper approximation result.

Proposition 2 (Local upper bound on the value function) *Let $(t_0, x_0) \in V$ be fixed. Let $(x, u) \in \mathrm{traj}_{t_0,x_0}$ be an admissible trajectory starting at x_0 at time t_0. Assume that $v \in \mathscr{C}^1([0, 1] \times X)$ satisfies condition (2) for a given $\varepsilon > 0$. Then $v^*(t, x(t)) \leq v(t, x(t)) + \varepsilon$ for all $t \in [t_0, 1]$. In addition, if v satisfies condition (1) then (x, u) is at most ε sub-optimal for problem (OCP): feasible with objective value at most ε greater than the optimal value.*

Proof Following similar integration arguments as in the proof of Proposition 1, using the first part of condition (2) yields:

$$\int_t^1 l(x(s), u(s))\, ds + v(1, x(1)) - v(t, x(t)) \leq (1 - t)\frac{\varepsilon}{2} \leq \frac{\varepsilon}{2}, \quad \forall t \in [t_0, 1],$$

and combining with the second part of condition (2),

$$\int_t^1 l(x(s), u(s))\, ds + l_T(x(1)) \leq l_T(x(1)) - v(1, x(1)) + v(t, x(t)) + \frac{\varepsilon}{2} \leq v(t, x(t)) + \varepsilon,$$

for all $t \in [t_0, 1]$. As the left hand side is an upper bound on $v^*(t, x(t))$, the first statement follows. In addition, if condition (1) is satisfied then we can use Proposition 1 at $(t, x(t))$ to obtain:

$$\int_t^1 l(x(s), u(s))\, ds + l_T(x(1)) \leq v^*(t, x(t)) + \varepsilon.$$

In particular, letting $t = t_0$ in the previous relation yields that (x, u) is at most ε-sub-optimal for problem (OCP). □

Again, a remarkable property of condition (2) is that it depends neither on the regularity of v^* nor on the existence of optimal trajectories and still provides powerful sub-optimality certificates. Note that Proposition 2 characterizes properties of v^* only along the specific chosen trajectory, whence the name "local" for this type of bounds.

4 Applications in Control

In this section, we consider applications in control and show how conditions (1) and (2) can be used to solve control problems.

The general methodology is to use conditions (1) and (2) as constraints in combination with additional constraints and linear objective functions depending on the application.

The reason why this is relevant and produces valid practical methods comes from the connection with Propositions 1 and 2. Depending on the problem at hand, definition of objective functions or addition of constraints allow to provide a systematic numerical scheme to solve the control problems we consider: approximating the region of attraction of a controlled system, solving optimal control and inverse optimal control problems, provided that they are described with polynomials and semi-algebraic sets (see also Sect. 5.3). All the material of this section is based on reformulation and simplification of the work presented in [14, 19, 26, 27].

4.1 Region of Attraction

The region of attraction is a subset of the domain of the value function, V in Definition 2, corresponding to a fixed initial time t_0. In other words, we are looking for the set X_0 of initial conditions, x_0, for which there exists an admissible trajectory starting in state x_0 at a given time t_0.

Definition 3 (*Region of attraction*) The region of attraction at time t_0, denoted by $X_0 \subset X$, is the set that satisfies

$$X_0 = \left\{x_0 \in X : \mathrm{traj}_{t_0,x_0} \neq \emptyset\right\},$$

where traj_{t_0,x_0} is the set of admissible trajectories as given in Definition 1. Following Definition 2, we have $\{t_0\} \times X_0 = V \cap [\{t_0\} \times X]$.

This exactly corresponds to the situation where $l = 0$ and $l_T = 0$ in (OCP). Indeed, in this case, v^* becomes the indicator of X_0 (equal to 0 on X_0 and $+\infty$ otherwise) and the optimal control problem is a feasibility problem.

Condition (1) becomes

$$\begin{aligned} \frac{\partial v}{\partial t}(x,t) + \frac{\partial v}{\partial x}(x,t) \cdot f(x,u) \geq 0 \quad & \forall (x,u,t) \in X \times U \times [0,1] \\ v(T,x) \leq 0 \quad & \forall x \in X_T \end{aligned} \tag{3}$$

and Proposition 1 has the following consequence.

Corollary 1 *If $v \in \mathscr{C}^1([0,1] \times X)$ satisfies condition (3) then $v(x_0,t_0) \leq 0$ for any $x_0 \in X_0$.*

Corollary 1 states that X_0 is contained in the zero sublevel set of v whenever v satisfies condition (3). However this is not sufficient to have a good approximation of X_0 and condition (3) is not strong enough to distinguish between accurate and rough sublevel set approximations of this type. In order to sort out accurate candidates v, a possibility is to search among all functions which satisfy condition (3) an "optimal" one, e.g. in the sense that it should be as greater than 0 as possible outside of X_0. Following [14], we introduce an additional decision variable $w \in \mathscr{C}(X)$. We will construct an optimization problem which ensures that w is non positive on X_0 and as close as possible to 1 on $X \setminus X_0$. This can be obtained by combining Corollary 1 with additional positivity constraints and a linear objective function. The following problem is a reformulation of problem (16) in [14].

$$\begin{array}{ll} \sup\limits_{v,w} & \int_X w(x)dx \\ \text{s.t.} & 0 \le \frac{\partial v}{\partial t} + \frac{\partial v}{\partial x} \cdot f \\ & 0 \le -v(T, \cdot) \\ & w(\cdot) \le v(\cdot, t_0) \\ & w \le 1. \end{array} \tag{4}$$

In problem (4), the first two constraints are exactly condition (3) and Corollary 1 ensures that $v(\cdot, t_0) \le 0$ on X_0. Therefore, the third constraint ensures that $w \le 0$ on X_0. The last constraint combined with the objective function allow to "choose" w as close as possible to 1 on $X \backslash X_0$. In general the supremum in (4) is not attained, but any candidate solution w, is such that its zero sublevel contains X_0 and remains close to it in a certain sense. Indeed it was shown in [14] that the supremum in (4) is equal to the volume of X_0 and this quantity can be approximated by hierarchies of semidefinite approximations which we describe in Sect. 5.

4.2 *Optimal Control*

In this section, we fix t_0 and x_0. As described in Sect. 3, condition (1) provides a global lower bound on v^*. However, the family of functions v which satisfy this condition is too large. For example, if $l \ge 0$ and $l_T \ge 0$, then $v = 0$ satisfies condition (1) and does not provide much insight regarding solutions of (OCP). Therefore, one should design a way to choose lower bounds of specific interest. In the (direct) optimal control problem, one is interested in the value $v^*(t_0, x_0)$. Hence an informal approach is to choose among all v that satisfy condition (1) one for which $v(t_0, x_0)$ is close to $v^*(t_0, x_0)$. Note that under condition (1) we already have $v(t_0, x_0) \le v^*(t_0, x_0)$ and hence it is sufficient to look for a function v such that $v(t_0, x_0)$ is as large as possible. This leads to the following optimization problem.

$$\begin{array}{ll} \sup\limits_{v} & v(t_0, x_0) \\ \text{s.t.} & 0 \le l + \frac{\partial v}{\partial t} + \frac{\partial v}{\partial x} \cdot f \\ & 0 \le l_T(\cdot) - v(T, \cdot). \end{array} \tag{5}$$

In general the supremum is not attained. Furthermore, for most reasonable practical situations, the value of the problem is exactly $v^*(t_0, x_0)$, providing a valid conceptual solution to the optimal control problem.

At this point a remark is in order. Solutions of problem (5) allow to approximate from below the value function v^*. In this respect they provide solutions of (OCP) because of their relations to v^* which is the value of specific interest. However, this approach does not give access to an optimal trajectory which achieves this optimal value. Indeed, without further assumptions, the existence of such an optimal trajectory is not guaranteed. In order to compute optimal trajectories, further conditions are required in combination with additional methods to search for optimal trajectories. When such a method is available, it is always possible to combine it with solutions of (5) by using condition (2) and Proposition 2 to certify the sub-optimality of the computed trajectory.

4.3 Inverse Optimal Control

In inverse optimal control the situation is somewhat reversed compared to direct optimal control. The Lagrangian is unknown but we are given a set of trajectories that should be optimal with respect to the unknown Lagrangian. So the goal is to find a Lagrangian for which the given trajectories are optimal. In Fig. 1, we display an informal description of this problem and its relation with the direct optimal control problem in the framework of positivity certificates. Briefly, the main goal is to infer a cost function (a Lagrangian) which can generate a set of given trajectories through an optimality process. The applications of this are twofold:

- Provide a tool for applications in which one assumes the existence of an optimality process behind decisions.
- Provide a modeling tool which could allow to summarize and reproduce the behaviour of observed systems.

In the rest of this section, we fix an admissible trajectory (x, u) starting from $x_0 \in X_0$ at time 0. We suppose that the state trajectory x as well as the control trajectory u are given and we look for candidate Lagrangians. The whole methodology naturally extends to an arbitrary number of trajectories. Actually, the higher the number of trajectories, the better and the more (physically) meaningful is the characterization

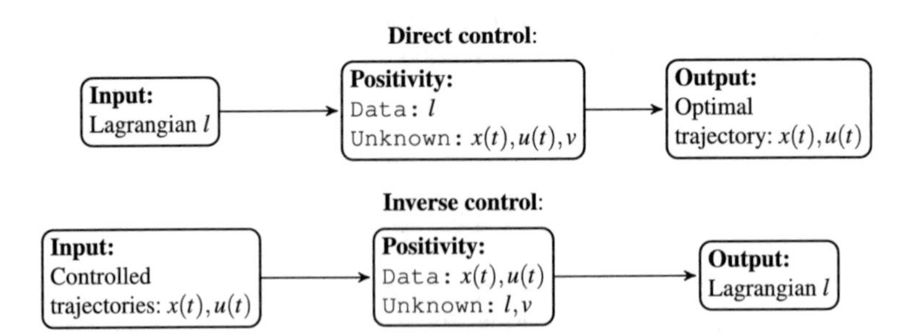

Fig. 1 Direct optimal and inverse optimal control flow chart. The dynamical system is described through the dynamics f, the state constraint set X, control constraint set U and terminal state constraint set X_T which are all given. We emphasize that the Lagrangian and the trajectories have symmetric roles for the direct and inverse problems. In particular, the output of the inverse problem is a Lagrangian

of the candidate Lagrangian that we are looking for. However for clarity of exposition, the approach is better understood when we consider a single given trajectory.

In order to provide a solution to the inverse problem, we combine conditions (1) and (2). The relevance of doing this comes from Proposition 2 which provides a sub-optimality certificate. In addition, we enforce $l_T = 0$ in order to simplify the problem. Among all potential certificates we look for the one that provides the smallest sub-optimality gap as described in Proposition 2. This leads to the following optimization problem.

$$\begin{aligned} &\inf_{\varepsilon,l,v} \ \varepsilon \\ &\text{s.t. } 0 \leq l + \tfrac{\partial v}{\partial t} + \tfrac{\partial v}{\partial x} \cdot f \\ &\quad\ \ 0 \leq l_T(\cdot) - v(T,\cdot) \\ &\quad\ \ \tfrac{\varepsilon}{2} \geq l(x(t),u(t)) + \tfrac{\partial v}{\partial t}(t,x(t)) + \tfrac{\partial v}{\partial x}(t,x(t)) \cdot f(x(t),u(t)) \ \forall t \in [0,1] \\ &\quad\ \ \tfrac{\varepsilon}{2} \geq l_T(x(T)) - v(1,x(1)). \end{aligned} \tag{6}$$

By Proposition 2, if l is a Lagrangian part of a feasible solution (ε, l, v) for problem (6), then the trajectory (x, u) is ε-sub-optimal for problem (OCP) with Lagrangian l. In other words, every feasible solution (ε, l, v) of (6) provides us with an ε-sub-optimality certificate for the trajectory (x, u).

However, this is not sufficient. Indeed, problem (6) always admits the trivial solution $(0, 0, 0)$ and it turns out that this solution is also valid from a formal point of view. Indeed, every admissible trajectory is optimal for the trivial null Lagrangian, and therefore, from the point of view of inverse optimality, the null Lagrangian is a valid (but not satisfactory) solution. To avoid trivial Lagrangians, additional constraints on l are needed. We will settle upon problem (6) as it highlights the main mechanism behind positivity in inverse optimal control and invite the reader to see [26, 27] for further discussions and more details about application in practical situations.

5 Duality, Hamilton–Jacobi–Bellman, SOS Reinforcement and Convergence

The results presented so far without much context are related to principles which have a long history in optimal control theory. In this section we mention a few of them and we also comment on how to use these results in practical contexts through the SOS hierarchy.

5.1 Occupation Measures

The constraints imposed in (1) have a *conic flavor* as they combine linear operators and positivity constraints. The space of continuous functions that are nonnegative on a given set form a convex cone. This cone admits a (convex) dual cone, see [4, Chap. IV] for a description of conic duality in Banach spaces. Representation results of Riesz type ensure that this dual cone can be identified with that of nonnegative measures on the same set. As is classical for duality in convex optimization:

- To the inequality constraints appearing in the conic optimization problem (1) are associated nonnegative *dual variables* in the dual conic optimization problem, and
- to the variables appearing in (1) are associated constraints on these dual variables.

The constraints in the dual problem describe a transport equation satisfied by the dual variables, more precisely the transport along the flow followed by admissible trajectories in Definition 1. These dual variables are called "occupation measures", see e.g. [34] for an accurate description and [11] for an extension to infinite horizons.

In other words, the dual counterpart of condition (1) allows to formally work with *generalized trajectories* instead of classical ones. Whence the name "relaxation" for this approach. Equivalently, one speaks of "weak formulation" of the optimal control problem (OCP) because the differential equation is replaced by a weaker constraint (the transport equation for occupation measures). One main benefit of working with weak formulations is that the question of attaining the infimum is solved, at least from a theoretical point of view, under weak conditions, e.g. compactness of the sets X and X_T. However, the relaxed problem is not equivalent to the original problem, and its optimal value may be smaller. But for most reasonable practical situations, there is no relaxation gap and the optimal values of both problems are the same [34, 35]. Although the use of occupation measures is much less popular than classical differential equations in the engineering community, it is classical in Markov processes and ergodic dynamical systems. Furthermore, understanding these dual aspects is crucial in the framework of positivity constraints that we describe.

5.2 Hamilton–Jacobi–Bellman Equation

Conditions (1) and (2) have the same flavor and structure as the well known Hamilton–Jacobi–Bellman (HJB) sufficient optimality conditions (see e.g. exposition in [2]). In fact, if we combine (1) with (2) with $\varepsilon = 0$, we recover exactly the same condition. This provides a certificate of optimality for a given trajectory. However, this condition is not necessary. Indeed if the value function v^* is not smooth (which is the case in most practical situations) then it is not possible to fulfill this condition in the classical sense. Whence the use of a relaxed condition involving $\varepsilon > 0$ that measures how far we are from the true optimality condition. Another possible workaround is the use of the elegant *viscosity solution* concept to define “solutions” of HJB equation [3]. This involves a lot more sophisticated mathematical machinery, far beyond the scope of this chapter.

5.3 SOS Reinforcement

Finally, conditions (1) and (2) are actually positivity constraints for functions. Moreover, all the examples presented in Sect. 4 consist of combining these constraints with additional constraints of the same type and linear objective functions. In full generality this type of constraint is not amenable to practical computation. In order to be able to use the results of this chapter to actually solve control problems, involving some practical “algorithm”, we need to enforce more structure on the objects we manipulate. A now widespread approach is to work with the following assumption.

Assumption 1 The dynamics f, the Lagrangian l and the terminal cost l_T are polynomials. Constraints set X, U and X_T are compact basic semi-algebraic sets.

Recall that a closed basic semi-algebraic set G can be defined by inequalities involving a finite number of polynomials $g_1, \ldots, g_q \in \mathbb{R}[X]$:

$$G = \{x : \; g_i(x) \geq 0, \; i = 1, \ldots, q\}. \tag{7}$$

Given a family of sum-of-squares (SOS) polynomials $p_0, p_1, \ldots, p_q$, hence nonnegative, it is direct to check that the following polynomial P *in Putinar form*

$$P = p_0 + \sum_{i=1}^{q} p_i g_i, \tag{8}$$

is nonnegative on G. Checking whether a given polynomial of degree $2d$ in n variables is a SOS reduces to solving a semidefinite program of size $\binom{n+d}{d}$. For moderate number of variables and degrees, this is thus amenable to efficient practical computation [33]. Actually dedicated software tools exist [21]. Hence under Assumption 1, if v is a polynomial then Condition (1) can be enforced by semidefinite constraints. This is of course an approximation and in fact the SOS constraints (7) are stronger than the original positivity constraints, whence the name “reinforcement”. But a counterpart of this approximation is that it is amenable to practical computation on moderate size problems which is not the case for general positivity constraints.

Conic convex duality also holds for semidefinite programs. In the present context of control, the dual variables associated with the SOS reinforcement of condition (1) are “moments” of the occupation measures discussed in Sect. 5.1. Hence the SOS approximation actually bears the name *moment-SOS* approximation, see [18] for a comprehensive treatment.

5.4 Convergence

The positivity certificate in Eq. (8) describes a family of nonnegative polynomials over the set G involving a family of SOS polynomials $\{p_i\}_{i=0}^q$. By increasing the degree allowed for these SOS polynomials p_i, one provides a hierarchy of increasing families of polynomials nonnegative on G. A relevant issue is:

What happens when we let the degrees of the SOS polynomials $\{p_i\}_{i=0}^q$ defining this hierarchy goes to infinity?

This issue is related to the question of the representation of nonnegative polynomials on compact basic semi-algebraic sets. Fortunately, powerful results from real algebraic geometry state that it is enough to work with certificates of the form of (8) [29]. This usually translates in global convergence results: replacing nonnegativity constraint in condition (1) by their SOS reinforcement and letting the degree of the SOS polynomials go to infinity is, in some sense, equivalent to the intractable constraints in condition (1). Applications of sufficient conditions to represent positive polynomials date back to [17] in static optimization and to [19] in optimal control; see also [6, 18] for a more recent overview. This methodology can be used for all the control problems described in Sect. 4 to provide converging hierarchies of semidefinite approximations [14, 19, 27], see also [12] for a detailed overview.

The size of the semidefinite programs which need to be solved grows extremely fast with the number of state and control variables and the degree of the SOS polynomials defining the hierarchy. It is worth emphasising that we are dealing with non-linear non-convex infinite-dimensional optimal control problems, arguably a broad class of difficult mathematical problems. Yet, despite the generality and the difficulty of the problems considered, this general methodology provides increasingly tight bounds with mathematically rigorous convergence guarantees. In terms of number of states and control, good bounds can be obtained at a moderate cost (say a few minutes of CPU time on a standard desktop PC) for problems with 4–5 states and 2–3 controls, see e.g. the experiments reported in [14]. With some additional engineering insights and programming skills, these techniques can be applied to larger size problems see e.g. [23] or [22] for a robotics example with 30 states and 14 controls.

6 Numerical Illustration

In this section we briefly describe numerical results obtained when applying the SOS reinforcement techniques of Sect. 5.3 to the abstract optimization problems of Sect. 4. We choose a simple but non trivial nonlinear system: Brockett's integrator. This system is of importance in humanoid robotics since it is equivalent to the nonholonomic Dubbins vehicle [31], a model of human walking [1], up to a change of variable [8]. The numerical results presented here relate to free terminal time optimal control problem, which is different from the fixed terminal time setting considered in this work. They still illustrate most important aspects of these simulations. Indeed, most of the ideas presented in Sects. 3 and 4 have direct equivalent in the free terminal time setting. In a nutshell, the terminal time in (OCP) is not fixed to be 1 but is a decision variable of the problem. In this case, the value function v^* as well as its lower approximations v can be chosen to be independent of time. The numerical examples of this section were originally presented in [14, 19, 26, 27]. All these examples were computed by combining the abstract infinite dimensional optimization problems of Sect. 4 with the SOS reinforcement techniques of Sect. 5.3.

6.1 Brockett's Integrator

Brockett's integrator is a 3-dimensional nonlinear system with two dimensional control. We set $X = \{x \in \mathbb{R}^3 : \|x\|_\infty \leq 1\}$, $U = \{u \in \mathbb{R}^2 : \|u\|_2 \leq 1\}$ and we let X_T be the origin in $\mathbb{R}^3$. The dynamics of the system are given by

$$f(x,u) = \begin{pmatrix} u_1 \\ u_2 \\ u_1 x_2 - u_2 x_1 \end{pmatrix}, \tag{9}$$

where the subscripts denote the corresponding coordinates. All the following examples are related to the minimum time to reach the origin under the previous dynamical constraints. The value function of this problem is known and described in [5, Theorems 1.36 and 1.41] and the corresponding optimal control is computed in [28, Corollary 1]. In what follows, $T(x)$ denotes the optimal time to reach the origin, starting at initial state x under the dynamical constraints (9).

6.2 Region of Attraction

For this application, the final time is set to 1 and initial time is set to 0. The region of attraction described in Sect. 4 is the set X_0 of initial states for which there exists a feasible trajectory reaching the origin in time less or equal to 1. In other words, it consists of the set of initial states x, for which $T(x) \leq 1$. This quantity is computable explicitly [5]. Combining the formulation in Eq. (4) with the SOS reinforcement technique described in Sect. 5.3, we get sublevel sets which are outer approximations of X_0. This is represented in Fig. 2 which compares the true region of attraction and its outer approximation in $\mathbb{R}^3$.

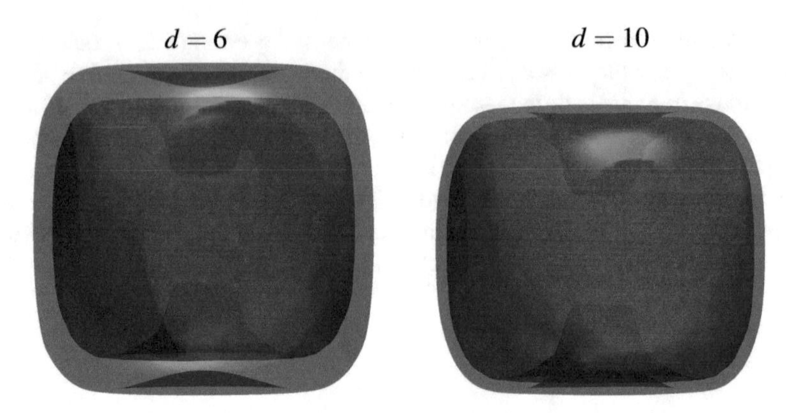

Fig. 2 Sublevel set outer approximations (light red, larger) to the region of attraction X_0 (dark red, smaller) for different degrees of the SOS reinforcement. d is the degree of the SOS polynomial in the reinforcement, for $d = 10$ the size of the corresponding SDP is 252 times 252. This was originally published in [14]

Table 1 Brockett's integrator, comparison of exact optimal time and SOS reinforcement. This was originally published in [19]. The cells represent different initial conditions: $x_1 = 1$ and $x_i \in \{1, 2, 3\}$ for $i = 2, 3$. The degree of the SOS polynomials of the reinforcement is 4 and the corresponding SDP is of size 126 times 126

SOS reinforcement			Optimal time		
1.7979	2.3614	3.2004	1.8257	2.3636	3.2091
2.3691	2.6780	3.3341	2.5231	2.6856	3.3426
2.8875	3.0654	3.5337	3.1895	3.1008	3.5456

6.3 Minimum Time Direct Optimal Control

For the direct optimal control problem, we are interested in the value of the optimal time $T(x)$. Following [19], we combine the formulation in (5) with the SOS reinforcement technique described in Sect. 5.3. As a result, we get a lower approximation of the optimal time $T(x)$. This is illustrated in Table 1 for different initial conditions: $x_1 = 1$ and $x_i \in \{1, 2, 3\}$, $i = 2, 3$. As expected, we obtain lower bounds on the optimal time which is known analytically. For these examples, the approximation is reasonably accurate.

6.4 Inverse Optimal Control

In this example, we are interested in recovering the minimum time Lagrangian (constant) from optimal trajectories which reach the origin in minimal time under dynamical constraints (9). These trajectories can be computed analytically [28]. As outlined

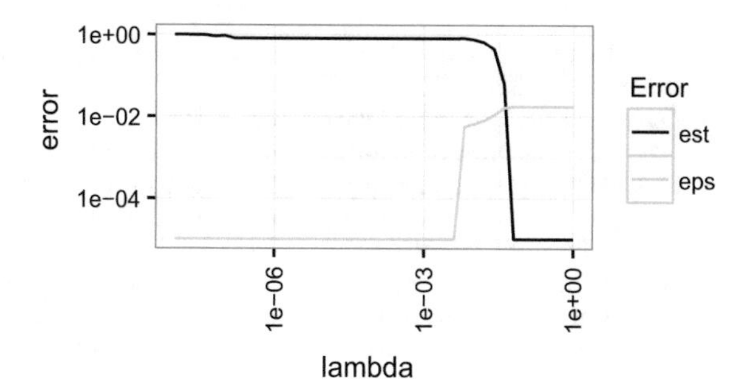

Fig. 3 Error versus a regularization parameter λ for the inverse optimal control problem. Estimation error (est) is a measure of distance between a candidate solution computed with (6) and the constant Lagrangian (normalized between 0 and 1). Epsilon error (eps) corresponds to the value of ε in (6). The Lagrangian is looked for among all 3-variate polynomials of degree up to 4 and the degree of the SOS polynomials in the reinforcement is 10 which means that the size of the matrix variables in the corresponding SDP is 252 times 252. These results were originally presented in [27]

in Fig. 1, the trajectories constitute an input of the inverse problem and the output is a Lagrangian function. In order to find this function, we follow the work of [26, 27] which combines the abstract problem in (6) with SOS reinforcement techniques described in Sect. 5.3 and additional constraints. We emphasize that the problem of Lagrangian identification is much less well posed than the direct optimal control problem and that accuracy of solutions highly depend on prior information about expected Lagrangians. In [26, 27], it is shown that the success of such a procedure requires careful normalization and prior knowledge enforcement (sparsity through a regularization term). We do not describe the details of the procedure here and refer to [26, 27] for more details. This formulation includes a regularization parameter denoted by λ. Figure 3 presents measures of inverse optimality accuracy, (the value of ε in (6)), and estimation accuracy (a distance to the constant Lagrangian, the true Lagrangian of minimum time optimal control), for various values of this parameters. The input is made of optimal time trajectories and Fig. 3 illustrates that the original Lagrangian can be recovered with a reasonable inverse optimality accuracy for some values of the regularization parameter close to 1.

Acknowledgements This work was partly supported by project 16-19526S of the *Grant Agency of the Czech Republic*, project ERC-ADG TAMING 666981,ERC-Advanced Grant of the *European Research Council* and grant number FA9550-15-1-0500 from the *Air Force Office of Scientific Research, Air Force Material Command.*

References

1. G. Arechavaleta, J.P. Laumond, H. Hicheur, A. Berthoz, An optimality principle governing human walking. IEEE Trans. Robot. **24**(1), 5–14 (2008)
2. M. Athans, P.L. Falb, *Optimal Control. An Introduction to the Theory and Its Applications* (McGraw-Hill, New York, 1966)

3. M. Bardi, I. Capuzzo-Dolcetta, *Optimal Control and Viscosity Solutions of Hamilton-Jacobi-Bellman Equations* (Springer, Berlin, 2008)
4. A. Barvinok, *A Course in Convexity* (AMS, Providence, 2002)
5. R. Beals, B. Gaveau, P.C. Greiner, Hamilton-Jacobi theory and the heat kernel on Heisenberg groups. Journal de mathématiques pures et appliquées **79**(7), 633–689 (2000)
6. G. Chesi, LMI techniques for optimization over polynomials in control: a survey. IEEE Trans. Autom. Control **55**(11), 2500–2510 (2010)
7. F.C. Chittaro, F. Jean, P. Mason, On inverse optimal control problems of human locomotion: stability and robustness of the minimizers. J. Math. Sci. **195**(3), 269–287 (2013)
8. D. DeVon, T. Bretl, Kinematic and dynamic control of a wheeled mobile robot. IEEE/RSJ Int. Conf. Intell. Robots Syst. (2007)
9. H.O. Fattorini, *Infinite Dimensional Optimization and Control Theory* (Cambridge Univ. Press, Cambridge, 1999)
10. K. Friston, What is optimal about motor control? Neuron **72**(3), 488–498 (2011)
11. V. Gaitsgory, M. Quincampoix, Linear programming approach to deterministic infinite horizon optimal control problems with discounting. SIAM J. Control Optim. **48**(4), 2480–2512 (2009)
12. D. Henrion, *Optimization on Linear Matrix Inequalities for Polynomial Systems Control*, Lecture notes of the International Summer School of Automatic Control (Grenoble, France, September 2014)
13. D. Henrion, A. Garulli (eds.), *Positive Polynomials in Control*, vol. 312, Lecture Notes on Control and Information Sciences (Springer, Berlin, 2005)
14. D. Henrion, M. Korda, Convex computation of the region of attraction of polynomial control systems. IEEE Trans. Autom. Control **59**(2), 297–312 (2014)
15. D. Henrion, J.B. Lasserre, Solving nonconvex optimization problems - how GloptiPoly is applied to problems in robust and nonlinear control. IEEE Control Syst. Mag. **24**(3), 72–83 (2004)
16. D. Hernández-Hernández, O. Hernández-Lerma, M. Taksar, The linear programming approach to deterministic optimal control problems. Applicationes Mathematicae **24**(1), 17–33 (1996)
17. J.B. Lasserre, Global optimization with polynomials and the problem of moments. SIAM J. Optim. **11**(3), 796–817 (2001)
18. J.B. Lasserre, *Moments, Positive Polynomials and Their Applications* (Imperial College Press, UK, 2010)
19. J.B. Lasserre, D. Henrion, C. Prieur, E. Trélat, Nonlinear optimal control via occupation measures and LMI relaxations. SIAM J. Control Optim. **47**(4), 1643–1666 (2008)
20. J.P. Laumond, N. Mansard, J.B. Lasserre, Optimality in robot motion: optimal versus optimized motion. Commun. ACM **57**(9), 82–89 (2014)
21. J. Löfberg, Pre-and post-processing sum-of-squares programs in practice. IEEE Trans. Autom. Control **54**(5), 1007–1011 (2009)
22. A. Majumdar, A.A. Ahmadi, R. Tedrake, Control and verification of high-dimensional systems via dsos and sdsos optimization, in *Proceedings of the 53rd the IEEE Conference on Decision and Control* (2014)
23. A. Majumdar, R. Vasudevan, M.M. Tobenkin, R. Tedrake, Convex optimization of nonlinear feedback controllers via occupation measures. Int. J. Robot. Res. **33**(9), 1209–1230 (2014)
24. K. Mombaur, A. Truong, J.P. Laumond, From human to humanoid locomotion-an inverse optimal control approach. Auton. Robots **28**(3), 369–383 (2010)
25. P.A. Parrilo, S. Lall, Semidefinite programming relaxations and algebraic optimization in control. Eur. J. Control **9**(2–3), 307–321 (2003)
26. E. Pauwels, D. Henrion, J.B. Lasserre, Inverse optimal control with polynomial optimization. IEEE Conf. Decis. Control (2014)
27. E. Pauwels, D. Henrion, J.B. Lasserre, Linear conic optimization for inverse optimal control. SIAM J. Control Optim. **54**(3), 1798–1825 (2016)
28. C. Prieur, Trélat, Robust optimal stabilization of the Brockett integrator via a hybrid feedback. Math. Control, Signals Syst. **17**(3), 201–216 (2005)

29. M. Putinar, Positive polynomials on compact semi-algebraic sets. Indiana Univ. Math. J. **42**(3), 969–984 (1993)
30. A.S. Puydupin-Jamin, M. Johnson, T. Bretl, A convex approach to inverse optimal control and its application to modeling human locomotion. Int. Conf. Robot. Autom. IEEE (2012)
31. P. Souères, J.P. Laumond, Shortest paths synthesis for a car-like robot. IEEE Trans. Autom. Control **41**(5), 672–688 (1996)
32. E. Todorov, Optimality principles in sensorimotor control. Nat. Neurosci. **7**(9), 907–915 (2004)
33. L. Vandenberghe, S.P. Boyd, Semidefinite programming. SIAM Rev. **38**(1), 49–95 (1996)
34. R. Vinter, Convex duality and nonlinear optimal control. SIAM J. Control Optim. **31**(2), 518–538 (1993)
35. R. Vinter, R. Lewis, The equivalence of strong and weak formulations for certain problems in optimal control. SIAM J. Control Optim. **16**(4), 546–570 (1978)

The Interplay Between Big Data and Sparsity in Systems Identification

O. Camps and M. Sznaier

Abstract Recent advances in distributed control, coupled with an exponential growth in data gathering capabilities, have made feasible a wide range of applications with potential to profoundly impact society, from safer self-aware environments and smart cities to enhanced model-based medical therapies. Yet, achieving this vision requires addressing the challenge of handling large amounts of very high dimensional data. In this chapter, we provide a tutorial showing how to exploit the inherent sparsity of the data, which is present in a large class of identification problems, to overcome the "curse of dimensionality". The concepts presented here extend traditional ideas from machine learning linking big data and sparsity, to challenging dynamic settings. In particular, we explore the connections between system identification and information extraction from large data sets, using as an example human activity analysis from video data.

1 Introduction

Recent advances in sensing, actuation, and data collection capabilities provide access to exponentially increasing amounts of data. The availability of this data opens up a wide range of applications that have the potential to profoundly impact society, with benefits ranging from safer, self-aware environments and smart cities, to enhanced model-based medical therapies. However, a major impediment to realize this vision stems from the amount of data and its high dimensionality.

This work was supported in part by NSF grants IIS–1318145 and ECCS–1404163; AFOSR grant FA9550-15-1-0392, and the Alert DHS Center of Excellence under Award Number 2013-ST-061-ED0001.

O. Camps · M. Sznaier (✉)
ECE Department, Northeastern University, Boston, MA 02115, USA
e-mail: msznaier@coe.neu.edu

O. Camps
e-mail: camps@coe.neu.edu

J.-P. Laumond et al. (eds.), *Geometric and Numerical Foundations of Movements*,
Springer Tracts in Advanced Robotics 117, DOI 10.1007/978-3-319-51547-2_7

The goal of this chapter is twofold: (i) to describe a set of techniques that exploit the inherent sparsity of the data, available in a large class of problems, to combat the *curse of dimensionality*; and (ii) discuss new research opportunities at the intersection of machine learning, dynamical system theory and sparse optimization to extend big data and sparsity ideas to dynamic settings.

The first part of the chapter illustrates the key role that machine learning and sparse optimization play while developing robust and computationally efficient system identification algorithms, capable of handling very large data sets. We start by showing that parsimonious models of Linear Time Invariant (LTI) systems can be identified from noisy, potentially incomplete, data by using a regularized atomic norm form. This approach leads to problems that can be efficiently solved by using a randomized version of the Frank-Wolfe algorithm, whose complexity scales linearly with the number of data points. Next, we address the harder case when the data is corrupted with outliers due to gross measurement errors or sensor failures. While this problem is generically NP-hard, we show that it can be reformulated as a regularized robust regression problem that recovers the exact solution, as long as the outliers are sparse. Lastly, we discuss the problem of identification of switched linear systems. These systems are particularly interesting because they are universal approximators [1, 2] and provide tractable approximations to general nonlinear control problems. Moreover, there is an intimate connection between identifying piece-wise affine systems and the problem of extracting actionable information from very large data streams [3]. Unfortunately, identification of switched linear systems is generically NP-hard, even in the absence of outliers. However, as shown here, it is possible to obtain tractable convex relaxations by using robust regression or sparse polynomial optimization methods that exploit the structure of the problem.

The chapter concludes by illustrating these techniques with several examples, including identification of a physical lightly damped structure for a civil engineering application and the problem of activity detection and analysis in video streams.

2 Notation

In this section we summarize, for ease of reference, the notation used in this chapter.

$\mathbb{R}, \mathbb{N}$	set of real number and non-negative integers
$\mathbb{D}$	unit disk in the complex plane
$\mathbf{x}, \mathbf{M}$	a vector in $\mathbb{R}^n$ (matrix in $\mathbb{R}^{m\times n}$)
$\mathbf{M}(:,\mathbf{j})$	j^{th} column of matrix $\mathbf{M}$.
$\Vert\mathbf{x}\Vert_2$	ℓ_2 norm of a vector: $\Vert\mathbf{x}\Vert_2^2 = \sum_i x_i^2$
$\Vert\mathbf{x}\Vert_{w,1}$	weighted ℓ_1 norm: $\Vert\mathbf{x}\Vert_{w,1} \doteq \sum \vert w_i x_i\vert$
$\Vert\mathbf{x}\Vert_0$	ℓ_0 quasi-norm, number of non-zero elements in $\mathbf{x}$
$\Vert\mathbf{x}\Vert_\infty$	ℓ_∞ norm, $\Vert\mathbf{x}\Vert_\infty \doteq \max_i \vert x_i\vert$
$\mathbf{M} \succeq \mathbf{N}$	$\mathbf{M} - \mathbf{N}$ is positive semidefinite
$\Vert\mathbf{A}\Vert_*$	Nuclear norm: $\Vert\mathbf{A}\Vert_* \doteq \sum \mathrm{svd}(\mathbf{A})$

conv($\mathcal{A}$) Convex hull of the set $\mathcal{A}$.
$|E|$ cardinality (e.g. number of elements) of the set E.

3 Identification of LTI Systems in the Presence of Missing Data via Atomic Norm Minimization

As indicated in the introduction, the goal of this chapter is to illustrate the ability of sparsification based techniques to provide computationally tractable solutions to hard problems. To this effect, we begin by considering the problem of identifying low order linear time invariant (LTI) models from noisy measurements. Specifically, we are interested in solving:

Problem 1 Given:

- N (noisy) samples of the time response y_t of an unknown plant G to a known input u_t:
$$y_t = (g * u)_t + \eta_t,\ t = 1, \ldots, N, \tag{1}$$
where $g(.)$ denotes the impulse response of the unknown plant and $*$ denotes convolution.
- A priori information about the peak value of the frequency response of the plant (e.g. $|G(e^{j\theta})| \leq 1$ for all $\theta \in [0, 2\pi)$), and its time constant ρ (e.g. the impulse response of the plant is known to decay as ρ^k), and
- Worst case bounds ϵ on the noise, that is $|\eta_t\|_{\ell_\infty} \leq \epsilon$

(i) Determine whether there exists a model of the form (1) with the desired time constant and peak frequency response that explains the observed experimental data, and, (ii) if so, find the coefficients of the lowest order model with this property.

In principle, the problem above can be solved by minimizing the rank of a suitable constructed matrix, subject to additional semidefinite constraints to enforce stability of the resulting model. Specifically, as shown in [4], Problem 1 is equivalent to the following constrained rank minimization:

$$\begin{gathered}\min_g \text{rank}\, \mathbf{H}(g) \text{ subject to:} \\ \begin{bmatrix} \mathbf{R}^{-2} & \mathbf{T}(g)^T \\ \mathbf{T}(g) & \mathbf{R}^2 \end{bmatrix} \succeq 0 \text{ and } |y_t - (\mathbf{T}(g)\mathbf{u})_t| \leq \epsilon,\ t = 1, \ldots, N\end{gathered} \tag{2}$$

where, for an infinite sequence g_i, the corresponding Hankel and Toeplitz matrices are defined as:

$$\mathbf{H}(g) \doteq \begin{bmatrix} g_1 & g_2 & \cdots & g_n & \cdots \\ g_2 & g_3 & \cdots & g_{n+1} & \cdots \\ \vdots & \vdots & \ddots & \vdots & \cdots \\ g_n & g_{n+1} & \cdots & g_{2n-1} & \cdots \\ \vdots & \vdots & \vdots & \vdots & \ddots \end{bmatrix}, \quad \mathbf{T}(g) \doteq \begin{bmatrix} g_1 & 0 & \ldots & 0 & \ldots \\ g_2 & g_1 & \ddots & 0 & \ldots \\ \vdots & \ddots & \ddots & 0 & \ldots \\ g_n & g_{n-1} & \ldots & g_1 & \ldots \end{bmatrix} \tag{3}$$

and where $\mathbf{R} = \text{diag}(1 \ \rho \ \ldots \ \rho^n \ \ldots)$. In principle, the problem above is very challenging due to the non-convex, non-differentiable objective function and the infinite dimensional constraints. However, a tractable convex relaxation can be obtained by using the nuclear norm as a surrogate for rank and considering truncated versions of the matrices $\mathbf{H}$, $\mathbf{T}$, leading to a semi-definite program (SDP) of the form:

$$\begin{aligned} &\min_g \|\mathbf{H}_n(g)\|_* \text{ subject to:} \\ &\begin{bmatrix} \mathbf{R}^{-2} & \mathbf{T}_n(g)^T \\ \mathbf{T}_n(g) & \mathbf{R}^2 \end{bmatrix} \succeq 0 \text{ and } |y_t - (\mathbf{T}_n(g)\mathbf{u})_t| \leq \epsilon, \ t = 1, \ldots, N \end{aligned} \tag{4}$$

where $n \gg N$. In turn, this problem can be solved using first order, ADMM type algorithms (see [4] for details). However, the computational complexity of this approach scales as $\mathcal{O}(\text{number of variables}^3)$, thus practically limiting its use to moderately large problems (few hundreds of data points). On the other hand, as briefly described below, recasting the identification problem into an atomic-norm constrained minimization leads to very efficient algorithms whose complexity scales linearly with the number of data points. In order to elaborate on these ideas, we briefly recall some key results on using atomic norms to promote sparsity.

3.1 Atomic Norms, Sparse Optimization and Low Order Models

The main motivation for using atomic norms is to obtain a sparse representation of a given object in terms of the elements of a given dictionary (the "atoms"). In particular, we will consider the case where this dictionary, $\mathcal{A}$, is centrally symmetric, that is, $a \in \mathcal{A} \Rightarrow -a \in \mathcal{A}$. Under this condition, we can assign to each point in space an "atomic norm" [5], $\|\mathbf{x}\|_{\mathcal{A}}$ defined as:

$$\|\mathbf{x}\|_{\mathcal{A}} = \inf\{t > 0 \ : \ \mathbf{x} \in t \ \text{conv}(\mathcal{A})\} \tag{5}$$

Alternatively, an equivalent definition is given by:

$$\|\mathbf{x}\|_{\mathcal{A}} = \inf\left\{\sum_{a \in \mathcal{A}} |c_a| \ : \mathbf{x} = \sum_{a \in \mathcal{A}} c_a a\right\} \tag{6}$$

Atomic norms play a key role when seeking sparse solutions to optimization problems of the form:

$$\min_x f(x) \text{ subject to } \|x\|_{\mathcal{A}} \leq \tau \tag{7}$$

where $f(x)$ is a smooth convex function, and τ is used to promote sparsity [5]. Note that (7) can be considered a constrained version of a regularized problem of the form:

$$\min_x f(x) + \lambda\|x\|_{\mathcal{A}} \tag{8}$$

For instance, in the case where the set of atoms consists of basis vectors and their symmetric images with respect to the origin, $\pm e_i \in \mathbb{R}^n$, the corresponding atomic norm is simply the ℓ_1 norm and the problem above reduces to the well known form $\min_x f(x) + \lambda\|x\|_1$. Similarly, a set of of atoms consisting of all unit norm rank-1 matrices leads to problems of the form $\min_x f(x) + \lambda\|x\|_*$. The advantage of the formulation (7) over (8) is that it can be solved using the following Frank-Wolfe type algorithm (see e.g. [6]), which has a convergence rate of $\mathcal{O}(\frac{1}{t})$.

Algorithm 1 Generic Frank-Wolfe algorithm to minimize a convex function over the τ-scaled atomic norm ball

1: $x_0 \leftarrow \tau a_0$ for arbitrary $a_0 \in \mathcal{A}$ ▷ Initialization
2: **for** $t = 0,1,2,3,...$ **do**
3: $\quad a_t \leftarrow \text{argmin}_{a\in\mathcal{A}} \langle \partial f(x_t), a\rangle$
4: $\quad \alpha_t \leftarrow \text{argmin}_{\alpha\in[0,1]} f(x_t + \alpha[\tau a_t - x_t])$
5: $\quad x_{t+1} \leftarrow x_t + \alpha_t[\tau a_t - x_t]$
6: **end for**

3.2 LTI Identification as an Atomic Norm Minimization

Recasting Problem 1 into an atomic norm framework requires identifying a suitable set of atoms. As shown in [7] one such set is given by: $\mathcal{A} = \mathcal{A}_1 \cup \mathcal{A}_2 \cup \mathcal{A}_3 \cup \mathcal{A}_4$, where:

$$\begin{aligned}
\mathcal{A}_1 &= \left\{\Psi_p(z) = \pm\frac{(1-|p|^2)}{2}\left(\frac{1}{z-p} + \frac{1}{z-p^*}\right) \ : \ p \in \mathbb{D}\right\} \\
\mathcal{A}_2 &= \left\{\Psi_p(z) = \pm\frac{(1-|p|^2)}{2}\left(\frac{-j}{z-p} + \frac{j}{z-p^*}\right) \ : \ p \in \mathbb{D}\right\} \\
\mathcal{A}_3 &= \left\{\Psi_p(z) = \pm 1\right\} \\
\mathcal{A}_4 &= \left\{\Psi_p(z) = \pm\frac{(1-|p|^2)}{z-p} \ : \ p \in [-\rho, \rho]\right\}
\end{aligned} \tag{9}$$

p^* denotes the complex conjugate of p, and where the normalization factor $1 - |p|^2$ guarantees that each Ψ_p has norm less or equal to 1. The intuition behind this choice of atoms is to try to express the impulse response of the unknown plant as the sum of impulse responses of first order systems (the same idea used in partial fraction expansions), with the specific choice for $\mathcal{A}_i$ motivated by the need to restrict the optimization to real numbers even in the case of oscillatory responses.

The set of atoms introduced above leads to the following convex relaxation of Problem 1:

$$\min_{g_i} \frac{1}{2} \sum_{t=1}^{N} \left(y_t - \sum_{j=1}^{t} g_j u_{t-j} \right)^2 \text{ s.t } \|\mathbf{g}\|_{\mathcal{A}} \leq \tau \tag{10}$$

$$= \min_{\mathbf{g}} \ \frac{1}{2} \|\mathbf{T}_{\mathbf{u}}^n \mathbf{g} - \mathbf{y}\|_{\ell_2}^2 \text{ s.t } \|\mathbf{g}\|_{\mathcal{A}} \leq \tau \tag{11}$$

where $\mathbf{g}$ denotes the (truncated) impulse response of the plant to be identified. Here the objective seeks to minimize the quadratic fitting error to the given experimental data, while the atomic norm constraint is used to promote sparsity. In principle, this problem can be solved using Algorithm 1. However, since the "dictionary" here is the set of impulse responses associated with poles in the (open) unit disk, it is infinite dimensional. Thus, finding the atom that leads to the steepest descent (step 3 in the algorithm) is far from trivial. In the case of well damped plants, this difficulty can be overcome by simply gridding the unit disk [8]. However, plants with poles close to the stability boundary require using dense grids, substantially increasing the computational complexity. To avoid dense griddings [9] proposed the randomized variant of Algorithm 1 shown in next page.

The main difference with Algorithm 1 is that here the search for the atom that provides the steepest descent has been replaced by a random search for an atom that just provides a descent direction. The main advantage of this strategy is that Algorithm 2 requires performing only inner products, resulting both in a substantial speed increase and reduction of memory requirements vis-a-vis rank minimization based algorithms. Further, it can be shown [9] that this algorithm retains the rate of convergence (albeit now in expected value) of its deterministic counterpart.

Algorithm 2 Randomized algorithm for atomic norm constrained minimization

1: Initialize $\mathbf{g_0} \leftarrow \tau\{\mathbf{a}_0\}$ for arbitrary $\mathbf{a}_0 \in \mathcal{A}$
2: **for** $k = 0,1,2,3,..., k_{max}$ **do**
3: Pick N poles uniformly distributed over $\mathbb{D}_\rho$, denote the set of these poles S_k
4: $\mathbf{a_k} \leftarrow \{\text{argmin}_{\mathbf{a} \in \mathcal{A}\{S_k\}} \langle \nabla f(\mathbf{g}_k), \mathbf{a} \rangle\}$
5: $\alpha_k \leftarrow \text{argmin}_{\alpha \in [0,1]} f(\mathbf{g_k} + \alpha[\tau \mathbf{a_k} - \mathbf{g_k}])$
6: $\mathbf{g_{k+1}} \leftarrow \mathbf{g_k} + \alpha_k[\tau \mathbf{a_k} - \mathbf{g_k}]$
7: **end for**

Remark 1 Missing data can be easily accommodated by the algorithm above by simple introducing a "selection" variable:

$$s_i = \begin{cases} 0 & \text{data at } t_i \text{ is missing} \\ 1 & \text{otherwise} \end{cases}$$

and replacing the objective in (10) with $\frac{1}{2}\|s_i(\mathbf{T}_u^n\mathbf{g} - \mathbf{y})\|_{\ell_2}^2$

Remark 2 Frequency domain experimental data can be handled by simply modifying the objective in (10) to:

$$\frac{1}{2}\|\mathbf{T}_\mathbf{u}^n\mathbf{g} - \mathbf{y}\|_{\ell_2}^2 + \frac{\lambda}{2}\|(\mathbf{G} - \boldsymbol{\Omega})\|_{\ell_2}^2 \tag{12}$$

where $\mathbf{G}$ and $\boldsymbol{\Omega}$ denote the frequency response of the unknown plant and the experimental measurements, respectively, and the parameter λ weights the relative importance of the time-domain versus the frequency domain fitting error.

4 Handling Outliers in LTI Identification

Many practical scenarios involve situations where the data is corrupted by outliers. Examples of these situations include sensor outages, data corrupted while transmitted over a wireless link or, in the case of computer vision applications, target occlusion. If the location of these outliers is known, then they can be handled as outlined in Remark 1 above. However, in most of the situations above, the location of these outliers is unknown. In these cases, it is of interest to find a model that interpolates the largest number of data points (the "inliers"), while, at the same time, identifying the outliers, leading to the following problem:

Problem 2 Given noisy input/output data $\{(u_t, y_t)_{t=t_0}^T\}$, find an ARX model of the form

$$y_t = \sum_{k=1}^{n_a} a_k y_{t-k} + \sum_{k=1}^{n_b} b_k u_{t-k} + \eta_t \tag{13}$$

that maximizes the number of points interpolated.

It is well known [10] that identification in the presence of outliers is generically NP hard. However, as we discuss next, efficient convex relaxations can be obtained by recasting the problem into either (i) a regularized robust regression form or (ii) a polynomial optimization form.

4.1 Identification with Outliers as a Robust Regression Problem

Define $\mathbf{r} = \left[a_1 \cdots a_{n_a}\ b_1 \cdots b_{n_b}\right]^T$ and $\mathbf{x}_t = \left[y_{t-1} \cdots y_{t-n_a}\ u_{t-1} \cdots u_{t-n_b}\right]^T$. In this context, Problem 2 can be compactly stated as finding a parameter vector $\mathbf{r}$ that maximizes the cardinality of the set $\mathcal{T} \doteq \{t : \left|y_t - \mathbf{r}^T\mathbf{x}_t\right| \leq \epsilon\}$. Equivalently, by introducing additional variables $\mathbf{r}_i \in \mathbb{R}^d$ the problem can be reformulated as:

$$\begin{array}{l} \mathbf{r}^* = argmin_{\mathbf{r},\mathbf{r}_i} \|\{\mathbf{r} - \mathbf{r}_i\}\|_0 \text{ subject to:} \\ |y_i - \mathbf{x}_i^T \mathbf{r}_i| \leq \epsilon,\ i = 1, \ldots, N \end{array} \tag{14}$$

where $\|\{\mathbf{r} - \mathbf{r}_i\}\|_0$ denotes the cardinality (e.g. number of non-zero vectors) of the sequence $\{\mathbf{r} - \mathbf{r}_i\}_{i=1}^N$. While this problem is still generically NP hard, a convex relaxation can be obtained using the fact that the convex envelope of the cardinality of a vector sequence $\{\mathbf{v}_i\}$ is given by $\|\{\mathbf{v}\}\|_{0,env} = \sum_i \|\mathbf{v}_i\|_\infty$ [11], leading to the following relaxation:

$$\begin{array}{l} \mathbf{r}_{env} = argmin_{\mathbf{r},\mathbf{r}_i} \sum_{i=1}^N \|\mathbf{r} - \mathbf{r}_i\|_\infty \text{ subject to:} \\ |y_i - \mathbf{x}_i^T \mathbf{r}_i| \leq \epsilon,\ i = 1, \ldots, N \end{array} \tag{15}$$

As shown in [12], under certain conditions,[1] $\mathbf{r}_{env} = \mathbf{r}^*$, the solution to (14). It is also worth noting that (15) is equivalent to:

$$\begin{array}{l} \min_{\mathbf{r},\eta} \sum_{i=1}^N \frac{|y_i - \mathbf{x}_i^T \mathbf{r} + \eta_i|}{\|\mathbf{x}_i\|_1} \\ \text{subject to } |\eta_i| \leq \epsilon,\ i = 1, \ldots, N \end{array} \tag{16}$$

that is, a traditional ℓ_1-regularized robust regression where the data points have been scaled. As shown in [12], this "self-scaling" property substantially reduces the effect of gross outliers on the estimation error.

4.2 A Moments Based Approach

The results of the previous section show that, under certain conditions, Problem 2 can be exactly solved by solving the convex relaxation (15). However, when these conditions fail, $\mathbf{r}_{env}$ may provide a poor approximation to $\mathbf{r}^*$. To avoid this difficulty, in this section we introduce an alternative approach based on polynomial optimization. This approach is guaranteed to recover the optimal solution $\mathbf{r}^*$, at the price of increased computational complexity. The idea is to associate to each pair $(y_i, \mathbf{x}_i)$ a binary variable $s_i \in \{0, 1\}$ that indicates whether the point is an inlier ($s_i = 1$)

[1] These conditions are related to the number of outliers and the minimum separation between the inlier and outliers subspaces.

or outlier ($s_i = 0$), allowing for recasting Problem 2 into the following polynomial optimization form:

$$\begin{aligned} p^* = \max_{s_j, \mathbf{r}} & \sum_{j=1}^{N_p} s_j \\ \text{s.t. } & |s_j(y_j - \mathbf{r}^T \mathbf{x}_j)| \le \epsilon s_j,\ s_j^2 = s_j,\ \forall_{j=1}^{N_p} \\ & \mathbf{r}^T \mathbf{r} = 1,\ \mathbf{r}(1) \ge 0 \end{aligned} \tag{17}$$

Here, the first constraint enforces that $(y_j, \mathbf{x}_j)$ is an inlier when $s_j \neq 0$ and is trivially satisfied otherwise; the second is simply a restatement of the fact that $s_j \in \{0, 1\}$, while the third and fourth constraints normalize the vector $\mathbf{r}$ and remove ambiguities. Clearly, since $s_i = 1 \iff (y_i, \mathbf{x}_i)$ is an inlier, the objective maximizes the number of inliers.

Problem (17) is a polynomial optimization problem and thus can be solved using the techniques outlined in the Appendix. Specifically, its first order moment relaxation has the following form:

$$\tilde{p}^* = \max_{\mathbf{M}_j} \sum_{j=1}^{N_p} m_j(s_j) \text{ subject to:} \tag{18}$$

$$\begin{aligned} & \mathrm{Tr}(\mathbf{Q}_+)\mathbf{M} \le 0,\ \mathrm{Tr}(\mathbf{Q}_- \mathbf{M}) \ge 0, \\ & \mathbf{M} \succeq \mathbf{0}, \mathbf{M}(1, 1) = 1 \end{aligned} \tag{19}$$

Here $m_j(s_j)$ denotes the moment variable associated with s_j, $\mathbf{M}$ denotes the moment matrix containing up to second-order moments of the variables $\mathbf{r}$ and s_i, $i = 1, \ldots, N_p$ and $\mathbf{Q}_+, \mathbf{Q}_-$ are matrices of the form

$$\mathbf{Q}_\pm = \text{block-diag } (\mathbf{Q}_{j,\pm}) \text{ with } \mathbf{Q}_{j,\pm} = \begin{bmatrix} 0 & 0 & 0 \\ \vdots & \vdots & \vdots \\ 0 & 0 & 0 \\ y_j \pm \epsilon & -\mathbf{x}_j^T & 0 \end{bmatrix}$$

Further, due to the block diagonal structure of $\mathbf{Q}_\pm$, the problem above exhibits the running intersection property (see Definition 2 in the Appendix), which allows for using the reduced complexity relaxation (49), leading to:

$$\tilde{p}^* = \max_{\mathbf{M}_j} \sum_{j=1}^{N_p} m_j(s_j) \text{ subject to:} \tag{20}$$

$$\begin{aligned} & \mathrm{Tr}(\mathbf{Q}_{j,+}\mathbf{M}_j) \le 0,\ \mathrm{Tr}(\mathbf{Q}_{j,-}\mathbf{M}_j) \ge 0, \forall_{j=0}^{N_p} \\ & \mathbf{M}_j \succeq \mathbf{0}, \mathbf{M}_j(1, 1) = 1, \forall_{j=0}^{N_p} \\ & \mathbf{M}_j(1 : n+1, 1 : n+1) = \mathbf{M}_0, \forall_{j=1}^{N_p} \end{aligned} \tag{21}$$

where $\mathbf{M}_j$ denotes the moment matrix containing up to second-order moments of the variables $\mathbf{r}$ and s_j, and where $\mathbf{M}_o$ denotes the moment matrix containing up to second order moments of the variables $\mathbf{r}$. In principle, the formulation above provides only a relaxation of the original problem. However, optimality of the solution can be enforced by exploiting a result, originally from [13], showing that a sufficient condition for the relaxation (20) to be exact is that $\text{rank}(\mathbf{M}_o) = 1$. Combining this result with the usual re-weighted heuristics for rank minimization leads to the following algorithm:

Algorithm 3 Identification in the presence of outliers via polynomial optimization

1: **Initialize:** $k = 0$, $\mathbf{W}^{(0)} = \mathbf{I}$, $0 < \delta \ll 1$, $k_{\max}$
2: **repeat**
3: solve

$$\begin{aligned} \{\mathbf{M}_j^{(k)}\} = \arg\min\ & \text{Tr}(\mathbf{W}^{(k)}\mathbf{M}_0) - \lambda \textstyle\sum_{i=1}^{N_p} s_i \\ \text{s.t. } & (21) \end{aligned} \tag{22}$$

4: update $\mathbf{W}^{(k+1)} = [\mathbf{M}_0^{(k)} + \sigma_2(\mathbf{M}_0^{(k)})]^{-1}$, $k = k + 1$;
5: **until** $\sigma_2(\mathbf{M}_0^{(k)}) < \delta\sigma_1(\mathbf{M}_0^{(k)})$ or $k > k_{\max}$.

It is worth emphasizing that in this algorithm, there are N_p semi-definite constraint, having fixed size $(n_r + 2 \times n_r + 2)$. Hence computational complexity scales linearly with the number of data points. To the best of our knowledge, this is the first algorithm for robust regression in the presence of outliers that exhibits this property.

5 Identifying Dynamical Graphical Models

So far, we have considered only unstructured models. However, in the past few years substantial attention has been devoted to the problem of identifying dynamical graphical models, represented by a directed graph structure $G = \{V, E\}$, where each node V corresponds to a given time series, and the edges E are linear shift invariant operators relating the values of these series at different time instants. The corresponding equations are given by

$$\begin{aligned} x_j(t) = \textstyle\sum_{i=1}^{n}\sum_{k=1}^{r} c_{j,i}(k)x_i(t-k) + \eta_j(t), \\ t \in [r+1, T],\ \ j = 1, \ldots, n \end{aligned} \tag{23}$$

where $x_j(.)$ denotes the time series at the jth node, $c_{j,i}(.)$ are the coefficients of an ARX model relating the present value of the time series at node j to the past values measured at node i, and $\eta_j(t)$ represents measurement noise. As briefly described below, these models, which appear in fields ranging from systems biology and chemistry to economics and video-analytics, can be identified from experimental data by recasting the problem into a super-atomic constrained minimization form. Note in passing that, unless a regularization criteria is added, the problem is ill posed, since an infinite number of topologies can explain a given set of finite, noisy observations. In this chapter, we will use "sparsity" to regularize the problem, reflecting the fact that usually the solution with the fewest number of edges is the correct one. Let

$$\begin{aligned}
\mathbf{x}_j &\doteq \left[x_j(T), \ldots, x_j(r+1)\right]^T \\
\boldsymbol{\eta}_j &\doteq \left[\eta_j(T), \ldots, \eta_j(r+1)\right]^T \\
\mathbf{c}_{j,i} &\doteq \left[c_{j,i}(1), \ldots, c_{j,i}(r)\right]^T \\
\mathbf{c}_j &\doteq \left[\mathbf{c}_{j,1}^T \ldots, \mathbf{c}_{j,n}^T\right]^T \\
\mathbf{C} &\doteq \left[\mathbf{c}_1, \ldots, \mathbf{c}_n\right] \\
\mathbf{X} &\doteq \left[\mathbf{x}_1, \ldots, \mathbf{x}_n\right] \\
\mathbf{H}_i &\doteq \begin{bmatrix} x_i(T-1) & x_i(T-2) & \ldots & x_i(T-r) \\ x_i(T-2) & x_i(T-3) & \ldots & x_i(T-r-1) \\ \vdots & \ldots & \ldots & \vdots \\ x_i(r) & \ldots & \ldots & x_i(1) \end{bmatrix} \\
\mathbf{H} &\doteq \left[\mathbf{H}_1 \ldots \mathbf{H}_n\right] \\
\boldsymbol{\Xi} &\doteq \left[\boldsymbol{\eta}_1, \ldots, \boldsymbol{\eta}_n\right]
\end{aligned}$$

With this notation, the equations describing the complete model can be written in compact form as:

$$\mathbf{X} = \mathbf{HC} + \boldsymbol{\Xi} \tag{24}$$

and the problem of interest here can be precisely stated as:

Problem 3 Given T measurements of n time series $x_i(t),\ i = 1, \ldots, n,\ t \in [1, T]$, and upper bounds ϵ and r on the noise level and edge model order, respectively, solve:

$$\begin{aligned} \min \textstyle\sum_i \|\{\mathbf{c}_i\}\|_0 \text{ s. t. (24) and } \|\boldsymbol{\eta}_i\|_2 \le \epsilon, \\ \forall i = 1, \ldots, n \end{aligned} \tag{25}$$

where $\mathbf{c}_i \in \mathbb{R}^r$ and $\|\{\mathbf{c}_i\}\|_0$ denotes the number of non-zero elements of the vector sequence $\mathbf{c}_i$.

Note that the objective function in this problem is precisely $|E|$, the number of edges in the graph, and that, due to its structure, the problem above decouples into n subproblems of the form:

$$\begin{aligned} &\min \|\{\mathbf{c}_i\}\|_0 \text{ s. t. } \|\boldsymbol{\eta}_j\|_2 \le \epsilon \text{ and} \\ &\mathbf{x}_j = \textstyle\sum_i \mathbf{H}_i \mathbf{c}_i + \boldsymbol{\eta}_j \end{aligned} \tag{26}$$

This is a (vector) sparsification problem similar to (14) and thus can be solved using a relaxation similar to (15). However, as we show next, a computationally attractive alternative can be obtained by expanding the concept of atomic norm introduced in Sect. 3.1 to encompass the case where it is desired to *block*-spasify a vector sequence.

5.1 Super Atoms and Block Sparsity

In this section we briefly discuss how to promote *block*-sparsity, rather than sparsity, using the concept of atoms, and present a computationally efficient algorithm to solve the resulting optimization problem.

Definition 1 ([14]) Assume that the set $\mathcal{A}$ can be partitioned into N centrally symmetric subsets $\mathcal{A}_i$ (the super-atoms), such that $\mathcal{A} = \cup_i \mathcal{A}_i$ and $\mathcal{A}_i \cap \mathcal{A}_j = \emptyset,\ i \neq j$ and associate to each super-atom $\mathcal{A}_i = \{a_{i,1}, ..a_{i,n_i}\}$ the matrix $\mathbf{A}_i$ having as its jth column $\mathbf{a}_{i,j}$, the coordinates of the atom $a_{i,j}$ in a suitable basis in X. Given a point $\mathbf{x} \in X$, its super-atomic norm is defined as:

$$\|\mathbf{x}\|_{s\mathcal{A}} \doteq \inf \left\{ \tau > 0 \colon \mathbf{x} = \sum_i (\tau \mathbf{A}_i)\mathbf{c}_i \text{ and } \sum_i \|\mathbf{c}_i\|_\infty = 1 \right\} \tag{27}$$

An alternative definition of the super atomic norm is given by:

$$\|\mathbf{x}\|_{s\mathcal{A}} = \min_{\mathbf{c}} \textstyle\sum_{i=1}^N \|\mathbf{c}_i\|_\infty \text{ s.t } \mathbf{x} = \sum_i \mathbf{A}_i \mathbf{c}_i \tag{28}$$

Since the convex envelope of the cardinality of a vector sequence $\{\mathbf{c}\}$, $\|\mathbf{c}_i\|_\infty \leq 1$ is given by [11]:

$$\|\{\mathbf{c}\}\|_{0,env} = \sum_i \|\mathbf{c}_i\|_\infty$$

it follows that, minimizing the super-atomic norm indeed promotes block-sparsity. Further, problems involving the minimization of a function subject to super-atomic norm constraints can be efficiently solved by using the following modification of Algorithm 1 [14]:

Algorithm 4 Convex minimization subject to super-atomic norm constraints

1: Data: set of super-atoms $\mathcal{A} = \{\mathcal{A}_1, \ldots, \mathcal{A}_i, \ldots\}$
2: Initialize $\mathbf{x}^{(0)} \leftarrow \tau \mathbf{a}$ for some arbitrary $\mathbf{a} \in \mathcal{A}$
3: **for** $k = 0,1,2,3,..., k_{max}$ **do**
4: $\quad L \leftarrow \arg\min_m \left\{ \min_{\|\mathbf{c}\|_\infty \leq 1} \langle \partial f(\mathbf{x}^{(k)}), \sum \mathbf{a}_{i,m} c_i \rangle \text{ s.t. } \mathbf{a}_{i,m} \in \mathcal{A}_m \right\}$
5: $\quad \mathbf{c} \leftarrow \arg \min_{\|\mathbf{c}\|_\infty \leq 1} \langle \partial f(\mathbf{x}^{(k)}), \sum \mathbf{a}_{i,L} c_i \rangle \text{ s.t. } \mathbf{a}_{i,L} \in \mathcal{A}_L.$
6: $\quad \mathbf{a} \leftarrow \sum_i \mathbf{a}_{i,L} c_i$
7: $\quad \alpha_k \leftarrow \text{argmin}_{\alpha \in [0,1]} f(\mathbf{x}^{(k)} + \alpha[\tau \mathbf{a} - \mathbf{x}^{(k)}])$
8: $\quad \mathbf{x}^{(k+1)} \leftarrow \mathbf{x}^{(k)} + \alpha_k [\tau \mathbf{a} - \mathbf{x}^{(k)}]$
9: **end for**

As shown in [14], when the super-atoms are centrally symmetric, explicit solutions to steps 4–6 of Algorithm 4 are given by

(i) Step 4: $L \leftarrow \arg\max_m \{\|[\partial f(\mathbf{z}^{(k)})]^T \mathbf{A}_m\|_1\}$

(ii) Step 5: $\mathbf{c} = -\operatorname{sign}([\partial f(\mathbf{z}^{(k)})]^T \mathbf{A}_L)$
(iii) Step 6: $\mathbf{a} \leftarrow \mathcal{A}_L \mathbf{c}$

where $\mathbf{A}_m$ denote the matrix having as columns the coordinates of $\mathbf{a}_{i,m}$, the elements of the super-atom $\mathcal{A}_m$.

5.2 *Sparse Graphical Model Identification as a Super-Atomic Norm Minimization*

The techniques discussed above can be used to solve Problem 3 by simply defining each super-atom as the collection of columns from the matrices $\mathbf{H}_i$, (e.g. a collection of vectors, each containing delayed measurements of the respective time-series):

$$\mathcal{A}_i = \{\mathbf{H}_i(:,t)\},\ t = 1, \dots r$$

each of these subproblems can be relaxed to a super-atomic norm minimizations of the form

$$\min \|\mathbf{z}\|_{s\mathcal{A}} \text{ subject to } \|\mathbf{x}_j - \mathbf{z}\|_2 \leq \epsilon \tag{29}$$

where $\mathbf{z} = \sum_i \mathbf{H}_i \mathbf{c}_i$. Finally, imposing soft, rather than hard constraints on the fitting error leads to:

$$\min \|\mathbf{x}_j - \mathbf{z}\|_2 \text{ subject to } \|\mathbf{z}\|_{s\mathcal{A}} \leq \tau \tag{30}$$

This is precisely a problem of the form discussed in Sect. 5.1 and can be efficiently solved using Algorithm 4.

Remark 3 Many practical scenarios require taking into account relatively rare external events, for instance to model interactions of the network with its environment. As proposed in [14], these interactions can be handled by adding at each node, a piecewise constant signal $u_j(\cdot)$, with a sparse derivative, modifying (23) to

$$\begin{aligned} x_j(t) = \textstyle\sum_{i=1}^{n}\sum_{k=1}^{r} c_{j,i}(k) x_i(t-k) + u_j(t) + \eta_j(t), \\ t \in [r+1, T],\ j = 1, \dots, n \end{aligned} \tag{31}$$

and the corresponding objective in (25) to

$$\min \|\{\mathbf{c}_i\}\|_0 + \lambda \|\{\boldsymbol{\Delta} u_j\}\|_0 \tag{32}$$

where $\boldsymbol{\Delta} u_j \doteq \left[u_j(2) - u_j(1), \dots, u_j(t) - u_j(t-1), \dots\right]$ and the parameter λ allows for trading-off graph versus input sparsity. The problem above can be solved using Algorithm 4 by simply adding the following super-atoms to the set $\mathcal{A}$:

$$\mathcal{A}_u = \frac{1}{\lambda}\{\mathbf{u}_1, \ldots, \mathbf{u}_T\} \tag{33}$$

where $\mathbf{u}_t$ is defined as the t-th column of a lower triangular matrix with $\{0, 1\}$ elements.

6 Semi-supervised Identification of Switched Systems in the Presence of Outliers

In the previous sections we have addressed identification of LTI systems. The goal of this section is to indicate how these techniques can be extended to switching systems. These systems are interesting in their own, since they appear in many scenarios (biological systems transitioning amongst different metabolic stages, human activity, physical systems with different operation modes, etc.) and as tractable approximations to more complex non-linear dynamics. For simplicity, we consider only single-input single-output systems, but extension to the MIMO case is straightforward. Specifically, we are interested in the following extension of Problem 2:

Problem 4 Given:

- A set of input/output data $\{(u_t, y_t)_{t=t_0}^T\}$ generated by an SARX model of the form

$$y_t = \sum_{k=1}^{n_a} a_k(\sigma_t) y_{t-k} + \sum_{k=1}^{n_b} b_k(\sigma_t) u_{t-k} + \eta_t \tag{34}$$

- A-priori information consisting of (i) a bound N_s on the number of subsystems (e.g. $\sigma_t \in \mathbb{N}_{N_s}$), (ii) a bound ϵ on the process noise η_t, (iii) additional information, such as N_{f_i}, the relative frequency of each submodel, point wise co-occurrences, constraints on the switching sequence, etc.

Find a set of coefficients $\{a_{k=1}^{n_a}(i), b_{k=1}^{n_b}(i)\}$, each associated with the submodel $G_i, \forall_{i=1}^{N_s}$, that maximizes the number of inliers.

As in Sect. 4, we will present two alternative approaches to solving the problem above, one based on robust-regression and the second based on polynomial optimization.

6.1 A Sparsification Based Approach

The idea underlying this approach (originally presented in [11]) is to find one submodel at a time, by successively finding a parameter vector $\mathbf{r}$ that makes $|y_t - \mathbf{x}_t^T \mathbf{r}| \leq \epsilon$ feasible for as many time instants t as possible. Once this model is found, the points explained by it are removed from the data set and the procedure is repeated until all data points are clustered. By considering at each stage, points not explained by the model as outliers, each parameter vector $\mathbf{r}_i$ can be found using the algorithm described in Sect. 4, leading to Algorithm 5. When the subspaces spanned by each subsystem are well separated, the recovery results in [12] guarantee that this approach will indeed find the correct set of models. On the other hand, if these conditions do not hold, due to its greedy nature, the algorithm can overestimate the number of subsystems required to explained the observed data. Nevertheless, consistent numerical experience shows that these instances are very rare, specially when the algorithm is combined with a re-weighted heuristics to enhance sparsity [11].

Algorithm 5 SARX ID via regularized regression

$t_0 \leftarrow \max(n_a, n_b)$, $N_1 \leftarrow \{t_0, \ldots, T\}$, $l \leftarrow 0$.
while $i < l$ **do**
  $l \leftarrow l + 1$
Find $\mathbf{r}_l$ by solving (14)
  $i \leftarrow 1$
  while $i < l$ **do**
    $K_{il} \leftarrow \{t \in N_i : |y_t - \mathbf{x}_t^T \mathbf{r}_l| \leq \epsilon\}$
    if $\#K_{il} > \#K_i$ **then**
      $\mathbf{r}_i \leftarrow \mathbf{r}_l$ and $l \leftarrow i$;
    end if
    $i \leftarrow i + 1$
  end while
  $K_l \leftarrow \{t \in N_l : |y_t - \mathbf{x}_t^T \mathbf{r}_l| \leq \epsilon\}$
  $N_{l+1} \leftarrow N_l \setminus K_l$
end while
return $s = l$ and $K_i, i = 1, \ldots, s$

6.2 A Moments Based Approach

As in Sect. 4.2, an alternative approach, with optimality certificates can be obtained by recasting the problem into a polynomial optimization form, by introducing a set of binary variables $s_{i,t}$ that indicate whether the submodel G_i is active at time instant t ($\sigma_t = i \Leftrightarrow s_{i,t} = 1$) and auxiliary variables $e_{i,t}$ (the fitting error of point t to model i), leading to:

$$
\begin{array}{l}
\max_{\mathbf{r}_i, e_{i,t}, s_{i,t}} \quad \sum_{t=t_o}^{T} \sum_{i=1}^{N_s} s_{i,t} \\
\text{subject to} \\
\mathbf{r}_i^T \mathbf{x}_t - e_{i,t} = 0 \\
||\mathbf{r}_i||_2 = 1 \\
|s_{i,t} e_{i,t}| \leq \epsilon s_{i,t} \\
s_{i,t}^2 = s_{i,t}, \ \sum_{i=1}^{N_s} s_{i,t} \leq 1 \\
\forall i = 1, \ldots, N_s, \forall t = t_o + n_a, \ldots, T
\end{array}
\tag{35}
$$

Here, $\mathbf{r}_i(j)$ is the jth entry of $\mathbf{r}_i$, and the last two constraints on $s_{i,t}$ guarantee that no more than one submodel is active at time instant t. As before, it can be easily shown that this problem satisfies the running intersection property. Thus, from the results in the Appendix, it follows that a convergent sequence of convex relaxations is given by:

$$
\begin{array}{l}
\max_{\mathbf{m}_N} \sum_{i,t} \mathbf{m}(s_{i,t}) \\
\text{subject to} \\
\quad \forall_{t=t_0}^{T} : \mathbf{M}_{t,N} \succeq 0 \\
\quad \forall_{t=t_0}^{T} : \mathbf{L}_{t,N} \succeq 0
\end{array}
\tag{36}
$$

where $m(s_{i,t})$ denotes the first order moments corresponding to the variables $s_{i,t}$ and $\mathbf{M}_{t,N}$ and $\mathbf{L}_{t,N}$ are the (truncated) moment and localizing matrices involving the moments of the variables $s_{i,t}$, $e_{i,t}$ and $\mathbf{r}_i$, $i = 1, .., N_s$. Further, as before, optimality can be guaranteed by introducing the rank constraint rank$(\{\mathbf{M}_r\} = 1)$, where $\mathbf{M}_r$ denotes the matrix containing up to second order moments of the variables $\mathbf{r}_i$. The resulting rank-constrained optimization can be solved by using a straightforward generalization of Algorithm 3.

Remark 4 In many practical scenarios, additional information is available about the system to be identified. Examples of these situations include knowledge about certain transitions are inhibited (common in biological applications), or co-occurrences (common in image processing and computer vision) where some of the data may be manually annotated, so that it is known that two given data points belong (or do not belong) to the same system. A salient feature of the approach above is its ability to incorporate these priors by imply imposing additional constraints on the variables $s_{i,t}$. For instance:

(i) submodel G_i is active for $f\%$ of the time $\iff \sum_{t=t_0+n_a}^{T} s_{i,t} = 0.01 f(T + 1 - t_0 - n_a)$;
(ii) the same submodel is active at time instants m and $n \iff s_{i,m} = s_{i,n}$, $\forall i = 1, \ldots, N_s$;
(iii) different submodels are active at time instants m and $n \iff s_{i,m} s_{i,n} = 0$, $\forall i = 1, \ldots, N_s$;
(iv) submodel i cannot be followed by submodel $j \iff s_{i,t} s_{j,t+1} = 0$, $\forall t$.

7 Model (In)Validation of Switched Systems

Model (in)validation is the dual problem of identification: given a model and experimental data, the goal here is to determine whether these are consistent. That is, whether or not the observed data (corrupted by noise) could have been generated by the model. Validating identified models against additional data is a key step before using these models for control synthesis. Additional applications of model (in)validation include fault detection and isolation, and, interestingly, to detect anomalies in time series, including abnormal human activity. While (in)validation of LTI models is a well understood problem (see for instance Chaps. 9 and 10 in [15]), the case of switched systems is considerably less developed. Nevertheless, as we show in the sequel, this problem can be addressed using the same polynomial optimization tools used for switched systems identification. Formally, the problem of interest here can be posed as determining whether a noisy input/output sequence could have been generated by a given model of the form:

$$\begin{aligned} \xi_t &= \textstyle\sum_{i=1}^{n_a} a_i(\sigma_t)\xi_{t-i} + \sum_{i=1}^{n_b} b_i(\sigma_t)u_{t-i} \\ y_t &= \xi_t + \eta_t, \ \ \sigma_t \in \{1, \ldots, s\}, \ \ \|\eta_t\|_\infty \leq \epsilon \end{aligned} \tag{37}$$

where y_t denotes the measured output corrupted by the noise η_t. Like the identification case, this problem is known to be generically NP-hard, due to the presence of noise and because the mode variable σ_t is not directly measurable. However, it is possible to obtain tractable relaxations by using sparsification and polynomial optimization tools. This can be done by noting that (37) holds if and only if there exist a set of "indicator" variables $s_{i,t}$ and admissible noise sequence η_t such that

$$\begin{aligned} & s_{i,t}\left(\mathbf{g_{i,t}} - \mathbf{h_i}\boldsymbol{\eta}_{t:t-n_a}\right) = \mathbf{0} \ \forall\, t \in [t_o + n_a, T] \\ & \text{subject to} \\ & \textstyle\sum_{i=1}^{n_s} s_{i,t} = 1 \\ & s_{i,t} \in \{0, 1\} \text{ and } \|\boldsymbol{\eta}_t\|_\infty \leq \epsilon \end{aligned} \tag{38}$$

where the notation was simplified by defining:

$$\begin{aligned} \mathbf{g_{i,t}} &\doteq a_1(i)y_{t-1} + \cdots + a_{n_a}(i)y_{t+n_a} \\ &-y_t + b_1(i)u_{t-1} + \cdots + b_{n_b}(i)u_{t-n_c} \\ \mathbf{h_i} &\doteq \left[-1\ a_1(i)\ \ldots\ a_{n_a}(i)\right] \\ \boldsymbol{\eta}_{t:t-n_a}^T &\doteq \left[\eta_t^T\ \ldots\ \eta_{t-n_a}^T\right]^T \end{aligned}$$

7.1 *Sparsification-Based Certificates*

It is easy to show that the above condition is equivalent to the feasibility of:

$$
\begin{aligned}
& s_{i,t}\mathbf{g_{i,t}} - \mathbf{h_i}\boldsymbol{\eta}_{i,t:t-n_a} = \mathbf{0} \\
& 0 \le s_{i,t} \le 1,\ \sum_i s_{i,t} = 1,\ \forall t \in [t_o + n_a, T] \\
& \|\boldsymbol{\eta}_{i,t:t-n_a}\|_\infty \le s_{i,t}\epsilon,\ \sum_{i=1}^{n_s} \boldsymbol{\eta}_{i,t:t-n_a} = \boldsymbol{\eta}_{t:t-n_a} \\
& \text{and} \\
& \|\mathbf{s}\|_o = T - t_o - n_a + 1
\end{aligned} \tag{39}
$$

where $\boldsymbol{\eta}_{i,t:t-n_a} \doteq s_{i,t}\boldsymbol{\eta}_{t:t-n_a}$ are auxiliary variables. Then, since all the constraints, except the last one, are convex, we can use a weighted ℓ_1 norm as proxy for cardinality [16, 17], to obtain a convex relaxation and (in)validation certificates using Algorithm 6.

Algorithm 6 Sparsification Based (In)Validation Certificates

Initialize: $k \leftarrow 0, \forall_{i=1}^{N_s} \forall_{t=n_a+t_0}^{T} : w_{i,t}^{(0)} \leftarrow 1$;
repeat
solve

$$
\begin{aligned}
&\min_{\mathbf{s},\eta} \quad \textstyle\sum_{i,t} w_{i,t}^{(k)} s_{i,t} \\
&\text{subject to (39)}
\end{aligned} \tag{40}
$$

update

$$
\begin{aligned}
& w_{i,t}^{(k+1)} \leftarrow (s_{i,t}^{(k)} + \delta)^{(-1)}, \forall i \forall t \\
& k \leftarrow k+1
\end{aligned}
$$

where $s_{i,t}^{(k)}$ denotes the optimal solution at the k-th iteration, and δ is a (small) regularization constant.
until convergence.

This algorithm can be used to efficiently compute convex (in)validation certificates – i.e. infeasibility of (40) – or to establish that the experimental data is compatible with the a priori models, when the solution satisfies $s_{i,j} \in \{0, 1\}$. However, it should be noted that these conditions are only sufficient. That is, they cannot explain the case when the relaxation admits a solution with non-integer elements. This case arises if, due to noise, some of the data points can be explained by more than one model, or if a model is invalid (it can be explained by a linear combination of other models). It is possible to obtain sharper certificates, albeit at a higher computational cost, by using moments to solve directly the following polynomial optimization problem (see [18]):

$$
\begin{aligned}
p^* = \ & \min_{\mathbf{s},\eta} \textstyle\sum_{i,t} s_{i,t}^2 \|\mathbf{g_{i,t}} - \mathbf{h_i}\boldsymbol{\eta}_{t:t-n_a}\|^2 \\
& \text{subject to} \\
& s_{i,t} = s_{i,t}^n \\
& \textstyle\sum_i s_{i,t}^2 = 1 \ \forall t \\
& \|\boldsymbol{\eta}_{t:t-n_a}\|_\infty \le \epsilon
\end{aligned} \tag{41}
$$

where the experimental data does not invalidate the model iff $p^* = 0$. It is easy to show that problem (41) exhibits the running intersection property. Thus, only the smaller moment matrices involving the moments of the variables $s_{1,t}, .., s_{n_s,t}, \eta_t, \eta_{t-1} \dots \eta_{t-n_a}$ need to be considered, leading to Algorithm 7 below.

Algorithm 7 Moments Based (In)Validation Certificates

$N \leftarrow 2;$

repeat

solve $p_m^* = \min_{\mathbf{m}} \sum_{i,t} l_{i,t}$ subject to:

$$\left.\begin{array}{l} \mathbf{M}_N(\mathbf{m_{t-n_a:t}}) \succeq 0 \\ \mathbf{L}_N(\mathbf{m_{t-n_a:t}}) \succeq 0 \end{array}\right\} \forall t \in [t_o + n_a, T]$$

update $N \leftarrow N + 1$

$$\textbf{until} \begin{cases} p_m^* > 0, \text{ or} \\ p_m^* = 0, \textbf{rank}[\mathbf{M}_N(\mathbf{m_{t-n_a:t}})] \\ = \textbf{rank}[\mathbf{M}_{N-1}(\mathbf{m_{t-n_a:t}})]; \text{ or} \\ N = T + 1. \end{cases}$$

If $p_m^* > 0$ the model is invalid, otherwise the data record is consistent with the a-priori assumptions.

---

8 Applications

Next, we illustrate the ideas presented in this tutorial with several application examples.

8.1 Identification of a Lightly Damped System

Consider the non-trivial problem of identifying of a very lightly damped two-degrees of freedom structure [19] from the time and frequency domain experimental data shown in Figs. 1 and 2. Using Algorithm 2 (Sect. 3) leads to a 6th order system that fits well the data, in approximately 2 s. For comparison, the ADMM approach in [4], while producing similar results, requires approximately 2500 s and the nonparametric approach in [19] led to a 19th order model.

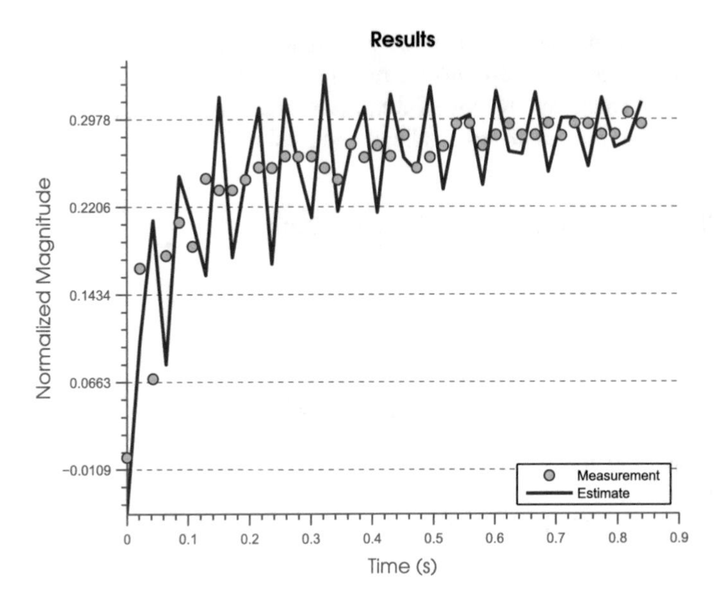

Fig. 1 Step response of a lightly damped structure

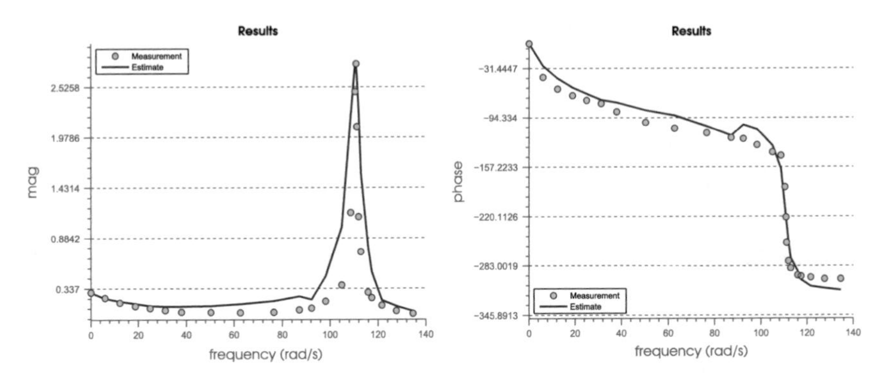

Fig. 2 Frequency response of the structure

8.2 Finding Causally Correlated Activities in Video Sequences

The goal of this section is to illustrate the use of dynamical graphical models to unveil causal relationships from time series generated by different agents. In particular, we consider here the first example from [20], which analyzed two video sequences (6 and 16) from the UT Human Interaction Data Set [21]. The specific time series used in this example are the trajectories of each individual's head, normalized to lie in the interval $[-1, 1]$. Figure 3 shows the result of using Algorithm 4 (Sect. 5), modified to take into account the existence of derivative-sparse exogenous inputs. As shown

Fig. 3 Sample Frames of the UT Sequences 6 and 16 showing the causally interacting groups identified using the approach outlined in Sect. 5

there, this algorithm successfully identified the interactions between agents in both sequences. In addition, the super-atomic norm based approach was, depending on the examples, 3 to 5 times faster than the ADMM approach proposed in [20].

8.3 Activity Analysis from Noisy Video Data

The goal of this application is to segment a video containing multiple activities into sub-activities, each characterized by an affine model. As described in Sect. 9, these models can then be used to recognize contextually abnormal activity. The experimental data used in this section (taken from [22]), illustrated in Fig. 4, consist of 55 frames extracted from a video sequence of a person walking, bending and resuming walking. To simulate a realistic scenario, several frames were corrupted with large amounts, consistent with a scenario where the data is corrupted by interference. In order to recast the segmentation problem into an identification form, the position of the center of mass of the person in each frame was modeled as the output of a switched affine system consisting of 2 first order submodels, and the system was identified using the algorithm outlined in Sect. 6.2. As shown in Fig. 5, this approach successfully segmented the sequence in the presence of outliers.

Fig. 4 Sample frames for the activity segmentation application

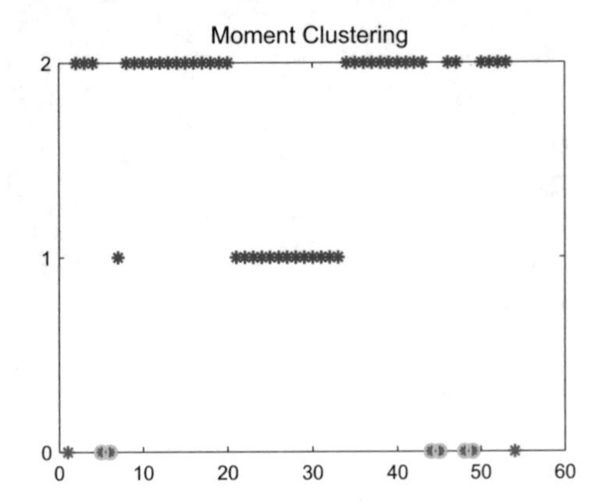

Fig. 5 Activity segmentation as a SARX identification problem with outliers (*red* stars and *green* circles denote the detected and true outliers, respectively)

9 Detection of Contextually Abnormal Activities

Next, we discuss the application of the ideas presented in this chapter to the problem of extracting actionable information from large data sets. Framing this problem using concepts from dynamical systems, allows us to exploit the tractable relaxations discussed above. This approach leads to scalable, computationally tractable algorithms, which can help making critical decisions based on dynamic information that is very sparsely encoded in the available data streams in real time. To this effect, we will consider the observed data as the output of a switched dynamical system, where jumps between systems indicate events that can be characterized by the parameters of the corresponding subsystems.

As an example, consider the problem of abnormal activity detection from video sequences which arises in surveillance systems of large public spaces. This is a very challenging problem since the video data usually contains many different activities. Thus, most machine learning based techniques must parse the data into sub-activities before they can detect an anomaly. Furthermore, the parsing can be very difficult if the individual segments are very short, spanning only a few frames. However, explicit parsing can be avoided by formulating the problem as a model (in)validation one: first, identifying models corresponding to "safe activities" and then detecting anomalies by using Algorithm 7 to invalidate the observed data against the set of trajectories that could have been generated by switching among these safe activities. This approach is illustrated by an example with three safe activities {*waiting, walking, running*}, described by the models (see [23])[2]

$$\begin{pmatrix} x_t \\ y_t \end{pmatrix} = \begin{pmatrix} 0.4747 & 0.0628 \\ -0.3424 & 1.2250 \end{pmatrix} \begin{pmatrix} x_{t-1} \\ y_{t-1} \end{pmatrix} + \begin{pmatrix} 0.5230 & -0.1144 \\ 0.3574 & -0.2513 \end{pmatrix} \begin{pmatrix} x_{t-2} \\ y_{t-2} \end{pmatrix} \quad \text{(walking)}$$

$$\begin{pmatrix} x_t \\ y_t \end{pmatrix} = \begin{pmatrix} 1 & 0 \\ 0 & 1 \end{pmatrix} \begin{pmatrix} x_{t-1} \\ y_{t-1} \end{pmatrix} \quad \text{(waiting)}$$

$$\begin{pmatrix} x_t \\ y_t \end{pmatrix} = \begin{pmatrix} 0.6058 & 0.0003 \\ 0.2597 & 0.8589 \end{pmatrix} \begin{pmatrix} x_{t-1} \\ y_{t-1} \end{pmatrix} + \begin{pmatrix} 0.3608 & 0.1853 \\ -0.2381 & 0.1006 \end{pmatrix} \begin{pmatrix} x_{t-2} \\ y_{t-2} \end{pmatrix} \quad \text{(running)}$$

where (x_t, y_t) are the coordinates of the centroid of the actor, and where transitions from *waiting* to *running* are not allowed, as shown in Fig. 6. The proposed approach successfully flagged the sequence in Fig. 7 as "anomalous", even though it all the sub-activities are safe activities, because it exhibits a forbidden transition.

10 Conclusions

A wide range of applications with potential for profound societal impact, such as self-aware and smart environments, have become feasible thanks to the ease of collecting data and recent developments in distributed control. However, achieving their full potential remains challenging due to the need for processing large amounts of very high dimensional data. As shown in this chapter, it is possible to exploit the inherent sparsity in the data, exhibited in a large class of identification problems, to overcome the "curse of dimensionality" by extending to dynamic settings ideas that were originally proposed in machine learning and polynomial optimization, often leading to algorithms that scale linearly with the number of data points. The potential of the techniques presented here was illustrated with several practical applications, including the problem of individual and group activity analysis from video data. Finally, a related issue not addressed in this chapter is the connection between non-linear systems identification and non-linear manifold embeddings. Interested

[2]These models were identified by considering the trajectories of the centroid of the person and using LP to find the coefficients that minimized the peak value of the fitting error. Note that both *walking* and *running* have two poles at 1, corresponding to constant velocity motion. However, the remaining two poles for the *running* model are complex conjugate, corresponding to oscillatory motion, while those in the *walking* model are real.

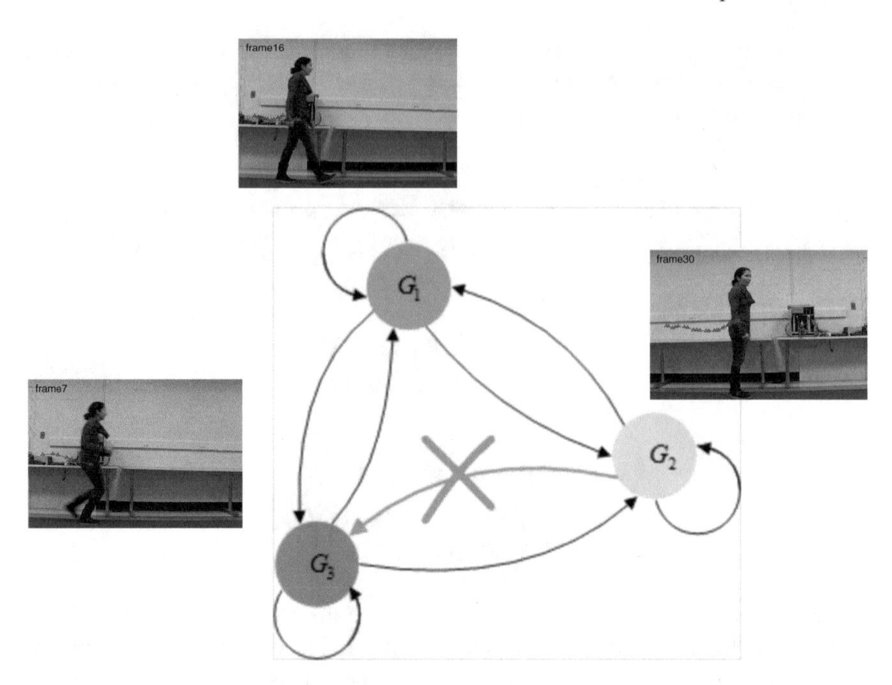

Fig. 6 A structurally constrained transition graph. One-step transitions from *waiting* to *running* are not allowed

Fig. 7 Anomalous behavior detection as a switched (in)validation problem. The video sequence is flagged as abnormal, even though it consists of "normal" activities (*walking,waiting,running*), since it contains an anomalous transition (*waiting* $\rightarrow$ *running*)

readers are directed to [24] where the authors find low order embeddings of dynamic data by recasting the problem as a Wiener system identification problem that can be efficiently solved using technique similar to the ones presented here.

Appendix: Polynomial Optimization

Many identification and model (in)validation problems can be framed as a (non-convex) constrained polynomial optimization problem of the form:

$$p_K^* := \min_{\mathbf{x} \in K} p(\mathbf{x}) \doteq \sum_{\alpha} p_\alpha \mathbf{x}^\alpha \tag{42}$$

where $\mathbf{x}^\alpha = x_1^{\alpha_1} x_2^{\alpha_2} \cdots x_n^{\alpha_n}$ and $K \subset \mathbb{R}^n$ is a compact semi-algebraic set defined by a collection of polynomial inequalities of the form $g_k(\mathbf{x}) \doteq \sum_\beta g_{k,\beta} \mathbf{x}^\beta \geq 0$, $k = 1, \cdots, d$. This problem is equivalent to the following convex, albeit infinite-dimensional, optimization [25]:

$$\tilde{p}_K^* := \min_{\mu \in \mathcal{P}(K)} \int p(\mathbf{x}) \mu(dx) := \min_{\mu \in \mathcal{P}(K)} \mathbf{E}_\mu [p(\mathbf{x})] = \min_{m_\alpha} \sum_\alpha p_\alpha m_\alpha \tag{43}$$

where $\mathcal{P}(K)$ is the space of probability measures supported on K, $\mathbf{E}_\mu$ denotes expectation with respect to μ, and where we have defined:

$$m_\alpha = \mathbf{E}_\mu(\mathbf{x}^\alpha) \doteq \int_K \mathbf{x}^\alpha \mu(dx) \tag{44}$$

As shown in [25], existence of a measure μ satisfying (44) is equivalent to positive semidefiniteness of the (infinite) moment $\mathbf{M}(\mathbf{m})$ and localizing $\mathbf{L}(g_k \mathbf{m})$ matrices. Thus, (43) is equivalent to:

$$p_K^* := \min_{m_\alpha} \sum_\alpha p_\alpha m_\alpha \text{ subject to } \mathbf{M}(\mathbf{m}) \succeq 0 \text{ and } \mathbf{L}(g_k \mathbf{m}) \succeq 0, \; k = 1, .., d \tag{45}$$

While the problem above is still infinite dimensional, a convergent sequence of finite dimensional approximations is given by

$$p_N^* := \min_{m_\alpha} \sum_\alpha p_\alpha m_\alpha \text{ subject to } \mathbf{M}_N(\mathbf{m}) \succeq 0 \text{ and } \mathbf{L}_N(g_k \mathbf{m}) \succeq 0, \; k = 1, .., d \tag{46}$$

where $\mathbf{M}_N$ and $\mathbf{L}_N$ are truncated versions of the matrices $\mathbf{M}(.)$ and $\mathbf{L}(.)$, given by:

$$\begin{aligned} &\mathbf{M}_N(\mathbf{m})(i, j) = m_{\alpha^{(i)} + \alpha^{(j)}}, \; \forall i, \; j = 1, \ldots, S_N \\ &\mathbf{L}_N(g_k \mathbf{m})(i, j) = \sum_\beta g_{k,\beta} m_{\beta + \alpha^{(i)} + \alpha^{(j)}}, \\ &\forall i, \; j = 1, \cdots, S_{N - \lceil \frac{\text{degree}\,(g_k)}{2} \rceil} \end{aligned} \tag{47}$$

where $S_N = \binom{N+n}{n}$ (e.g. the number of moments in $\mathbb{R}^n$ up to order N) and the moments have been arranged according to a grevlex ordering. Further, it can be shown that as N increases, p_N^* in (46) monotonically increases to p_K^* from below. The necessary and sufficient conditions to guarantee the equivalence between (45) and (42) are either:

- positive semi-definiteness of the infinite dimensional matrices $\mathbf{M}_N$ and $\mathbf{L}_{N-\lceil \frac{\text{degree}(g_k)}{2} \rceil}$ as N increases to infinity; or
- for a finite N, the flat extension property holds [25], that is, for $\mathbf{M}_N \succeq 0$, $\mathbf{L}_{N-\lceil \frac{\text{degree}(g_k)}{2} \rceil} \succeq 0$, and $\text{rank}(\mathbf{M}_N) = \text{rank}(\mathbf{M}_{N-\lceil \frac{\text{degree}(g_k)}{2} \rceil})$.

Exploiting Sparsity: The problems considered in this chapter exhibit a special sparse structure that can be exploited to reduce the computational complexity entailed in solving (42).

Definition 2 Consider problem (42) and let $I_k \subset \{1, \ldots, n\}$ be the set of indices of variables such that each $g_k(\mathbf{x})$ contains variables only from some I_k. Assume that the objective function $p(\mathbf{x})$ can be partitioned as $p(\mathbf{x}) = p_1(\mathbf{x}) + \ldots + p_l(\mathbf{x})$ where each p_k contains only variables from I_k. Problem (42) satisfies the running intersection property if there exists a reordering $I_{k'}$ of I_k such that for every $k' = 1, \ldots, l-1$:

$$I_{k'+1} \cap \bigcup_{j=1}^{k'} I_j \subseteq I_s \quad \text{for some} \quad s \leq k' \tag{48}$$

As shown in [26], for problems satisfying the running intersection property, the sequence of approximations (45) can be replaced by a hierarchy of semidefinite programs of smaller size:

$$\begin{aligned} p_N^* = \min_{\mathbf{m}} & \textstyle\sum_{j=1}^{l} \sum_{\alpha(j)} p_{j,\alpha(j)} m_{\alpha(j)} \\ \text{s.t.} \quad & \mathbf{M}_N(\mathbf{m}_{I_k}) \succeq 0, k = 1, \ldots, l, \\ & \mathbf{L}_N(g_k \mathbf{m}_{I_k}) \succeq 0, k = 1, \ldots, l, \end{aligned} \tag{49}$$

where $p_{j,\alpha(j)}$ is the coefficient of the $\alpha(j)^{th}$ monomial in the polynomial p_j, $\mathbf{M}_N(\mathbf{m}_{I_k})$ denotes the moment matrix, and $\mathbf{L}_N(g_k \mathbf{m}_{I_k})$ is the localizing matrix for the subset of variables in I_k. Note that, for a given N, (49) involves matrices containing $O(\kappa^{2N})$ variables, where κ is the maximum cardinality of I_k, rather than $O(n^{2N})$. Since in the problems considered in this chapter $\kappa \ll n$ this leads to substantial complexity reduction.

References

1. E. Sontag, Nonlinear regulation: the piecewise linear approach. IEEE Trans. Autom. Control **26**, 346–358 (1981)
2. L. Breiman, Hinging hyperplanes for regression, classification and function approximation. IEEE Trans. Inf. Theory **39**, 999–1013 (1993)
3. M. Sznaier, Compressive information extraction: a dynamical systems approach, in *Proceedings of the 16th IFAC Symposium on Systems Identification (SYSID 2012), plenary talk* (2012), pp. 1559–1568
4. M. Sznaier, M. Ayazoglu, T. Inanc, Fast structured nuclear norm minimization with applications to set membership systems identification. IEEE Trans. Autom. Control **59**(10), 2837–2842 (2014)

5. V. Chandrasekaran, B. Recht, P.A. Parrilo, A.S. Willsky, The convex geometry of linear inverse problems. Found. Comput. Math. **12**(6), 805–849 (2012)
6. A. Tewari, P. Ravikumar, I.S. Dhillon, Greedy algorithms for structurally constrained high dimensional problems, in *Advances in Neural Information Processing Systems* vol. 24 (2011), pp. 882–890
7. B. Yilmaz, C. Lagoa, M. Sznaier, An efficient atomic norm minimization approach to identification of low order models, in *2013 IEEE CDC* (2013), pp. 5834 – 5839
8. P. Shah, B.N. Bhaskar, G. Tang, B. Recht, Linear system identification via atomic norm regularization, in *2012 IEEE 51st Annual Conference on Decision and Control (CDC)* (2012), pp. 6265–6270
9. B. Yilmaz, C. Lagoa, M. Sznaier, An efficient atomic norm minimization approach to identification of low order models, in *52nd IEEE CDC* (2013), pp. 5834–5839
10. B.K. Natarajan, Sparse approximate solutions to linear systems. SIAM J. Comput **24**(2), 227–234 (1995)
11. N. Ozay, M. Sznaier, C.M. Lagoa, O.I. Camps, A sparsification approach to set membership identification of switched affine systems. IEEE Trans. Autom. Control **57**(3), 634–648 (2012)
12. Y. Wang, C. Dicle, M. Sznaier, O. Camps, Self scaled regularized robust regression, in *2015 IEEE CVPR* (2015), pp. 3261–3269
13. Y. Cheng, Y. Wang, O. Camps, M. Sznaier, Subspace clustering with priors via sparse quadratically constrained quadratic programming, in *2016 CVPR* (to appear)
14. Y. Wang, O. Camps, M. Sznaier, A super-atomic norm minimization approach to identifying sparse dynamical graphical models, in *2016 American Control Conference* (to appear)
15. J. Chen, G. Gu, *Control Oriented System Identification, An $\mathcal{H}_\infty$ Approach* (Wiley, New York, 2000)
16. E.J. Candes, M. Wakin, S. Boyd, Enhancing sparsity by reweighted l1 minimization. J. Fourier Anal. Appl. **14**(5), 877–905 (2008)
17. M. Fazel, H. Hindi, S.P. Boyd, Log-det heuristic for matrix rank minimization with applications to hankel and euclidean distance matrices, in *Proceedings of American Control Conference 2003*, vol. 3 (AACC, 2003), pp. 2156–2162
18. Y. Cheng, Y. Wang, M. Sznaier, N. Ozay, C. Lagoa, A convex optimization approach to model (in)validation of switched arx systems with unknown switches, in *51 IEEE CDC* (2012), pp. 6284–6290
19. T. Inanc, M. Sznaier, P.A. Parrilo, R.S. Sanchez Pena, Robust identification with mixed time/frequency domain experiments and parametric/nonparametric models: theory and an application. IEEE Trans. Control Syst. Tech. **9**(4), 608–617 (2001)
20. M. Ayazoglu, B. Yilmaz, M. Sznaier, O. Camps, Finding causal interactions in video sequences, in *2013 IEEE International Conference on Computer Vision (ICCV)* (IEEE, 2013), pp. 3575–3582
21. M.S. Ryoo, J.K. Aggarwal, UT-Interaction Dataset, ICPR contest on Semantic Description of Human Activities (SDHA) (2010). http://cvrc.ece.utexas.edu/SDHA2010/Human_Interaction.html
22. N. Ozay, C. Lagoa, M. Sznaier, Set membership identification of switched linear systems with known number of subsystems. Automatica **51**, 180–191 (2015)
23. N. Ozay, M. Sznaier, C. Lagoa, Model (in)validation of switched ARX systems with unknown switches and its application to activity monitoring, in *Proceedings of 2010 IEEE Conference on Decision and Control (CDC)* (2010), pp. 7624–7630
24. F. Xiong, Y. Cheng, O. Camps, M. Sznaier, C. Lagoa, Hankel based maximum margin classifiers: a connection between machine learning and wiener systems identification, in *Proceedings of 2013 IEEE CDC* (2013)
25. J.B. Lasserre, Lasserre. Global optimization with polynomials and the problem of moments. SIAM J. Optim. **11**(3), 796–817 (2001)
26. J.B. Lasserre, Convergent sdp-relaxations in polynomial optimization with sparsity. SIAM J. Optim. **17**(3), 822–843 (2006)

Part III
Foundation of Human Movement

Inverse Optimal Control as a Tool to Understand Human Movement

Katja Mombaur and Debora Clever

Abstract In this paper, we discuss numerical foundations and computational results for inverse optimal control of human locomotion based on human motion capture data. The task of inverse optimal control is to identify the precise underlying objective function that is optimized in an observed motion. The presented methods can cope with partial and imprecise measurements of the state variables which is typically the case for motion capture recordings. We investigate human walking and running motions on different levels of detail and consequently different underlying models which all have their own motivation depending on the question asked. Whole-body models are used to explore the mechanisms of motions on joint level, while simple models describing the subject as a single entity can be used to describe overall locomotion behavior. At an intermediate level, template models describe some relative motions of bodies while maintaining simplicity and computational efficiency. Results will be presented for all model types and different walking tasks. We also show for some of them how the identified objective functions can be used to generate new waking motions for humanoid robots in novel scenarios.

1 Introduction

Movement is a central aspect of our life, since it represents an important way to interact with the world. The human body is a remarkably complex system capable of a wide range of movements such as walking and running on different terrains and at different speeds, grasping and manipulating complex objects or interacting with other humans. Gaining a fundamental understanding of the human body and its movements has long been an important research topic in biomechanics, sports science, physiology, neuroscience, computer animation and humanoid robotics. The human body is at the same time redundant and under-actuated. Redundancy in this

---

K. Mombaur (✉) · D. Clever
ORB, IWR, Heidelberg University, Berliner Str. 45, 69120 Heidelberg, Germany
e-mail: katja.mombaur@iwr.uni-heidelberg.de

D. Clever
e-mail: debora.clever@iwr.uni-heidelberg.de

J.-P. Laumond et al. (eds.), *Geometric and Numerical Foundations of Movements*, Springer Tracts in Advanced Robotics 117, DOI 10.1007/978-3-319-51547-2_8

context means that there usually is an infinite number of ways to perform a given motion task, e.g. walk from A to B, since many degrees of freedom of the human body are involved and different combinations of joint trajectories lead to the same goal. Underactuation means that only internal degrees of freedom - the joints - are actuated by muscles, but the overall position and orientation of the human body in space only results indirectly from the combination of the internal joints' actions and the interaction of the body with the environment (e.g. foot contact with the ground, touching a wall or a hand rail or holding an object).

How does the central nervous system control human movement? How does the motor system activate the more than 600 highly interdependent muscles to control the over 200 mechanical degrees of freedom of the body? How is the redundancy issue solved? Which is the way we choose to walk from A to B? How do we distinguish natural from unnatural movement?

It is a common assumption that motions of humans and animals are performed in an optimal way due to evolution, learning and training [4, 5]. Optimization effects can be found on the mechanical properties of the execute movements, but also in the closed loop sensory motor system [53]. As a logical consequence of this optimizing property of nature, from a mathematical perspective human movements can be formulated as optimal control problems. Researchers from different fields have used optimal control approaches to generate or synthesize realistic anthropomorphic movements, e.g. in human movement studies to generate optimal walking and running [1, 18, 22, 50], in computer animation to generate motions for human shapes and legged fantasy characters [21] and in robotics to compute motions for specific humanoid robots [7, 10, 31]. Here the optimization criterion has been predefined along with the underlying dynamic model, and the optimal motions are computed as the result of a (forward) optimal control problem, i.e. an optimization problem with a dynamic process model as a constraint.

In this paper, we are addressing the opposite or inverse problem which is called the inverse optimal control problem. Given a specific human movement for which motion capture data is available, and a defined model used for its description, which is the underlying objective function that gives rise to this movement? The are special types of human movement for which this question is easy to answer since the objective function corresponds to the voluntary goal of the human subjects. This is particularly true in sports, e.g. in sprinting (maximization of speed) or in long or hight jump (maximization of jump width or height). However in most situations this question is far from trivial since motion control is performed in a less conscious way. What is the optimization criterion of everyday motions such as walking, opening a door or shaking hands? It can be assumed that for most types of motion, not a single objective function is optimized but rather a combined criterion of several elementary functions.

Objective functions have been identified for some particular motions. Flash and Hogan study the effects of the minimization of jerk in reaching motions [20], and Park and Levine investigate different optimality criteria in running motions [46]. Berret et al. [8] have identified combined criteria in reaching movements. Mainprice et al. study the concepts of optimality in collaborative manipulation tasks [38]. However, there is no commonly identified criterion for human movement since everything depends on the specific motion task and situation, and partly also on the specific subject.

In this paper, we discuss the use of inverse optimal control for the example of human locomotion. We are interested in human locomotion from different perspectives:

- The whole-body perspective, which considers detailed rigid multi-body system models of the human body and investigates optimality of the motion on joint level;
- The perspective of the overall locomotion trajectory selection and its optimality criterion considering only the position and orientation of the subject as a whole in the plane of locomotion, i.e. only external degrees of freedom (DOF);
- The intermediate perspective which considers template models with some internal DOFs and partial dynamics of the system and studies optimality on this level.

We will give several examples of successful inverse optimal control computations in all cases and identify future research directions. In all cases, we used motion capture data as reference which typically only provides partial and imprecise information about the state and control variables of the system.

Inverse optimal control problems are challenging since they require the solution of an identification problem inside and optimal control problem. We discuss different methods for the numerical solution of this class of problems.

There is always a discussion to what extent optimality criteria of particular movement tasks are generalizable between different subjects and to what extent the criteria define the individual style of a subject. The approach discussed in this paper can be used to answer this question since inverse optimal control problems can be solved for different data sets - either using data for many subjects simultaneously to study average behavior or just using data for one subject to describe individual behavior and perform a comparison afterwards. In Sect. 3.1, we address this issue in more detail.

Once optimality criteria are identified for a particular class of motions, they are expected to provide useful insights into the nature of the motion itself. In addition, they are of big importance for the use in robotics and computer graphics. Using this same criterion and applying it to the corresponding dynamic model of a robot or an animated character it is possible to generate new motions (also for new situations) in the same style as the ones considered in the inverse optimal control problem. However, some special care has to be taken to properly handle kinematic and dynamic constraints of the robots which typically differ from those of humans. This problem is usually less relevant for computer graphics.

The paper is organized as follows. Section 2 gives an introduction to the mathematical formulation and numerical solution approaches for inverse optimal control problems. In Sect. 3, we give different examples of inverse optimal control applied to locomotion tasks seen from the different perspectives: whole body models describing level ground running motions with 25 DOF in 3D and walking motions with 16 DOF in 2D, overall walking trajectory selection tasks to a defined point for single entity models in the plane, as well 3D template walking models for different walking scenarios such as step stones, stairs and level ground. Section 4 presents options for the transfer of these results to humanoid robots and discusses the necessity for adapted transfer rules. In Sect. 5, we finally give some conclusions and perspectives on future research in this area.

2 The Inverse Optimal Control Problem

The practical relevance of the inverse optimal control problem has been discussed in the introduction. The purpose of this section is to show how the problem of solving a parameter estimation problem within an optimal control problem will be transferred into an appropriate mathematical formulation. We also discuss different numerical solution approaches.

2.1 Mathematical Problem Formulation

The mechanical model of the locomotion system in form of a system of differential equations is a first important part of the model formulation. We do not discuss this in detail in this paper due to reasons of space, but only give a verbal description of the model underlying the examples in Sect. 3. But it is crucial to choose a model that corresponds to the desired description accuracy since an approximation of a measured trajectory can only be as good as the underlying model permits. This accuracy concerns the choice of degrees of freedom of the model as well as the determination of model parameters correctly describing the recorded subject(s).

As a basis for the formulation of the unknown objective function inside the inverse optimal control problem, we assume that we are able to establish a set of reasonable independent base objective functions $\Psi_i(t)$. Typically there are some expert guesses in the literature in biomechanics or medicine for the different movement tasks, stating that some specific physical quantity e.g. mechanical energy, jerk, joint torques, variability of end position, stability etc. might be optimized. Other base functions can be generated by independent reflections about the movement possibilities and potential intentions of the subject. Such formulations can also take into account metabolic effects and other internal processes of the human body. From a mathematical perspective the use of mathematical base functions, such as Fourier series would also seem reasonable, however they are much less interesting from the human movement science or robotics perspective, since they do not have any physical meaning. The relative contributions of all base functions $\Psi_i(t)$ to the overall objective function can be expressed by the respective weight factors α_i, which are the unknown variables to be determined by the inverse optimal control problem. The choice of good basis functions is the second crucial step in the formulation of the inverse optimal control problem, since any optimal solution can only get as close to the measured trajectory

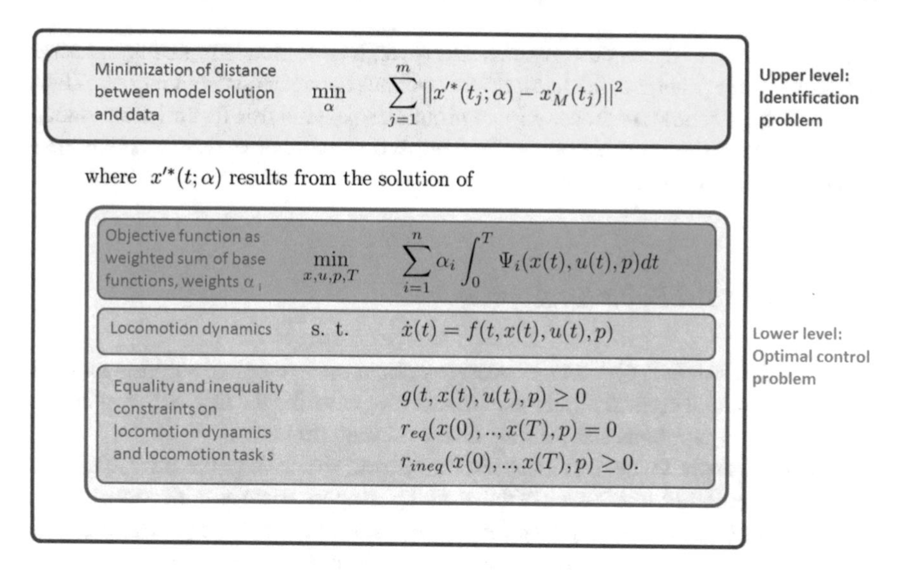

Fig. 1 Inverse optimal control. Bi-level problem with a parameter identification problem on the upper and an optimal control problem on the lower level (List of symbols: α weighting vector of objective, x'^* (subset of) optimal state, x'_M reference data, u control, p model parameters, T final process time, n number of basis criteria, Ψ_i basis criteria, g path constraints, r_{eq} point equality constraints, r_{ineq} point inequality constraints, m number of data points)

as possible within the range of its base functions. It should be taken care that they are as complete as possible but redundancy has to be avoided. This may not be trivial for complex problems since global as well as local redundancy may occur, and it should be carefully performed. No general rule can be given for the required number and type of base functions since the problem is highly nonlinear and the answer depends on the specific system and solutions.

The formulation of the inverse optimal control problem is shown in Fig. 1.

This problem is a bi-level optimization problem. This structure results naturally from the two problems that have to be solved simultaneously. In the upper level, we aim to minimize the distance between the measured motion and the computed one by optimizing over the vector of weight parameters α. In the lower level, we solve a (forward) optimal control problem for the current iterate of α in order to compute the solution x^* to evaluate the objective function of the higher level problem.

It should be noted that the distance between recorded and computed solution is usually not evaluated based on a full knowledge of the states. We therefore use the notation x' instead of x to show that either only a subset of the states is measured or the measurements concern variables which are not direct state variables but allow to draw conclusions about state variables.

Another thing that should be noted is that we have here - for clarity of presentation - only given the problem formulation for a single phase, i.e. for a problem which uses just one set of dynamic equations. However, locomotion is typically described as a multi-phase problem which implicitly switches between different models. Using appropriate numerical methods, it is no problem to extend this formulation to locomotion models with multiple phases, and this has actually been done to solve many of the examples discussed in Sect. 3.

2.2 *Numerical Solution Approaches*

In this section, we will give a brief overview about some possible solution methods for inverse optimal control problems. We will also describe the link between inverse optimal control problems and reinforcement learning problems.

A straightforward way is to maintain the bi-level structure of the inverse optimal control problem and address each of the levels by an appropriate optimization method. We have proposed a method for nonlinear inverse optimal control problems in [42] and recently implemented a new version of this method. All the results we are showing here have been computed with this kind of method.

In the lower level an efficient approach to solve optimal control problems is required. These methods have to handle the fact that state and control variables are functions in time, i.e. infinite dimensional variables. Direct - also called first-discretize-then-optimize-methods - use a discretization of the control variables. For the treatment of states, collocation or shooting methods can be applied. While collocation only considers states at discrete points, multiple shooting keeps only the states at so called multiple sooting points in the optimization problem, but the full dynamics of the system are simultaneously treated by an integrator communicating with the optimization problem. These discretizations and parameterizations together result in large and structured nonlinear optimization problem. We have chosen to use the direct multiple shooting optimal control code MUSCOD [9, 34] which is able to efficiently solve this task. For the identification in the upper level a solution approach based on non-differentiable optimization approaches seems to be promising since the function evaluation of this problem involves a whole solution of the lower level optimal control problem. This function can therefore not be expected to satisfy the usual smoothness assumptions. Derivative information of this function could only be generated in a black-box finite difference way. The derivative-free optimization codes COBYLA (Constrained Optimization BY Linear Approximation) and BOBYQA (Bound Optimization BY Quadratic Approximation) by Michael Powell [47, 48] are an excellent choice. They are both suitable to handle bounds on the free parameters, where COBYLA even allows for more general constraints on these parameters. We have chosen to use them (in the original form, and for the reimplementation in the version implemented in NLOPT) since they perform particularly well in this context. Other options we tested are e.g. Nelder-Mead simplex methods [44] which are however much slower.

In the mathematical community, there is a big interest to resolve the natural bi-level structure of inverse optimal control problems and to develop solution methods based on a reformulation of the original problem as one-level problem treating it as a so-called MPEC (mathematical program with equilibrium constraints). Here the lower level optimal control problem is replaced by the corresponding first order optimality conditions, also called Karush-Kuhn-Tucker (KKT) conditions which essentially state that the first order derivatives of the Lagrangian function (taking into account objective function and constraints) have to be zero. These are then formulated as constraints of the higher level parameter estimation problem [37]. A lot of work is performed on the theory of MPECs formulating appropriate optimality conditions and constraint qualifications (e.g. [16, 54]). If this approach is applied in the context of direct optimal control methods (or first-discretize-then-optimize-methods), the optimality conditions must be formulated for the discretized optimal control problem. One example for an applied method of this type is given in [3] which is based on a state discretization by collocation and a solution of the resulting nonlinear programming problem (NLP) by an interior point method, and which is applied to study human arm movement. In [24] the authors propose an alternative approach in which the KKT conditions are formulated for an optimal control problem discretized by a direct multiple shooting technique and which solves the NLP by sequential quadratic programming (SQP). The method is used to investigate walking motions of cerebral palsy patients [23] and has also been used by us to analyzes simple walking models [13].

In [36] pioneering work has been performed in the area of physics-based character animation addressing a problem similar to the optimal control problem stated here: instead of the objective function, which is assumed to be known, they identify unknown model parameters from motion capture data using a nonlinear inverse optimization technique.

Inverse optimal control is related to learning control [6]. Like inverse optimal control, learning systems assume the existence of both an optimization objective and an estimation of performance index [2]. While such a performance index is evaluated by direct numerical optimization in inverse optimal control, in learning control the index can be evaluated on the basis of physical experiences. In [35] and [17] optimization and policy learning is combined to find optimal policies from initial human demonstrations and practice to demonstrate the improved generalization ability compared to learning from human demonstration alone.

3 Identifying Optimality Criteria of Human Locomotion on Different Levels

In this section, we give an overview of several inverse optimal control computations we have performed so far in the context of human locomotion studies. As Fig. 2 shows, we have investigated whole-body models, template models, and overall locomotion trajectory tasks. As also shown, the time segments studied in the different cases vary - from the study of single periodic steps over sequences of steps to entire locomotion maneuvers.

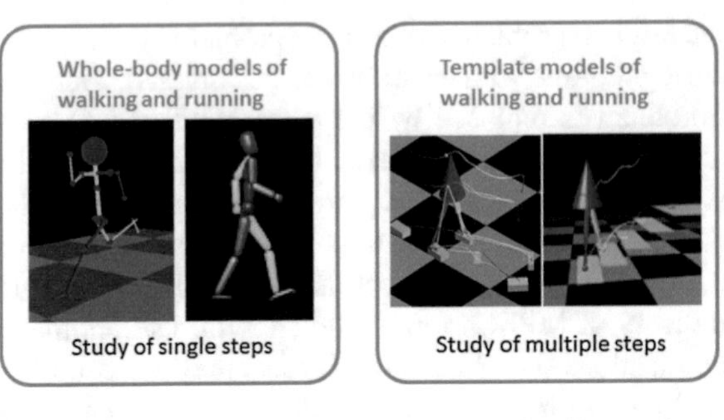

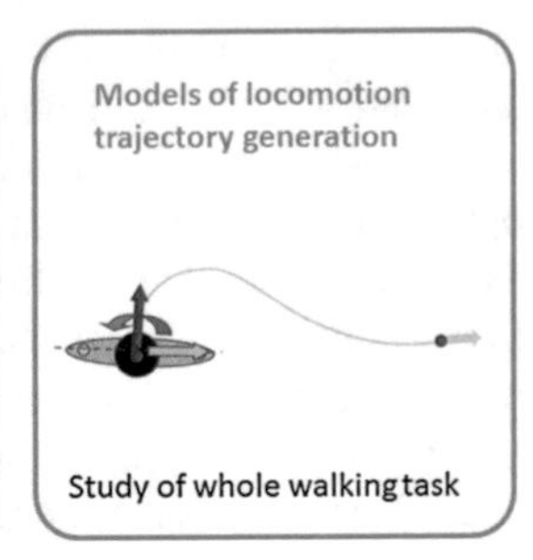

Fig. 2 Three different perspectives for the study of human locomotion: whole-body models, template models, and models of locomotion trajectory generation

3.1 Optimality Criteria for Whole-Body Motions on Joint Level

In this section, we briefly present results achieved for the study of motions by means of whole-body models. These contain the study of a 3D level ground running motions for one subject as well as a study of 2D level ground walking performed for several subjects. Both problems are multi-phase problems since the models go through different phases with different contacts with the environment.

Whole-body 3D running model

This section summarizes the inverse optimal control studies we have performed on whole-body models of running at a moderate pace. More detailed information about these results can be found in [41].

The human running model consists of 12 rigid bodies, namely thighs, shanks, feet, upper arms and lower arms, as well as two combined bodies for pelvis + lower trunk and upper trunk + head. The model has 25 degrees of freedom (DOF) in flight - 6 global DOF associated with the position and orientation of the pelvis and 19 internal DOF related to internal joint angles. The system is equipped with torque actuators at each of the 19 internal DOF replacing the action of the human muscles. For geometry and inertia parameters, anthropometric table data by de Leva [15] is used and adjusted to the subject's height and weight. The model describes human-like forefoot running, i.e. there is no flat foot ground contact but only point-like contact with the ball of foot. This assumption is very realistic for high speed sprinting which we studied with a similar model in [50], and for slower running speeds it still holds for the type of runners who tend to run on the forefoot. In the inverse optimal control case we have studied slower running motions at 10 km/h for which tread-mill data has been collected by our collaboration partners at the University of Rennes using a Vicon system at 100 Hz and 43 markers (see [41]). The fit between modeled and recorded motion here is achieved on the level of marker positions: marker locations on the model are computed as functions of the model variables and the distances to the

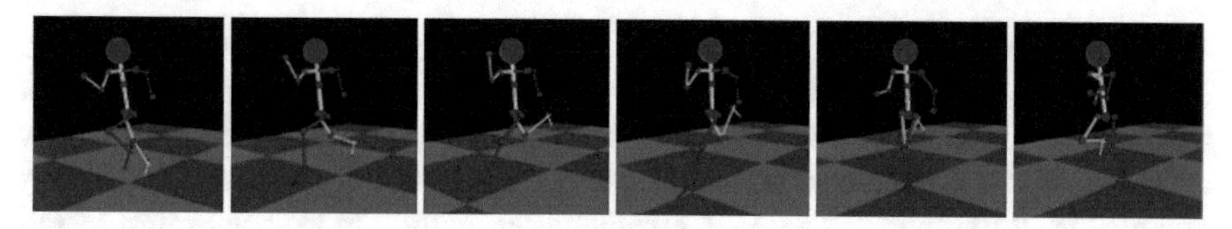

Fig. 3 3D human model performing the running motion under investigation

Table 1 Weight factors of base objective function contributions for 3D runner

Hip torques	Knee torques	Ankle torques	Trunk torques	Shoulder torques	Elbow torques	Head rotation	Step length
0.56	3.7	4.0	9.1	9.1	5.8	435	337
10.7				3.0		300	
0.0				0.0			

real marker locations are minimized in the upper level of the inverse optimal control problem.

Running motions consist of a sequence of alternating flight phases and single-leg contact phases. Since we assume that the running motion considered in this study is periodic and symmetric, i.e. right and left steps are assumed to be identical, we can cut down the study of running to the study of one step to which appropriate periodicity constraints (including the shift of sides) are formulated (Fig. 3).

As discussed above, an important prerequisite for any inverse optimal control computation is the establishment of a basis of potential objective functions which then form the base functions $\Psi_i(x(t), u(t), p)$ for the lower level problem. These functions are context specific and should reflect the current expert guesses on the problem under investigation. For running at controlled slow to medium speed (jogging), we postulate that runners typically are capable to set their speed to the desired value - which mathematically can be formulated as a constraint - and then adjust their running style which results from the solution of the optimization problem. For this mode, we formulate as potential base functions:

- minimization of all 19 joint torques (squared) - these receive individual weights for each internal DOF, but corresponding left and right torques have the same weights. This results in 10 different inverse optimal control parameters α_i.
- minimization of head motions, compare [49] (e.g. velocities squared) - resulting in two parameters α_i
- maximization of stride length - i.e. one more term.

Step time and phase times are not explicitly minimized or maximized, however it should be noted that they are of course free variables of the problem and are indirectly influenced by the objective functions listed above. The identified weights (using the inverse optimal control code with BOBYQA in the upper level) are shown in (Table 1).

A fundamental drawback of the model used in this study is the small number of DOF in the back and neck area. These missing mechanical DOF make it hard to follow the recorded trajectories precisely which also blurs the subsequent inverse optimal control analysis. It also makes the formulation of the head related criterion difficult since here the head is rigidly connected to the upper trunk. This drawback is addressed in our more recent whole-body models for walking presented in [18, 19] and also used in the next section. Corresponding running models with more DOF are currently being developed. The distance minimization on marker level makes it hard to analyze which DOF and which functions might contribute to improving the fit, and a distance minimization on the level of joint angles might facilitate the interpretation.

Whole-Body 2D Walking Model

To analyze walking motions taking into account changing contacts with the ground, we consider a planar whole-body human model that consists of 14 rigid segments connected by 13 rotational joints: a pelvis, two trunk segments, a head, two thighs, two shanks, two feet, two upper arms and two lower arms. Including a floating base with 3 DOFs this results in a full-body model with 16 planar DOF. We assume that motions of the human body are driven by torques in the joints that summarize the action of all related muscle forces. Contacts of the feet with the ground are described by additional constraints. Discontinuities due to impacts which are assumed to be fully inelastic are computed based on conservation of angular momentum equations. As before, we assume human walking to be periodic and symmetric and therefore we can model a sequence of several steps by just one step with proper periodic state constraints on the boundaries. Foot contact in walking is more complex than in running, since the possibilities of flat foot, heel only and toe only contact have to be considered, and as a consequence, each of these steps is divided in four continuous phases with varying contact sets and two discontinuities at heel and hallux touch down. All phase durations are left free and phase switching conditions are specified implicitly and therefore the identification procedure allows to identify optimality criteria which are not only related to motion in space but which are also related to timing issues.

We consider seven different optimality criteria as base functions. Four criteria are devoted to the minimization of squared joint torques in different parts of the body (legs, arms, hip, head+torso) one is meant for head stabilization, one is maximizing the step length, and the last one is maximizing the step frequency. All criteria are scaled such that they have approximately the same order of magnitude. As a pre-study, we have investigated the effect on the model of each of these functions separately. Not surprisingly, the motion maximizing the step frequency results in the fastest step and the motion maximizing the step length results in the largest step. More interesting is the observation that a minimization of squared joint torques in the legs results in a bend posture whereas a minimizing the squared torques in the arms results in an overstretched back. For a more detailed definition based on formulas and a visualization of the different elementary motions we refer to [14].

We focus on unconstrained and straight human walking on level ground (i.e. no obstacles, predefined footholds or curves), using motion capture data collected by our KoroiBot project partners from CIN, Tübingen, Germany (M. Giese and team) and publicly available in the KoroiBot motion capture data base, set up by our project partner KIT, Karlsruhe, Germany (T. Asfour and team) [32, 39]. The data have been collected with a Vicon system at a frame rate of 100 Hz. We consider six different

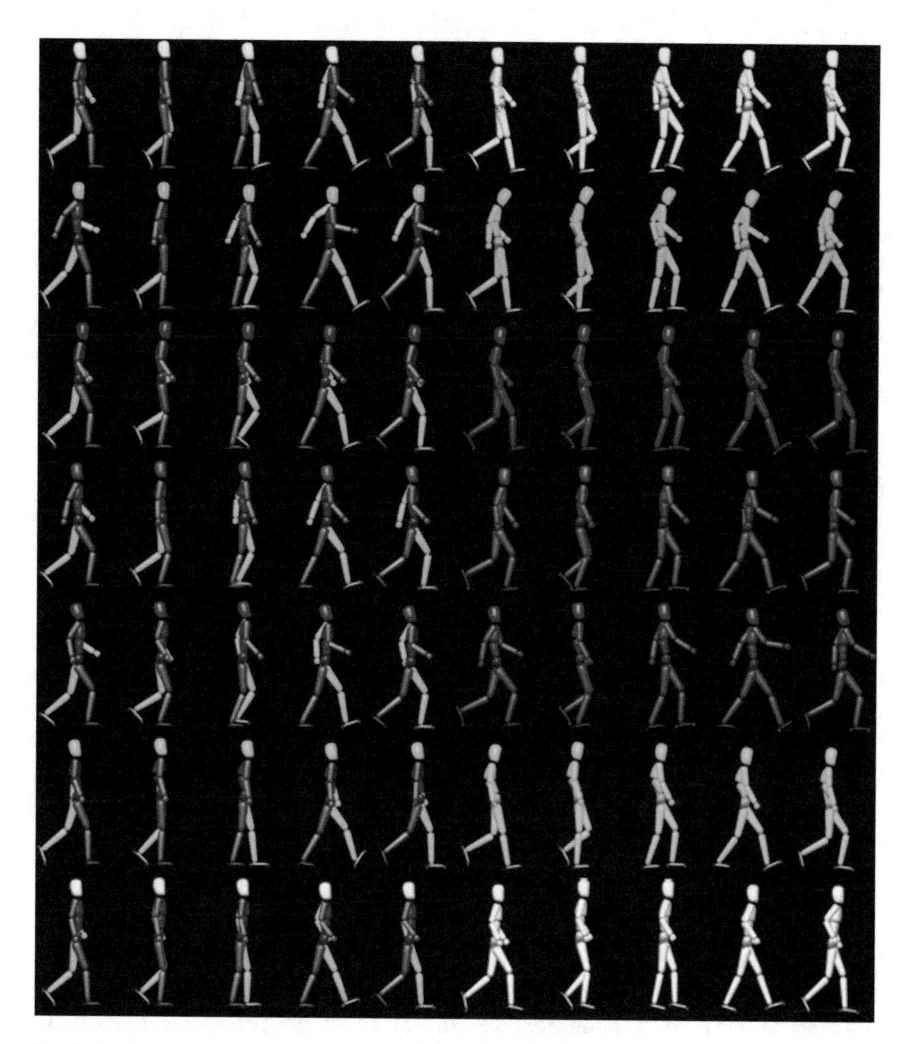

Fig. 4 Optimal motions for the identified objective function weights and the corresponding reference motions. Computed motions are shown by multi-colored walkers (*left*), their corresponding reference motion by the single-colored walkers (corresponding to the *head color*) on the *right*. Each motion is visualized as a sequence of five screen-shots, arranged next to each other. A corresponding video is available online: http://orb.iwr.uni-heidelberg.de/ftp/CleverEtAl_IOC_wholebody

Table 2 Weight factors of base objective function contributions for different subjects

Subject	Hip torques	Head and torso torques	Leg torques	Arm torques	Head stab.	Step length	Frequency
S1	1.61	1.53	1.52	0.30	0.71	1.08	0.25
S2	1.44	1.55	1.53	0.51	0.35	1.33	0.29
S3	0.73	0.41	1.75	0.66	0.59	2.64	0.23
S4	1.59	0.63	0.80	0.76	1.62	1.13	0.48
S5	0.87	1.66	1.17	0.95	0.73	0.88	0.74
S6a	1.54	0.77	1.18	1.44	0.79	1.12	0.16
S6b	1.90	1.51	1.47	0.51	0.38	0.86	0.38

subjects with significantly different total height. From each of the trials (one per subject, two for subject 6) we extract the first quasi periodic step on the right leg, see Fig. 4, right (single colored walkers). As distance minimization on the level of joint angles facilitates the interpretation of inverse optimal control results, 3D marker positions are transformed via the MMM framework [52] to quasi-measured positions and joint angles for the DOF of our planar walking model to which then a fit of the modeled motion is performed.

The identified objective weights are presented in Table 2 and the resulting motion in Fig. 4, left (multi-colored walkers). The head color of the multi-colored walkers (computed motions) coincides with the color of the single-colored walkers representing the corresponding reference motion. Computations here have been performed using COBYLA in the upper level. The fitting results are very good, except for subject 5, where the phase transition time of phase 1 is reproduced less accurate than for the others.

It is not surprising that for the different walking styles of the different subjects no unique combination of weights can be found. However, looking at the correlation matrix all entries show a correlation coefficient between 0.7 and 1.0. This means that even though we can not identify *the* objective function for all walking motions, we can observe that there exists a significant correlation between the individual ones.

3.2 *Optimality Criteria for Locomotion Trajectory Selection*

In this section we study the generation of the overall locomotion trajectory, i.e. how persons choose to move from starting point and orientation A to a given end point and orientation B. For this question, we can take a high level perspective at walking which results in a very simplified model representing the person only as one single body described by its global position and orientation in the walking plane. All details about internal joint movements are ignored for this purpose. The understanding of the natural human locomotion trajectory selection is important for many applications in computer animation, humanoid robotics, and human interaction behavior, see

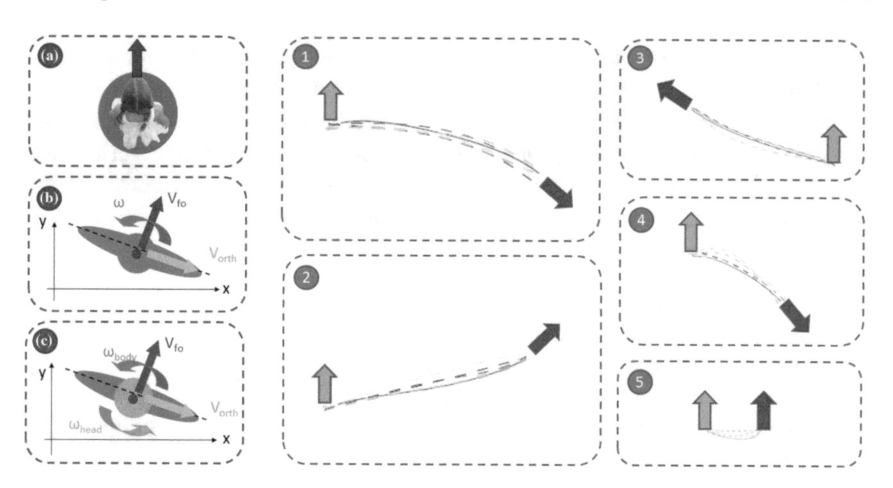

Fig. 5 The *left part* of the figures shows a person and her orientation in the walking plane (**a**), the locomotion model used for inverse optimal control consisting of one body only (**b**), and the extended model with a body and a separate head (**c**). The *right part* of the figure shows the five different locomotion tasks (1–5, *green*: start, *red*: target) investigated in inverse optimal control with *dashed lines* for the data of the 5 subjects and a *solid red line* for the modeled optimal trajectory

Table 3 Weight factors of base objective functions for locomotion trajectory generation

Time	Forward acceleration	Orthogonal acceleration	Rotational acceleration	Orientation to target
1.0	1.2	1.7	0.7	5.2

e.g. [25, 33, 45]. Also ignoring the joint level actions there still is redundancy in locomotion since there is an infinite number of ways to go from A to B.

Human and human-like locomotion is omnidirectional in a partial way - which means that in contrast to wheeled systems, humans are in principle capable to move in all different directions in the walking plane, however they have a clear preference for walking in forward direction due to perception and lower limb anatomy. These capabilities and preferences have a direct influence on the locomotion trajectory selection. It can be observed that for longer trajectories humans always tend to move in forward direction while for shorter trajectories also sidewards, backward or oblique steps are taken. Sometimes humans turn on the spot, but it is more likely that they do this at the beginning than at the end or in the middle of a motion. There also seems to be a clear preference to look towards the goal early on. Can this behavior also be described as an optimization problem in the same sense as walking generation on joint level?

In [42] we have investigated the optimality criteria of the overall locomotion trajectory generation by inverse optimal control. The state variables used for this model are its position and orientation in the walking plane, as well as the corresponding velocities and rotational velocities, respectively (see Fig. 5 (left)). The model in this case does only describe the kinematic relationships in the model and no dynamics. There are also no directly produced forces and torques related to the state variables in this model since they represent exactly the unactuated DOF of the human body, and forces and torques only appear indirectly by the interaction of the external limbs with the environment. The control variables used in this model are the corresponding accelerations in forward, sidewards and rotational direction. We have postulated the following base functions for inverse optimal control:

- minimization of total time from A to B
- minimization of forward acceleration squared
- minimization of sidewards (orthogonal) acceleration squared
- minimization of rotational acceleration squared
- minimization of difference between current orientation and direction to goal squared.

We have formulated the inverse optimal control problem simultaneously for five different subjects all performing motions to five different target positions and orientations (see Fig. 5 (right)) and used the bilevel formulation with BOBYQA as solver in the upper level. We have identified weights for the base functions above, which allowed to explain all 25 motions at the same time (See Table 3). The last term in the objective function describes an origin-target asymmetry while all others represent symmetric properties. With symmetry we mean in this case that the direction of motion on a given path does not play a role: the terms related to timing and accelerations would produce the same path for walking from A to B as for walking from B to A, just the timing would be exactly inversed. For the last term however, the walking direction on the path is a crucial factor, and it therefore plays an important role to explain why people are not walking back on their paths. Intuition as well as the resulting clearly non-zero weight tell us that this asymmetry is important. The dependency of this asymmetry on different parameters was investigated in more detail in [51].The dominant reason for this asymmetry is obviously perception. In fact, we do not want to turn our body towards the target but we want to *see* the target which then induces the orientation of the body, closely following the orientation of the head. In order to investigate this connection between head and trunk orientation further, we have developed an extension of the above model with a second segment for the head sitting on top of the body segment. i.e. they share the same position, but have different orientations [26]. While the orientation of the head is assumed to be controlled, the orientation of the trunk follows closely via a passive spring-damper-like element. We have identified parameters of the model based on experiments in [26], but the solution of the inverse optimal control problem for this extended model is subject of current research.

3.3 Optimality Criteria for Walking Motions of Template Models

Besides the two previously presented levels of modeling locomotion (whole-body and locomotion trajectory) so called template models are an interesting alternative. Template models describe an abstract version of the embodiment under investigation and define motion on the basis of some relevant quantities, e.g. the center of mass, the foot placement, the phase timing. This offers the possibility to study both, motion and locomotion trajectories over several steps on the one hand and internal body characteristics such as swing foot trajectories, torso orientation and phase timing on the other hand. Furthermore, due to their abstract architecture, template models are suitable to describe motion for very different embodiments and are therefore a good candidate for transfer models between humans and humanoids.

In this section, we consider the 3D dynamical template model that we have introduced in [11, 12] to describe and analyze human locomotion. The model consists of two legs with prismatic joints combined with damped series elastic actuators (SEA) instead of knees, two point-masses as feet and a reactive mass as upper body. Motion is described on the basis of center of mass trajectory, foot trajectories and upper body orientation. We consider a particularly interesting constrained environment, namely a step stone scenario, which is also very difficult, because it combines several steps of different step length, direction and duration for which common optimality criteria have to be found.

In Sect. 4 we then use the same template model with robot specific parameters to generate similar walking motions for different robots in a new environment.

The objective function to be determined is parameterized by twelve basis function, that depend on the states, the controls, and one optimality parameter which defines the optimal ratio of double and single support:

- minimization of the SEA actuation in the stance foot,
- minimization of the SEA actuation in the swing foot,
- minimization of hip torque of the swing foot,
- minimization of angular momentum in x-direction,
- minimization of angular momentum in y-direction,
- minimization of vertical center of mass oscillations,
- minimization of absolute swing foot velocity,
- minimization of the planar distance between the foot position at touch down and the capture point,
- minimization of the periodicity gap in center of mass velocities,
- minimization/maximization of overall single support duration,
- minimization of absolute swing foot velocity at touch down,
- tracking of ratio between sub-sequent double and single support phase to constant but unknown parameter.

Table 4 Weight factors of base objective function contributions for 3D template model

act. stance	act. swing	hip tor.	AM x	AM y	CoM osc.	foot vel.	CP	peri	ss dur.	vel. TD	ratio_α	ratio_p
1.0	1.3	0.8	0.92	0.93	0.91	0.91	1.35	1.28	0.51	0.98	0.1	1.61

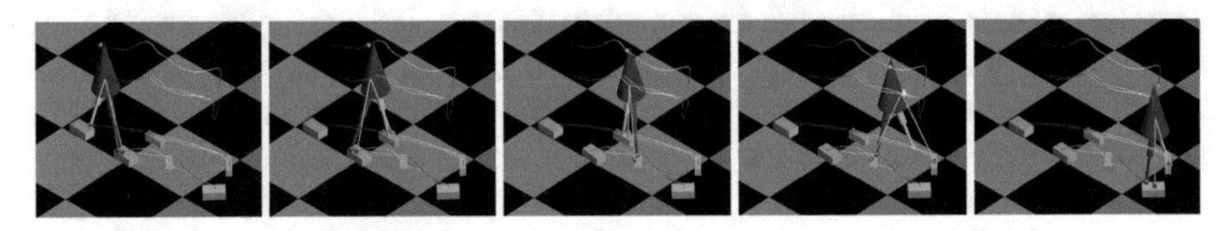

Fig. 6 Identification of optimality criteria for a 3D template model. The identified motion of interest is shown by the 3D template model. The reference motion (extracted from motion capture data) is sketched by *light colored lines* as legs and a small ball as head. The head trajectory of the motion (which is not an explicit state) is only plotted for the reference solution and is meant to simplify the comparison of the torso orientation. A corresponding video is available online: http://orb.iwr.uni-heidelberg.de/ftp/CleverMombaur_IOC_RSS2016

All criteria are scaled such that they have approximately the same order of magnitude. For a more detailed definition including the mathematical formulas we refer the reader to [12].

For the identification of optimality criteria, or to be more precise, of the optimal weights and optimality criteria related parameters, we rely on a least square fit of the model trajectories of the center of mass (CoM), feet, torso orientation, and phase durations to the corresponding reference trajectories and reference times.

As in the previous section, we use motion capture and subject specific data, recorded in Tübingen and published in the KoroiBot motion database, but here of course data for step stone scenarios is used. The reference motion starts, when the rear leg is about to lift off from the first step stone while the front leg is already on the second tread. The motion ends after four full steps, when the front leg is on the last step stone and the rear leg is about to take off again.

The weights and parameters optimized by inverse optimal control (based on BOBYQA), are given in Table 4.

With the identified weights, the average deviation between optimized and reference motion is reduced to quantities in the order of 10^{-2} which means that phase durations, feet trajectories, and x-y coordinates of the center of mass are reconstructed to a satisfying degree. The torso orientation and the vertical CoM velocities have the greatest relative difference. However, this is due to their small scaling and can be changed by adjusting the corresponding weight vector.

The optimal motion for the identified objective function, together with indicators of the reference motion is visualized as motion sequence in Fig. 6.

Fig. 7 Model flexibility. Due to its reduced and abstract architecture the 3D template model is suitable to explain human and humanoid walking over several steps in very different scenarios

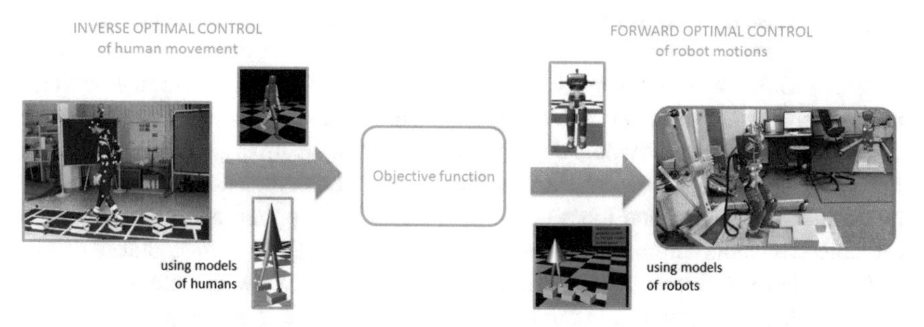

Fig. 8 From inverse optimal control to optimal control: The aim is to transfer optimality principles identified for human locomotion to humanoid robots in order to generate robot locomotion. This takes the different dynamic properties of humans and robots into account since objective functions are applied to different models

Due to its abstract and reduced architecture this 3D template model is suitable to explain human and humanoid walking over several steps in very different scenarios, as walking on level ground, on slopes, up and down stairs or even around curves, see Fig. 7.

4 Use of Inverse Optimal Control Results for Robots

One important application area for the results of inverse optimal - besides gaining a deeper understanding of human movement - is robotics. The identified optimality criteria for human motions can be used for a generation of new walking motions for a robot, as shown in Fig. 8. The principal idea is to apply these criteria to a model of the robot at the same level as the model used for identification, and to formulate appropriate boundary conditions to describe the specific walking task. This approach is particularly advantageous over a direct transfer of trajectories since it can handle differences in embodiments and environments in a natural way. Furthermore, the approach of identifying weights for physically meaningful basis optimality criteria even allows for an intuitive handling of the differences with respect to kinematic and dynamic bounds even though some care has to be taken to address them properly.

A first attempt to transfer optimality criteria for motions from humans to robots has been made for the optimization of locomotion trajectories discussed in Sect. 3.2, see also [42]. The optimal solutions generated for the human model never bring the human to any of its position, velocity or acceleration constraints in forward, orthogonal or rotational limits, since we are only considering regular walking motions, so we are observing an unconstrained optimum. However, if the same optimality criteria are applied to the corresponding model of a humanoid robot (in this case HRP-2), limits - as for example in the velocities - are very quickly reached. This has complex consequences: if the limit e.g. of the forward velocity is reached, this does not only mean that the forward velocity can not be further increased, but also that the same criterion which in the quasi unconstrained case produced a motion essentially moving in forward direction may now also produce the emergence of an orthogonal component and the robot would end up moving in oblique direction where the human moved forward. This of course changes the nature of the motion completely and must be avoided by appropriate measures. The general idea applied here was to modify the objective function parameters α_i such that the resulting solution automatically stays inside the velocity bounds without reaching them. We were able to achieve a similar balance between time minimization and the fast angle change term on one side and the terms minimizing acceleration of the other side by applying a simple scaling rule: we estimated that humans are about four times as fast as the HRP-2 robot, and therefore scale the weight parameters $\alpha_1 - \alpha_3$ for the robot by this factor. The optimal control problem with the modified parameters has then been solved, which lead to quite human-like locomotion trajectories for HRP-2, as shown in [42].

As mentioned before, due to their abstract architecture, template models are a suitable candidate to be used for a transfer between humans and humanoids. To run a computed robot motion on the real platform, the computed trajectories serve as control variables of the robot specific whole-body motion generation tools. In the following, we present an example for such a procedure based on the analyzed step stone scenario (see Sect. 3) and two very different humanoid robots. The first robot, the iCub platform of Heidelberg University (HeiCub), has 21 degrees of freedom and consists of an upper body, a hip and two legs [28, 40]. It does neither have arms nor a head and is therefore quite similar to the considered template model. It has a total height of 0.97 m and a total mass of 26 kg. The second robot, which is the HRP2-2 robot mentioned above, is a full humanoid with 36 degrees of freedom. It has a total height of 1.54 m and a total mass of 58 kg [29, 30]. We set up two different step stone scenarios, one for each robot. On the level of template models, differences between the three embodiments (human model, HeiCub, HRP-2) and the three step stone scenarios can be directly taken into account by adjusting the model parameters, model constraints and environment constraints.

Performing motion transfer based on optimality, it is desirable to transfer the identified weights of human optimality criteria as directly as possible to the humanoid objective function. However, as robots have much harder kinematic and dynamic constraints than humans (with respect to joint velocities and accelerations, joint torque etc.), a direct transfer of objective weights usually drives the robot model to some of its bounds. Those active constraints can change the nature of the motion significantly, as we have already seen above in the case of locomotion trajectory optimization. Therefore, it is necessary to identify transfer rules in terms of weight re-scaling which prevent the model to be stuck on a certain bound which was not active in the human model but lead to some balancing with respect to the constraints.

For the first robot model (HeiCub) the only modification we introduce is a division of the time minimization weight by two, to take into account the ratio describing how much slower the robot moves than the human. In addition to a decrease of the weight for time minimization, for the second robot model (HRP-2) we exploit the fact that in previous experiments (using a standard pattern generator) the robot HRP-2 has successfully crossed the considered step stone scenario with a ratio of double to single support of $1/7$ and include this knowledge in the corresponding parameter. In scenario 1 the topology of the step stones is quite similar to the human example, except for the smaller distances and the fact, that the robot has to start the motion with the other leg. Due to the characteristics of the template model this issue does not require any additional modifications, neither in the dynamics of the model, nor in the constraints, nor in the objective. In the second example, HRP-2 is asked to manage steps with a step height of 5 cm which is twice as much as the highest step in the training step stone scenario. This environmental difference is taken into account by dividing the weight for swing foot SEA actuation by two such that it allows for appropriate leg shortening which is necessary to step on a 5 cm higher step stone.

Considering those transfer rules, we now use the same optimal control model which we have used in the lower level of the inverse optimal control approach, but adjust model parameters and constraints to the robot properties, for more details see [12]. The results of these computations are optimal trajectories for the center of mass, the feet, the torso angles and the phase durations for the two robot models, see Fig. 9. Note, that the computed motions are not meant to be run open loop on the robot. Rather they substitute existing methods (e.g. based on the table cart model) to generate the input for the robot specific whole body motion generation tools.

The situation gets even more challenging for whole-body models. The optimality criteria identified in Sect. 3.1 could now easily be used to generate motions in new situations for models of humans, also with different parameter distributions. This might be very interesting for the fields of computer animation and biomechanical movement studies. However it gets more difficult if robots are concerned, since humans and robots have not only very different dynamic characteristics in terms

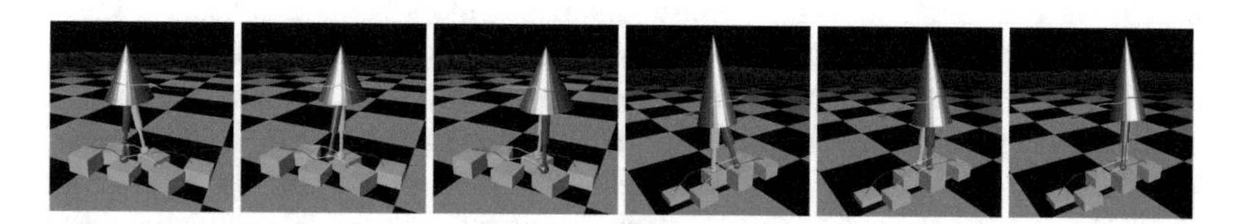

Fig. 9 Robot template motions as result of an optimal control problem in terms of CoM and feet trajectories, torso orientation and phase durations. Objective weights, model parameters and bounds are adjusted to the constraints of the two robots HeiCub (*red*) and HRP-2 (*gray*). A corresponding video is available online: http://orb.iwr.uni-heidelberg.de/ftp/CleverMombaur_IOC_RSS2016

of geometry and mass distribution (which would be taken care of by the different dynamic models used), but also very different kinematic and dynamic constraints. These are not only caused by hardware limitations, e.g. motors which are weaker than muscles and therefore produce smaller torques or smaller joint range or velocity limits, but also by the control software, which e.g. heavily constrains the ZMP. The effect of this can be clearly seen when comparing results of minimizing joint torques for a human model [18] and for an HRP-2 model [31]. The behaviors are fundamentally different especially if the ZMP bounds for the robot are introduced. It is of course not a very satisfying approach to put a lot of effort into the solution of the inverse optimal control problem to identify the objective function and then use some heuristics in order to adjust the function for the new system. We therefore have started a detailed study on defining the similarity of motions for models with different dynamic and kinematic constraints. The study consists in an online survey on the perceived similarity of motions which can be accessed on [43] and the following analysis. We expect to be able to derive general rules for similarity which will then serve as a basis for the formulation of transfer rules between walking trajectories on joint level, and also between optimality criteria.

5 Conclusions and Outlook

In this paper, we have discussed inverse optimal control as a tool to identify optimality criteria of human movement, based on the fundamental hypothesis that everyday and well trained human movement is optimal in some sense. The mathematical and numerical foundations of the algorithms have been presented. We have also given an overview of previous inverse optimal control studies of human walking and running motions. Human locomotion is a very diverse movement task and can take place in many different terrains and situations, as studied extensively in the European project KoroiBot.

The presented inverse optimal control studies help to shed light on human locomotion from different perspectives. Whole-body dynamic models of walking and running allow to study optimality of motions on joint level by considering the complex nonlinear kinematic and dynamic interactions of the relevant segments of the human body, typically looking at single steps in the cyclic or close to cyclic movements. On the other end of the spectrum, single entity kinematic models are used to study the overall locomotion trajectory selection considering the entire walking maneuver from origin to target. In between those two, template models provide a simple description of walking behavior while allowing to perform limited studies on the kinematics and dynamic interaction of segments and can be applied to single steps, but also to sequences of many steps. Along with the model descriptions also the possibilities to formulate objective functions change. Corresponding base objective functions have been formulated in each case, and the inverse optimal control task generally consisted in identifying the correct weights, i.e. the contribution of each base function to the overall objective underlying a measurement. For all three perspectives we have presented results giving insights into the optimality properties of the movement. We also have treated different walking scenarios such as walking and running on flat ground, on step stones and stairs, in straight lines an on arbitrary locomotion trajectories.

Besides contributing to the knowledge of human walking, such identified optimality criteria can be used to generate human walking movements in computer animations or in biomechanical or rehabilitation studies, but they are also very interesting for walking generation for humanoid robots. We have discussed some of the research performed to transfer identified optimality criteria for humans to humanoid robots based on the different modeling levels, including the challenges of adjusting the weights in a systematic way such that differences in kinematic and dynamic inequality constraints between humans and robots are considered.

There are still many open questions and ongoing research in the context of inverse optimal control of human movement and its transfer to robots.

We are currently working on systematic studies of walking motions based on 2D whole body models and their 3D extensions (see [19]). The goal is to have a full comparison of optimality criteria for walking, not only for different subjects, as shown in Sect. 3.1, but also for different walking scenarios, also including stairs, slopes and step stones in addition to level ground. We also plan to include our walking models with compliance modulation in the joints [27] in the inverse optimal control studies. Another possible path to pursue is the inclusion of muscle models in the whole-body models for inverse optimal control which however would represent a significant boost of complexity. Also the walking investigations based on template models will be continued for all kinds of terrains, since in particular the type of template model used here presents a very good compromise between simplicity and consideration of dynamic intersegmental effects. We plan to include further objective function terms related to efficiency, stability and timing in the future.

The most pressing issue, however, is the development of appropriate transfer rules for trajectories and optimality criteria from humans to humanoids taking into account the different kinematic and dynamic limits (i.e. inequality constraints). This question can not be answered by the research on retargeting performed in computer animation, since it does not address the feasibility issue, but in general only the difference in kinematic and sometimes dynamic segment parameters between different embodiments (forming part of the equality constraints). As mentioned in the previous section, we have developed a similarity survey tool [43], with which we are currently evaluating the dominant factors for perceived similarity between motions of models with different kinematic and dynamic limitations. This will serve as a basis to define appropriate norms that allow to define closeness of two motions - which can not be identical due to different constraints - in the sense of this perceived similarity measure.

Acknowledgements The research leading to these results has received funding from the EU seventh Framework Program (FP7/2007-2013) under grant agreement no 611909 (KoroiBot), the German Excellence Initiative and the French ANR project Locanthrope. We thank the Simulation and Optimization group of H.G. Bock at Heidelberg University for providing the optimal control code Muscod-II.

References

1. M. Ackermann, A.J. van den Bogert, Optimality principles for model-based prediction of human gait. J. Biomech. **43**(6), 1055–1060 (2010)
2. N. Aghasadeghi, T. Bretl, Maximum entropy inverse reinforcement learning in continuous state spaces with path integrals, in *Proceedings of IEEE/RSJ IROS* (2011)
3. S. Albrecht, C. Passenberg, M. Sobotka, A. Peer, M. Buss, M. Ulbrich, Optimization criteria for human trajectory formation in dynamic virtual environments. in *Haptics: Generating and Perceiving Tangible Sensations, LNCS* (2010)
4. R.M. Alexander, The gaits of bipedal and quadrupedal animals. Intern. J. Robot. Res. **3**(2), 49–59 (1984)
5. R.M. Alexander, *Optima for Animals* (Princeton University Press, New Jersey, 1996)
6. C.G. Atkeson, S. Schaal, Learning control in robotics. IEEE Robot. Autom. Mag. **17**, 20–29 (2010)
7. C.G. Atkeson, C. Liu, Trajectory-based dynamic programming, in *Modeling, Simulation and Optimization of Bipedal Walking Cognitive Systems Monographs*, vol 18 (Springer, Berlin Heidelberg, 2013), pp. 1–15
8. B. Berret, E. Chiovetto, F. Nori, T. Pozzo, Evidence for composite cost functions in arm movement planning: an inverse optimal control approach. PLoS Comput. Biol. **7**(10) (2011)
9. H.G. Bock, K.-J. Plitt, A multiple shooting algorithm for direct solution of optimal control problems, in *Proceedings of the 9th IFAC World Congress, Budapest*, (International Federation of Automatic Control, 1984), pp. 242–247
10. T. Buschmann, S. Lohmeier, M. Bachmayer, H. Ulbrich, F. Pfeiffer, A collocation method for real-time walking pattern generator, in *Proceedings of the IEEE-RAS International Conference on Humanoid Robots* (2007)
11. D. Clever, K. Mombaur, A new template model for optimization studies of human walking on different terrains, in *2014 14th IEEE-RAS International Conference on Humanoid Robots (Humanoids)*, (IEEE, 2014), pp. 500–505
12. D. Clever, K. Mombaur, An inverse optimal control approach for the transfer of human walking motions in constrained environment to humanoid robots, in *Robotics: Science and Systems (RSS)* (2016)
13. D. Clever, K. Mombaur, On the relevance of common humanoid gait generation strategies in human locomotion - an inverse optimal control approach, in *Modeling, Simulation and Optimization of Complex Processes - HPSC 2015*, ed. by X.P. Hoang, R. Rannacher, J. Schlöder, H.G. Bock (Springer, Heidelberg, 2016) (to appear)
14. D. Clever, R.M. Schemschat, M.L. Felis, K. Mombaur, Inverse optimal control based identification of optimality criteria in whole-body human walking on level ground, in *Proceedings of International Conference on Biomedical Robotics and Biomechatronics (BioRob2016)* (2016)
15. P. De Leva, Adjustments to Zatsiorsky-Seluyanov's segment inertia parameters. J. biomech. **29**(9), 1223–1230 (1996)
16. S. Dempe, N. Gadhi, Necessary optimality conditions for bilevel set optimization problems. Glob. Optim. **39**(4), 529–542 (2007)
17. A. Dörr, N. Ratliff, J. Bohg, M. Toussaint, S. Schaal, Direct loss minimization inverse optimal control, in *Proceedings of Robotics Sciece and Systems (RSS)* (2015)
18. M.L. Felis, K. Mombaur, Synthesis of full-body 3-D human gait using optimal control methods, in *IEEE International Conference on Robotics and Automation (ICRA 2016)* (2016)
19. M.L. Felis, K. Mombaur, A. Berthoz, An optimal control approach to reconstruct human gait dynamics from kinematic data, in *IEEE/RAS International Conference on Humanoid Robots (Humanoids 2015)* (2015), pp. 1044–1051
20. T. Flash, N. Hogan, The coordination of arm movements: an experimentally confirmed mathematical model. J. Neurosci. **5**, 1688–1703 (1984)
21. T. Geijtenbeek, M. van de Panne, A.F. van der Stappen, Flexible muscle-based locomotion for bipedal creatures. ACM Trans. Graph. **32**(6) (2013)

22. H. Geyer, H. Herr, A muscle-reflex model that encodes principles of legged mechanics produces human walking dynamics and muscle activities. IEEE Trans. Neural Syst. Rehabil. Eng. **18**(3), 263–273 (2010)
23. K. Hatz, Efficient Numerical Methods for Hierarchical Dynamic Optimization with Application to Cerebral Palsy Gait Modeling. Ph.D. thesis, University of Heidelberg (2014)
24. K. Hatz, J.P. Schlöder, H.G. Bock, Estimating parameters in optimal control problems. SIAM J. Sci. Comput. **34**(3), 1707–1728 (2012)
25. H. Hicheur, Q.-C. Pham, G. Arechavaleta, J.-P. Laumond, A. Berthoz, The formation of trajectories during goal-oriented locomotion in humans I: a stereotyped behaviour. Eur. J. Neurosci. **27**(8), 2376–2390 (2007)
26. M. Horn, M. Sreenivasa, K. Mombaur, Optimization model of the predictive head orientation for humanoid robots, in *IEEE/RAS International Conference on Humanoid Robots (Humanoids 2014)* (2014)
27. Y. Hu, K. Mombaur, Analysis of human leg joints compliance in different walking scenarios with an optimal control approach, in *IFAC International Workshop on Periodic Control Systems (PSYCO 2016)* (2016)
28. Y. Hu, K. Mombaur, F. Nori, Using optimal control to generate squat motions for the humanoid robot iCub with SEA, in *Proceedings of Dynamic Walking* (2015)
29. S. Kajita, T. Nagasaki, K. Kaneko, K. Yokoi, K. Tanie, A running controller of humanoid biped HRP-2LR, in *ICRA* (2005)
30. K. Kaneko, F. Kanehiro, S. Kajita, K. Yokoyama, K. Akachi, T. Kawasaki, S. Ota, T. Isozumi, Design of prototype humanoid robotics platform for HRP, in *2002 IEEE/RSJ International Conference on Intelligent Robots and Systems*, vol. 3 (IEEE, 2002), pp. 2431–2436
31. K.H. Koch, K. Mombaur, P. Souères, Studying the effect of different optimization criteria on humanoid walking motions, in *Simulation, Modeling, and Programming for Autonomous Robots*, Lecture Notes in Computer Science, ed. by I. Noda, N. Ando, D. Brugali, J.J. Kuffner, vol. 7628 (Springer, Berlin Heidelberg, 2012), pp. 221–236
32. KoroiBot Motion Capture Database. https://koroibot-motion-database.humanoids.kit.edu/ (2016) Last visited, May 2016
33. J.P. Laumond, G. Arechavaleta, T.-V.-A. Truong, H. Hicheur, Q.-C. Pham, A. Berthoz, The words of the human locomotion, in *Proceedings of 13th International Symposium on Robotics Research (ISRR-2007)* (Springer Star Series, 2007)
34. D.B. Leineweber, I. Bauer, H.G. Bock, J.P. Schlöder, An efficient multiple shooting based reduced SQP strategy for large-scale dynamic process optimization - Part I: theoretical aspects (2003), pp. 157 – 166
35. S. Levine, V. Koltun, Guided policy search, in *ICML* (2013)
36. C.K. Liu, A. Hertzmann, Z. Popovic, Learning physics-based motion style with inverse optimization. ACM Trans. Graph. (SIGGRAPH 2005) **24**(3), 1071 (2005)
37. Z.-Q. Luo, J.-S. Pang, D. Ralph, *Mathematical Programs with Equilibrium Constraints* (Cambridge University Press, Cambridge, 1996)
38. J. Mainprice, R. Hayne, D. Berenson, Predicting human reaching motion in collaborative tasks using inverse optimal control and iterative re-planning, in *2015 IEEE International Conference on Robotics and Automation (ICRA)*, (IEEE, 2015), pp. 885–892
39. C. Mandery, Ö. Terlemez, M. Do, N. Vahrenkamp, T. Asfour, The KIT whole-body human motion database, in *IEEE International Conference on Advanced Robotics (ICAR 2015)* (2015), pp. 329–336
40. G. Metta, L. Natale, F. Nori, G. Sandini, D. Vernon, L. Fadiga, C. Von Hofsten, K. Rosander, M. Lopes, J. Santos-Victor et al., The iCub humanoid robot: an open-systems platform for research in cognitive development. Neural Netw. **23**(8), 1125–1134 (2010)
41. K. Mombaur, A.H. Olivier, A. Crétual, Forward and inverse optimal control of bipedal running, in *Modeling, Simulation and Optimization of Bipedal Walking, Cognitive Systems Monographs*, vol. 18 (Springer, Berlin Heidelberg, 2013), pp. 165–179
42. K. Mombaur, A. Truong, J.-P. Laumond, From human to humanoid locomotion an inverse optimal control approach. Auton. Robots **28**(3), 369–383 (2010)

43. Motion Similarity Study. https://orb.iwr.uni-heidelberg.de/ratingapp/similarity/ (2016) Last visited, May 2016
44. J.A. Nelder, R. Mead, A simplex method for function minimization. Comput. J. **7**, 308–313 (1965)
45. J. Ondřej, J. Pettré, A.-H. Olivier, S. Donikian, A synthetic-vision based steering approach for crowd simulation. ACM Trans. Graph. **29**(4), 123:1–123:9 (2010)
46. T. Park, S. Levine, Inverse optimal control for humanoid locomotion, in *Robotics Science and Systems-Workshop on Inverse Optimal Control and Robotic Learning from Demonstration* (2013)
47. M.J.D. Powell, A direct search optimization method that models the objective and constraint functions by linear interpolation, in *Advances in Optimization and Numerical Analysis* (Springer, Heidelberg, 1994), pp. 51–67
48. M.J.D. Powell, The BOBYQA algorithm for bound constrained optimization without derivatives. Report No. DAMTP 2009/NA06, Centre for Mathematical Sciences, University of Cambridge, UK, 2009
49. T. Pozzo, A. Berthoz, L. Lefort, Head stabilization during various locomotor tasks in humans. Exp. Brain Res. **82**(1), 97–106 (1990)
50. G. Schultz, K. Mombaur, Modeling and optimal control of human-like running. Trans. Mechatron. **15**(5) (2010)
51. M. Sreenivasa, K. Mombaur, J.P. Laumond, Walking paths to and from a goal differ: on the role of bearing angle in the formation of human locomotion paths. PLOS ONE **10**(4) (2015)
52. Ö. Terlemez, S. Ulbrich, C. Mandery, M. Do, N. Vahrenkamp, T. Asfour, Master motor map (MMM) framework and toolkit for capturing, representing, and reproducing human motion on humanoid robots, in *IEEE/RAS International Conference on Humanoid Robots (Humanoids 2014)* (2014), pp. 894–901
53. E. Todorov, Optimality principles in sensorimotor control. Nat. Neurosci. **7**(9), 907–915 (2004)
54. J.J. Ye, Necessary and sufficient optimality conditions for mathematical programs with equilibrium constraints. J. Math. Anal. Appl. **307**, 350–369 (2005)

Versatile Interaction Control and Haptic Identification in Humans and Robots

Yanan Li, Nathanael Jarrassé and Etienne Burdet

Abstract Traditional industrial robot controllers are typically dedicated to a specific task, while humans always interact with new objects yielding unknown interaction forces and instability. In this chapter, we examine the neuromechanics of such contact tasks. We develop a model of the necessary adaptation of force, mechanical impedance and planned trajectory for stable and efficient interaction with rigid or compliant surfaces of different structures. Simulations demonstrate that this model can be used as a novel adaptive robot controller yielding versatile control in representative interactive tasks such as cutting, drilling and haptic exploration, where the robot acquires a model of the geometry and structure of the surface along which it is moving.

1 Introduction

Current industrial robots generally work in well-controlled environments and are mostly involved in non-contact tooling tasks such as welding and gas cutting, where the robot does not physically come in contact with the tooled object. However,

---

This work was funded in part by the European Community under the grants EU-FP7 PEOPLE-ITN-317488-CONTEST, ICT-601003 BALANCE, ICT-611626 SYMBITRON, and EU-H2020 ICT-644727 COGIMON.

Y. Li · E. Burdet (✉)
Imperial College of Science, Technology and Medicine, London SW7 2AZ, UK
e-mail: e.burdet@imperial.ac.uk

Y. Li
e-mail: yanan.li@imperial.ac.uk

E. Burdet
School of Mechanical and Aerospace Engineering, Nanyang Technological University, Singapore, Singapore

N. Jarrassé
CNRS, Sorbonne University, UPMC Univ Paris 06, UMR 7222, ISIR INSERM, Paris U1150 Agathe-ISIR, France
e-mail: jarrasse@isir.upmc.fr

J.-P. Laumond et al. (eds.), *Geometric and Numerical Foundations of Movements*,
Springer Tracts in Advanced Robotics 117, DOI 10.1007/978-3-319-51547-2_9

new robotic applications require interaction with unknown environments and human beings [1]. Such interactions can generate instability and large forces, e.g. when working with a tool.

Surface exploration is one of the few applications where current robot manipulators make purposeful contact with the external objects. Surface exploration is performed either using tactile sensing [2–6] or aided by vision [7, 8] but utilises minimal contact forces. It aims at either forming a 3D model of the object [9–11] or determining the texture of the object's surface [12], for which dedicated sensors and controls have been proposed [13, 14]. Thus, the control of the robot during non-contact tool tasks and surface exploration is relatively easy as it only deals with the robot kinematics and dynamics (which are assumed to be known), and can use specialised sensors.

In contrast, the challenge in controlling contact tool tasks like polishing, carving, cutting, drilling or writing is to follow a given surface profile while maintaining a significant force on the surface. A contact-tool task is influenced by the dynamics of the robot, and also by that of the tooled surface, which is normally unknown. Furthermore, the presence of a large contact force makes the task highly unstable to disturbances caused by friction, irregularities in the surface or noise in the robot motor output. The robot must maintain the right impedance to counter these instabilities. As contact tooling usually involves penetration of the object's surface, vision cannot help in determining its irregularities and variations.

When the tooling task and environment are well known, specific controllers can be implemented to perform the task. For example, a drilling task may be achieved with stiff position control or impedance control [15], though the task instability will require careful tuning of impedance. Also, exploration of simple surfaces may be achieved using hybrid control [16] by controlling position tangential to the surface and force normal to it, but a stable interaction requires an accurate model. However, *the challenge is to develop a controller able to handle a variety of contact tasks with minimal prior knowledge about the tooled surface while interacting with the unknown and possibly inhomogeneous material.*

Compared to current robots, *humans are very versatile* and interact with all kinds of objects without requiring specialised controllers. To illustrate this, humans can learn to compensate for novel force fields in a few trials [17–19]. In order to understand the way humans deal with unstable dynamics typical of tool use, studies investigated how subjects adapt arm movements in negative damping [20] and negative stiffness [21, 22] environments. The results revealed that humans automatically learn to activate muscles in order to compensate for the environment force and instability [23, 24]. Computational modelling of this human motor adaptation capability gave rise to the first controller able to simultaneously adapt force and mechanical impedance in the presence of unknown dynamics [25, 26]. A theoretical analysis of this novel adaptive controller and robotic implementations demonstrated how it can deal with unstable situations typical of tool use and gradually acquire a desired stability margin [26].

Humans can explore unknown objects with the hand even blindly without any difficulty. How to realise such versatile, flexible control with a robot that has to deal with rigid or soft objects of unknown shape? If the planned trajectory enters the surface, the human-like robotic control behaviour such as in [25, 26] will tend to increase interaction force until the actuator limits are reached or the arm or environment breaks. To address this issue, we proposed in the robotic adaptive behaviour of [27, 28] that the planned trajectory is deformed to comply to the rigid object and lower the contact force. Interestingly, recent studies [29, 30] provide evidence that a similar adaptation is occurring when humans interact with a stiff environment.

In this chapter, we take the opportunity offered by these recent studies to shed light on the manner in which humans interact with unknown objects, suggesting how they modify the planned trajectory in order to comply with rigid environments and combine force and trajectory adaptation. We develop and demonstrate a robotic model of the concurrent adaptation of impedance, force and planned trajectory in humans necessary to deal with contact tasks. We show that our computational model has similar adaptive properties as exhibited by humans [29, 30]. We then simulate the corresponding adaptive robot behaviour in typical contact tasks of drilling, cutting and polishing, thus developing haptic identification of an unknown surface. Finally, we outline the frontier developments of our interdisciplinary approach of neuroscience and robotics to interaction control in robotics, haptic sensing and human-robot interaction.

2 Neuromechanical Control During Transport

In this section, we analyse the dynamics of a human carrying out arm movements while interacting with the environment. The nomenclatures used in this chapter are summarised in Table 1. To carry out a reaching movement, the nervous system will generate a *motor command* u that will activate muscles and generate a force F at the endpoint of the arm x. In general $F(u, x)$ is a nonlinear but smooth function of the neural activation u and the *state of the musculoskeletal system* x. Furthermore, muscles and neural feedback together yield a spring-like response to mechanical perturbations, both at static positions [31] and during movement [32]. We express these perturbation dynamics by linearising the endpoint force F along the undisturbed trajectory x_u:

$$F(u, x) \equiv FF(u, x_u) + K(u, x_u)\,\varepsilon\,, \quad \varepsilon \equiv e + \delta(u, x_u)\,\dot{e}\,, \quad e \equiv x_u - x \qquad (1)$$

where $FF(u, x_u)$ is *feedforward force*, $K(u, x_u) \equiv (\partial F/\partial x)_{x=x_u}$ *endpoint stiffness* and $K(u, x_u)\delta(u, x_u)$ *damping*.

By planning a *reference trajectory* $x_r \equiv x_u$ along which the feedforward is computed and relative to which the spring force $K(u, x_r)(x_r - x) + K(u, x_r)\delta(u, x_r)(\dot{x}_r - \dot{x})$ is exerted, Eq. (1) can be considered as a model of how the nervous system controls the endpoint force to interact with the environment. To facilitate the discussion of this controller, let us simplify Eq. (1) to

Table 1 Nomenclature

Symbol	Description
F	Endpoint force
u	Motor command
x	Actual trajectory
x_u	Undisturbed trajectory
e	Position error
ε	Tracking error
FF	Feedforward force
K	Endpoint stiffness
δ, δ_E	Ratio of damping over stiffness
x_r	Reference trajectory
K_0	Minimal endpoint stiffness to ensure stability
K_e	Stiffness to compensate for the environment
Q_F, Q_K, Q_x	Learning rates for force, stiffness, trajectory
γ	Time constant of effort minimisation
F_E	Interaction force
K_E	Environment stiffness
x_0	Surface without interaction
F_d	Desired interaction force

$$F(u, x_r) \equiv FF(u, x_r) + K(u, x_r)(x_r - x)\,. \tag{2}$$

Is the force control redundant? To increase F, it would be a-priori possible to use a larger feedforward force FF, increase stiffness K, or modify x_r. However, the redundancy due to the stiffness term is only apparent. We could show, by studying reaching arm movements in unstable dynamics typical of tool use [21–23, 33], that humans automatically learn to activate muscles which compensate for the environment instability independently of the applied force. For instance, considering Eq. (2), stiffness can be modelled as

$$K(u, x_r) \equiv K_0(u, x_r) + K_e(u, x_r) \tag{3}$$

where $K_0(u, x_r)$ ensures motion stability in free motion, and $K_e(u, x_r)$, which increases monotonically with the torque magnitude [34], is adapted to compensate for the stiffness of the environment [35]. This means that mechanical impedance, which can be characterised by stiffness, viscosity and inertia, regulates our interaction with the environment independently of the applied force [21]. Thus, it cannot be used to control the desired force.

On the other hand, the representation of force in Eq. (2) is redundant, as FF and Kx_r could be combined in an infinite number of ways to generate a desired force F.

For instance, one could set $FF \equiv 0$ and regulate the force using x_r. This corresponds to a form of the *equilibrium point trajectory hypothesis*, which was popular until the 1990s. Interestingly, according to this hypothesis the central nervous system would only need to plan the movement kinematics. The movement would then be carried out by muscles through shifting the reference position along the equilibrium point trajectory without needing to model the movement dynamics. As during arm reaching, the hand movement is stereotyped along a straight line between start and end points [36], therefore the brain could rely on these points to plan motion [37]. However, the relatively low stiffness value estimated during arm reaching movements identified in [38] means that the reference trajectory would become complex and would need to widely deviate from the actual movement in order to produce sufficiently large forces to move the arm. This contradicts the simplifying idea of the equilibrium point trajectory hypothesis and makes it unattractive.

A simple way to identify the parameters of Eq. (1) or (2) for transport movements is to use the undisturbed trajectory as a reference trajectory of the spring-like response: $x_r \equiv x_u$. As a consequence, the feedforward term FF should produce the movement dynamics and compensate for the forces experienced during movement. An advantage of such *feedforward control* is that the movement dynamics are not affected by the delays of sensory feedback. Various nonlinear adaptive controllers have been proposed in computational neuroscience and control theory to identify the feedforward term FF during movements [39, 40]. These algorithms can be used to model the force adaptation in novel dynamics as well as the generalisation to movements different from the trained movements [18, 19], as was demonstrated in [25].

Traditional nonlinear adaptive controllers use constant feedback gains. In contrast, humans can control mechanical impedance independent of the applied force [21, 33, 41]. By observing the changes of muscle activation during repeated arm movements in unstable dynamics, we could develop a model of human motor adaptation encompassing force and impedance adaptation [24, 25] that reproduces various properties of human motor adaptation and generalisation. In this model, we consider repeated or periodic movements, and the *update laws for force and impedance*

$$\begin{aligned} FF^{j+1} &\equiv FF^{j} + Q_F(\varepsilon^{j} - \gamma FF^{j}), \quad \gamma < 1 \\ K^{j+1} &\equiv max\{K_0, K^{j} + Q_K[\varepsilon^{j}\,(x^{j})^T - \gamma K^{j}]\} \\ \varepsilon &\equiv (x_u - x) + \delta(\dot{x}_u - \dot{x}), \quad \delta > 0 \end{aligned} \tag{4}$$

that minimise the tracking error ε and effort [26]. Here j is the iteration number (of either movement trial or period), Q_F and Q_K are learning rates, and γ is the decay to minimise effort. The terms $Q_F\,\varepsilon^j$ and $Q_K\,\varepsilon^j\,(x^j)^T$ aim at reducing tracking error ε. x^j appears in the impedance adaptation because Kx corresponds to the stiffness force that is adapted. This computational model has given rise to a novel nonlinear adaptive control behaviour for robots that was analysed using Lyapunov theory, and demonstrated on the DLR LWR manipulator and on variable impedance actuators [26].

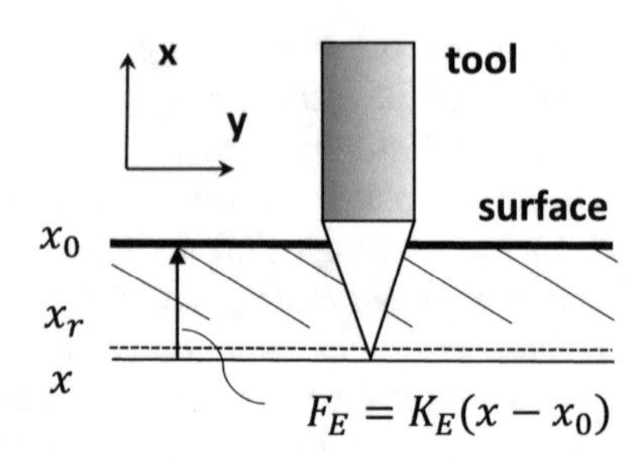

Fig. 1 Surface deformation coordinates. x is the actual position of the tool, x_0 the surface without interaction and x_r the reference position for the restoring force

3 Control of Contact Tasks

While humans constantly make contact with objects in order to carry out activities of daily living, few studies have investigated the control of contact tasks. Recently, the group of Mussa-Ivaldi examined reaching movements along a convex obstacle [29] and the adaptation of force control [30]. *Contact tasks*, in which an object's surface is detected by tactile sensors and proprioception (e.g. through an impact), consist of moving the hand along an object while applying a contact force F_d (Fig. 1). This *desired force* will be counterbalanced by the *interaction force* F_E with the environment. For instance, assuming an elastic environment characterised by stiffness K_E,

$$F_E \equiv K_E(x - x_0)\,, \tag{5}$$

where x_0 is the position of the object surface without interaction.

To apply the desired force F_d and let FF provide the actual movement dynamics, it is necessary to adapt x_r. We therefore introduce the *trajectory adaptation update law*

$$x_r^{j+1} \equiv x_r^j + Q_x(F_d - F^j)\,, \tag{6}$$

that minimises the error between the actual interaction force and the desired one, where Q_x is the learning rate and F^j is given in Eq. (2). This update law converges when $F_d = F = FF + K\,(x_r - x)$. From the error dynamics developed in [26, 42], together with the force and impedance adaptation of Eq. (4), we then have $F_d = F_E$, which indicates that the desired interaction force between the human and the environment is maintained.

The adaptive controller of Eqs. (2), (4), (6) yields simultaneous adaption of force, impedance and reference trajectory to achieve stable and efficient control of contact tasks. This controller is discussed here only in task space and the transformation between task and joint spaces can be carried out by usual means as described e.g. in [43]. The reference trajectory is adapted automatically to maintain a desired interaction force with the environment. Through force and impedance adaptation, the actual trajectory is guaranteed to track the (adapted) reference trajectory. Moreover, force and impedance adaptations result in just-as-needed feedforward force and impedance, corresponding to minimisation of the control effort.

These properties are in line with a human's interaction control. In non-contact tasks, motion control is similar to tracking a planned trajectory by compensating for the disturbance due to the interaction with the environment and reducing the control effort when the interaction weakens or disappears, as described in Sect. 2. In a contact task, this interaction may lead to an undesired large force, so that the original reference trajectory has to be adapted to maintain a desired interaction force. Through interaction, the adaptive controller further enables the robot/human to identify the geometry (i.e. the adapted reference trajectory) and impedance characteristics of the environment. The rigorous analysis of this adaptive controller can be found in [42], which also discusses the conditions that need to be satisfied for achieving the above properties and how to choose the controller's parameters.

4 Adaptation of Force and Trajectory

The adaptation of feedforward force and impedance of Eq. (4) was illustrated extensively in [25, 26]. To understand how humans deal with contact tasks and illustrate the adaptation of applied force and reference trajectory, let us first simulate the experiment of [30]. In this experiment, subjects were required to push in the forward direction against environments of various stiffnesses produced by a haptic interface. The subjects had to match the desired force profile

$$F_d(t) = 5(1 - \cos(\pi t)) \, [N], \quad 0 \leq t \leq 1s \tag{7}$$

displayed on a computer monitor by pushing appropriately on the interface. The interface was controlled to render a viscoelastic environment characterised by the force

$$F_E = K_E(x + \delta_E \, \dot{x}) \tag{8}$$

where x is relative to the spring rest position 0. The subjects first had to exert a force against a rigid surface. Figure 2a shows that the subjects were able to match the desired force profile. Then, the surface became compliant (b) and the subjects adapted to this novel environment (c), which was tested in catch trials with the rigid environment (d).

The results of force field adaptation studies [17, 18] suggest that humans form a model of the interaction force and compensate for it. If neuromechanical control would consist only of this feedforward force, approximately the same interaction force should be measured when the interface switches to a more compliant environment. However, a significantly lower interaction force was measured by the interface (Fig. 2b). This mismatch can be easily interpreted when considering the spring-like properties at the endpoint of the arm, as expressed e.g. in Eq. (2). As the first trial in the compliant condition exhibits a reduced force, this means that $x_r(t) < x(t)$, and it is practical to assume $x_r(t) \approx 0$, corresponding to the surface learned during the

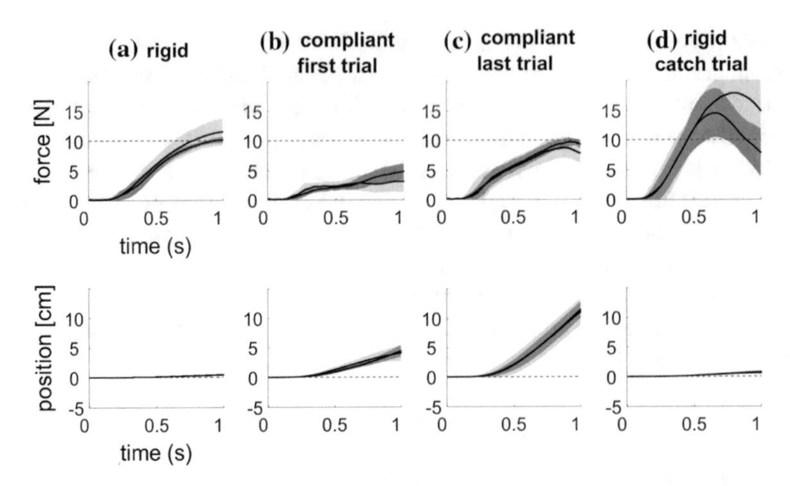

Fig. 2 Results of the force control task from [30] (reproduced with permission). After subjects learned to push an increasing force profile on a rigid surface (**a**), this surface became compliant (**b**), and the subjects had to adapt contact to this new environment (**c**). **d** exhibits after-effects of this learning

previous trials. During subsequent trials in the compliant environment, the subjects adapted their control, the force converged to the desired profile and the trajectory penetrated the surface deeper trial after trial.

We simulate this experiment using pointmass dynamics $m\ddot{x}$ with $m = 4\,\text{kg}$ and the adaptive controller of Eqs. (2), (4), (6). The rigid environment is characterised by $K_E = -1000\,\text{N/m}$, $\delta_E = 1\,\text{s}$ and the compliant environment by $K_E = -100\,\text{N/m}$, $\delta_E = 1\,\text{s}$. The control and learning parameters used for simulation are $\delta(t) = 10\,\text{s}$, $K_0 = 100\,\text{N/m}$, $\gamma(t) = \frac{10^{-4}}{1+1000\|\varepsilon\|}$, $Q_K = 10$, $Q_F = 15$, $Q_x = 2.5/10^4$.

Simulation results are shown in Fig. 3. Figure 3a exhibits that the desired force is achieved in the case of a rigid surface. Figure 3b illustrates that when the surface suddenly becomes compliant, the desired force is not achieved because of the trajectory control component. However, the trajectory iteratively moves forward and interaction force increases. After learning, the reference trajectory has adapted to penetrate the surface and the desired interaction force is reached again. Note that while the same desired force is achieved as in Fig. 3a, the reference trajectory changes due to the compliant surface in Fig. 3b. Figure 3c illustrates the after-effects of learning; when the surface becomes rigid again, the interaction force surpasses the desired force.

These results are similar to the behaviour observed in human experiments (Fig. 2). Note that both force/impedance adaptation and trajectory adaptation are involved in the evolution: the latter adapts the reference trajectory to achieve the desired force, while the former adapts feedforward force and impedance to track the updated reference trajectory.

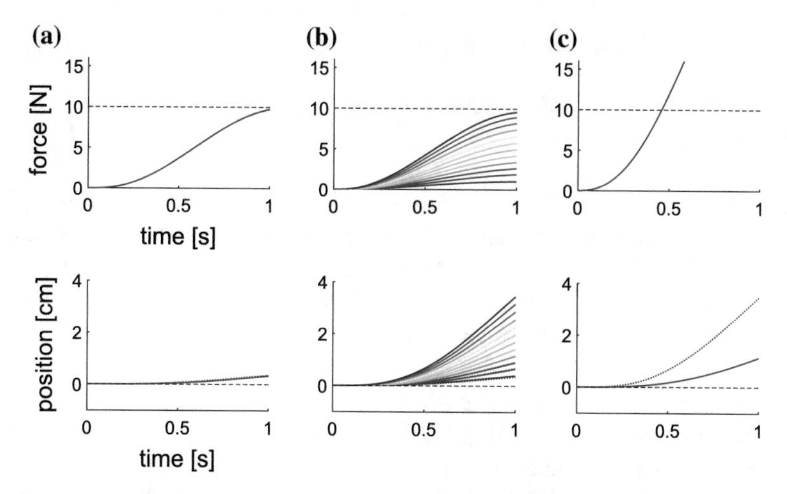

Fig. 3 Simulation of concurrent adaption of force, impedance and trajectory as in the experiment of Fig. 2. *Top panels* show the interaction force and *bottom panels* the actual trajectory (*solid*) and updated reference trajectory (*dotted*): **a** after learning in a rigid environment; **b** in a compliant environment (plotted from *blue to red* in every 16 trials, with the actual trajectories almost superimposed on the respective reference trajectories); and **c** exposed to a rigid environment after learning the compliant environment

5 Haptic Identification

Using the above model of motor adaptation, we can also simulate the haptic exploration observed when human subjects carry out reaching movements along surfaces of different stiffnesses [29]. We simulate a 1 s long arm movement ahead of the body, $D = 0.1$ m in the forward direction, with a smooth planned trajectory

$$\begin{bmatrix} x_r(t) \\ y_r(t) \end{bmatrix} = \begin{bmatrix} x(0) \\ y(0) \end{bmatrix} + t^3(10 - 15t + 6t^2) \begin{bmatrix} 0 \\ D \end{bmatrix}, \tag{9}$$

$$\begin{bmatrix} x(0) \\ y(0) \end{bmatrix} \equiv \begin{bmatrix} 0.4385 \\ 0.4494 \end{bmatrix} \text{m}\,. \tag{10}$$

The interaction force is generated according to

$$F_E = \begin{cases} K_E(t) \begin{bmatrix} x(t) - x_0(t) \\ y(t) - y_0(t) \end{bmatrix} & \text{if } x(t) \geq x_0(t) \\ \begin{bmatrix} 0 \\ 0 \end{bmatrix} & \text{otherwise.} \end{cases} \tag{11}$$

The surface rest position is set as the smooth sine wave

$$\begin{bmatrix} x_0(t) \\ y_0(t) \end{bmatrix} = \begin{bmatrix} x(0) - 0.0083(1 + \sin(20\pi(y(t) - y(0)) - \frac{\pi}{2}) \\ y(t) \end{bmatrix} \tag{12}$$

where $x(0)$ and $y(0)$ are defined in Eq. (10). The stiffness matrices of a compliant, medium and rigid surface K_E are $-200I_2$, $-800I_2$ and $-2000I_2$ N/m respectively, where $I_2 \equiv [1, 0; 0, 1]$ is the 2×2 identity matrix. The desired interaction force is set as $F_d(t) = [-3e^{-(5(t-\frac{T}{2}))^2}\ 0]^T N$. The following control and learning parameters are used: $\delta(t) = 10$ s, $K_0 = 50I_2$ N/m, $\gamma(t) = \frac{10^{-4}}{1+1000\|\varepsilon\|}$, $Q_K = I_2$, $Q_F = 10I_2$, $Q_x = 2/10^4 I_2$.

Simulation results are shown in Fig. 4. The resulting trajectory adaptation is in line with experimental results in [29]: the reference trajectory conforms to a rigid surface, but is little influenced by a compliant surface. The resulting forces illustrate the underlying reason: a large interaction force is generated from the interaction with a stiff surface, and trajectory adaptation reduces this force to track the desired interaction force. The trajectory does not change much with a compliant surface as the small interaction force is achieved soon. The trajectory conformation is in between when the surface stiffness is medium. After-effect trajectories (i.e. when the surface disappears after learning) exhibit the converse effect. After learning along a surface, the trajectories move in the direction into the surface as a desired interaction force is expected to be maintained. Moreover, the adaptation is found to involve two processes: first, force and impedance adaptation compensate for the interaction and leads to slightly larger force, then the trajectory adapts to reduce the force and achieve the desired force.

The results of this and the previous sections show that the model of Sect. 3 predicts the adaptation of force and trajectory observed when humans carry out movements in contact with rigid and compliant environments [29, 30]. To further understand the role of trajectory adaptation in human-like learning behaviour, we simulate this learning when trajectory adaptation is frozen. We see in Fig. 4d that trajectories after learning come closer to the initial straight reference trajectory, such that the interaction force becomes even larger. Without trajectory adaptation the reference trajectory is tracked, at the cost of a possibly large force and surface deformation. These results contrast to those of Fig. 4c where trajectory adaptation prevented too large interaction force and surface deformation.

In above simulations, the initial reference trajectory was set inside the surface. Let us now consider the case that the initial reference trajectory lies outside the surface. In this case a small non-zero interaction force e.g. $F_d(t) = [-2\ 0]^T N$ is required to follow the surface. In particular, let us consider the adaptation along the $1s$ long rotated initial reference trajectory

$$\begin{bmatrix} x_r(t) \\ y_r(t) \end{bmatrix} = \begin{bmatrix} x(0) \\ y(0) \end{bmatrix} + t^3(10 - 15t + 6t^2) \begin{bmatrix} -D\, s\theta \\ D\, c\theta \end{bmatrix} \tag{13}$$

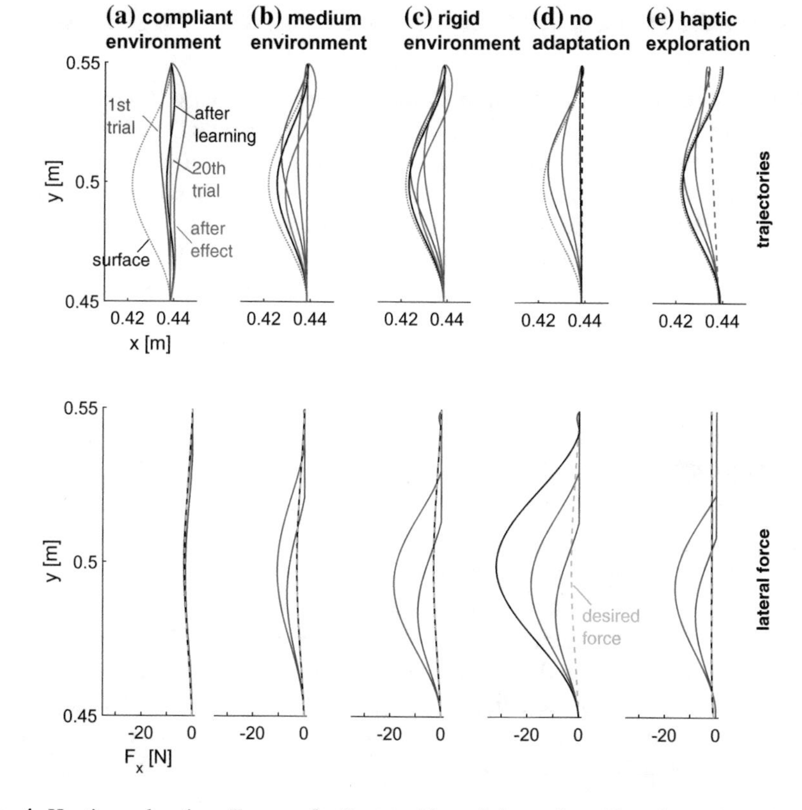

Fig. 4 Haptic exploration. *Top panels*: Rest position of the surface (*dotted grey line*), before-learning reference (*dashed blue line* covered by *solid red line*) and actual (*solid blue line*) trajectories, reference (*dashed red line*) and actual (*solid red line*) trajectories in the 20th trial, reference (*dashed black line*) and actual (*solid black line*) trajectories in the last trial and after-effect actual trajectory (*solid pink line*). *Bottom panels*: Desired interaction force (*dotted green line*), before-learning interaction force (*solid blue line*), interaction forces in the 20th trial (*solid red line*) and in the last trial (*solid black line*)

where $x(0)$ and $y(0)$ are defined in Eq. (10), $s\theta \equiv \sin(3°)$ and $c\theta \equiv \cos(3°)$. The interaction force is generated by Eq. (11) with stiffness $K_E = -2000 I_2$ N/m. The same control and learning parameters are used as above in this section.

Figure 4e shows that after adaptation both the reference and actual trajectories are within the surface and there is a small distance between them and the rest position of the surface, corresponding to the interaction force which is close to the desired force. While some parts of the initial reference trajectory lie outside of the surface and other parts inside, surface identification is achieved for the whole trajectory after adaptation. These results demonstrate how the human-like adaptive behaviour leads to haptic exploration, using a desired non-zero interaction force. Note that with a positive desired interaction force the reference trajectory will drift in the direction opposite to the surface.

6 Versatile Human-Like Interaction Behaviour for Robots

Can the above model of interaction control in humans be used to control a robot carrying out typical contact tasks? In this section, we simulate three typical contact tooling tasks to test the abilities of human-like interaction control. We use the algorithm of Eqs. (2), (4), (6) in the robot task space. The tool is modelled as a point mass in two degrees-of-freedom, with y in the forward direction and x normal to the surface as is shown in Fig. 1. The tool stiffness is assumed to be much more rigid than the material being tooled, such that there is no deformation of the tool during interaction. In all tasks the default learning parameters are $Q_F = 0.005$, $Q_K = 7$, $\gamma = 0.005$, $Q_x = 0.01$.

6.1 *Cutting*

Cutting is similar to milling or carving. It is difficult to perform with force-based control that would require thorough knowledge of the object geometry and mechanical structure in order to determine the force level required for the cut. This task may be achieved using impedance control [15], but the gains would have to be tuned to each surface. In particular, during cutting some amount of tool stiffness is required to counter friction, but too high rigidity can get the tool stuck in the material, such as when meeting a knot while sawing wood. In contrast, the above human-like control automatically adjusts the stiffness as required for each surface. Cutting requires:

- a constant feedforward force to cut against the material;
- minimal contact force (normal to the direction of cut) to avoid excessive penetration of the surface;
- adjustment of stiffness so as to maintain stability against the irregularities in the tooled material.

In our simulation, the surface to cut is modelled such that an unknown threshold force is required to penetrate it with the tool (in the x direction). Once inside, large damping opposes further penetration and forward movement. When the tool reaches the desired cutting depth, a high frequency and low amplitude sinusoidal vertical (i.e. along x) force simulates the effect of the material's internal texture. Some dry friction (proportional to the force applied normally by the tool) along the forward movement requires the force to reach a threshold level for the tool to start moving in that direction. The robot is required to make a cut of $d = 3$ cm depth on a surface. Figure 5 shows the normal force, position and stiffness of the robot simulated in the cutting task with our controller. At the beginning of the task, the reference (red trace in Fig. 5b) is set at the required depth below the surface corresponding to the required depth of cut. The reference adaptation Eq. (6) is switched off for this operation.

First, the robot automatically increases the contact force (Fig. 5a) and penetrates to the required depth. It then proceeds with a horizontal cut, during which the contact

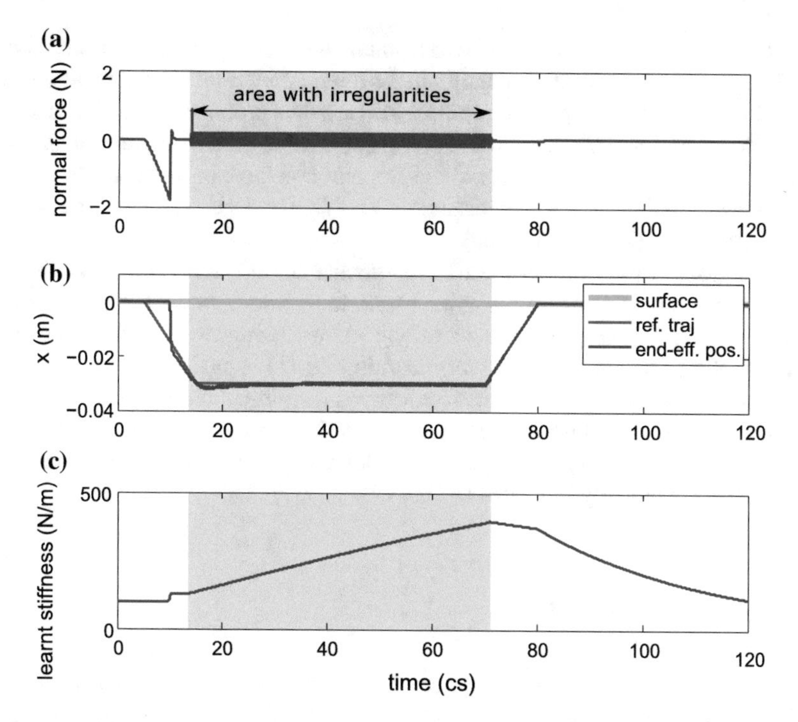

Fig. 5 Cutting task simulation. This figure shows, along the x forward movement direction: **a** the contact force; **b** the task reference (in *red*) and position of the tool (*blue*); **c** stiffness of the tool along x. Irregularities in the surface were simulated (*orange region*) by a sinusoidal external force

force (normal to the cut) is reduced to 0. In the presence of irregularities (marked in Fig. 5a), the robot automatically increases its stiffness (Fig. 5c) with little increase in contact force (note the small perturbations in the orange region in Fig. 5b). In the absence of irregularities, and after the end of the cutting operation, the robot automatically reduces stiffness.

6.2 Drilling

While drilling is similar to cutting, it usually involves a larger contact force on the surface. Furthermore, drilling is inherently unstable as the heavy drill (and the robot) supported at the end of a narrow tool tip will amplify any noise. Thus, a drilling task requires:

- constant force and minimal stiffness in the drilling direction;
- stability against the large, random perturbations perpendicular to the drilling direction.

While force control could be used to maintain the drilling force, additional control would be required to maintain stability in the lateral direction. Hybrid position/force control [44] or hybrid impedance control [45] are thus probably suited for this operation. However, the controller would still require some additional control along the drilling direction so as to monitor the movement and increase the controlling force in the presence of obstacles (for instance, a knot in wood would temporarily require more force in the drilling direction).

Our controller provides all these requirements automatically. In our simulation, the surface is modelled as for the cutting task. In addition, the tool encounters sinusoidal horizontal force perturbations (along y) simulating the vibrations generated by the drilling process. The trajectory learning Eq. (6) is again switched off in this operation. We see in Fig. 6e that the feedforward learning in the algorithm adapts the movement force as required and reduces the position error and hence the feedback gain (Eq. (4)), making the drill compliant along x (Fig. 6f). At the same time, the algorithm automatically increases stiffness along y.

6.3 Surface Exploration

This task, similar to polishing, requires the robot to glide along unknown surfaces, with controlled e.g. constant force. This operation needs force control normal to the unknown surface and position control tangential to it. At the same time, the robot could also acquire information about the object geometry and texture. Some hybrid force-position or force-impedance control appears to be suitable strategies for this task, but would require an additional surface estimation algorithm in order to modulate the control dimensions dependent on the object surface as it explores it. With its ability to adapt the reference trajectory (Eq. (6)) and to identify stiffness during movement (Eq. (4)), our algorithm inherently possesses capabilities to observe the surface geometry and texture.

In our simulation, the surface is composed of a rising and then falling ramp, an area similar to a "macro velcro" (modelled as a high frequency and low amplitude sinusoidal position perturbation) and finally a large sinusoidal profile. Some dry friction (proportional to the normal force applied on the surface) was again added along the forward horizontal direction, along with some damping resistive force when the tool is in contact with the surface. To implement exploration with our algorithm, a rough reference trajectory is initially set below the unknown surface (see start of simulation in Fig. 7b, where the red trace starts below green) similar to a cutting task.

The adapted reference follows the surface (Fig. 7b) and can be used to estimate the surface normal and tangent at any time instance. Note that the normal force (Fig. 7a) is kept at a low value throughout the movement while the stiffness (Fig. 7c) increases only when there are surface irregularities.

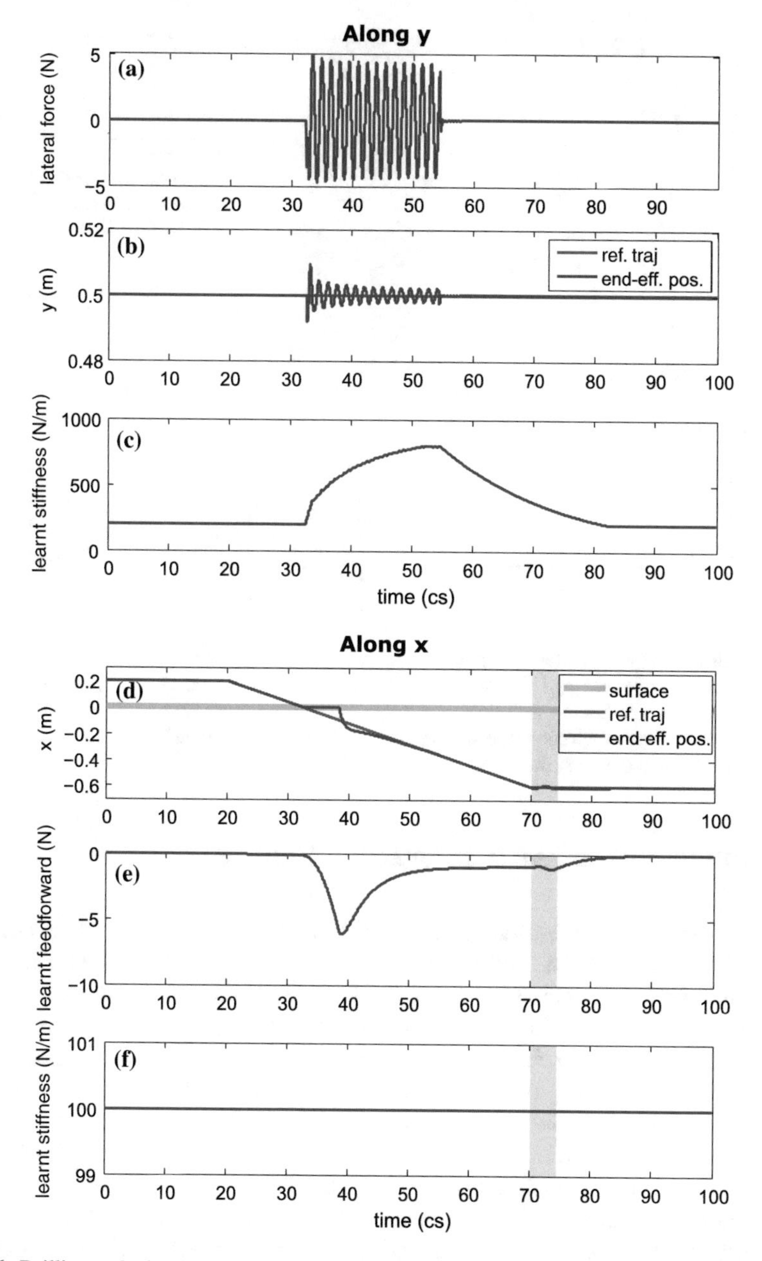

Fig. 6 Drilling task simulation. Along y: In the presence of perturbations lateral to the drilling direction (**a**), the robot increases stiffness only along this direction (**c**) to keep the tool vibrations low (*blue* trace in **b**) and close to the reference (*red* trace in **b**). Once the vibrations decrease, stiffness is reduced to a low value again. Along x: position (**d**), feedforward force (**e**) and tool stiffness (**f**) in the drilling direction. Note that a low stiffness is maintained while feedforward force is learnt to achieve the task. In the presence of an internal obstacle (like a wood knot, during the orange period), the feedforward is automatically increased to counter this, before returning to a lower value

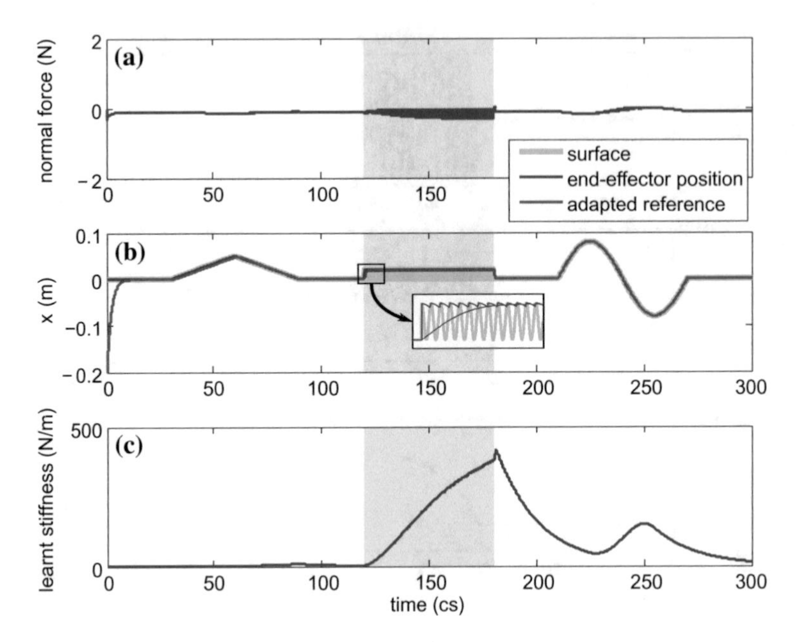

Fig. 7 Surface exploration task simulation. The robot explores an unknown surface and maps it (*red* trace in **b**) while moving (*blue* trace in **b**). The contact force (along x) during the movement is maintained (**a**) at a minimum value while stiffness along x increases (**c**) only in the presence of a rough texture (*orange region*) on the surface. The stiffness changes could be thus used to identify the texture of the explored surface

7 Outlook: Frontiers in Interaction Control

In this chapter, we have used an interdisciplinary approach of neuroscience and robotics to investigate interactive control in humans and robots. Robotics and control theory were used as tools to investigate human sensorimotor control and develop a computational model of neuromechanical control in humans, which in turn yielded a novel adaptive robot behaviour. We have described recent advances enabled by such human robotics approach [46] for adapting the interaction with soft and rigid surfaces, as well as the resulting haptic identification. The simulation of representative interactive tasks of a robot in contact with the environment demonstrated the versatility and efficiency of the novel adaptive behaviour. Future research may investigate how such methodologies can be used to build expectation about future action in robots and how such cognitive motor memory is built in humans. In the following paragraphs, let us outline three other frontiers in interaction control whose solutions require both human motor control experiments and the development of suitable robotic strategies.

Transitions between transport and contact. Many actions such as opening a door or parts insertions involve transport and contact phases. We have analysed how to control both of these phases in Sects. 2 and 3, but have not dealt with the transition between them. Arguably the detection of a contact and corresponding consideration of a desired contact force by dedicated sensors would arise gradually, thus addressing the continuity issue for slow transitions. However, further studies are required when the transition involves an impact. In order to understand how the human central nervous system learns to control the limbs in such a transition task, we have recently investigated how subjects perform an insertion task [47]. We could observe how the subjects modify the movement velocity and the impedance during movement to minimise time, effort and the impact force. Such learning is not addressed by the supervised learning techniques described in this chapter, but could be addressed by reinforcement learning techniques developed in recent years, e.g. [48].

Haptic interaction using tactile and force information. Haptic sensing used for objects manipulation requires the integration of signals from both tactile sensing and proprioception. However, research in tactile sensing on the one hand, and in sensorimotor control using proprioception and vision information on the other were pursued in parallel, with little communication between these two communities. What role does/can tactile information play in characterising the environment in which the human or robotic limb interacts during movements? Behavioural experiments with humans should be designed to examine how the nervous system combines tactile sensing and proprioception, and how robots should use and integrate these two sources of information.

Human-like interaction with robots. Human motor control and robotics have investigated the interaction with time-invariant dynamical systems, e.g. fixed force fields. In contrast, interacting with another human or any autonomous agent involves time-varying dynamics. How do humans interact with each other, e.g. to carry a table together, perform interpersonal physical therapy, or to assist an infant or an elderly in walking? Experiments with pairs of human subjects who have to carry out a collaborative task whilst being mechanically connected have been carried out for over a decade [49]. These studies could show that physical coupling enables pairs or dyads to improve in joint tasks [50–52], and promote interacting individuals to negotiate specialised roles [51, 53, 54]. Computational modelling of these results is needed to reveal the underlying mechanism and enable human-like collaboration algorithm for robots. The interactive motor control framework of [55] may serve as a starting point for such developments, that could be extended to encompass the estimation of the partner's state and motion intention.

Acknowledgements The authors thank Atsushi Takagi for editing the text.

References

1. N. Jarrassé, V. Sanguineti, E. Burdet, Slaves no longer: review on role assignment for human-robot joint motor action. Adapt. Behav. **22**(1), 70–82 (2014)
2. P. Dario, M. Bergamasco, A. Sabatini, Sensing body structures by an advanced robot system, in *Proceedings of the IEEE International Conference on Robotics and Automation (ICRA)* (1988), pp. 1758–1763

3. K. Pribadi, J.S. Bay, H. Hemami, Exploration and dynamic shape estimation by a robotic probe. IEEE Trans. Syst. Man Cybern. **19**(4), 840–846 (1989)
4. S. Sestili, A. Starita, Learning objects by tactile perception, in *Proceedings of the IEEE International Engineering in Medicine and Biology Society (EMBC)* (1989), pp. 896–897
5. S.A. Stansfield, Haptic perception with an articulated, sensate robot hand. Robotica **10**(6), 497–508 (1992)
6. A.M. Okamura, M.R. Cutkosky, Haptic exploration of fine surface features, in *Proceedings of the IEEE International Conference on Robotics and Automation (ICRA)* (1999), pp. 2930–2936
7. A.M. Okamura, M.R. Cutkosky, Feature-guided exploration with a robotic finger, in *Proceedings of the IEEE International Conference on Robotics and Automation (ICRA)* (2001), pp. 589–596
8. H.T. Tanaka, K. Kushihama, N. Ueda, S. Hirai, A vision-based haptic exploration, in *Proceedings of the IEEE International Conference on Robotics and Automation (ICRA)* (2003), pp. 3441–3448
9. A. Bierbaum, M. Rambow, T. Asfour, R. Dillmann, A potential field approach to dexterous tactile exploration of unknown objects, in *Proceedings of the IEEE-RAS International Conference on Humanoid Robots* (2008), pp. 360–366
10. F. Mazzini, D. Kettler, S. Dubowsky, J. Guerrero, Tactile robotic mapping of unknown surfaces: an application to oil well exploration, in *IEEE International Workshop on Robotic and Sensors Environments (ROSE)* (2009), pp. 80–88
11. D.R. Faria, R. Martins, J. Lobo, J. Dias, Probabilistic representation of 3D object shape by in-hand exploration, in *Proceedings of the IEEE/RSJ International Conference on Intelligent Robots and Systems (IROS)* (2010), pp. 1560–1565
12. W.W. Mayol-Cuevas, J. Juarez-Guerrero, S. Munoz-Gutierrez, A first approach to tactile texture recognition. in *Proceedings of the IEEE International Conference on Systems, Man, and Cybernetics* (1993), pp. 4246–4250
13. P.A. Schmidt, E. Mael, R.P. Wurtz, A sensor for dynamic tactile information with applications in human- robot interaction and object exploration. Robot. Auton. Syst. **54**(12), 1005–1014 (2006)
14. D. Kraft, A. Bierbaum, M. Kjaergaard, J. Ratkevicius, A. Kjaer-Nielsen, C. Ryberg, H. Petersen, T. Asfour, R. Dillmann, N. Kruger, Tactile object exploration using cursor navigation sensors, in *Proceedings of the IEEE WorldHaptics* (2009), pp. 296–301
15. N. Hogan, Impedance control: an approach to manipulation. J. Dyn. Syst. Meas. Control Trans. ASME **107**(1), 1–24 (1985)
16. O. Khatib, A unified approach for motion and force control of robot manipulators: the operational space formulation. IEEE J. Robot. Autom. **3**(1), 43–53 (1987)
17. J.R. Lackner, P. Dizio, Rapid adaptation to coriolis-force perturbations of arm trajectory. J. Neurophysiol. **72**(1), 299–313 (1994)
18. R. Shadmehr, F.A. Mussa-Ivaldi, Adaptive representation of dynamics during learning of a motor task. J. Neurosci. **74**(5), 3208–3224 (1994)
19. M.A. Conditt, F. Gandolfo, F.A. Mussa-Ivaldi, The motor system does not learn the dynamics of the arm by rote memorization of past experience. J. Neurophysiol. **78**, 554–560 (1997)
20. T.E. Milner, C. Cloutier, Compensation for mechanically unstable loading in voluntary wrist movement. Exp. Brain Res. **94**(3), 522–532 (1993)
21. E. Burdet, R. Osu, D.W. Franklin, T.E. Milner, M. Kawato, The central nervous system stabilizes unstable dynamics by learning optimal impedance. Nature **414**(6862), 446–449 (2001)
22. A. Kadiallah, G. Liaw, M. Kawato, D.W. Franklin, E. Burdet, Impedance control is selectively tuned to multiple directions of movement. J. Neurophysiol. **106**(5), 2737–2748 (2011)
23. D.W. Franklin, G. Liaw, T.E. Milner, R. Osu, E. Burdet, M. Kawato, Endpoint stiffness of the arm is directionally tuned to instability in the environment. J. Neurosci. **27**(29), 7705–7716 (2007)
24. D.W. Franklin, E. Burdet, K.P. Tee, R. Osu, C.M. Chew, T.E. Milner, M. Kawato, CNS learns stable, accurate, and efficient movements using a simple algorithm. J. Neurosci. **28**(44), 11165–11173 (2008)

25. K.P. Tee, D.W. Franklin, M. Kawato, T.E. Milner, E. Burdet, Concurrent adaptation of force and impedance in the redundant muscle system. Biol. Cybern. **102**(1), 31–44 (2010)
26. C. Yang, G. Ganesh, S. Haddadin, S. Parusel, A. Albu-Schaeffer, E. Burdet, Human-like adaptation of force and impedance in stable and unstable interactions. IEEE Trans. Robot. **27**(5), 918–30 (2011)
27. G. Ganesh, A. Albu-Schaeffer, M. Haruno, M. Kawato, E. Burdet, Biomimetic motor behavior for simultaneous adaptation of force, impedance and trajectory in interaction tasks, in *Proceedings of the IEEE International Conference on Robotics and Automation (ICRA)* (2010), pp. 2705–2711
28. G. Ganesh, N. Jarrassé, S. Haddadin, A. Albu-Schaeffer, E. Burdet, A versatile biomimetic controller for contact tooling and haptic exploration, in *Proceedings of the IEEE International Conference on Robotics and Automation (ICRA)* (2012), pp. 3329–3334
29. V.S. Chib, J.L. Patton, K.M. Lynch, F.A. Mussa-Ivaldi, Haptic identification of surfaces as fields of force. J. Neurophysiol. **95**(2), 1068–1077 (2005)
30. M. Casadio, A. Pressman, S. Mussa-Ivaldi, Learning to push and learning to move: the adaptive control of contact forces. Front. Comput. Neurosci. **9**(118) (2015)
31. F.A. Mussa-Ivaldi, N. Hogan, E. Bizzi, Neural, mechanical, and geometric factors subserving arm posture in humans. J. Neurosci. **5**(10), 2732–2743 (1985)
32. J. Won, N. Hogan, Stability properties of human reaching movements. Exp. Brain Res. **107**(1), 125–126 (1995)
33. D.W. Franklin, E. Burdet, R. Osu, M. Kawato, T.E. Milner, Functional significance of stiffness in adaptation of multijoint arm movements to stable and unstable dynamics. Exp. Brain Res. **151**(2), 145–157 (2003)
34. H. Gomi, R. Osu, Task-dependent viscoelasticity of human multijoint arm and its spatial characteristics for interaction with environments. J. Neurosci. **18**(21), 8965–8978 (1998)
35. K.P. Tee, E. Burdet, C.M. Chew, T.E. Milner, A model of endpoint force and impedance in human arm movements. Biol. Cybern. **90**, 368–375 (2004)
36. P. Morasso, Spatial control of arm movements. Exp. Brain Res. **42**, 223–227 (1981)
37. T. Flash, The control of hand equilibrium trajectories in multi-joint arm movements. Biol. Cybern. **57**(4–5), 257–274 (1987)
38. H. Gomi, M. Kawato, Equilibrium-point control hypothesis examined by measured arm stiffness during multijoint movement. Science **272**(5258), 117–120 (1996)
39. J.-J.E. Slotine, W. Li, *Applied Nonlinear Control* (Prantice-Hall, 1991)
40. E. Burdet, A. Codourey, L. Rey, Experimental evaluation of nonlinear adaptive controllers. IEEE Control Syst. Mag. **18**(2), 39–47 (1998)
41. R. Osu, E. Burdet, D.W. Franklin, T.E. Milner, M. Kawato, Different mechanisms in adaptation to stable and unstable dynamics. J. Neurophysiol. **90**(5), 3255–3269 (2003)
42. Y. Li, E. Burdet, Dynamic analysis of simultaneous adaptation of force, impedance and trajectory (2016). arXiv:1605.07834 [cs.RO] (also submitted)
43. R.M. Murray, Z. Li, S.S. Sastry, *A Mathematical Introduction to Robotic Manipulation* (CRC, 1994)
44. M.H. Raibert, J.J. Craig, Hybrid position/force control of manipulators. J. Dyn. Syst. Meas. Contr. **103**(2), 126–133 (1981)
45. R.J. Anderson, M.W. Spong, Hybrid impedance control of robotic manipulators. IEEE J. Robot. Autom. **4**(5), 549–556 (1982)
46. E. Burdet, T.E. Milner, D.W. Franklin, *Human Robotics: Neuromechanics and Motor Control* (MIT press, 2013)
47. G. de Magistris, Etude et conception de la commande de mannequins virtuels dynamiques pour l'évaluation ergonomique des postes de travail. Ph.D. thesis, Université Pierre et Marie Curie Paris VI (2013)
48. E. Theodorou, J. Buchli, S. Schaal, A generalized path integral control approach to reinforcement learning. J. Mach. Learn. Res. **11**, 3137–3181 (2010)
49. N. Sebanz, H. Bekkering, G. Knoblich, Joint action: bodies and minds moving together. Trends Cognit. Sci. **10**, 70–76 (2006)

50. C. Basdogan, C.-H. Ho, M.A. Srinivasan, M. Slater, An experimental study on the role of touch in shared virtual environments. ACM Trans. Comput.-Hum. Interact. **7**, 443–460 (2000)
51. K. Reed, M. Pershkin, Haptically linked dyads: are two motor-control systems better than one? Psychol. Sci. **17**, 365–366 (2006)
52. G. Ganesh, R. Osu, M. Kawato, E. Burdet, Two is better than one: physical interactions improve motor performance in humans. Sci. Rep. **4**, 3824 (2014)
53. R.P.R.D. van der Wel, G. Knoblich, N. Sebanz, Let the force be with us: Dyads exploit haptic coupling for coordination. J. Exp. Psychol. Hum. Percept. Perform. **37**, 1420–1431 (2011)
54. A. Melendez-Calderon, V. Komisar, E. Burdet, Interpersonal strategies for disturbance attenuation during a rhythmic joint motor action. Physiol. Behav. **147**, 348–358 (2015)
55. N. Jarrassé, T. Charalambous, E. Burdet, A framework to describe, analyze and generate interactive motor behaviors. PLoS ONE **7**, e49945 (2012)

The Variational Principles of Action

Karl Friston

Abstract This chapter provides a theoretical perspective on action and the control of movement from the point of view of the free-energy principle. This variational principle offers an explanation for neuronal activity and ensuing behavior that is formulated in terms of dynamical systems and attracting sets. We will see that the free-energy principle emerges when considering the ensemble dynamics of biological systems like ourselves. When we look closely what this principle implies for the behavior of systems like the brain, one finds a fairly straightforward explanation for many aspects of action and perception; in particular, their (approximately Bayesian) optimality. Within the Bayesian brain framework, the ensuing dynamics can be separated into those serving perceptual inference, learning and behavior. Variational principles play a key role in what follows; both in understanding the nature of self-organizing systems but also in explaining the adaptive nature of neuronal dynamics and plasticity in terms of optimization—and the process theories that mediate optimal inference and motor control. A special focus of this chapter is the pre-eminent role of heteroclinic cycles in providing deep and dynamic (generative) models of the sensorium; particularly the sensations that we generate ourselves through action. In what follows, we will briefly rehearse the basic theory and illustrate its implications using simulations of action (handwriting)—and its observation.

1 Introduction

The premise we will pursue is that the brain is trying to optimize something (specifically variational free-energy), using a generalized gradient descent to perform this optimization. In other words, one can understand neuronal dynamics as optimizing a quantity through the method of steepest descent that can be described with a (complicated) set of ordinary differential equations. It is these equations that give rise to purposeful movement that have been described in previous chapters. In what

K. Friston (✉)
The Wellcome Trust Centre for Neuroimaging, University College London, Queen Square, London WC1N 3BG, UK
e-mail: k.friston@ucl.ac.uk

J.-P. Laumond et al. (eds.), *Geometric and Numerical Foundations of Movements*, Springer Tracts in Advanced Robotics 117, DOI 10.1007/978-3-319-51547-2_10

follows, we will see how the optimization of free-energy leads naturally to optimal action and perception. Crucially, the nature of this optimization rests on an internal, forward or generative model of the world that it navigates. This model includes prior beliefs about the causal structure and dynamics in the world, which constrain both perception and action. This adds a second layer of dynamics, reflecting our prior expectations about the trajectories of states and their attractors in our environment. This chapter focuses on itinerant (wandering) dynamics and how movement can be understood in terms of prior beliefs about sensorimotor trajectories. In particular, we will look at action-observation in the context of handwriting and how it rests on stable heteroclinic channels. This is one of many examples of how itinerant dynamics are embedded in generative models of the sensorium. It is particularly relevant in the context of this book, given we appeal again and again to variational principles. Here, we disclose their fundamental role in shaping action and perceptual inference.

This chapter comprises two parts. In the first, we provide a didactic overview of the free-energy principle, motivating it from basic principles. We will consider the underlying imperative that applies to all biological agents; namely to conserve themselves by minimizing surprise – and how this calls upon the minimization of variational free-energy. We then unpack the free-energy principle in terms of action and perception. This leads to active inference that subsumes perceptual inference the sort considered by the Bayesian brain hypothesis. We illustrate the key aspects of this treatment with a few selected examples and conclude by thinking about the timescales over which the dynamics of free-energy minimization may be manifest. The second part of this chapter presents a particular example in greater detail. This example considers handwriting in terms of itinerant expectations about sequences of movements. Not only does it provide a plausible account of sensorimotor execution but touches upon the cognitive neuroscience of action-observation and how we represent ourselves and others.

2 The Free-Energy Principle

In recent years, there has been growing interest in free-energy formulations of brain function [13, 22], not just from the neuroscience community, where has caused some puzzlement [69] but from fields as far apart as psychotherapy [7] and social politics [35]. The free-energy principle has been described as a unified brain theory [40] and may have broader implications for how we interact with our environment. This section describes the origin of the free-energy formulation, its underlying premises and the implications for how we represent and interact with the world. Table 1 provides a glossary of the quantities that we will be dealing with and the appendix offers a technical overview of how the free energy principle applies to the brain.

The free-energy principle is a simple postulate that has complicated ramifications. It says that all agents or biological systems (like us) must minimize free-energy. This postulate is closely related to Hamilton's law of Least Action and the celebrated H-theorems in statistical physics [49]. The principle was originally formulated as

Table 1 Generic variables and quantities in the free-energy formation of active inference, under the Laplace assumption (i.e., generalized predictive coding)

Variable	Description
$m \in \mathcal{M}$	**Generative model or agent:** In the free-energy formulation, each agent or system is taken to be a model of the environment in which it is immersed. The model specifies the form of the process generating predictions of sensory signals
$a \subset A$	**Action:** These variables are states of the world that correspond to the movement or configuration of an agent (i.e., its effectors)
$\tilde{s}(t) = s \oplus s' \oplus s'' \oplus \ldots \in S$	**Sensory signals:** These generalized sensory signals or samples comprise the sensory states, their velocity, acceleration and temporal derivatives to high order. In other words, they correspond to the trajectory of an agent's sensations
$\mathcal{L}(\tilde{s} \mid m) = -\ln p(\tilde{s} \mid m)$	**Surprise:** This is a scalar function of sensory samples and reports the improbability of sampling some signals, under a generative model of how those signals were caused. It is sometimes called (sensory) surprisal or self-information. In statistics it is known as the negative log-evidence for the model
$H(S \mid m) \propto \int dt \mathcal{L}(\tilde{s}(t) \mid m)$	**Entropy:** Sensory entropy is, under ergodic assumptions, proportional to the long-term time average of surprise
$\mathcal{G}(\tilde{s}, \psi) = -\ln p(\tilde{s}, \psi \mid m)$	**Gibbs energy:** This is the negative log of the density specified by the generative model; namely, surprise about the joint occurrence of sensory samples and their causes
$\mathcal{F}(\tilde{s}, \tilde{\mu}) = \mathcal{G}(\tilde{s}, \tilde{\mu}) + \frac{1}{2} \ln \mid \mathcal{G}_{\tilde{\mu}\tilde{\mu}} \mid$ $\geq \mathcal{L}(\tilde{s} \mid m)$	**Free-energy:** This is a scalar function of sensory samples and a recognition density, which upper bounds surprise. It is called free-energy because it is the expected Gibbs energy minus the entropy of the variational density. Under a Gaussian (Laplace) assumption about the form of the variational density, free-energy reduces to the simple function of Gibbs energy shown in Fig. 2
$\mathcal{S}(\tilde{s}, \tilde{\mu}) = \int dt \mathcal{F}(\tilde{s}, \tilde{\mu})$ $\geq H(S \mid m)$	**Free-action:** This is a scalar functional of sensory samples and a variational density, which upper bounds the entropy of sensory signals. It is the time or path integral of free-energy
$q(\psi) = \mathcal{N}(\tilde{\mu}, C)$ $\tilde{\mu} = \mu \oplus \mu' \oplus \mu'' \oplus \ldots$ $C = \mathrm{G}_{\tilde{\mu}\tilde{\mu}}^{-1}$	**Variational density**: This is also known as an ensemble or recognition density and becomes (approximates) the conditional density over hidden causes of sensory samples, when free-energy is minimized. Under the Laplace assumption, it is specified by its conditional expectation and covariance

(continued)

Table 1 (continued)

Variable	Description
$\Psi = \{\mathbf{u}, \varphi\}$ $\psi = \{u, \varphi\}$ $u = \{x, v\}$ $\varphi = \{\theta, \gamma\}$	**True (bold) and hidden (italics) causes:** These quantities cause sensory signals. The true quantities exist in the environment and the hidden homologues are those assumed by the generative model of that environment. Both are partitioned into time-dependent variables and time-invariant parameters
$\theta \subset \varphi \subset \psi$	**Hidden parameters:** These are the parameters of the mappings (e.g., equations of motion) that constitute the deterministic part of a generative model
$\gamma \subset \varphi \subset \psi$	**Log-precisions:** These parameters control the precision (inverse variance) of fluctuations that constitute the random part of a generative model
$x(t) = x^{(1)} \oplus x^{(2)} \oplus x^{(3)} \dots$	**Hidden states:** These hidden variables encode the hierarchical states in a generative model of dynamics in the world
$v(t) = v^{(1)} \oplus v^{(2)} \oplus v^{(3)} \dots$	**Hidden causes:** These hidden variables link different levels of a hierarchical generative model
$g(x^{(i)}, v^{(i)}, \theta)$ $f(x^{(i)}, v^{(i)}, \theta)$	**Deterministic mappings:** These are equations at the i-th level of a hierarchical generative model that map from states at one level to another and map hidden states to their motion within each level. They specify the deterministic part of a generative model
$\omega^{(i,v)}$ $\omega^{(i,x)}$	**Random fluctuations:** These are random fluctuations on hidden causes and the motion of hidden states. Gaussian assumptions about these fluctuations furnish the probabilistic part of a generative model
$\tilde{\Pi}^{(i,v)} = R^{(i,v)} \otimes \Pi(\gamma^{(i,v)})$ $\tilde{\Pi}^{(i,x)} = R^{(i,x)} \otimes \Pi(\gamma^{(i,x)})$	**Precision matrices:** These are the inverse covariances among (generalized) random fluctuations on the hidden cases and motion of hidden states
$R^{(i,v)}$ $R^{(i,x)}$	**Roughness matrices:** These are the inverse of a matrix encoding serial correlations among (generalized) random fluctuations on the hidden cases and motion of hidden states
$\tilde{\varepsilon}^{(i,v)} = \tilde{v}^{(i-1)} - \tilde{g}^{(i)}$ $\tilde{\varepsilon}^{(i,x)} = \mathcal{D}\tilde{x}^{(i)} - \tilde{f}^{(i)}$	**Prediction errors:** These are the prediction errors on the hidden causes and motion of hidden states evaluated at their current conditional expectation
$\xi^{(i,v)} = \tilde{\Pi}^{(i,v)}\tilde{\varepsilon}^{(i,v)}$ $\xi^{(i,x)} = \tilde{\Pi}^{(i,x)}\tilde{\varepsilon}^{(i,x)}$	**Precision-weighted prediction errors:** These are the prediction errors weighted by their respective precisions

a computational account of perception [22] that borrows heavily from statistical physics and machine learning [19, 38, 50]. However, its explanatory scope includes action and behavior [25] and may be linked, at a fundamental level, to our very existence [24]). In brief, the free-energy principle takes well-known statistical ideas and applies them to problems in population (ensemble) dynamics and self-organization [1, 36, 42, 54]. In applying these ideas, many aspects of our brains, how we perceive and the way we act become understandable as necessary and self-evident attributes of biological systems [9, 39]). To see this consider the following problem:

How, in a changing and unpredictable world, do biological agents resist a natural tendency to disorder and thermodynamic equilibrium? All the physics that we know, such as the fluctuation theorem (which generalizes the second law of thermodynamics; [16]), suggests that random fluctuations in our environment will ultimately change our physical states to the point we cease to exist (i.e., we should gently decompose or evaporate). And yet, biological systems seem to violate these laws, maintaining precise physiological states for long periods of time [4]. In other words, they occupy a small number of states with a high probability and avoid a large number of other states. In short, they appear to resist thermodynamic imperatives. Mathematically, we can summarize this remarkable capacity by saying biological agents maintain a low entropy bound on the distribution over states that they could occupy. Entropy is just the average surprise (aka surprisal or self information) or negative log probability of an agent being in a particular state (see Table 1). In short, the question we need to address is how biological systems minimize their average surprise. Surprise here just means something unexpected, like tripping and falling in the street. One might think that exotic phenomena from theories of pattern-formation and self-organization may provide a sufficient explanation for the emergence of orderly (unsurprising) state-transitions. However, they do not. These patterns certainly have beautiful structures that unfold over short periods of time; but self-organization *per se* cannot explain the ability of biological agents to avoid surprise indefinitely. However, there is a solution that is almost tautological in its simplicity:

The solution lies in noting that surprise in ensemble dynamics is exactly the same as the (negative log) evidence for a model in statistics: $\mathcal{L} = -\ln p(s \mid m)$ (see Table 1). The conceptual link between surprise and log-evidence rests on assuming that every agent or person is a model of their environment or, more specifically, the sensory data to which they are exposed. This means that to minimize average surprise (entropy), each agent should maximize the evidence for its model of sensory exchanges with the world. Model optimization of this sort is a solved problem in statistics and machine learning (e.g., [47, 50]). In fact, most forms of statistical inference rest on comparing the evidence for one model relative to another, given some data.

So what does this mean for our brains? It suggests that we are obliged to optimize our model of the world through evolution, neurodevelopment and learning. In other words, we are statistical machines that make inferences about the world, given the (sensory) data available to us. The idea that we are *inference machines* is very old and was most clearly articulated by the renowned physicist [71]. Indeed, perception has been explicitly equated with hypothesis testing [34] and the brain has been referred to as a Helmholtz machine [13]. More recent incarnations of this idea appear as the Bayesian brain hypothesis [43, 46] and are instantiated in schemes like predictive coding [52, 57]. All these explanations borrow from Helmholtz's idea that the brain makes inferences about its sensations. A large body of work in theoretical neuroscience provides a plausible and compelling account of perception and the architecture of the wet-ware (brain) required to make these inferences. The ensuing perspective on biological systems says something quite profound: It says that all biological organisms can be regarded as a model of the environmental niche (econiche) they inhabit. In this sense, each species represents the product of evolutionary model optimization and each phenotype (including our brain) is a physical model or transcription of causal structure in its econiche. However, we have overlooked one small problem: Optimizing models is not easy and, in most situations, evaluating surprise or model evidence is an intractable problem. This is where variational free-energy comes in:

Free-energy was introduced (in the context of quantum physics) by Feynman [19] to solve the sort of difficult integration problems inherent in computing model evidence. It has been exploited in statistics and machine learning (e.g., [53]) as a very efficient way of measuring and maximizing model-evidence (i.e., minimizing surprise). The idea is quite simple, instead of trying to minimize something that cannot be measured, one simply creates a bound that can be measured, which is always bigger than the unknown quantity. One then minimizes the unknown quantity by minimizing the bound. So, what is this bound? In physics and statistics it is variational free-energy (recent statistical treatments of evolution consider a related quantity called free-fitness; [64]). Its construction is simple (See Fig. 1): The free-energy bound is constructed by adding a non-negative (Kullback–Leibler divergence) quantity to surprise. The clever thing is that adding this term renders the free-energy easily measurable. This Kullback–Leibler divergence measures the difference between two probability distributions; the first is called a variational density and is an arbitrary probability distribution used to create the bound. The second is the posterior or conditional density on the causes of our sensations (for example the presence of an object in our field of view). The posterior density is the probability of causes after seeing their consequences. Minimizing the bound reduces the difference between the variational and the posterior density. When they are identical, free-energy becomes surprise or negative log-evidence. This means to evaluate surprise, we have to make (Bayesian) inferences about what caused our sensations. This is the Bayesian brain hypothesis, where minimizing free-energy entails Bayes-optimal perception. In short, free-energy converts an intractable mathematical integration problem into a simple optimization problem. This statistical device furnishes another important perspective on how we, as organisms, work. It suggests that we minimize surprise by optimizing an upper bound on surprise. In other words, everything we do can be cast in terms of optimization. This is self-evidently true in many contexts, certainly in fields like reinforcement learning and economics [8, 12, 58, 68] but also fields like evolutionary biology, where adaptive fitness is optimized.

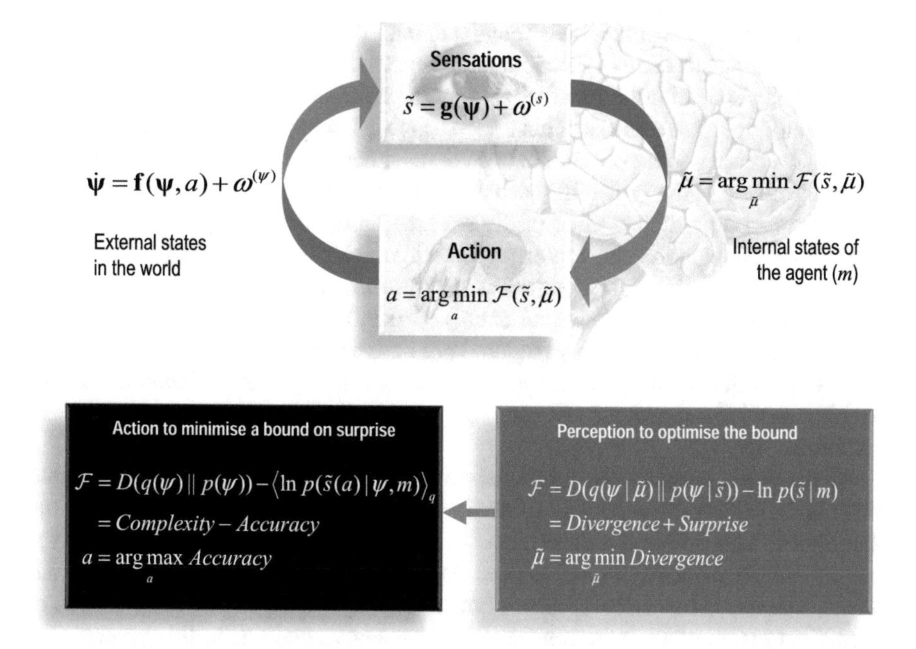

Fig. 1 **The free-energy principle**. This schematic shows the dependencies among the quantities that define the free-energy of an agent or brain, denoted by m. These include, its generalized internal states $\tilde{\mu}(t)$ and sensory signals $\tilde{s}(t)$ (generalized states include their generalized motion; i.e., velocity, acceleration *etc.*.). The environment is described by equations, which specify the motion of its states ψ, when depend on action $a(t)$. Both internal brain states and action minimize free-energy $\mathcal{F}(\tilde{s}, \tilde{\mu})$, which is a function of sensory input and the internal states. Internal states encode a variational density $q(\psi \mid \tilde{\mu})$ on the causes of sensory input. These comprise states of the world and the amplitude of random fluctuations $\omega(t)$. The *lower* panels provide the key equations behind the free-energy formulation. The *right* equality shows that optimizing brain states, with respect to the internal states, makes the variational density an approximate conditional density on the causes of sensory input. Furthermore, it shows that free-energy is an *upper* bound on surprise. This is because the first term of the equality is a divergence between the variational density and the true conditional or posterior density. Because this divergence can never be less than zero, minimizing free-energy renders it a proxy for surprise. At the same time, the variational density becomes the posterior density. The *left* equality shows that action can only reduce free-energy by selectively sampling sensory data that are predicted under the variational density

2.1 The Bayesian Brain

The Bayesian brain hypothesis makes complete sense in this context. If our imperative is to reduce surprise, then we need to have some reference or expectations against which to measure surprise. These expectations depend upon a model of the world and its current state. The probabilistic state of the world we infer is the variational density above (Fig. 1) and – when things are working properly – corresponds to the true but unknown posterior density. In the brain, this variational density (or more precisely, its sufficient statistics like its mean or expectation) may be encoded by neuronal activity or connection strengths among different parts of the brain. This leads to an understanding of perceptual inference and learning as changing synaptic activity and connectivity respectively, to minimize free-energy.

There are many schemes that have been proposed to implement this optimization. Among the more popular is predictive coding. Under some simplifying assumptions about the shape of the probability densities involved, the free-energy reduces to the sum of squared prediction error (see Fig. 2). In short, minimizing free-energy corresponds to reducing prediction errors. The hierarchical scheme depicted in Fig. 3 represents a fairly plausible architecture that the brain might use to suppress prediction errors and thereby reduce free-energy. Crucially, this scheme is based upon a gradient descent on free-energy (squared prediction error) and, as such, can be cast as a set of ordinary differential equations. It is these equations of motion that we suppose provide a model for neuronal dynamics that will be used in the second part of this chapter.

In summary, surprise cannot be measured directly but we can induce a bound on surprise called variational free-energy and reduce this bound by optimizing the activity and connectivity in our brains. This renders free-energy approximately the same as surprise and obliges us to make Bayesian inferences about the state of our world. The implementation of this optimization may rest upon the minimization of prediction errors of the sort considered by predictive coding. In this context, the gradient descent on free-energy (prediction errors) provides a plausible account or process theory for synaptic activity (perceptual inference) and synaptic efficacy (perceptual learning). An important aspect of this optimization is the proper estimation of the precision (inverse variance or uncertainty) associated with prediction errors. In the generalized predictive coding scheme of Fig. 3, we consider this precision to be encoded by synaptic gain, which has to be optimized in exactly the same way as synaptic activity (encoding expected states of the world) and synaptic efficacy (encoding the coupling among these states). The role of precision or synaptic gain will become important later when we consider the difference between action and action-observation later. The scheme described in Fig. 3 has been used to explain many different aspects of perceptual learning and inference in psychophysics and psychology. Figure 4 shows an example of perceptual categorization using simulated bird songs. However, perceptual *inference and learning does not itself reduce surprise*; it just reduces the difference between free-energy and surprise. To understand how surprise *per se* is reduced, we have to consider action and the active sampling of sensory data.

3 Active Inference

So far, we have seen that perception can be understood as furnishing a proxy for surprise, in the sense that perception reduces the divergence between the variational density and the true conditional density over hidden states causing sensations. In doing this, it makes free-energy a tighter bound or better approximation to surprise.

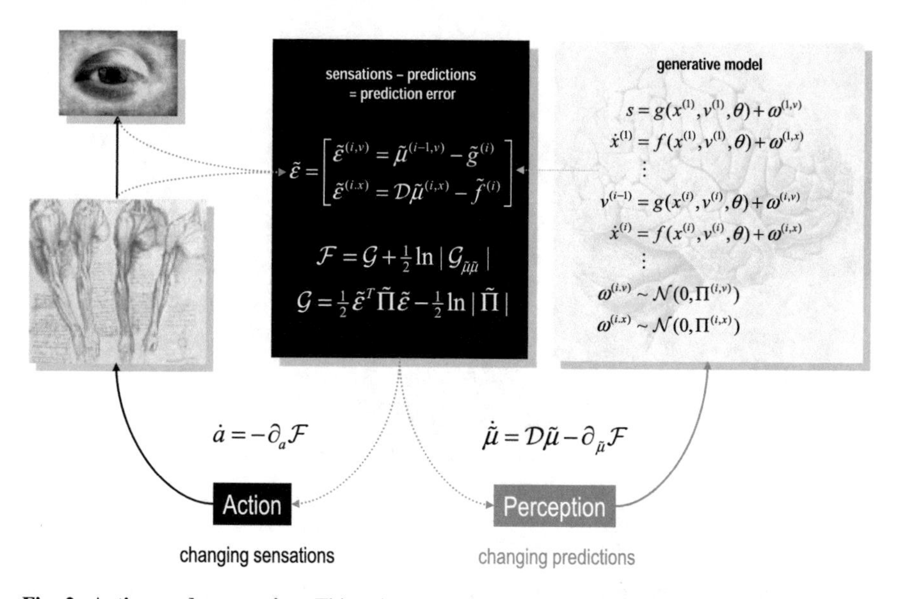

Fig. 2 Action and perception. This schematic illustrates the bilateral role of free-energy (i.e., prediction error) in driving action and perception: **Action**: Acting on the environment by minimizing free-energy enforces a sampling of sensory data that is consistent with the current representation (i.e., changing sensations to minimize prediction error). This is because free-energy is a mixture of complexity and accuracy (the first expression for free-energy in Fig. 1). Crucially, action can only affect accuracy. This means the brain will reconfigure its sensory epithelia to sample inputs that are predicted by its representations; in other words, to minimize prediction errors. The equation above action simply states that action performs a gradient decent on (i.e., minimizes) free-energy. **Perception**: Optimizing free-energy by changing the internal states that encode the variational density makes it an approximate posterior or conditional density on the causes of sensations. This follows because free-energy is surprise plus a Kullback–Leibler divergence between the variational and conditional densities (the second expression for free-energy in Fig. 1). Because this difference is non-negative, minimizing free-energy makes the variational density an approximate posterior probability. This means the agent implicitly infers or represents the causes of its sensory samples in a Bayes-optimal fashion. At the same time, the free-energy becomes a tight bound on surprise that is minimized through action. The equation above perception describes a gradient decent in a moving frame of reference for generalized states and accumulates gradients over time for the parameters. **Prediction error**: The equations show that the free-energy comprises a (Gibb's) energy $\mathcal{G}(t)$, which is effectively the (precision weighted) sum of squared prediction error. This error contains the sensory prediction error and other differences that mediate empirical priors on the motion of hidden states. The predictions rest on a generative model of how sensations are caused. These models have to explain complicated dynamics on continuous states with hierarchical or deep causal structure. An example of one such generic model is shown on the *right*. **Generative model**: Here $g^{(i)}$ and $f^{(i)}$ are continuous nonlinear functions of (hidden) causes and states, parameterized by $\theta \subset \psi$ at the i-th level of a hierarchical dynamic model. The random fluctuations $\omega^{(i,u)} : u \in x, v$ play the role of observation noise at the sensory level and state-noise at higher levels. Hidden causes $v^{(i)} \subset \psi$ link hierarchical levels, where the output of one level provides input to the next. Hidden states $x^{(i)} \subset \psi$ link dynamics over time and lend the model memory. Gaussian assumptions about the random fluctuations specify the likelihood of the model and furnish empirical priors in terms of predicted motion. These assumptions are encoded by the precision or inverse variance of the random fluctuations on hidden causes and the motion of hidden states; $\Pi^{(i,v)}$ and $\Pi^{(i,x)}$, respectively. These depend on precision parameters $\gamma \subset \psi$. The associated message-passing scheme implementing perception is shown in the next figure. Here, $\mathcal{D}$ is a temporal derivative operator that acts on generalized states

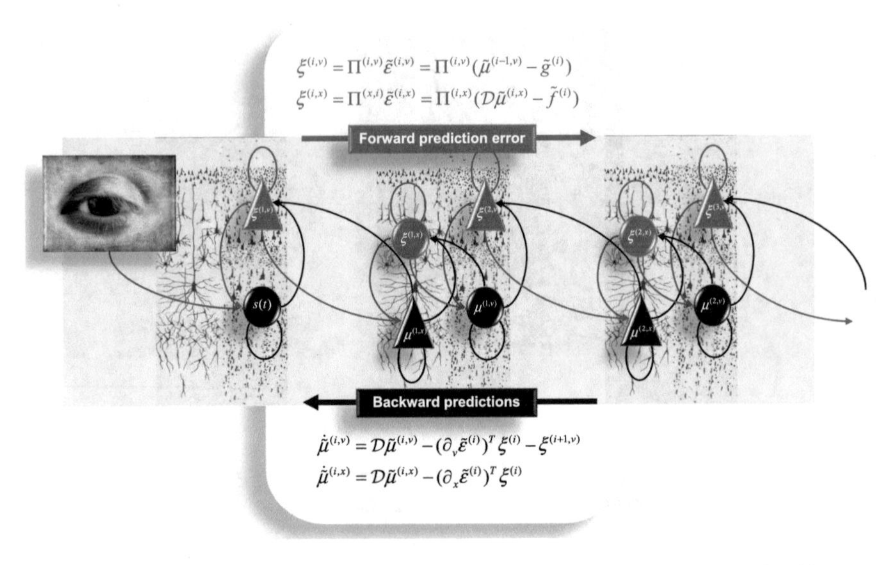

Fig. 3 Hierarchical message-passing in the brain. The schematic details a neuronal architecture that optimizes the conditional expectations of causes in hierarchical models of sensory input of the sort illustrated in the previous figure. It shows the putative cells of origin of forward driving connections that convey prediction-error from a *lower* area to a higher area (*red arrows*) and nonlinear backward connections (*black arrows*) that construct predictions [22, 52]. These predictions try to explain away (inhibit) prediction-error in *lower* levels. In this scheme, the sources of forward and backward connections are superficial and deep pyramidal cells (*triangles*) respectively, where state-units are *black* and error-units are *red*. The equations represent a generalized gradient descent on free-energy using the generative model of the previous figure**. Predictions and prediction-error**: If we assume that synaptic activity encodes the conditional expectation of states, then recognition can be formulated as a gradient descent on free-energy. Under Gaussian assumptions, these recognition dynamics can be expressed compactly in terms of precision weighted prediction-errors $\xi^{(i,u)} : u \in x, v$ on the causal states and motion of hidden states (at level i of the hierarchy). The ensuing equations suggest two neuronal populations that exchange messages; causal or hidden state-units encoding expected states and error-units encoding prediction-error. Under hierarchical models, error-units receive messages from the state-units in the same level and the level above; whereas state-units are driven by error-units in the same level and the level *below*. These provide *bottom-up* messages that drive conditional expectations $\mu^{(i,u)} : u \in x, v$ towards better predictions to explain away prediction-error. These *top-down* predictions correspond to $g(\tilde{\mu}^{(i,u)})$ that are specified by the generative model. This scheme suggests the only connections that link levels are forward connections conveying prediction-error to state-units and reciprocal backward connections that mediate predictions. Note that the prediction errors that are passed forward are weighted by their precision. This tells us that precision may be encoded by the postsynaptic gain or sensitivity or error units, which also has to be optimized: see [26] for further details

Next, we consider how action can actually reduce surprise. In brief, we can minimize prediction error in one of two ways: We can either change our expectations or predictions (perception) or we can change the things that are predicted (action). This perspective suggests that we should selectively sample data (or place ourselves in relation to the world) so that we experience what we expect to experience. In other

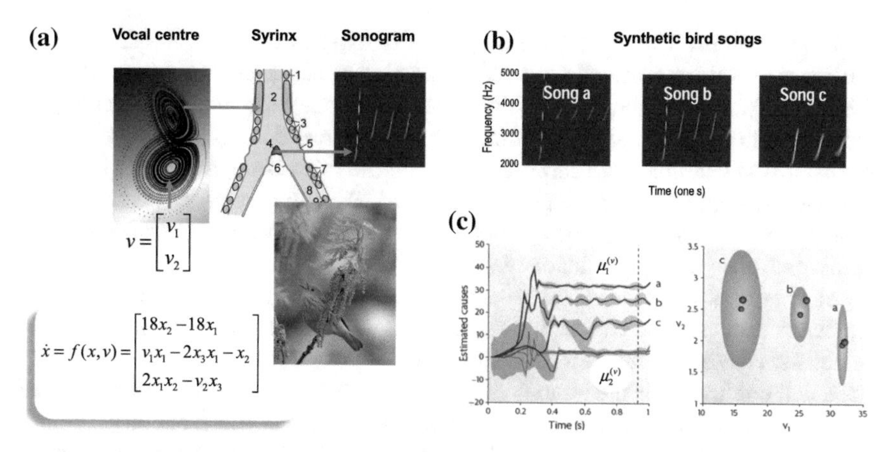

Fig. 4 Birdsongs and perceptual categorization. **a** The generative model of birdsong used in this simulation comprises a Lorenz attractor with two control parameters (or hidden causes) (v_1, v_2), which, in turn, delivers two control parameters to a synthetic syrinx to produce 'chirps' that were modulated in amplitude and frequency (an example is shown as a sonogram). Simulated chirps were presented to a synthetic bird to see if it could infer the hidden causes and thereby categorize the song. This entails minimizing free-energy by changing the conditional expectations of the control parameters. Examples of this perceptual inference or categorization are shown on the *right*. **b** Three simulated songs are shown (*upper panels*) in sonogram format. Each comprises a series of chirps whose frequency and number fall progressively from song **a** to song **c**, as a causal state (known as the Raleigh number; v_1 in the *left panel*) is decreased. **c** The graph on the *left* depicts the conditional expectations of the hidden causes, shown as a function of peristimulus time for the three songs. It shows that the causes are identified after about 600 milliseconds with high conditional precision (90 % confidence intervals are shown in *grey*). The graph on the *right* shows the conditional density on the causes shortly before the end of peristimulus time (i.e., the *dotted line* in the *left panel*). The *small dots* correspond to conditional expectations and the *grey* areas correspond to the 90 % conditional confidence regions. Note that these encompass the true values (*large dots*) that were used to generate the songs. These results illustrate the nature of perceptual categorization under the inference scheme in Fig. 3: Here, recognition corresponds to mapping from a continuously changing and chaotic sensory input to a fixed point in perceptual space

words, we will act upon the world to ensure that our predictions come true [23]. This is exactly the sort of behavior that we were trying to explain at the beginning; namely, how do biological systems avoid surprising exchanges with the environment?

It is fairly easy to show that the only part of free-energy that can be changed by action is sensory prediction error. This simple fact provides a nice explanation for how we interact with the world at a number of levels. First, in biological terms, it suggests that our muscles are wired to cancel sensory prediction errors. We are all familiar with this as a reflex: If I stretched the muscles in your leg by tapping the tendons below your knee, then they respond by contracting to cancel the unpredicted stretch-receptor signals. This reflects a basic functional architecture in movement (motor) control; whereby movements are elicited by prediction errors about the position of limbs – this is the classical motor reflex. If we generalize this view of how the brain controls our bodies, then peripheral motor or muscle systems are enslaved to fulfill predictions. This means we only have to expect or predict an action and it will be executed automatically – an old idea dating back to ideomotor theories of the 19th-century. The resulting perspective implies a curious yet compelling relationship between action and perception: on the one hand, perception optimizes predictions so that action can minimize surprise, while, on the other hand, our motor behavior is prescribed entirely by perceptual predictions. If action and perception work in synergy, we will navigate our econiche, never straying from well trodden paths, eluding surprise (and potential danger).

At a more abstract level, the selective sampling of sensory data we expect to encounter may provide a metaphor for the way we live. This is particularly true of scientists, who spend most of their life designing experiments to gather data they hope will confirm their predictions (hypotheses). It is precisely this imperative that underlies variational explanations for the way we sample data from our visual world, with saccadic eye movements [18, 29, 73]. Indeed, one could regard any phenotype as garnering evidence for its own existence. This brings us back to the notion that each individual is a model of its environment – a model that has to be continually affirmed by actively sampling from that environment. So far, we have only considered action as supplying further evidence for internal models of how the world works. Is this sufficient to explain behaviors such as goal-seeking, exploration and innovation? Not quite. To conclude this summary we will look at the fundamental role of prior expectations in shaping predictions and behavior.

3.1 Polices and Priors

Clearly, if each individual is adapted or optimized to their own environment, either at an evolutionary level or on a day-to-day basis in terms of learning and inference, the expectations of each individual must differ. Furthermore, we must inherit some aspect of these expectations, such that the physical form encoding each generation's model of its econiche is conserved (e.g., the way that the brain is wired). This speaks to the important role of innate or prior expectations about how and what we will sample from the world. For example, the fact we have eyes belies the fact our environment is bathed in light; and we avoid the dark because we expect to see things. This perspective touches on situated and embodied cognition [74] and the notion that we adapt environment to fulfill our expectations [5, 45]).

In the free-energy formulation, prior expectations are a key determinant of behavior and are an integral part of an individual's model. These priors may not be very complicated but can have a profound effect on what we expose ourselves to. For example, one could cast innately rewarding states (e.g. being sated or warm) as states we expect to encounter and that we are least likely to avoid. In brief, if an agent expects to move when, and only when, it is not in a rewarding state, then action will fulfill this prior and remove it from costly (non-rewarding states). This means that the probability of finding an agent in a non-rewarding state is much smaller than finding it in a rewarding state. This is precisely the low entropy distribution of states we want to explain and can be accounted for by one prior belief: "I will move unless rewarded". Simulations of this implicit policy produce remarkably intentional and adaptive behaviors that can solve benchmark problems in optimal control theory (like the mountain car problem: see Fig. 5 for an example). Heuristically, agents in these simulations move through the space of their states as if they were in a medium with negative viscosity or friction. This means that in most parts of state-space they speed up, until they find states they, a priori, believe they should occupy. At this point the viscosity becomes positive and the agent slows down to exploit the state it expects to be in. The trick here is to formulate prior expectations about movement through state space in terms of cost or loss-functions, of the sort considered in reinforcement and value learning. In this example, cost (negative reward) controls the viscosity the agent believes it will encounter at different points in state space. Viscosity is positive only in low cost or rewarding regions and it is these regions that agents populate. This is one example of a generic link between active inference and optimal decision theory. Briefly, the complete class theorem suggests that all admissible decision rules are equivalent to a Bayesian decision policy given some prior beliefs and a cost-function [55]. This means that any decision is optimal for a Bayesian generative model and cost-function or it is not rational. In fact, the implicit equivalence between priors and cost-functions means that we can recast any cost-function as a prior belief.

The example in Fig. 5 illustrates this by incorporating the cost-function into prior beliefs about motion through state-space. It also illustrates a simple mechanism for generating itinerant or wandering exploration of state-space. The basic idea is that unattractive or surprising fixed points destroy themselves by being rendered unstable. We have referred to this as autovitiation (see [23] for details). This sort of prior on state-transitions (a policy) provides a simple explanation for foraging behavior in ethology and, to a certain extent, addresses the exploitation-exploration trade-off in game theory and economics [10, 41].

3.1.1 Epistemic (Intrinsic) and Pragmatic (Extrinsic) Priors

The very existence of the exploitation and exploration trade-off speaks to the epistemic value of certain behaviors in reducing uncertainty or expected surprise in the future. A generic formulation of the prior beliefs about action that covers both epistemic (information seeking) and pragmatic (goal seeking) behavior is afforded in terms of *expected free energy*. In brief, if it is sufficient for organisms to minimize variational free energy, then it is sufficient for them to have prior beliefs that they will minimize expected free energy – via their action. In other words, actions are a priori more likely if they reduce the expected free energy of future states. It is relatively straightforward to show that expected free energy can be decomposed or carved in a number of ways that make sense in relation to established criteria for adaptive behavior. For example, one can express expected free energy in terms of information gain or Bayesian surprise plus expected utility – where utility is defined as log probability of preferred outcomes. Conversely, one can express expected free energy in terms of ambiguity and risk [32].

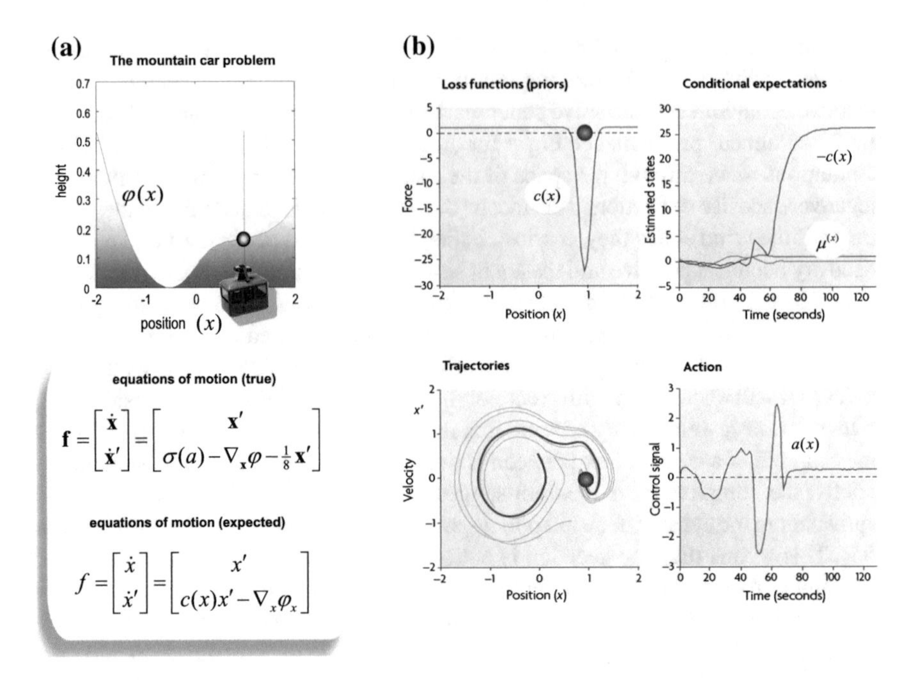

Fig. 5 **Solving the mountain car problem with prior expectations**. **a** This Figure shows how paradoxical but adaptive behavior (e.g. moving away from a target to ensure it is secured later) emerges from simple priors on the motion of hidden states in the world. The Figure shows the landscape or potential energy function (with a minimum at position $x = -0.5$) that exerts forces on a mountain car. The car is shown at the target position on the hill at $x = 1$, indicated by the ball. The true and expected equations of motion of the car are shown below. Crucially, at $x = 0$ the force on the car cannot be overcome by the agent, because a squashing function $-1 \leq \sigma(a) \leq 1$ is applied to action to prevent it being greater than one. This means that the agent can only access the target by starting halfway up the *left* hill to gain enough momentum to carry it up the other side. **B.** The results of active inference under priors that destroy fixed points outside the target domain. The priors are encoded in a loss or cost function $c(x)$ (*upper left*), which acts like negative friction. When 'friction' is negative the car expects to go faster. The inferred hidden states (*upper right*) show that the car explores its landscape until it encounters the target, when friction increases (i.e., cost decreases) dramatically to prevent the car from escaping the target (by falling down the hill). The ensuing trajectory is shown in the *lower left*. The *paler lines* provide exemplar trajectories from different trials, with different starting positions. In the real world, friction is constant. However, the car 'expects' friction to change as it changes position, thus enforcing exploration or exploitation. These expectations are fulfilled by action (*lower right*)

The imperative to minimize expected free energy provides a useful explanation for saccadic searches of the visual environment [28] and has been used to model a variety of tasks in psychology and behavioral economics: for example, waiting games [24] the urn task and evidence accumulation [20], trust games from behavioral economics [51, 62], addictive behavior [63], two-step maze tasks [32] and the mountain car problem considered in Fig. 5 [28]. It has also been used in the setting of computational fMRI [61]. Given the importance of prior beliefs, one might ask where they come from.

The answer is implicit in their evolutionary motivation; in that they can be specified genetically and elaborated through (Bayes-optimal) learning. This explains how one generation tells the next what is valuable (expected, preferred or characteristic of a phenotype), without having to prescribe the details of how to attain preferred states. This is a nice aspect of the free-energy formulation because it connects dynamics at different levels or scales. For example, the same free-energy is minimized by inferring things about someone on the phone and by the evolution of our ancestors. The only difference is that the long-term average of free-energy is optimized by evolution, development and learning, whereas perception minimizes free-energy over short time scales. Interestingly, the long-term average or path-integral of energy is called *action* in physics. This means the free-energy principle is just an example of Hamilton's principle of least *action*.[1]

3.2 Summary

In conclusion, we have reviewed the free-energy principle in terms of explaining how self-organizing adaptive and biological systems manage to resist a tendency to disorder. When we unpack this principle, we see that it accommodates both perception and action, while embedding the action-perception cycle in an evolutionary context. We have seen that the underlying imperative of all biological systems can be expressed as minimizing (a free-energy bound) on surprise; and that surprise, self-evidently, depends upon predictions. These predictions can be constrained by prior expectations (that will minimize free energy), which allow our behavior to be optimized by evolution and neurodevelopment (learning). In the next section, we will apply these ideas to understand how agents emit sequences of movements or action. We will focus on handwriting, noting that the same principles should apply to any structured and sequential pattern of behavior. This example has been chosen to highlight the central role of itinerant dynamics in furnishing prior expectations about action and concomitant perception.

4 Action and Its Observation

In this section, we describe a generative model of handwriting and then apply the fee-energy scheme of the previous section to simulate the emergent neuronal dynamics and behavior. To create these simulations, all we have to do is specify a generative model. This model and (generalized) sensations define the free-energy, which determines the dynamics of action and neuronal states encoding the conditional expec-

[1] We have used italics to distinguish *action* (integral of energy) from action (enacted by the agent).

tations of hidden states in the world. Action and perception are prescribed by the equations in Fig. 2, which simulate neuronal and behavioral responses respectively.

$$
\begin{aligned}
\dot{\tilde{\mu}} &= \mathcal{D}\tilde{\mu} - \partial_{\tilde{\mu}}\mathcal{F}(\tilde{s}(a), \mu) \\
\dot{a} &= -\partial_a \mathcal{F}(\tilde{s}(a), \mu) \\
\dot{\boldsymbol{\Psi}} &= \mathbf{f}(\boldsymbol{\Psi}, a) + \omega^{(\psi)} \\
\tilde{s} &= \mathbf{g}(\boldsymbol{\Psi}) + \omega^{(s)}
\end{aligned}
\tag{1}
$$

The first equation represents a generalized or instantaneous gradient descent on free-energy for the conditional expectations of hidden states causing sensory input (i.e., neuronal activity). The first term represents their expected generalized motion, while the second is simply the gradient of the free-energy with respect to the expectations. The reason that this is a *generalized* descent is that it is formulated in generalized coordinates of motion, such that the first term augments and anticipates the descent so that it becomes effectively instantaneous. Representing states in generalized coordinates of motion $\tilde{\mu} = (\mu, \mu', \mu'', \ldots)$ means that each state comprises its current value, velocity, acceleration, jerk and so on. This means that the generalized motion $\mathcal{D}\tilde{\mu} = (\mu', \mu'', \mu'' \ldots)$ just involves shifting generalized expectations to the left: see [23] for details.

The second equality is the equivalent gradient descent for action. Both of these equations rest upon the free-energy, which is a function of sensory information and current expectations. This function depends upon a generative model, which is specified completely by equations of motion of the hidden states and a function mapping hidden states to sensory signals (see Fig. 2). This means all we have to do to simulate action and perception is to specify the equations of the generative model and then solve or integrate Eq. 1 over time.

The second pair of equalities describes how true (but hidden) states in the world evolve and generate sensations, where both are subject to random fluctuations. The dependencies among hidden states, sensory states, expectations and action mean that sensations and action constitute a *Markov blanket* that separates hidden (external) states from expected (internal) states (see also Fig. 1). The very existence of this Markov blanket can be used to show that the free energy principle is necessarily true for any ergodic system; including those that possess random dynamical attractors and therefore attain *nonequilibrium steady-state*. This provides a theoretical back story for the free energy principle, as a fundament of biological self organisation [24, 31].

In what follows, we describe the generative model that will be used for the remainder of this chapter. We have chosen this model because it entails the sort of itinerant dynamics found in real-world movements, such as reading and walking.

4.1 Itinerant Dynamics and Attractors

Our agent was equipped a simple hierarchical model of its sensorium based on a Lotka–Volterra system. The particular form of this model has been discussed previously as the basis of putative speech decoding [44]. Here, it is used to model a stable heteroclinic channel [59] encoding successive locations to which the agent expects its arm to move. The resulting trajectory was contrived to simulate synthetic handwriting.

A stable heteroclinic channel is a particular form of itinerant trajectory or orbit that revisits a sequence of (unstable) fixed points. In our model, there are two sets of hidden states, which we will associate with two levels of a hierarchical model. The first set $x^{(2)} \in \mathbb{R}^{6\times 1}$ corresponds to the state-space of a Lotka–Volterra system. This is an abstract (attractor) state-space, in which a series of attracting points are visited in succession. The second set $x^{(1)} = \{x_1, x_2, x_1', x_2'\}$ corresponds to the (angular) positions and velocities of two joints in (two dimensional) physical space. The dynamics of hidden states at the first level embody the agent's prior expectation that the arm will be drawn to a particular location, $v^{(1)} = g(x^{(2)})$ specified by the attractor states of the second level. This is implemented simply by placing a (virtual) elastic band between the tip of the arm and the attracting location. The hidden states basically draw the arm's extremity (finger) to a succession of locations to produce an orbit or trajectory, under classical Newtonian mechanics. We chose the locations so that the resulting trajectory looked like handwriting. These hidden states generate both proprioceptive and visual (exteroceptive) sensory data: The proprioceptive data are the angular positions and velocities of the two joints $x^{(1)}$, while the visual information was the location of the arm in physical (Cartesian) space $\{\ell_1, \ell_1 + \ell_2\}$, where $\ell_2(x^{(1)})$ is the displacement of the finger from the location of the second joint $\ell_1(x^{(1)})$ (see Fig. 6 and Table 2).

Crucially, because this generative model generates two (proprioceptive and visual) sensory modalities, the solutions to Eq. 1 implement Bayes-optimal multisensory integration. However, because action is also trying to reduce prediction errors, it will move the arm to reproduce the expected trajectory (under the constraints of the motor plant). In other words, the arm will trace out a trajectory prescribed by the itinerant priors (to cancel proprioceptive prediction errors). This closes the loop, producing autonomous self-generated sequences of behavior of the sort described below. Note that the real world does not contain any attracting locations or elastic bands: The only causes of observed movement are the self-fulfilling expectations encoded by the itinerant dynamics of the generative model. In short, hidden attractor states essentially prescribe the intended movement trajectory, because they generate predictions that action fulfils. This means expected states encode conditional percepts (concepts) about latent abstract states (that do not exist in the absence of action), which play the role of intentions. We now describe the model formally.

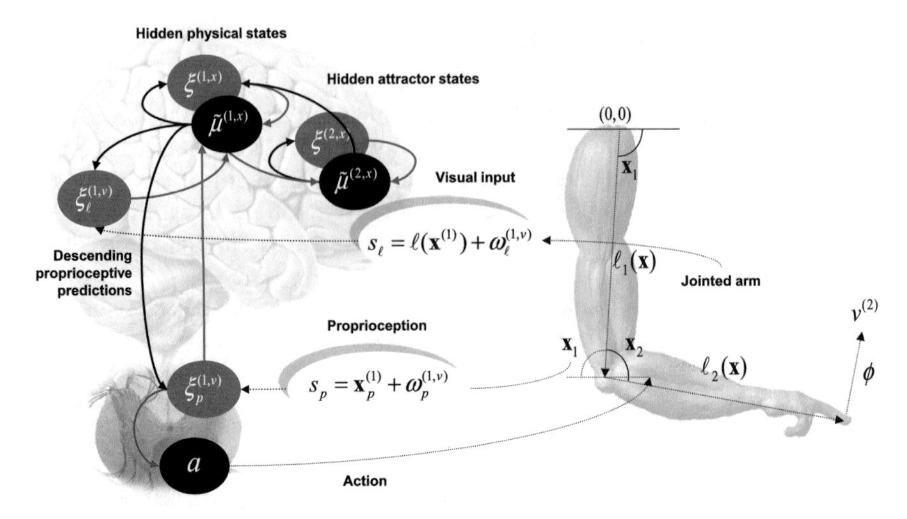

Fig. 6 Simulating self-generated movement. This schematic details a simulated (mirror neuron) system and the motor plant that it controls (*left* and *right* respectively). The *right panel* depicts the functional architecture of the supposed neural circuits underlying active inference. The *red ellipses* represent prediction error-units (neurons or populations), while the *black ellipses* denote state-units encoding conditional expectations about hidden states of the world (for simplicity, we have omitted hidden causes). The hidden states are split into two hierarchical levels: the *higher* abstract attractor states (that supports stable heteroclinic orbits) and *lower* physical states of the arm (angular positions and velocities of the two joints). *Red arrows* are forward connections conveying prediction errors and *black arrows* are backward connections mediating predictions. Motor commands are emitted by the *black* units in the ventral horn of the spinal cord. Note that these just receive prediction errors about proprioceptive states. These, in turn, are the difference between sensed proprioceptive input from the two joints and descending predictions from optimized representations in the motor cortex. The two jointed arm has a state space that is characterized by two angles, which control the position of the finger that will be used for writing in subsequent figures

4.2 The Generative Model

The model used in this section concerns the movements of a two-joint arm. When simulating active inference, it is important to distinguish between the agent's generative model and the actual dynamics generating sensory data. To make this distinction clear, we will use bold for true equations and states, while those of the generative model will be written in italics. Proprioceptive input corresponds to the angular position and velocity of both joints, while the visual input corresponds to the location of the extremities of both parts of the arm. This means the mapping from hidden states to sensory consequences is:

$$\mathbf{g}^{(1)} = g^{(1)} = \begin{bmatrix} x^{(1)} \\ \ell_1(x^{(1)}) \\ \ell_1(x^{(1)}) + \ell_2(x^{(1)}) \end{bmatrix} \tag{2}$$

Table 2 Variables and quantities specific to the writing example of active inference (see main text for details)

Variable	*Description*
$x^{(2)} \in \mathbb{R}^{6\times 1}$	**Hidden attractor states:** A vector of hidden states that specify the current location towards which the agent expects its arm to be pulled
$x^{(1)} \in \mathbb{R}^{4\times 1}$	**Hidden effector states:** Hidden states that specify the angular position and velocity of the i-th joint of a two-jointed arm
$\ell_1(x^{(1)}) \in \mathbb{R}^{2\times 1}$ $\ell_2(x^{(1)}) \in \mathbb{R}^{2\times 1}$	**Joint locations:** Locations of the end of the two arm parts in Cartesian space. These are functions of the angular positions of the joints
$v^{(1)} = g(x^{(2)}) \in \mathbb{R}^{2\times 1}$	**Attracting location:** The location towards which the arm is drawn. This is specified by the hidden attractor states
$\phi(x^{(1)}, v^{(1)}) \in \mathbb{R}^{2\times 1}$	**Newtonian Force:** This is the angular force on the joints exerted by the attracting location
$A \in \mathbb{R}^{6\times 6} \subset \theta$	**Attractor parameters:** A matrix of parameters that govern the (sequential Lotka–Volterra) dynamics of the hidden attractor states
$L \in \mathbb{R}^{2\times 6} \subset \theta$	**Cartesian parameters:** A matrix of parameters that specify the attracting locations associated with each hidden attractor state

We will ignore the complexities of inference on retinotopically mapped visual input and assume the agent has direct access to the locations of the arm in visual space. The kinetics of the arm conforms to Newtonian laws, under which action forces the angular position of each joint. Both joints have an equilibrium position at ninety degrees; with inertia $m_i \in 8, 4$ and viscosity $\kappa_i \in 4, 2$, giving the following equations of motion for the hidden states

$$\mathbf{x}^{(1)} = \begin{bmatrix} \mathbf{x}_1 \\ \mathbf{x}_2 \\ \mathbf{x}_1' \\ \mathbf{x}_2' \end{bmatrix} \mathbf{f}^{(1)} = \begin{bmatrix} \mathbf{x}_1' \\ \mathbf{x}_2' \\ (a_1 + \mathbf{v}_1 - \frac{1}{4}(\mathbf{x}_1 - \frac{\pi}{2}) - \kappa_1 \mathbf{x}'_1)/m_1 \\ (a_2 + \mathbf{v}_2 - \frac{1}{4}(\mathbf{x}_2 - \frac{\pi}{2}) - \kappa_2 \mathbf{x}'_2)/m_2 \end{bmatrix} \tag{3}$$

However, the agent's empirical priors on this motion have a very different form. Its generative model assumes the finger is pulled to a (goal) location $v^{(1)}$ by a force $\phi(t)$, which implements the virtual elastic band above:

$$x^{(1)} = \begin{bmatrix} x_1 \\ x_2 \\ x'_1 \\ x'_2 \end{bmatrix} f^{(1)} = \begin{bmatrix} x'_1 \\ x'_2 \\ (\phi^T \ell_2 \ell_2^T O \ell_1 - \frac{1}{16}(x_1 - \frac{\pi}{2}) - \kappa_1 x'_1)/m_1 \\ (\phi^T O \ell_2 - \frac{1}{16}(x_2 - \frac{\pi}{2}) - \kappa_2 x'_2)/m_2 \end{bmatrix} \tag{4}$$

$$\ell_1 = \begin{bmatrix} \cos(x_1) \\ \sin(x_1) \end{bmatrix} \quad \ell_2 = \begin{bmatrix} -\cos(-x_2 - x_1) \\ \sin(-x_2 - x_1) \end{bmatrix} \quad O = \begin{bmatrix} 0 & -1 \\ 1 & 0 \end{bmatrix}$$

$$\phi = \tfrac{1}{2}(v^{(1)} - \ell_1 - \ell_2)$$

The (moving) target location is specified by the second level of the hierarchy as a nonlinear (softmax) function of the hidden attractor states.

$$\begin{aligned} v^{(1)} &= g(x^{(2)}) = Ls(x^{(2)}) \\ f^{(2)} &= A\sigma(x^{(2)}) - \tfrac{1}{8}x^{(2)} + \mathbf{1} \\ \sigma(x_i) &= \tfrac{1}{1+e^{2x_i}} \quad s(x_i) = \tfrac{e^{2\alpha_i}}{\sum_j e^{2x_j}} \end{aligned} \tag{5}$$

Heuristically, these equations of motion mean that the agent thinks that changes in its world are caused by the dynamics of attractor states on an abstract (conceptual) space. The currently active state selects a location $v^{(1)}$ in the agent's physical (Cartesian) space, which exerts a force $\phi(t)$ on its finger. The equations of motion in Eq. 4 pertain to the resulting motion of the arm in Cartesian space, while Eq. 5 mediates the attractor dynamics driving these movements.

The (Lotka–Volterra) form of the equations of motion for the hidden attractor states ensures that only one has a high value at any one time and imposes a particular sequence on the underlying states. Lotka–Volterra dynamics basically induce competition among states that no state can win. The resulting winnerless competition rests on the (logistic) function $\sigma(x^{(2)})$, while the sequence order is determined by the elements of the matrix

$$A = \begin{bmatrix} 0 & -\frac{1}{2} & -1 & -1 & \cdots \\ -\frac{3}{2} & 0 & -\frac{1}{2} & -1 & \\ -1 & -\frac{3}{2} & 0 & -\frac{1}{2} & \ddots \\ -1 & -1 & -\frac{3}{2} & 0 & \\ \vdots & & \ddots & & \ddots \end{bmatrix} \tag{6}$$

Each attractor state has an associated location in Cartesian space, which draws the arm towards it. The attracting location is specified by a mapping from attractor space to Cartesian space, which weights different locations,

$$L = \begin{bmatrix} 1 & 1.1 & 1.0 & 1 & 1.4 & 0.9 \\ 1 & 1.2 & 0.4 & 1 & 0.9 & 1.0 \end{bmatrix}$$

with a softmax function $s(x^{(2)})$ of the attractor states. The location parameters were specified by hand but could, in principle, be learnt as described in [23, 25]. The inertia and viscosity of the arm were chosen somewhat arbitrarily to reproduce realistic writing movements over 256 time bins, each corresponding to roughly eight milliseconds (i.e., a second). Unless stated otherwise, we used a log-precision of four for sensory noise and eight for random fluctuations in the motion of hidden states.

Figure 7 shows the results of integrating Eq. 1, using the generative model above. The top right panel shows the hidden states embodying Lotka–Volterra dynamics (the hidden joint states are smaller in amplitude). These generate predictions about the position of the joints (upper left panel) and consequent prediction errors that drive action. Action is shown on the lower right panel and displays intermittent forces that move the joint to produce a motor trajectory. This trajectory is shown on the lower left in visual space over time. This trajectory or orbit is translated as a function of time to reproduce handwriting. Although this is a pleasingly simple way of simulating an extremely complicated motor trajectory, it should be noted that this agent has a very limited repertoire of behaviors; it can only reproduce this sequence of graphemes, and will do so *ad infinitum*.

In summary, we have covered the functional architecture of a generative model whose autonomous (itinerant) expectations prescribe complicated motor sequences through active inference. This rests upon itinerant dynamics (stable heteroclinic channels) that can be regarded as a formal prior on abstract causes in the world. These are translated into physical movement through classical Newtonian mechanics, which correspond to the physical states of the model. Action tries to fulfill predictions about proprioceptive inputs and is enslaved by autonomous predictions, producing realistic behavior. These trajectories are both caused by neuronal representations of abstract (attractor) states and cause those states in the sense that they are conditional expectations. Closing the loop in this way ensures a synchrony between internal expectations and external outcomes.

An interesting technical issue – that follows from formulating kinetics in generalized coordinates of motion – is that prior beliefs about the amplitude of random fluctuations enter free energy in a way that is consistent with minimum variance heuristics in the motor control literature [37]. This follows from the fact that smooth trajectories can be characterized by precise (low variance) velocities and accelerations. In short, if an agent believes its generalised motion is precise, it will execute smooth movements of the sort seen empirically.

In the next section, we will make a simple change which means that movements are no longer caused by the agent. However, we will see that the conditional expectations about attractor states are relatively unaffected, which means that they still anticipate observed movements. We conclude with this example because it illustrates nicely the potential role of itinerant dynamics in explaining some of the higher cognitive aspects of brain function. Our focus here is on emulating the electrophysiological phenomenology of the mirror neuron system; in particular, the fact that certain neurons in the ventral premotor cortex and inferior parietal cortex respond not only to the execution of particular movement primitives but also when these movements are observed in other agents [14, 21, 33, 60].

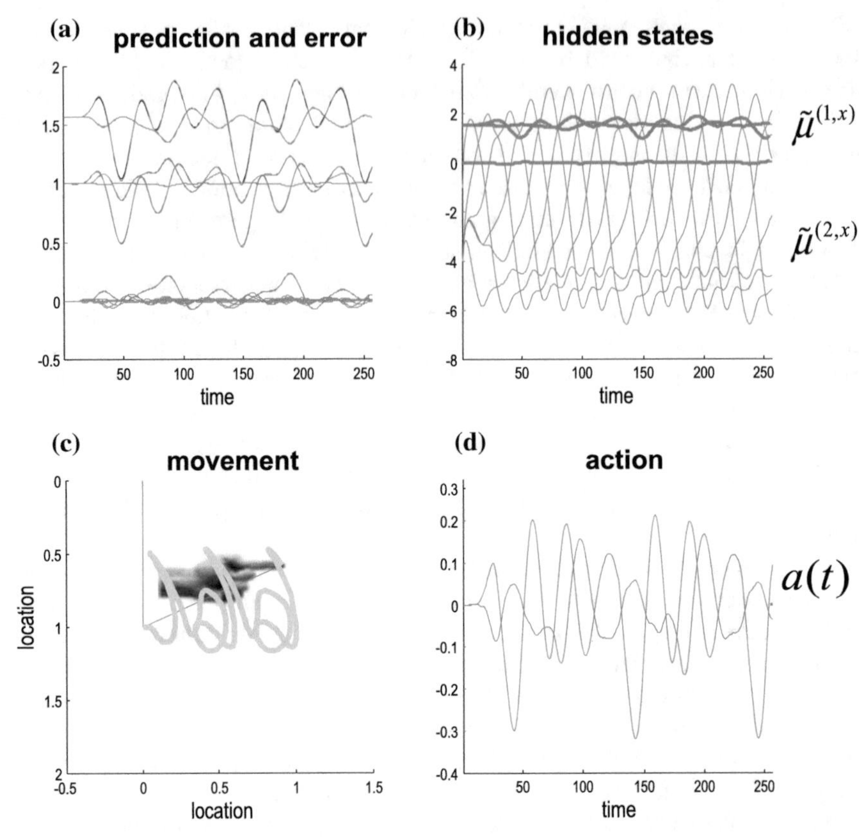

Fig. 7 Itinerant dynamics and active inference. This figure shows the results of simulated action (writing), under active inference, in terms of conditional expectations about hidden states of the world (*upper right*), consequent predictions about sensory input (*upper left*) and the ensuing behavior (*lower left*) that is caused by action (*lower right*). The autonomous dynamics that underlie this behavior rest upon the expected hidden states that follow Lotka–Volterra dynamics. These are the thinner lines in the *upper right panel*. The hidden physical states (*thicker lines*) have smaller amplitudes and map directly on to the predicted proprioceptive and visual signals (shown on the *left*). The visual locations of the two joints are shown above the predicted joint positions and angular velocities that fluctuate around zero. The *dotted lines* correspond to prediction error, which shows small fluctuations about the prediction. Action tries to suppress this error by 'matching' expected changes in angular velocity through exerting forces on the joints. These forces are shown on the *lower right*. The subsequent movement of the arm is traced out on the *lower left*; this trajectory has been plotted in a moving frame of reference so that it looks like synthetic handwriting (e.g., a succession of 'j' and 'a' letters). The straight lines on the *lower left* denote the final position of the two jointed arm and the hand icon shows the final position of its finger

4.3 Action-Observation

The simulations above were repeated but with one small but important change. Basically, we reproduced the same movements but the proprioceptive consequences of action were removed, so that the agent could see but not feel the arm moving. From the agent's perspective, this is like seeing an arm that looks like its own arm but does not generate proprioceptive input (i.e., the arm of another agent). However, the agent still expects the arm to move with a particular itinerant structure and will try to predict the trajectory with its generative model. In this instance, the hidden states still represent itinerant dynamics (intentions) that govern the motor trajectory but these states do not produce any proprioceptive prediction errors and therefore do not result in action. Crucially, the perceptual representation still retains its anticipatory or prospective aspect and can therefore be taken as a perceptual representation of intention, not of self, but of another. We will see below that this representation is almost exactly the same under action-observation as it is during action.

Practically speaking, to perform these simulations, we simply recorded the forces produced by action in the previous simulation and replayed them as exogenous forces (real causes in Eq. 2) to move the arm. This change in context (agency) was modeled by down-weighting the precision of proprioceptive signals. This is exactly the same mechanism that we have used previously to model attention [17]. In this setting, reducing the precision of proprioceptive prediction errors prevents them from having any influence on perceptual inference (i.e., the agent cannot feel changes in its joints). Furthermore, action is not compelled to reduce these prediction errors because they have no precision. In these simulations, we reduced the log-precision of proprioceptive prediction errors from eight to minus eight. To illustrate the key results of these simulations of action-observation, in relation to simulated action, we recorded the activity of units encoding hidden attractor states and examined and their relationship to observed movements:

4.4 *Place-Cells, Itinerancy and Oscillations*

It is interesting to think about the attractor states as representing trajectories through abstract representational spaces (cf., the activity of place cells; [6, 56, 70]). Figure 8 illustrates the sensory or perceptual correlates of units representing expected attractor states. The left hand panels show the activity of one (the fourth) hidden state unit under action, while the right panels show exactly the same unit under action-observation. The top rows show the trajectories in visual space, in terms of horizontal and vertical displacements (grey lines). The black dots correspond to the time bins in which the activity of the hidden state unit exceeded an amplitude threshold of two arbitrary units. They key thing to take from these results is that the activity of this unit is very specific to a limited part of Cartesian space and, crucially, a particular trajectory through this space. The analogy here is between directionally selective place-cells of the sort studied in hippocampal recordings: In tasks involving goal-directed, stereotyped trajectories, the spatially selective activity of hippocampal cells depends on the animal's direction of motion [3]. A further interesting connection with hippocampal dynamics is the prevalence of theta rhythms during action: "Driven either by external landmarks or by internal dynamics, hippocampal neu-

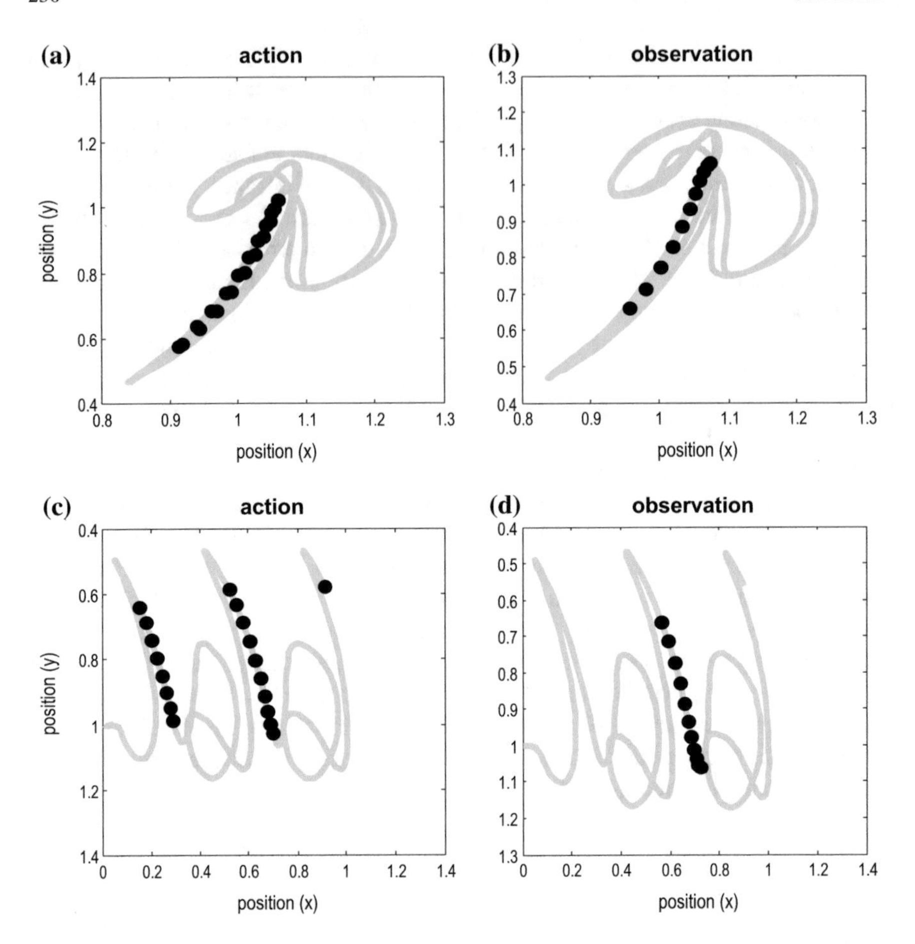

Fig. 8 Simulating action-observation. These results illustrate the sensory or perceptual correlates of units encoding expected hidden (attractor) states. The *left hand panels* show the activity of one (the fourth attractor) hidden state-unit under action, while the *right panels* show exactly the same unit under action-observation. The *top rows* show the trajectory in visual space in terms of *horizontal* and *vertical* position (*grey lines*). The *dots* correspond to the time bins during which the activity of the state-unit exceeded an amplitude threshold of two arbitrary units. They key thing to take from these results is that the activity of this unit is very specific to a limited part of visual space and, crucially, a particular trajectory through this space. Notice that the same selectivity is seen under action and observation. The implicit direction selectivity can be seen more clearly in the *lower panels*, in which the same data are displayed but in a moving frame of reference to simulate writing. They key thing to note here is that this unit responds preferentially when, and only when, the motor trajectory produces a down-stroke, but not an up-stroke

rons form sequences of cell assemblies. The coordinated firing of these active cells is organized by the prominent "theta" oscillations in the local field potential (LFP): place cells discharge at progressively earlier theta phases as the rat crosses the respective place field (phase precession)" [66]. Quantitatively, the dynamics of the hidden state-units in Fig. 7 (upper left panel) show quasiperiodic oscillations in the (low) theta range. The notion that quasiperiodic oscillations may reflect stable heteroclinic channels is implicit in many treatments of episodic memory and spatial navigation, which "require temporal encoding of the relationships between events or locations" [15], and may be usefully pursued in the context of active inference under itinerant priors.

Notice that the same 'place' and 'directional' selectivity is seen under action and observation (Fig. 8 right and left columns). The direction selectivity can be seen more clearly in the lower panels, in which the same data are displayed but in a moving frame of reference (to simulate writing). They key thing to note here is that this unit responds preferentially when, and only, when the motor trajectory produces a down-stroke, but not an up-stroke. There is an interesting dissociation in the firing of this unit under action and action-observation: during observation the unit only starts responding to down-strokes *after* it has been observed once. This reflects the finite amount of time required for visual information to entrain the perceptual dynamics and establish veridical predictions.

5 Conclusion

In this chapter, we have tried to show that many aspects of action, perception and high-level (cognitive) inference are consistent with (Bayes-optimal) active inference under the free-energy principle. Put simply, the brain does not represent intended motor acts or the perceptual consequences of those acts separately. The constructs represented in the brain are both intentional and perceptual: They are amodal inferences about the states of the world generating sensory data that have both sensory and motor correlates, depending upon the context in which they are made. The predictions generated by these representations are modality-specific, prescribing both exteroceptive (e.g., visual) and interoceptive (e.g., proprioceptive) predictions, which action fulfils. The functional segregation of motor and sensory cortex could be regarded as a hierarchical decomposition, in the brain's model of its world, which provides predictions that are primarily sensory (e.g. visual cortex) or proprioceptive (motor cortex). If true, this means that high level representations can be used to furnish predictions in either visual or proprioceptive modalities, depending upon the context in which those predictions are called upon. This picture of functional anatomy fits well with the juxtaposition of somatosensory and motor cortex – and speaks to the participation of the motor and premotor cortex to sensory information flow through the thalamus [67].

In one sense, this conclusion takes us back to very early ideas concerning the nature of movements and intentions. The notion of an ideomotor reflex or response was introduced in the 1840s by the Victorian physiologist and psychologist William Benjamin Carpenter. The ideomotor response (reflex) refers to the process whereby a thought or mental image induces reflexive or automatic movements, often very small and potentially outside awareness. Active inference formalizes this idea and suggests that all movements are prescribed by mental images that correspond to prior beliefs about what will happen next. These priors are inherently dynamic and itinerant. This suggests that our exchanges with our environment are constrained to an exquisite degree by local and global brain dynamics; and that these dynamics have been carefully crafted by evolution, neurodevelopment and experience to optimize behavior. In short, our variational treatment is quintessentially enactivist [2, 65].

Acknowledgements The Wellcome trust funded this work. I would also like to thank Daniel Bennequin for invaluable help in formulating these ideas.

Appendix

This appendix provides a brief technical overview of how the free energy principle applies to neuronal dynamics. In this setting, the states of the brain (e.g., the activity of neurons and other systems that are crucial for its function, such as glial cells), are viewed as encoding the sufficient statistics of probability measures on hidden states of the external world. In this view, the main quantities are probabilities measures, denoted by $p(\tilde{s}, \psi \mid m)$, on the product $S \times \Psi$ of possible values of (generalized) sensory states and hidden states, under a particular model m. Time plays a hidden but fundamental role in this formalism, in the sense that (S, Ψ) are path spaces, and $(\tilde{s}, \psi)$ are points in manifolds that depend on time. Particular attention is required by this point in Bayesian modelling [11, 48]).

The underlying premise is that the sufficient statistics $\tilde{\mu}$ and the induced probability $q(\psi \mid \tilde{\mu})$ evolve to maximize the marginal likelihood or model evidence:

$$p(\tilde{s} \mid m) = \int d\psi p(\tilde{s}, \psi \mid m) \tag{A.1}$$

However, this marginalization is generally intractable. The main simplification rests on replacing the difficult marginalization in A.1, by the practically easier problem of minimizing free energy:

$$\mathcal{F} = E_q[\mathcal{G}(\tilde{s}, \psi)] - H[q(\psi \mid \tilde{\mu})] \tag{A.2}$$

where $H[q(\psi \mid \tilde{\mu})]$ denotes the entropy of the probability law $q(\psi \mid \tilde{\mu})$ on Ψ, and the first term is the Gibbs internal energy $\mathcal{G}(\tilde{s}, \psi) = -\ln p(\tilde{s}, \psi \mid m)$ expected under $q(\psi \mid \tilde{\mu})$. It is easy to show that the concavity of the logarithmic function on $]0, \infty[$ implies:

$$\mathcal{F}(\tilde{s}, \tilde{\mu}) \geq -\ln p(\tilde{s} \mid m) \tag{A.3}$$

Then, the minimization of free energy with respect to the sufficient statistics $\tilde{\mu}$ affords a constraint on the good direction for the maximization of the marginal likelihood or model evidence. In general, the problem is further simplified, for instance by an Ansatz of mean-field approximation, or by reducing to a belief propagation algorithm; c.f., [72].

References

1. W.R. Ashby, Principles of the self-organizing dynamic system. J. Gen. Psychol. **37**, 125–128 (1947)
2. D.H. Ballard, D. Kit, C.A. Rothkopf, B. Sullivan, A hierarchical modular architecture for embodied cognition. Multisensory Res. **26**, 177 (2013)
3. F.P. Battaglia, G.R. Sutherland, B.L. McNaughton, Local sensory cues and place cell directionality: additional evidence of prospective coding in the hippocampus. J. Neurosci. **24**(19), 4541–4550 (2004)
4. C. Bernard, *Lectures on the phenomena common to animals and plants*. Trans Hoff HE, Guillemin R, Guillemin L, Springfield (IL): Charles C Thomas (1974). ISBN 978-0398028572
5. G. Buason, N. Bergfeldt, T. Ziemke, Brains, bodies, and beyond: competitive co-evolution of robot controllers, morphologies, and environments. Genet. Program. Evolvable Mach. **6**(1), 25–51 (2005)
6. N. Burgess, C. Barry, J. O'Keefe, An oscillatory interference model of grid cell firing. Hippocampus **17**(9), 801–812 (2007)
7. R.L. Carhart-Harris, K.J. Friston, The default-mode, ego-functions and free-energy: a neurobiological account of Freudian ideas. Brain **133**(Pt 4), 1265–1283 (2010)
8. C.F. Camerer, Behavioural studies of strategic thinking in games. Trends Cogn. Sci. **7**(5), 225–231 (2003)
9. A. Clark, Whatever next? Predictive brains, situated agents, and the future of cognitive science. Behav. Brain Sci. **36**, 181–204 (2013)
10. J.D. Cohen, S.M. McClure, A.J. Yu, Should I stay or should I go? How the human brain manages the trade-off between exploitation and exploration. Philos. Trans. R Soc. Lond. B Biol. Sci. **362**(1481), 933–942 (2007)
11. F. Colas, J. Diard, P. Bessiere, Common Bayesian models for common cognitive issues. Acta Biotheoretica **58**, 191–216 (2010)
12. N.D. Daw, K. Doya, The computational neurobiology of learning and reward. Current Opinion Neurobiol. **16**(2), 199–204 (2006)
13. P. Dayan, G.E. Hinton, R.M. Neal, The Helmholtz machine. Neural Comput. **7**, 889–904 (1995)
14. G. Di Pellegrino, L. Fadiga, L. Fogassi, V. Gallese, G. Rizzolatti, Understanding motor events: a neurophysiological study. Exp. Brain Res. **91**, 176–80 (1992)
15. G. Dragoi, G. Buzsáki, Temporal encoding of place sequences by hippocampal cell assemblies. Neuron **50**(1), 145–157 (2006)
16. D.J. Evans, A non-equilibrium free-energy theorem for deterministic systems. Mol. Phys. **101**, 1551–1554 (2003)
17. H. Feldman, K.J. Friston, Attention, uncertainty, and free-energy. Front Hum. Neurosci. **4**, 215 (2010)
18. M. Ferro, D. Ognibene, G. Pezzulo, V. Pirrelli, Reading as active sensing: a computational model of gaze planning during word recognition. Front. Neurorobotics **4**, 1 (2010)
19. R.P. Feynman, *Statistical Mechanics* (Benjamin, Reading, 1972)
20. T.H. FitzGerald, P. Schwartenbeck, M. Moutoussis, R.J. Dolan, K. Friston, Active inference, evidence accumulation, and the urn task. Neural Comput. **27**, 306–328 (2015)
21. L. Fogassi, P.F. Ferrari, B. Gesierich, S. Rozzi, F. Chersi, G. Rizzolatti, Parietal lobe: from action organization to intention understanding. Science **308**, 662–667 (2005)
22. K.J. Friston, A theory of cortical responses. Philos. Trans. R. Soc. Lond. B Biol. Sci. **360**, 815–836 (2005)

23. K. Friston, The free-energy principle: a unified brain theory? Nat. Rev. Neurosci. **11**(2), 127–138 (2010)
24. K. Friston, Life as we know it. J. R. Soc. Interface **10**, 20130475 (2013)
25. K. Friston, J. Daunizeau, S. Kiebel, Active inference or reinforcement learning? PLoS One **4**(7), e6421 (2009)
26. K.J. Friston, J. Daunizeau, J. Kilner, S.J. Kiebel, Action and behavior: a free-energy formulation. Biol. Cybern. **102**(3), 227–260 (2010)
27. K. Friston, K. Stephan, B. Li, J. Daunizeau, Generalised filtering. Math. Probl. Eng. **2010**, 621670 (2010)
28. K. Friston, R. Adams, R. Montague, What is value-accumulated reward or evidence? Front. Neurorobotics **6**, 11 (2012)
29. K. Friston, R.A. Adams, L. Perrinet, M. Breakspear, Perceptions as hypotheses: saccades as experiments. Front. Psychol. **3**, 151 (2012)
30. K. Friston, P. Schwartenbeck, T. FitzGerald, M. Moutoussis, T. Behrens, J. Raymond, R.J. Dolan, The anatomy of choice: active inference and agency. Front. Hum. Neurosci. **7**, 598 (2013)
31. K. Friston, M. Levin, B. Sengupta, G. Pezzulo, Knowing one's place: a free-energy approach to pattern regulation. J. R. Soc. Interface **12** (2015)
32. K. Friston, F. Rigoli, D. Ognibene, C. Mathys, T. Fitzgerald, G. Pezzulo, Active inference and epistemic value. Cogn. Neurosci. 1–28 (2015)
33. V. Gallese, A. Goldman, Mirror-neurons and the simulation theory of mind reading. Trends Cogn. Sci. **2**, 493–501 (1998)
34. R.L. Gregory, Perceptions as hypotheses. Phil. Trans. R. Soc. Lond. B **290**, 181–197 (1980)
35. M. Grist, Changing the Subject. RSA. www.thesocialbrain.wordpress.com, pp. 74–80 (2010)
36. H. Haken, *Synergetics: An Introduction. Non-Equilibrium Phase Transition and Self-Organization in Physics, Chemistry and Biology*, 3rd edn. (Springer, Heidelberg, 1983)
37. C.M. Harris, D.M. Wolpert, Signal-dependent noise determines motor planning. Nature **394**, 780–784 (1998)
38. G.E. Hinton, D. van Cramp, Keeping neural networks simple by minimizing the description length of weights, in *Proceedings of COLT-93* (1993), pp. 5–13
39. J. Hohwy, The Self-Evidencing Brain. Noûs (2014)
40. G. Huang, Is this a unified theory of the brain? New Sci. Mag. (2658) (2008)
41. S. Ishii, W. Yoshida, J. Yoshimoto, Control of exploitation-exploration meta-parameter in reinforcement learning. Neural Netw. **15**(4–6), 665–687 (2002)
42. S. Kauffman, *The Origins of Order: Self-Organization and Selection in Evolution* (Oxford University Press, Oxford, 1993)
43. D. Kersten, P. Mamassian, A. Yuille, Object perception as Bayesian inference. Annu. Rev. Psychol. **55**, 271–304 (2004)
44. S.J. Kiebel, K. von Kriegstein, J. Daunizeau, K.J. Friston, Recognizing sequences of sequences. PLoS Comput. Biol. **5**, e1000464 (2009)
45. D. Kirsh, Adapting the environment instead of oneself. Adapt. Behavr. **4**(3/4), 415–452 (1996)
46. D.C. Knill, A. Pouget, The Bayesian brain: the role of uncertainty in neural coding and computation. Trends Neurosci. **27**(12), 712–719 (2004)
47. D. Kropotova, D. Vetrovb, General solutions for information-based and Bayesian approaches to model selection in linear regression and their equivalence. Pattern Recognit. Image Anal. **19**(3), 447–455 (2009)
48. O. Lebeltel, P. Bessière, Basic concepts of Bayesian programming, in *Probabilistic Reasoning and Decision Making in Sensory-Motor Systems* (Springer, 2008), pp. 19–48
49. E.M. Lifshitz, L.P. Pitaevskii, *Physical Kinetics* (Pergamon, London, 1981)
50. D.J.C. MacKay, Free-energy minimization algorithm for decoding and cryptoanalysis. Electron. Lett. **31**, 445–447 (1995)
51. M. Moutoussis, N.J. Trujillo-Barreto, W. El-Deredy, R.J. Dolan, K.J. Friston, A formal model of interpersonal inference. Front. Hum. Neurosci. **8**, 160 (2014)

52. D. Mumford, On the computational architecture of the neocortex. II. The role of cortico-cortical loops. Biol. Cybern. **66**, 241–51 (1992)
53. R.M. Neal, G.E. Hinton, A view of the EM algorithm that justifies incremental, sparse, and other variants', in *Learning in Graphical Models*, ed. by M.I. Jordan (Kluwer Academic Publishers, Dordrecht, 1998), pp. 355–368
54. G. Nicolis, I. Prigogine, *Self-Organization in Non-Equilibrium Systems* (Wiley, New York, 1977)
55. D.W. North, A tutorial introduction to decision theory. IEEE Trans. Syst. Sci. Cybern. **4**(3), 200–210 (1968)
56. J. O'Keefe, Do hippocampal pyramidal cells signal non-spatial as well as spatial information? Hippocampus **9**(4), 352–364 (1999)
57. R.P. Rao, D.H. Ballard, Predictive coding in the visual cortex: a functional interpretation of some extra-classical receptive field effects. Nat. Neurosci. **2**, 79–87 (1998)
58. R.A. Rescorla, A.R. Wagner, A theory of Pavlovian conditioning: variations in the effectiveness of reinforcement and nonreinforcement, in *Classical Conditioning II: Current Research and Theory*, ed. by A.H. Black, W.F. Prokasy (Appleton Century Crofts, New York, 1972), pp. 64–99
59. M. Rabinovich, R. Huerta, G. Laurent, Neuroscience. Transient dynamics for neural processing. Science **321**, 48–50 (2008)
60. G. Rizzolatti, L. Craighero, The mirror-neuron system. Ann. Rev. Neurosci. **27**, 169–192 (2004)
61. P. Schwartenbeck, T.H. FitzGerald, C. Mathys, R. Dolan, K. Friston, The dopaminergic midbrain encodes the expected certainty about desired outcomes. Cereb. Cortex **25**, 3434–3445 (2015)
62. P. Schwartenbeck, T.H. FitzGerald, C. Mathys, R. Dolan, M. Kronbichler, K. Friston, Evidence for surprise minimization over value maximization in choice behavior. Sci. Rep. **5**, 16575 (2015)
63. P. Schwartenbeck, T.H. FitzGerald, C. Mathys, R. Dolan, F. Wurst, M. Kronbichler, K. Friston, Optimal inference with suboptimal models: addiction and active Bayesian inference. Med. Hypotheses **84**, 109–117 (2015)
64. G. Sella, A.E. Hirsh, The application of statistical physics to evolutionary biology. Proc. Natl. Acad. Sci. USA **102**(27), 9541–9546 (2005)
65. A.K. Seth, Interoceptive inference, emotion, and the embodied self. Trends Cogn. Sci. **17**, 565–573 (2013)
66. W.E. Skaggs, B.L. McNaughton, M.A. Wilson, C.A. Barnes, Theta phase precession in hippocampal neuronal populations and the compression of temporal sequences. Hippocampus **6**, 149–172 (1996)
67. S.M. Sherman, R.W. Guillery, *Exploring the Thalamus and its Role in Cortical Function* (MIT Press, Cambridge, 2006)
68. R.S. Sutton, A.G. Barto, Toward a modern theory of adaptive networks: expectation and prediction. Psychol. Rev. **88**(2), 135–170 (1981)
69. C. Thornton, Some puzzles relating to the free-energy principle: comment on Friston. Trends Cogn. Sci. **14**(2), 53–54 (2010); author reply 54–55
70. M. Tsodyks, Attractor neural network models of spatial maps in hippocampus. Hippocampus **9**(4), 481–489 (1999)
71. H. von Helmholtz, Concerning the perceptions in general, in *Treatise on physiological optics*, vol. III, 3rd edn. (1866) (translated by J.P.C. Southall, 1925 Opt. Soc. Am. Section 26, reprinted New York: Dover, 1962)
72. Y. Weiss, Correctness of local probability propagation in graphical models with loops. Neural Comput. **12**, 1–41 (2000)
73. R.H. Wurtz, K. McAlonan, J. Cavanaugh, R.A. Berman, Thalamic pathways for active vision. Trends Cogn. Sci. **5**, 177–184 (2011)
74. T. Ziemke, Introduction to the special issue on situated and embodied cognition. Cogn. Systs. Res. **3**(3), 271–274 (2002)

Modeling of Coordinated Human Body Motion by Learning of Structured Dynamic Representations

Albert Mukovskiy, Nick Taubert, Dominik Endres, Christian Vassallo, Maximilien Naveau, Olivier Stasse, Philippe Souères and Martin A. Giese

Abstract The modeling and online-generation of human-like body motion is a central topic in computer graphics and robotics. The analysis of the coordination structure of complex body movements in humans helps to develop flexible technical algorithms for movement synthesis. This chapter summarizes work that uses learned structured representations for the synthesis of complex human-like body movements in real-time. This work follows two different general approaches. The first one is to learn spatio-temporal movement primitives from human kinematic data,

---

J.P. Laumond et al. (Eds.): Geometric and Numerical Foundations of Movements, Springer STAR Series, 2016.

A. Mukovskiy (✉) · N. Taubert · M.A. Giese (✉)
Section for Computational Sensomotorics, Department of Cognitive Neurology, Hertie Institute for Clinical Brain Research & Center for Integrative Neuroscience, University Clinic Tübingen, Otfried-Müller Str. 25, 72076 Tübingen, Germany
e-mail: albert.mukovskiy@medizin.uni-tuebingen.de

N. Taubert
e-mail: nick.taubert@uni-tuebingen.de

M.A. Giese
e-mail: giese@uni-tuebingen.de

D. Endres
Theoretical Neuroscience Group, Section for General and Biological Psychology, Department of Psychology, University of Marburg, Gutenbergstr. 18, 35032 Marburg, Germany
e-mail: dominik.endres@uni-marburg.de

C. Vassallo · M. Naveau · O. Stasse · P. Souères
Gepetto Lab, LAAS/CNRS, Université de Toulouse, Av. du Colonel Roche 7, 31400 Toulouse, France
e-mail: christian.vassallo@laas.fr

M. Naveau
e-mail: maximilien.naveau@laas.fr

O. Stasse
e-mail: ostasse@laas.fr

P. Souères
e-mail: philippe.soueres@laas.fr

J.-P. Laumond et al. (eds.), *Geometric and Numerical Foundations of Movements*, Springer Tracts in Advanced Robotics 117, DOI 10.1007/978-3-319-51547-2_11

and to derive from this Dynamic Movement Primitives (DMPs), which are modeled by nonlinear dynamical systems. Such dynamical primitives are then coupled and embedded into networks that generate complex human-like behaviors online, as self-organized solutions of the underlying dynamics. The flexibility of this approach is demonstrated by synthesizing complex coordinated movements of single agents and crowds. We demonstrate that Contraction Theory provides an appropriate framework for the design of the stability properties of such complex composite systems. In addition, we demonstrate how such primitive-based movement representations can be embedded into a model-based predictive control architecture for the humanoid robot HRP-2. Using the primitive-based trajectory synthesis algorithm for fast online planning of full-body movements, we were able to realize flexibly adapting human-like multi-step sequences, which are coordinated with goal-directed reaching movements. The resulting architecture realizes fast online planing of multi-step sequences, at the same time ensuring dynamic balance during walking and the feasibility of the movements for the robot. The computation of such dynamically feasible multi-step sequences using state-of-the-art optimal control approaches would take hours, while our method works in real-time. The second presented framework for the online synthesis of complex body motion is based on the learning of hierarchical probabilistic generative models, where we exploit Bayesian machine learning approaches for non-linear dimensionality reduction and the modeling of dynamical systems. Combining Gaussian Process Latent Variable Models (GPLVMs) and Gaussian Process Dynamical Models (GPDMs), we learned models for the interactive movements of two humans. In order to build an online reactive agent with controlled emotional style, we replaced the state variables of one actor by measurements obtained by real-time motion capture from a user and determined the most probable state of the interaction partner using Bayesian model inversion. The proposed method results in highly believable human-like reactive body motion.

Keywords Dynamic movement primitives · Animation · Machine learning · Gaussian process latent variable model · Gaussian process dynamical model · Navigation · Walking pattern generator · Goal-directed movements · Motor coordination · Action sequences

1 Introduction

The generation of realistic human movements in reactive fashion is a difficult task with high relevance for computer graphics and robotics. An especially challenging task in this domain is the online-synthesis of complex behaviors that consist of sequences of individual actions, which adapt to continuously changing environmental constraints.

The whole body movements of humans and animals are organized in terms of muscle synergies or movement primitives [4, 17]. Such primitives characterize the coordinated involvement of subsets of the available degrees of freedom in different actions. An example is the coordination of periodic and non-periodic components of the full-body movements during reaching while walking, where behavioral studies reveal a mutual coupling between these components [8, 12, 47, 68]. The realism and human-likeness of synthesized movements in robotics and computer graphics can be improved by taking such biological constraints into account [15, 18, 73].

In this chapter we present two learning-based frameworks that make such biological properties applicable to the realtime synthesis of human-like movements in technical systems, one that is based on *Dynamic Movement Primitives (DMPs)*, and another one that exploits unsupervised Bayesian learning methods.

The chapter is organized into three main sections. Section 2 introduces a framework that approximates complex human movements by combining learned dynamic movement primitives. Highly adaptive coordinated full-body movements, and even the coordination of the movements of multiple agents, can be generated online by networks of such dynamic primitives, which are mutually coupled. Section 3 discusses how the same methods can be exploited for the movement planning of humanoid robots. We present an architecture that embeds such an online synthesis model into a control architecture of a real humanoid robot, which is based on model predictive control. The proposed solution ensures the dynamic balance of the robot, so that it is prevented from falling, while realizing highly flexible online planning of movements. The last Sect. 4 introduces a completely different approach for the learning-based representation of reactive human movements, which is based on Bayesian machine learning methods for dimension reduction and model inversion. Space constraints allow us only to give the outline of these different approaches, and we refer to the cited original publications with respect to many technical and mathematical details.

2 Modeling of Human Movements Based on Learned Primitives

Human full-body movements involve typically a large number of degrees of freedom. It has been a classical idea in biological motor control that such complex body movements might be composed from lower-dimensional control units, often referred to as *movement primitives* or *synergies*. Substantial work in motor control has been dedicated to the identification of such primitives from kinematic and EMG data, applying unsupervised learning techniques for dimension reduction [14, 31, 71]. Different techniques have been applied, including Principle Component Analysis (PCA), Independent Component Analysis (ICA), or more sophisticated methods that include time shifts of the superpositioned components. Such methods approximate a set of time-dependent signals by a superposition of learned source functions, which have been interpreted as movement primitives or (muscle) synergies.

Work in computer graphics shows that the accurate approximation of motion capture data from complex full-body movements using PCA requires typically more than 8 principal components (e.g. [70]). In the following section we describe a method that often leads to more compact representations with less components or primitives.

Such compact representations are important especially if parameters of the learned models have to be interpreted, e.g. in order to characterize motion styles [65]. Compact models tend to concentrate the data variance on a low number of interpretable parameters. Compact primitive-based representations are also beneficial if they are embedded into control systems or dynamic architectures for the online generation of motion. In this case, the number of primitives determines the dimensionality of the underlying system dynamics, and systems with lower dimensionality often are easier to control and more robust against perturbations.

In the following, after reviewing some related methods in Sect. 2.1, we give first a short introduction in the method that we apply to learn primitives from trajectory data (Sect. 2.2). The resulting kinematic primitives are given by basis functions or trajectories, which by appropriate combination can approximate complex joint angle trajectories. We then discuss how from such kinematic primitives dynamic movement primitives can be constructed that generate the learned trajectories online (cf. Sect. 2.3). These dynamic primitives are nonlinear dynamical systems that produce the learned basis trajectories as stable solutions. In the following Sect. 2.4 we demonstrate how such learned dynamical generative models can be augmented by controllers that make the behaviors adaptable, realizing for example navigation through space or the control of step length or emotional style. It is demonstrated that the developed approach is suitable for the online generation of quite complex coordinated behaviors, either of single agents or even of whole crowds of agents that execute coordinated collective behaviors. In Sect. 2.5 we discuss finally, how such complex generative dynamical models can be designed, guaranteeing the robustness of their solutions. Contraction Theory, a special type of mathematical stability analysis, which is especially applicable to nonlinear systems which are composed of many components, makes it possible to ensure that the desired behavior is the only stable solution of the resulting nonlinear dynamical system.

2.1 Related Work

The synthesis of the kinematics of sequences of human full-body movements has been treated extensively in computer graphics [42]. The prominent classical approach for the synthesis of human motion in computer graphics is the adaptive interpolation between motion-captured example actions, which is typically realized off-line [2, 22, 23, 93]. Other approaches are based on learned low-dimensional parameterizations of whole body motion that are embedded in mathematical frameworks for the online generation of motion [9, 27, 43, 44, 66, 70, 88]. In addition, a variety of methods for the segmentation of action streams into individual actions have been proposed, where the models for individual actions can be adapted online in order to fulfill additional constraints, such obstacle avoidance or the correct positioning of end-effectors [16, 28, 34, 62, 67]. Only very few of these works have focused on the modeling of the flexible coordination of groups of degrees of freedom, similar to synergies in biological systems [70, 75].

2.2 Approximation of Human Movement Data by Anechoic Mixtures

Many standard approaches use principle component analysis (PCA) or independent component analysis (ICA) for the reduction of the dimensionality of motion data. A set of trajectories is represented as a linear combination of a limited number of basis components or source functions. In our work we used a more sophisticated mixture model for the approximation of the joint angle trajectories that contains time shifts for the superposed components or sources [11, 56]. This model is known from acoustics as *anechoic mixture* and superposes source functions s_j that are temporally shifted with the time delays τ_{ij} in order to to approximate a set of trajectories $\xi_i(t)$. The corresponding model is characterized mathematically by the equation

$$\underbrace{\xi_i(t)}_{\text{angles}} = m_i + \sum_j w_{ij}\, \underbrace{s_j\left(t - \tau_{ij}\right)}_{\text{sources}} + \text{noise} \tag{1}$$

The parameters w_{ij} specify the mixing weights, and the variables m_i signify constant offsets (means) of the approximated trajectories. Learning of an anechoic mixture model requires the estimation of these parameters, the source functions s_j, and the delays τ_{ij}. In our case, the trajectories were given by the angle trajectories of 17 joints expressed as quaternions. We have shown in previous work for different classes of human movements that this anechoic mixture model results in very accurate approximations of complex human movent data, often with as few as 3–4 source functions, and typically with factor 2 less sources than classical approaches using PCA or ICA [50, 60].

2.3 Online Synthesis by Networks of Dynamic Primitives

The discussed mixture model can be applied for an off-line analysis and synthesis of classes of trajectories. Movement types or styles can be characterized by the mixing weights (and delays) of the model, and the movement can be analyzed using these weights as features. In addition, novel movement trajectories can be generated off-line by specifying or interpolating these parameters, and using the Eq. (1) as a generative model [65]. However, this approach is not sufficient for applications that require an online synthesis of complex movements.

In order to make the learned structured model applicable for real-time synthesis we associated each learned source function (*kinematic primitive*) with an associated dynamic primitive [19, 29]. The dynamic primitives are defined by dynamical systems whose stable solutions approximate the learned source functions. Each dynamical primitive is defined by a *canonical dynamical system*, which has an attractor solution with well-defined mathematical properties. We used limit cycle oscillators (Andronov-Hopf oscillators) for the approximation of periodic source functions, and

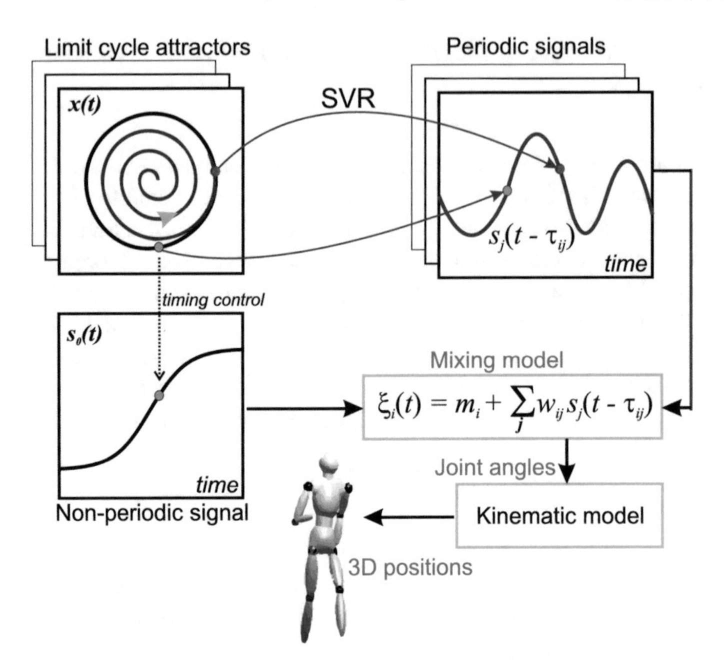

Fig. 1 Architecture for the online synthesis of body movements using dynamic primitives. The solutions $x_j(t)$ of a canonical dynamical systems (limit cycle oscillators) are mapped by Support Vector Regression (SVR) onto the values of the periodic source functions $s_j(t)$. In addition, a non-periodic source function $s_0(t)$ is constructed from these solutions. From these online generated source functions joint angle trajectories are computed using the learned anechoic mixing model

a ramp-like solution, which is derived from the state of a limit cycle oscillator, for the non-periodic ones. We then learned nonlinear functions that map the state spaces of the canonical dynamics onto the values of the source functions $s_j(t)$ using Support Vector Regression (SVR) [10]. Figure 1 shows an overview of the developed architecture for real-time synthesis.

By insertion of couplings between the different canonical dynamical systems it is possible to synchronize their dynamics, so that the corresponding source signals are evolving in synchrony. Such couplings can be used either to model coordinated behavior between the movement primitives within a single agent, or by introduction of couplings between the dynamics of multiple agents, for the simulation of coordinated interactive behavior of multiple agents.

2.4 Style Morphing and Navigation

The proposed method for the online generation of body motion trajectories can be combined with a dynamic variations of motion style. For this purpose multiple

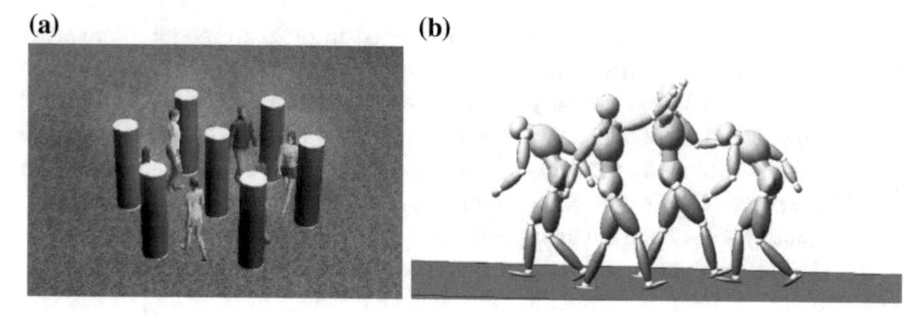

Fig. 2 **a** Reactive online control of locomotion. Agents avoid the obstacles (poles) and other agents in the scene. Trajectories are generated by morphing between steps with different length, and curvatures of the walking path (*left*, *straight*, *right*), where blending weight are controlled by a navigation dynamics that controls the heading direction dependent on obstacle and goal positions. **b** Folk dance of two couples, one forming a bridge, and the other crouching beneath it. The behavior is fully self-organized, where the behavior of the agents depends on the relative positions with respect to the other agents

examples of the same motion were motion-captured that realize different styles, and intermediate styles were generated by online interpolation (motion blending). For thus purpose, we linearly interpolated the average angles m_i, the mixing weights w_{ij}, and the delays τ_{ij} of mixing models that were learned from training trajectories representing different motion styles. (See [21, 50] for further details.)

The blending weights were modulated by controllers that depend on task parameters, such as the position or orientations of agents in the scene, distances of agents to goal points, etc. One example is the generation of walking steps that realize locomotion along curved paths by morphing between straight and curved walking steps to the right or to the left. In this case, the morphing weights of the three walking patterns were determined by a controller that determines the heading direction dependent on obstacles and desired goal points (cf. [74, 91]). Likewise, movements with different emotional styles can be generated by blending between models that realize the same motion with different emotional styles, or steps with different length can be generated by morphing between long and short steps.

We worked out an application of this approach for the simulation of locomoting and navigating agents. Blending weights were controlled by a simplified version of a dynamic navigation model that had been applied successfully in robotics before [74], and which we extended by inclusion of a prediction of expected collision points with obstacles in order to make the navigation behavior more human-like [59]. The heading direction is controlled by a nonlinear first-order differential equation that depends on the actual positions of the agents and of obstacles in the scene. (See [21, 59] for details.)

An example for the navigation behavior that can be simulated with the described architecture is illustrated in Fig. 2a, where six agents avoid the obstacles and the other agents. **Demo Movie I**[1] shows this example and other obstacle avoidance

[1] http://tinyurl.com/hvwv9ra.

scenarios. The same method can also be exploited in order to model interactions between multiple agents that realize more complex behaviors, which integrate periodic and non-periodic movement primitives. An example is shown in Fig. 2b that shows a figure from a folk dance that requires one couple of agents to walk beneath a bridge that is formed by the arms of another couple. Both both couples take turn and change places. This whole complex behavior with highly human-like appearance was completely self-organized using only 10 prototypical movements (normal walk, walks with left or right arm lifted, crouching walk, left and right forward turnings with two different angular velocities, left and right backward stepping turns). See also **Demo Movie I**. The proposed approach thus can be used to simulate highly complex full-body coordination patterns, and even patterns that include multiple agents. The underlying architecture is very simple, consisting only of a low-dimensional nonlinear dynamical system and some linear and nonlinear mappings. This makes it possible to generate the behavior, even of larger groups of agents in realtime.

2.5 Dynamic Stability Design Exploiting Contraction Theory

Effectively, the proposed method synthesizes desired motion trajectories online by generating them as stable solutions of a complex dynamical system, which can be characterized as a 'network' of dynamic movement primitives. The elements of such networks are highly nonlinear: the canonical dynamics, the mappings from the state space of the canonical state variables to the source functions, and the kinematic relationship between the joint angles and the behavior of the agent in the external space. This raises the question whether for such systems any guarantees can be given that the desired behavior is the only stable solution of the system. This question is of particular importance because for nonlinear systems, and even more for complex ones, multiple stable solutions may exist.

An interesting control-theoretical approach for the analysis of the stability of composite dynamical systems that consist of coupled nonlinear elements is *Contraction Theory (CT)* [45]. We were able to show that this method is suitable to guarantee the stability of highly coordinated behaviors of crowds of locomoting avatars, where our dynamical models included the full complexity and nonlinearity that is generated by the body articulation of the locomoting agents.

The question of the dynamic stability of the created behaviors has been rarely addressed in traditional work on crowd animation. Like in our work, some approaches also tried to learn rules of interactive behaviors from human crowds [13, 41, 58], while other approaches tried to optimize the interaction within crowds by numerical optimization of appropriate cost functions (e.g. [25]). Most existing approaches for the control of group motion in computer graphics neglect the effects of the body articulation during locomotion on the control dynamics [36, 54, 63]. Another field that typically pays attention to the dynamic stability of solutions is control theory. Some work in this area has studied the temporal and spatial self-organization of crowds, typically assuming highly simplified and partly even linear agent models (e.g. [57, 72]). This shows that for more detailed models of the agent dynamics systematic methods for the design of a stable system dynamics are largely lacking.

Contraction Theory is a special type of nonlinear stability analysis that has been introduced by J.-J. Slotine and coworkers [45, 64, 89] The special property of this framework, which makes it possible to simplify the analysis of complex composite systems, is that it permits to transfer stability results from parts to composite systems. In general, such a transfer is not possible for nonlinear dynamical systems, which typically renders the analysis of composite nonlinear dynamical systems impossible, even for moderately-sized systems.

Opposed to the classical approach for stability analysis that computes first a stationary solution and then linearizes about it, Contraction Theory analyzes differences between trajectories with different initial conditions. If these differences vanish exponentially over time, all solutions converge towards a single trajectory, independent from the initial states. In this case, the system is called *contracting*, and at the same time is globally asymptotically stable. More specifically, for a general dynamical system of the form

$$\dot{\mathbf{x}} = \mathbf{f}(\mathbf{x}, t) \tag{2}$$

assume that $\mathbf{x}(t)$ is one solution of the system, and $\tilde{\mathbf{x}}(t) = \mathbf{x}(t) + \delta\mathbf{x}(t)$ a neighboring one with a different initial condition. The function $\delta\mathbf{x}(t)$ is also called *virtual displacement*. With the Jacobian of the system $\mathbf{J}(\mathbf{x}, t) = \frac{\partial \mathbf{f}(\mathbf{x},t)}{\partial \mathbf{x}}$ it can be shown [45] that any nonzero virtual displacement decays exponentially to zero over time if the symmetric part of the Jacobian $\mathbf{J}_s = (\mathbf{J} + \mathbf{J}^T)/2$ is uniformly negative definite, denoted as $\mathbf{J}_s < 0$. This implies that it has only negative eigenvalues for all relevant state vectors $\mathbf{x}$ (within a contraction region). In this case, it can be shown that the norm of the virtual displacement decays at least exponentially to zero, for $t \to \infty$. If the virtual displacement is small enough, one can also prove the inequality: $||\delta\mathbf{x}(t)|| \leq ||\delta\mathbf{x}(0)||\, e^{\int_0^t \lambda_{\max}(\mathbf{J}_s(\mathbf{x},s))\, ds}$. This implies that the virtual displacements decay with a *convergence rate* (inverse timescale) that is bounded from below by the quantity $\rho_c = -\sup_{\mathbf{x},t} \lambda_{\max}(\mathbf{J}_s(\mathbf{x}, t))$, where $\lambda_{\max}(.)$ signifies the largest (negative) eigenvalue.

Contraction analysis can be generalized to systems with individual non-contracting directions (partial contraction) [89]. This is important, for example, for limit cycle oscillators, where the directions tangential to the stable oscillatory solution are non-contracting, but the system is contracting in all other directions orthogonal (transversally) to these trajectories. Contraction analysis can be applied to hierarchically coupled systems [45], where the systems on higher hierarchy levels do not feed back into the lower levels. Such systems can be shown to be contracting if each component system is contracting for all bounded inputs. In addition, one can derive constraints for the coupling between two contracting systems that are reciprocally connected (i.e. in a non-hierarchical forward-backward fashion) that guarantee that the resulting system also is contracting. This makes it possible to design contracting systems from contracting system components, by appropriate design of hierarchical and reciprocal connections of the modules. We applied this framework to a simplified model of the dynamics that generates coordinated behavior of crowds.

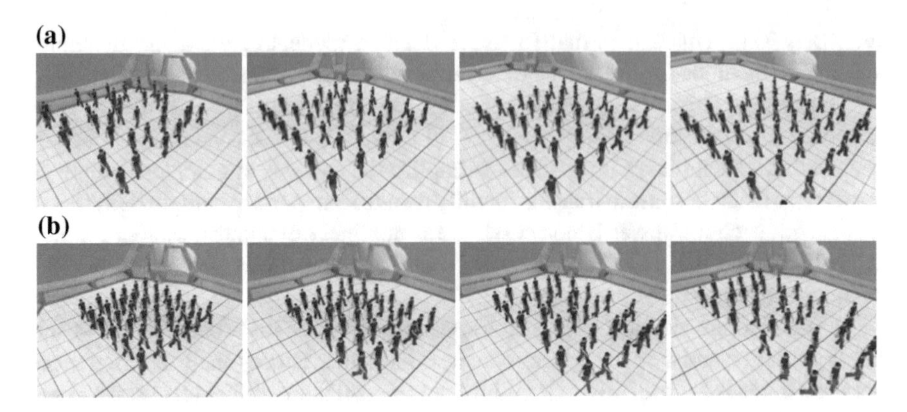

Fig. 3 Self-organized reordering of a crowd. Control dynamics affects direction, row and column distances, and gait phases. **a** When the sufficient contraction conditions of the system dynamics are satisfied the agents organize into an ordered formation where all agents synchronize their steps. **b** For a violation of the contraction conditions the behavior becomes unstable, and the agents diverge and do not synchronize their behaviors. See [52]

In order to apply Contraction Theory for the stability analysis of locomoting crowds, we used a model that integrated the following control levels: (i) Control of heading direction (as described before); (ii) step-size control by morphing between long and short steps; (iii) control of the gait phase in order to achieve a synchronization between all agents; and (iv) control of step frequency by adaptation of the frequency parameters of the limit cycle oscillators. (See [21, 51, 61] for further details.) The resulting dynamics can be approximated by a simplified nonlinear differential equation system that depends on a nonlinear function that describes the relationship between the propagation speed of the characters and the corresponding state variable of the canonical system. This approximative system dynamics is accessible for an application of tools from Contraction Theory. This allows to derive sufficient contraction conditions that ensure that the generated behavior is stable and that no other attractors of the system dynamics exist. (Further details about this analysis are laid out in [51, 52].)

Figure 3a shows a crowd with 36 avatars generated with a dynamics that fulfills the derived contraction conditions. By self-organization the group evolves into a spatially ordered configuration with a synchronization of gait phase, and step frequency. This behavior is robustly approached from different initial conditions and placements of the agents within the scene. Figure 3b shows the situation of the relevant contraction condition is violated. In this case, the crowd diverges and the dynamics becomes unstable. This example demonstrates the applicability of CT for stability design even for systems that model quite complex coordinated behaviors. (See also **Demo Movie II**[2]).

[2]http://tinyurl.com/jxgpptb.

3 Planning of Movements for Humanoid Robots

Standard approaches for kinematic planning in robotics model complex sequential activities by concatenations of elementary motions, each one accomplishing a specific sub-task. Differing from this, skilled human behavior is highly predictive, and behaviors are adapted to task constraints even far in the future. An example for this is the *maximum end-state comfort principle* [69] that has been demonstrated for the human coordination of walking and reaching [37, 92]:

It seems desirable to transfer such flexible human-like planning strategies to robots, e.g. for the generation of locomotion behaviors that are coordinated with hand or arm actions. The mathematical framework presented in Sect. 2 is suitable for the modeling of such highly predictive coordinations strategies. For this purpose, the desired behavior of the robot is synthesized online by a network of dynamic primitives, exploiting the architecture described in the previous section and specifying a virtual kinematic trajectory that the robot should follow. However, real robots are associated with additional constraints, e.g. for the joint angles or realizable torques. In addition the behavior of bipedal walking robots has to ensure specific constraints to ensure dynamic balance, in order to prevent the robot from falling. This part of the chapter describes how the framework presented in Sect. 2 can be embedded as online motion planning system in the control architecture of the humanoid robot HRP-2.

The core of this control architecture is a Walking Pattern Generator (WPG), which is based on nonlinear Model Predictive Control [55]. The underlying algorithm is based on a simplified model of a bipedal walker and synthesizes a dynamically feasible behavior of the legs that prevents the robot from falling. This lower body motion is then combined with the desired motion of the arms, correcting the lower body motion by a special *Dynamic Filter* [94], in order to ensure that the overall behavior is always dynamically feasible and thus realizable on the real robot without falling. We demonstrated the functionality of this architecture for the example of coordinated walking and reaching. The developed system models flexible and and very human-like behaviors for the online replanning after perturbations of the behavior, which realizes the *maximum end-state comfort principle* of human motor control.

Compared to a direct computation of dynamically feasible multi-step movements using optimal control approaches (c.f. [33]), our method is characterized by a much lower computational complexity. Optimal control methods using a accurate model of the robot require typically hours of computation time for the generation of multi-step sequences that ensure that the robot does not fall. The same goal can be achieved with our method with a computational complexity that is of the same order as the one of standard real time-capable WPG algorithms [26].

3.1 Related Work

Some work in prioritized control and stack-of-task approaches in the synthesis of trajectories from training trajectories [16, 75]. In robotics, numerous architectures which combine walking and grasping have been proposed that are not directly inspired by human behavior [1, 7, 35, 78]. Human-inspired frameworks for the decomposition of human reach-to-grasp movements into sequential actions were proposed in [48, 76]. An algorithm for the computation of optimal stance locations with respect to the reaching target within a dynamical systems approach was proposed in [20]. In [96] a task priority approach was applied for the integration of several sub-tasks, including stepping, hand motion, and gaze control. Other work has exploited global path planning in combination with walking pattern generators (WPGs) [32] in order to generate collision-free dynamically stable gait paths. A first attempt to transfer human reaching movements to humanoid robots by using motion-primitives was proposed in [79].

3.2 Drawer Opening Task

Human motor sequences have been shown to be highly predictive. Our implementation of such predictive strategies on a humanoid robot is based on recent study on the coordination of walking and reaching in humans [37]. Participants had to walk towards a drawer and to grasp an object, which was located at different positions in the drawer. Participants optimized their behavior already multiple steps before the object contact, consistent with the hypothesis of *maximum end-state comfort* during the reaching action [68, 92]. This implies that the steps prior to the reaching were modulated in a way that optimized the distance for the final reaching action in a way that simplified the reaching and grasping.

The initial distance from the drawer and the position of the object inside it were varied in the data set. The participants walked towards a drawer, opened it with their left hand and reached for an object inside the drawer with their right hand [49] (see Fig. 4). Each recorded sequence included three subsequent actions: (1) a normal walking step; (2) a shortened step with the left-hand reaching towards the drawer. This step showed a high degree of adaptability, and its length was typically adjusted in order to create an optimum distance from the drawer for the final reaching movement (consistent with the *maximum end-state comfort hypothesis*); (3) the drawer opening combined with the reaching for the object while standing. **Demo Movie III**[3] shows an example for the recorded human behavior.

[3]http://tinyurl.com/he3dhb2.

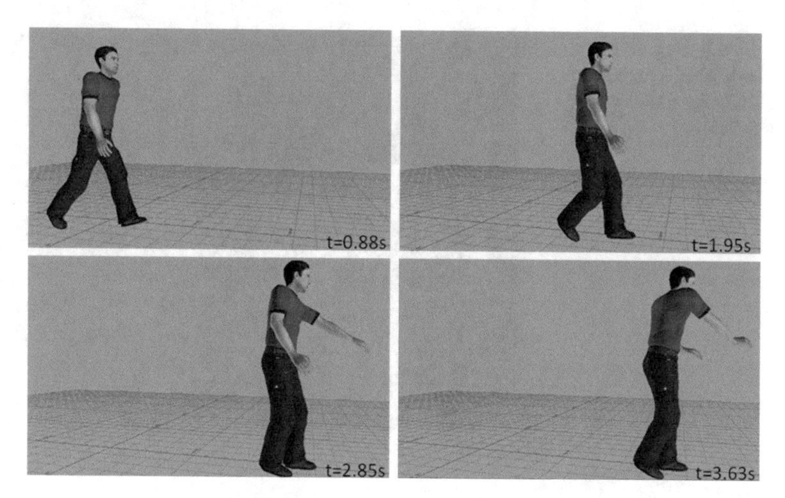

Fig. 4 Important intermediate postures from the human behavior: step with initiation of reaching, standing while opening the drawer, and reaching for the object

3.3 Adaptive Model of the Kinematics of Multi-step Sequences

In order to make the recorded motion capture data useful for a transfer of the behavior to the robot, it was retargeted on a kinematic model of the human-sized humanoid robot HRP-2 using the commercial software MotionBuilder. During retargeting the feet positions of the HRP-2 were constrained to level ground, and the step sizes were reduced proportionally to the height of the robot. This made the joint angle ranges compatible with the ones that can be realized by the robot. The data was split, separating the stored pelvis trajectories (pelvis position and pelvis direction angles in the horizontal plane), and the upper body trajectories, approximating the human trajectories by a kinematic model of the HRP-2. The pelvis position trajectories were also rescaled in order to match the maximally admissible propagation speed limit of the HRP-2 (0.5 m/s). In addition, corrections were applied to the pelvis and trunk yaw-angle trajectories. Figure 5 shows a comparison between an original human and the retargeted pose, illustrated using the corresponding avatar models. (See also **Demo Movie IV**.[4])

In order to model step sequences with a human-like coordination of the periodic walking and the non-periodic reaching behavior, we approximated the training data by anechoic mixtures (see Sect. 2.2). For this specific application, we used a step-wise regression approach. We introduced a total of five sources in order to model the three different actions (steps) within the sequence. The first action is the normal walking gait cycle. After extraction of the mean value and non-periodic component from this step, we approximated the remaining residuals by anechoic mixtures with

[4]http://tinyurl.com/j8qnbtp.

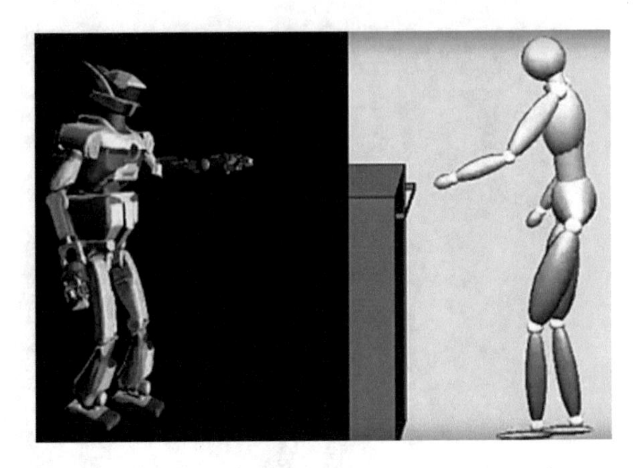

Fig. 5 Retargeting of the movements from a human to the unconstrained skeleton of the HRP-2 robot

three sources, applying a modified demixing algorithm that constrained all time delays belonging to the same source function and joint angle to be equal across all trials. These additional constraints make it necessary to introduce more sources in order to reach the same approximation quality, but significantly simplify the motion morphing. The second highly adaptive step was approximated by the same sources, and the remaining residuals were modeled by introducing two additional periodic sources. The same set of five sources were then used to model the last action.

In order to control the styles of the actions online, we learned nonlinear mappings between task parameters (step length and duration) and the weights of the source functions in our mixing model, applying Locally Weighted Linear Regression (LWLR) [3, 49]. For the synthesis of multi-step sequences the step lengths was computed from the actual estimated target distance. Based on the training data, we computed the achievable step ranges. Additional steps were automatically introduced if the target could not be reached within three steps. For the second step, the step length was adjusted in order to realize a maximum comfort distance for reaching, and the planning distance of the other steps was adjusted accordingly. A more detailed description of the algorithms for the smooth interpolation of the weights of the kinematic primitives at the transition points between the different steps is given in [49].

For the learned parameters the system generates very natural-looking coordinated three-step sequences for total goal distances between 2.3 and 3 m, which were not included in the training data set. This is illustrated in Fig. 6. When the specified goal distance exceeds this interval, the system automatically introduces additional gait steps, adapting the behavior for goal distances above 3 m. Clips illustrating the highly flexible synthesis of multi-step sequences are shown in **Demo Movie V**.[5] Figure 7 shows the highly adaptive online replanning if the goal (drawer) jumps away while the agent is approaching it, requiring the introduction of an additional step.

---

[5] http://tinyurl.com/gktjxre.

Fig. 6 Two synthesized behaviors for two conditions with different initial distance of the character from the drawer. Both distances were not present in the training data set. Adopted from [49]

3.4 Embedding in the Robot Control Architecture

The algorithm described in Sect. 3.3 generates trajectories for human-like coordinated behavioral sequences. However, these sequences are not guaranteed to result in dynamically stable behavior of the robot, and the robot just may fall due to a loss of dynamic balance. To solve this problem, we integrated the described algorithm for the online planning of multi-step-sequences with the control architecture of the HRP-2 robot that ensures the dynamic feasibility of the executed behaviors. An overview of the developed architecture is given in Fig. 8. The online planning module is called 'Kinematic Pattern Synthesis' in the figure. The planned gait cycle trajectory is transmitted to a Walking Pattern Generator (WPG) that is based on model predictive control, which computes the foot placements and the trajectory of the Zero Moment Point (ZMP) for the current step from the desired Center of Mass (CoM) velocity and the pelvis angular velocity of the planned gait cycle [55]. It can

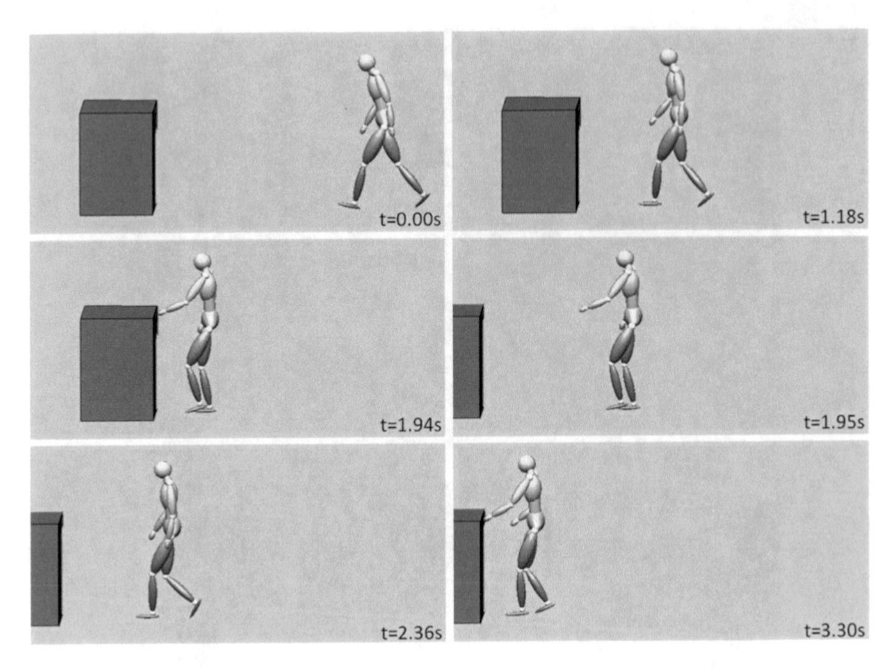

Fig. 7 Online perturbation experiment. The goal (drawer) jumps away while the agent is approaching it, requiring online replanning of the multi-step sequence. The algorithm introduces automatically an action of type 2 (short step) in order to adjust for the increased distance to the goal

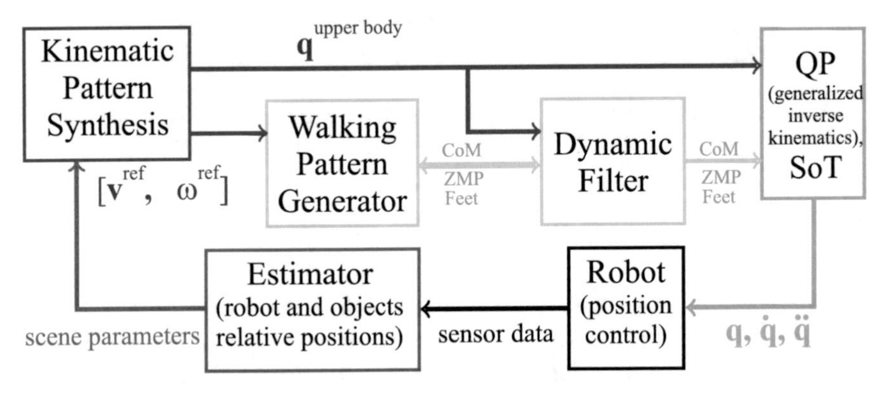

Fig. 8 Control architecture of HRP-2 humanoid robot. The online kinematic pattern synthesis module is linked to a Walking Pattern generator, which computes foot placements and the ZMP trajectory (see text). The Dynamic Filter corrects the walking trajectory dependent on the joint angles of the upper body. Both, gait control parameters and upper body joint angles are integrated in a generalized kinematic planning module using a Stack-of-Task (SOT) approach, which computes the control torques for the robot. The variables $[\mathbf{v}^{ref}, \boldsymbol{\omega}^{ref}]$ signify the linear and angular velocity of CoM, and $\mathbf{q}^{upperbody}$ are the upper-body joint trajectories computed from the kinematic pattern synthesis. The variables $\mathbf{q}$, $\dot{\mathbf{q}}$, $\ddot{\mathbf{q}}$ are the generalized position, velocity and acceleration vectors computed by the Stack of Tasks (SoT) approach

be shown that the gait of the robot is dynamically stable if the projection of the Zero Moment Point to the ground plane is within the support polygon on the floor, which surrounds the feet that are in contact with the ground [86].

The generated preplanned CoM and ZMP trajectories are corrected, taking into account the planned upper-body joint angles by a Dynamic Filter (DF), which operates in closed-loop together with the WPG. Both, the planned CoM and ZMP trajectories, and the upper-body joint angles are then combined in an inverse kinematics module that implements 'Stack-of-Task' approach (SoT) [46, 77]. This module outputs angular trajectories for legs and upper-body and ensures that the executed behavior respects the dynamic stability constraints of the robot, at the same time approximating the desired behavior of the upper body, as far as possible. These resulting trajectories $\mathbf{q}(t)$ can then feasibly be realized by the low-level controllers of HRP-2 robot.

During motion execution, the real-world environmental and task parameters and the current state of the robot are fed back to the kinematic planner, closing the control loop for adaptive interaction in the real world. For the successful realization of the system it is important to retrain the primitives on example trajectories that are feasible for the robot, which are generated by a robot physics simulator.

3.5 *Experiments on the Robot*

The synthesis architecture was first tested by simulating 'open-loop' control, using the OpenHRP simulator to realize a physical model of HRP-2 robot. In the open loop simulations the robot replays the training movements, but does not create online adapted movements with adjusted step sizes and sequences dependent on the distance of the robot from the reaching target. In the simulations, the robot starts from the parking position and makes a transition to a normal step. At the end of this step the pelvis velocities (propagation and angular) were determined and used as initial conditions for the generation of a three-action sequence. At the end of the last action a spline interpolation of pelvis angular and positional coordinates was used to change the robots state back to the parking position (introducing two additional steps on the spot). A snapshot of the executed behavior is shown in Fig. 9. Examples of full three- and four-action sequences are provided in **Demo Movie VI**.[6]

As final step of this validation, the architecture was also tested using the real HRP-2 robot (Fig. 10). The behavior could be successfully realized, maintaining the balance of the robot. Examples of the corresponding behaviors of the real HRP-2 robot are provided in **Demo Movie VI**.

After these tests of the 'open loop behavior' of the system, without an adaptation of step and reaching parameters using the movement planning algorithm, we tested the full system including such online planning in extensive simulations using the OpenHRP simulator. As result, we found that the proposed architecture really works

[6]http://tinyurl.com/jxwmwnt.

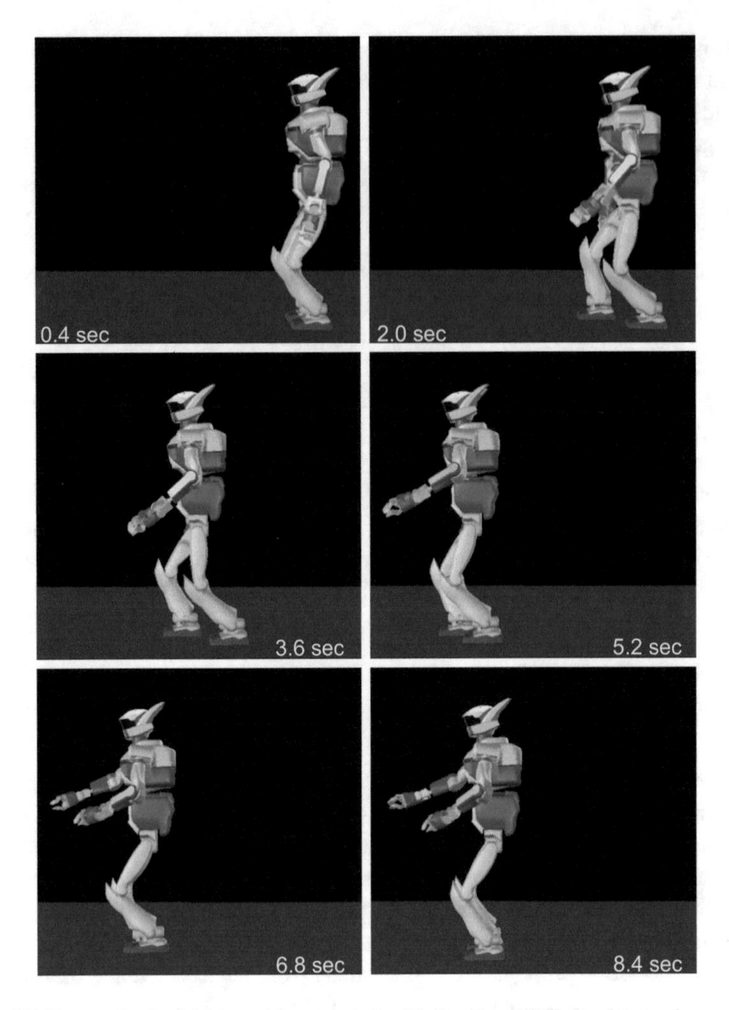

Fig. 9 Off-line synthesised trajectories generated with the OpenHRP simulator using a realistic physical model of the robot

robustly also in the case of online adaptation and replanning. In addition, we tested our architecture in comparison with a simpler machine learning-based approach, where one learns the output trajectories $\mathbf{q}(t)$ from many training examples that produce dynamically behavior of the robot, and where one tries to interpolate between them using learning techniques. It turns out that this simplistic strategy works for only a subset of the training trajectories and fails completely for the generation of adaptive behavior online planning of new adapted step sizes and reaching movements [53].

An example of the quantitative validation of the method is shown in Fig. 11, that illustrates the ground reaction forces (maximal normal force of the feet over the whole action sequence, based on 30 simulations with parameters that different

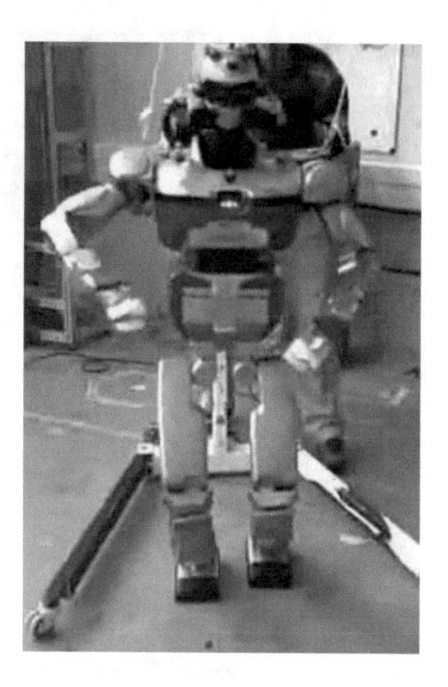

Fig. 10 Real HRP-2 robot performing walking-reaching sequences at LAAS-CNRS

from the training data). The maximum admissible ground reaction force for the real HRP-2 is 800 N. The figure compares the peak forces for trajectories directly created by the WPG without approximation of human behavior, the results from the naïve machine learning approach, and the ones obtained with our method. For the synthesis methods, the figure compares the results of the reconstruction of the training trajectories, using different numbers of source functions of the anechoic mixing model, and the case with an optimum number of sources with an inference of novel step sizes and reaching distances in the closed-loop system that includes online planning. For the naïve machine learning approach except for the case of 9 source functions, the force limit of the robot gets violated. Even with this optimum number of sources, the force limit is violated when the system is operating in closed loop. Consequently, the robot falls sometimes during the execution of such behaviors [53]. Contrasting with this result, for our methods the peak ground reaction forces remain always in the feasible region, and they are extremely similar to the ones when the movement was directly planed using the WPG without training to approximate human behavior. In is remarkable that even for the most difficult case, the closed-loop inference of adaptive behavior, the ground reaction forces do not significantly increase. Similar behavior is observed for other critical mechanical parameters, like the joint torques. (See [53] for details.) This demonstrates that the special form of the integration of the planning algorithm in the control system if critical in order to obtain dynamically feasible behavior of the robot that prevents it from falling.

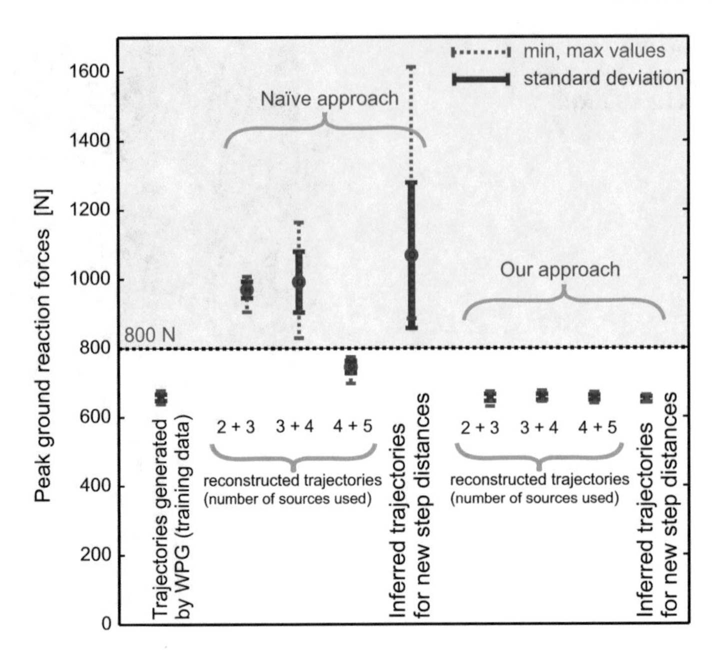

Fig. 11 Peak ground reaction forces obtained for simulated test trials. Comparison of three different synthesis methods: WPG: trajectories generated with the WPG without approximation of human behavior; naïve ML: interpolation of feasible trajectories using machine learning methods; and with our method. In addition, the figure compares the results for the resynthesis of the training trajectories with different numbers of sources, and the full closed-loop behavior with an adaptive synthesis of novel step sizes and reaching distances. (Blue error bars indicate mean and standard deviation, and red lines indicate the ranges between minimum and maximum values)

4 Probabilistic Model for the Online Synthesis of Stylized Reactive Movements

In the last section of this chapter we describe a completely different approach for the generation of reactive complex body movements that exploits state-of-the-art Bayesian approaches in machine learning. We applied this approach in order to simulate a reactive avatar in Virtual Reality (VR) that reacts to the movements of the user with gradually controlled emotional style. Reactive motions are generated by a dynamical extension of hierarchical Gaussian Process Latent Variable Model (GPLVM). (See [80, 81] for details.) This probabilistic model includes latent variables that encode the emotional style of the executed actions, where these variables can be adjusted at run-time. We have verified by psychophysical experiments that this method generates human motion that is almost indistinguishable from real human trajectories. In addition, it allows to control precisely and continuously the emotional style of the executed actions [82, 83]. This makes the developed method interesting for many applications, including experiments in neuroscience and psychology, computer graphics, and for the realization of human-machine interactions.

4.1 Related Work

The modeling of emotional styles is a classical problem in computer graphics (e.g. [6, 90]). A variety of statistical motion models have been proposed for style interpolation [6, 30], the editing of motions styles [38], and for the analysis and synthesis of human motion data in general (e.g. [9]). However, many of these techniques result either in off-line models that cannot react in real-time to external inputs, such as other characters in the scene, or they are strongly simplified, resulting in movements that are not completely believable when compared with real human motion.

More recent approaches have tried to learn highly accurate models of human motion in an unsupervised manner from motion capture data bases. A very successful approach has been the use of Gaussian Process Latent Variable Models (GPLVMs), a nonlinear dimension reduction technique. GPLVMs have been applied in computer graphics for the modeling of kinematics and motion interpolation [24], for the realization of inverse kinematics [42], and for the learning of low-dimensional dynamical models [95]. A related approach are Gaussian Process Dynamical Models (GPDM), a method that uses the same framework for the learning on nonlinear dynamical systems that generate highly realistic human motion [88]. In our previous work [80] we introduced a dynamical mapping similar to a GPDM in a hierarchical generative model to learn the dynamics of stylized interactive movements and interpolate between them. The major problem of these models for real-time synthesis is the associated computational cost, which requires additional approximations, such as the introduction of sparsified representations, to accomplish synthesis in real-time.

4.2 Probabilistic Model for Interactive Movements

In order to learn a generative model for two-person interactions we use motion capture data from couples of actors that executed interactive behaviors, such as handshakes or high-five movements with different emotional styles [82, 83].

The learned probabilistic model is depicted in Fig. 12. It has a hierarchical structure and consists of three levels, an *agent layer* which encodes the kinematics (joint angles) of either agent, an *interaction layer* that encodes the interaction between the two agents on the level of individual time points (frames), and a *dynamics layer* that encodes a dynamic sequence of states in the interaction layer. Along the hierarchy a strong dimension reduction is realized, with a reduction from 159 joint angles to a two-dimensional latent space at the agent layer, and a further reduction from four to three dimensions in the interaction layer. The dynamics that is modeled by the dynamics layer runs in a three-dimensional state space. The whole model can be interpreted as a probabilistic graphical model, and inference techniques for such models can be applied to determine the state of the latent variables [5]. Specifically, we used a maximum-a-posteriori approximation to determine the most probable settings of the latent $\mathbf{x}_t^j$ and $\mathbf{i}_t$ (see Fig. 12) In the following, the individual layers are described in more detail.

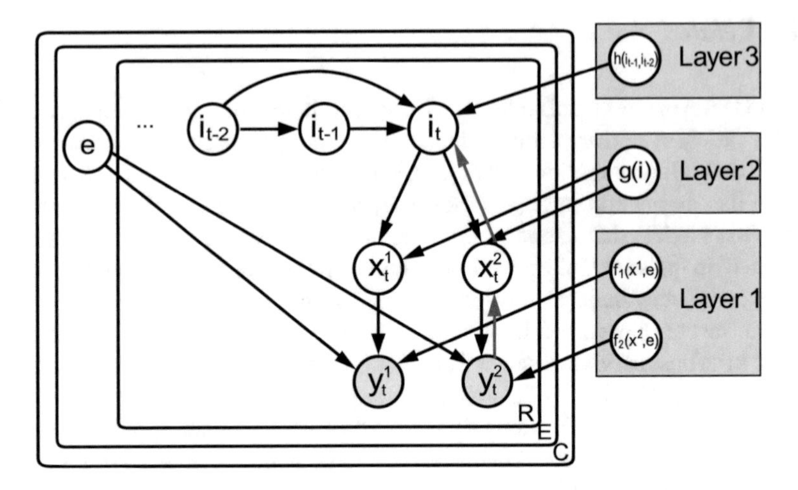

Fig. 12 *Hierarchical probabilistic model for interactive movements*. The graphical model comprises three layers. At the *bottom* are the observable joint angle vectors $\mathbf{y}_t^j$ of each actor $j \in \{1, 2\}$ and point in time t. The means of the $\mathbf{y}_t^j$ are generated by the latent states $\mathbf{x}_t^j$ of the agent layer and the emotion style variable $\mathbf{e}$ via functions $f_j(\mathbf{x}_t^j, \mathbf{e})$ drawn from Gaussian processes. The $\mathbf{x}_t^j$ have a much smaller dimensionality than the $\mathbf{y}_t^j$. The means of the $\mathbf{x}_t^j$ of both actors are generated by a function $g(\mathbf{i}_t)$ from a yet lower-dimensional interaction-layer state $\mathbf{i}_t$, whose time evolution is controlled by a mapping $h(\mathbf{i}_{t-2}, \mathbf{i}_{t-1})$ in the dynamics layer. Both $g()$ and $h()$ are drawn from Gaussian processes, too. The plates denote the assumption of replicated independent identically distributed (i.i.d.) draws across R trials per E many emotions from C couples of actors. For details, see text. Figure adopted from [82]

The **agent layer** approximates, separately for the two agents $j \in \{1, 2\}$, a set of training trajectories by nonlinear dimensionality reduction using a GPLVM. For this purpose, we learn a nonlinear mapping from a two-dimensional latent variable $\mathbf{x}_t^j$ and emotional style $\mathbf{e}$ onto the 159-dimensional joint angle vectors $\mathbf{y}_t^j$. The nonlinear functions f_j that realize this mapping are drawn from a Gaussian process with a composite kernel that combines a radial basis functions (RBF) kernel for the joint angle variables, and a linear kernel for an additional style variable $\mathbf{e}$ that controls the emotional style of the movements. This defines a multi-factor model [87], where the kernel function for the GPLVM is constructed by a product of different kernel functions for motion and style. In addition, we engineered a special prior that promotes factorization of the latent variables into motion dimensions and style dimensions during learning via back constraints [40], expressing the approximately periodic nature of the movements (i.e. the avatar begins and ends a trial in approximately the same pose). This step stabilizes the highly ill-posed factorization problem in a way that results in relatively simple manifolds representing the data in the latent space. Mathematically, the mapping from latent into joint angle space is given by the equation (time index t omitted):

$$\mathbf{y}^j = f_j(\mathbf{x}^j, \mathbf{e}) + \varepsilon_j, \;\; f_j(\mathbf{x}^j, \mathbf{e}) \sim GP(\mathbf{0}, k_Y^j([\mathbf{x}^j, \mathbf{e}], [(\mathbf{x}')^j, \mathbf{e}'])), \;\; j \in \{1, 2\}, \tag{3}$$

where k_Y^j is an appropriate kernel function, and where ε_j is isotropic Gaussian noise. The linear kernel component makes it possible to morph easily along this dimension. A GPLVM can be seen as a nonlinear extension of probabilistic dual PCA that learns a low-dimensional latent space and a mapping from this space to the data space. Unlike PCA, this mapping is nonlinear. It turned out that already two latent dimensions plus a dimension per emotion were sufficient to achieve a highly accurate approximation of the data. See [80, 83] for further details.

The latent variable of the agent layer represents the behavior of the two agents as a trajectory in a four-dimensional space. The dimensionality of this high-dimensional trajectory is further reduced in the **interaction layer**, which learns a mapping from a three-dimensional latent space (variable $\mathbf{i}$) onto this trajectory. This mapping is again realized by a GPLVM that is trained with the aforementioned data basis. Consequently, the individual points of the latent space of the interaction layer are mapped by the two lower layers of the model onto a pair of postures of both agents for each moment of the evolving interaction. The temporal evolution of the interaction corresponds to a three-dimensional trajectory $\mathbf{i}_t$ in the latent space of this layer.

This time course is modeled by the **dynamics layer** by learning of an autonomous dynamical system that generates this trajectory as stable solution. This dynamical system was modeled using a Gaussian Process Dynamical Model (GPDM) [88], which can be interpreted as a nonlinear generalisation of an Autoregressive (AR) model in time series analysis. Mathematically, this model is defined by the equations:

$$\begin{aligned} \mathbf{i}_t &= h(\mathbf{i}_{t-1}, \mathbf{i}_{t-2}) + \boldsymbol{\xi}, \\ h(\mathbf{i}_{t-1}, \mathbf{i}_{t-2}) &\sim GP(\mathbf{0}, k_h([\mathbf{i}_{t-1}, \mathbf{i}_{t-2}], [\mathbf{i}_{\tau-1}, \mathbf{i}_{\tau-2}])) \end{aligned} \tag{4}$$

where the function h is drawn from a Gaussian process with the kernel function k_h, and where $\boldsymbol{\xi}$ signifies Gaussian noise.

The described model allows for the generation of pair interactions with controllable emotional style by variation of the emotion parameter $\mathbf{e}$. The accuracy of the synthesized movements was validated in psychophysical experiments that show that the generated motion is perceptually almost indistinguishable from real motion capture data from human pair interactions [80].

4.3 Inference for the Generation of Reactive Movements

The described generative probabilistic model can be exploited for the simulation of the interactive behaviors of reactive human virtual agents, who react to the movements of a real human user in a human-like fashion. For this purpose, we exploited the fact that probabilistic generative models can be 'inverted' by conditioning. More precisely, using standard techniques [5], such models allow to make inference of unobserved nodes in the network, dependent on given (observed) information on a subset of the nodes. For this application, we strongly simplified the model for one agent and modeled only its hand position $\mathbf{y}^2$. We then replaced the corresponding random variables by online motion capture data from the user of the system. It is then possible to infer the distributions of all other nodes in the network by conditioning on the available hand position information. Maximizing the joint probability of the latent variables and the observed hand trajectory, one can then in principle find the most probable values of all other variables in the probabilistic network. Specifically, this allows to determine the most probable joint angles of the other agent (agent 1) that correspond best to the observed trajectories of the user. The resulting way of inferring a likely posture sequence from a generative probabilistic model, using the observations as constraints to find the most probable trajectory can be interpreted as a special form of 'style-based inverse kinematics' [24].

The straightforward practical implementation of this idea suffers from a very high computational cost, like many naïve Bayesian machine learning approaches. The inversion of the described probabilistic model using straight-forward methods results in a system that is way too slow for real-time synthesis of reactive movements. In order to solve this problem, we implemented in addition the following two approximations: (1) Direct mappings from the data variables to the latent variables were explicitly learned and modeled by Gaussian process regression. This is much faster than to determine the latent variables by conditioning on the observed input, since we only need to evaluate a Gaussian process prediction. We are learning these direct mappings from latent/observed pairs during an off-line training phase. (2) To learn the model from large data sets we applied sparse approximations techniques, which approximate the data manifolds in latent space by a small set of inducing points, resulting in much fewer effective model parameters and computational cost for the evaluation of the kernel-dependent functions: This approximation makes computational cost per learning step effectively linear in the number of data-points, as opposed to cubic for the exact solution [39]. With these two additional approximations, which simplify both learning and latent inference, a speed increase of more than factor 100 was obtained, making the resulting architecture suitable for the real-time synthesis of interactive behaviors.

4.4 Application Results

The proposed architecture was tested with different types of interactive human movements. One data set consisted of 'high fives' with four emotional styles (neutral, happy, angry and sad) executed by different actors. A total of 105 motion-captured trajectories was learned, which were performed on a imaginary 3×3 spatial grid for hand contact positions.

The synthesized movements look so natural that human observers were not able to distinguish them from original motion-captured trajectories, the method passes

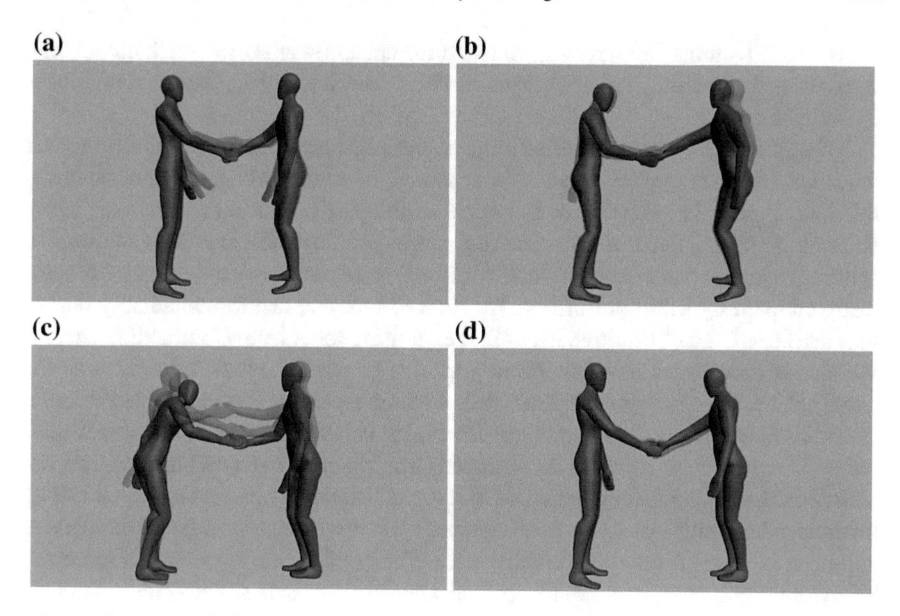

Fig. 13 *Motion sequences of synthesized emotional handshakes*. **a** neutral, **b** angry, **c** happy and **d** sad. Different emotions are associated with different postures and ranges of joint movements

effectively the 'Turing test of computer graphics' [80]. Demonstration movies are also provided as part of **Demo Movie VII**.[7]

For testing of the reactive movement generation the architecture was embedded in a virtual reality setup that is described in detail in [83]. The underlying animation pipeline integrated a Vicon (Nexus) motion capture system, the game engine Ogre 3D, and the proposed learning-based architecture. The reconstruction took place in real-time with Ogre rendering at a frame rate of 68 fps. The functioning of this system, including the variation of emotional style, are also shown in **Demo Movie VII**. A second data set for which the novel architecture was tested were handshakes, which were executed with different emotional styles. Snapshots from the generated stylized motions are shown in Fig. 13. Movies can be found in **Demo Movie VII**.

5 Conclusions and Future Work

In this chapter we reviewed two approaches that approximate complex human full-body motion by structured models that can be embedded in architectures that require a real-time synthesis of complex human motion. Opposed to the many available methods for the off-line synthesis of human motion, online synthesis requires an embedding of the synthesis process into dynamical systems that can be integrated in

[7] http://tinyurl.com/j3d9xtk.

control architectures. We have shown two different approaches how such models can be generated, exploiting concepts from machine learning and the theory of nonlinear dynamical systems.

The first approach approximated human joint angle trajectories by highly compact (anechoic) mixture models. The resulting source functions were then synthesized online by mapping the solutions of canonical nonlinear dynamical systems onto them, defining a special form of dynamic movement primitive (DMP). This allowed to synthesize highly complex coordinated full-body movements by networks of dynamically coupled dynamic primitives. We showed that one can systematically design the stability of such 'primitive networks' exploiting tools from Contraction Theory. We also demonstrated how this framework can be used to model complex coordinated behaviors of individual agents, and of whole crowds of interacting individuals. In addition, we demonstrated that this method is suitable for the online planning of multi-step sequences which are coordinated with arm movements in humanoid robots in real-time, accomplishing dynamically feasible behaviors on a real humanoid robot including the control of dynamically stable walking. The advantage of the chosen approach is that it is computationally more efficient than the synthesis of the same behaviors using straight-forward optimal control approaches, since the computational complexity of the underlying optimization problems with the presently available computational power would not permit an adaptive planning of such multi-step sequences in real-time.

The second approach for the learning of such real-time capable synthesis models was based on established methods in Bayesian machine learning. We demonstrated that an approach that learns a hierarchical ('deep') architecture by combining GPLVMs and GPDMs was suitable for the synthesis of highly natural-looking human movements. In addition, the resulting probabilistic graphical model can be inverted (Bayesian model inversion), i.e. conditioned on observable data. This allowed us to learn interactions from pairs of actors, and to use these learned models then to generate online the maximally probable reactive behavior of a virtual agent that responds directly to a human, whose movements were motion-captured online.

To make this system working in real-time required a substantial amount of engineering work, due to the high computational cost of the chosen Bayesian machine learning approach. This shows that it is a non-trivial step to make such methods work in real-world applications, especially with real-time constraints. It seems likely that it will be even less trivial to embed such methods in complex control architectures, such as the one shown in Fig. 8. This illustrates limitations of these popular approaches which cannot ignored when dealing with real technical control systems. An advantage of the described probabilistic architectures is that they can be integrated with other probabilistic systems, e.g. in computer vision or pattern analysis.

Another interesting challenge is to link the discussed hierarchical probabilistic architectures to spatial movement primitives that, similar to the source functions discussed in Sect. 2, allow the modeling of separately coordinated clusters of degrees of freedom. First work in this direction has been successfully performed [84, 85], and it seems an exciting avenue for future research to see how far such approaches can be extended in the context of real-world problems.

Acknowledgements The work supported by EC FP7 under grant agreements FP7-611909 (Koroibot), H2020 ICT-644727 (CogIMon), FP7-604102 (HBP), PITN-GA-011-290011 (ABC), DFG GI 305/4-1, DFG GZ: KA 1258/15-1, DFG IRTG-GRK 1901 'The brain in action', BMBF, FKZ: 01GQ1002A, and DFG SFB/TRR 135 Cardinal Mechanisms of Perception, project C06.

References

1. A. Ajoudani, J. Lee, A. Rocchi, M. Ferrati, E.M. Hoffman, A. Settimi, D.G. Caldwell, A. Bicchi, N.G. Tsagarakis, A manipulation framework for compliant humanoid COMAN: application to a valve turning task, in *14th IEEE-RAS International Conference on Humanoid Robots (Humanoids)* (2014), pp. 664–670
2. O. Arikan, D.A. Forsyth, J.F. O'Brien, Motion synthesis from annotations. ACM Trans. Gr. SIGGRAPH '03 **22**(3), 402–408 (2003)
3. C.G. Atkeson, A.W. Moore, S. Schaal, Locally weighted learning. A.I. Review **11**, 11–73 (1997)
4. N.A. Bernstein, *The Coordination and Regulation of Movements* (Pergamon Press, New York, 1967)
5. C.M. Bishop, *Pattern Recognition and Machine Learning* (Springer, Berlin, 2007)
6. M. Brand, A. Hertzmann, Style machines, in *Proceedings of SIGGRAPH Conference* (2000), pp. 183–192
7. M. Brandao, L. Jamone, P. Kryczka, N. Endo, K. Hashimoto, A. Takanishi, Reaching for the unreachable: integration of locomotion and whole-body movements for extended visually guided reaching, in *In Proceedings of 13th IEEE-RAS International Conference on Humanoid Robots (Humanoids)* (2013), pp. 28–33
8. H. Carnahan, B.J. McFadyen, D.L. Cockell, A.H. Halverson, The combined control of locomotion and prehension. Neurosci. Res. Commun. **19**, 91–100 (1996)
9. J. Chai, J.K. Hodgins, Performance animation from low-dimensional control signals. ACM Trans. Gr. SIGGRAPH '05 *24*(3), 686–696 (2005)
10. C.-C. Chang, C.-J. Lin, *LIBSVM: A Library for Support Vector Machines* (2001). Software available at http://www.csie.ntu.edu.tw/~cjlin/libsvm
11. E. Chiovetto, A. d'Avella, D. Endres, M.A. Giese, A unifying algorithm for the identification of kinematic and electromyographic motor primitives, in *Bernstein Conference* (2013)
12. E. Chiovetto, M.A. Giese, Kinematics of the coordination of pointing during locomotion. PLoS One **8**(11), e79555 (2013)
13. W. Daamen, S.P. Hoogendoorn, Controlled experiments to derive walking behaviour. Eur. J. Trans. Infrastruct. Res. **3**(1), 39–59 (2003)
14. A. d'Avella, E. Bizzi, Shared and specific muscle synergies in neural motor behaviours. Proc. Natl. Acad. Sci. USA **102**(8), 3076–3081 (2005)
15. S. Degallier, L. Righetti, S. Gay, A.J. Ijspeert, Towards simple control for complex, autonomous robotic applications: combining discrete and rhythmic motor primitives. Auton. Robots **31**(2–3), 155–181 (2011)
16. A.W. Feng, Y. Xu, A. Shapiro, An example-based motion synthesis technique for locomotion and object manipulation. Proc. ACM SIGGRAPH **I3D**, 95–102 (2012)
17. T. Flash, B. Hochner, Motor primitives in vertebrates and invertebrates. Current Opinion Neurobiol. **15**(6), 660–666 (2005)
18. A. Fod, M.J. Mataric, O.C. Jenkins, Automated derivation of primitives for movement classification. Auton. Robots **12**(1), 39–54 (2002)
19. A. Gams, B. Nemec, L. Zlajpah, M. Wächter, A.J. Ijspeert, T. Asfour, A. Ude, Modulation of motor primitives using force feedback: Interaction with the environment and bimanual tasks, in *In Proceedings of IEEE/RSJ International Conference on Intelligent Robots and Systems (IROS 2013)* (2013), pp. 5629–5635

20. M. Gienger, M. Toussaint, C. Goerick, Whole-body motion planning building blocks for intelligent systems, in *Motion Planning for Humanoid Robots*, ed. by K. Harada (Springer, Berlin, 2010), pp. 67–98
21. M.A. Giese, A. Mukovskiy, A. Park, L. Omlor, J.J.E. Slotine, Real-time synthesis of body movements based on learned primitives, in *Statistical and Geometrical Approaches to Visual Motion Analysis*. LNCS, vol. 5604, ed. by D. Cremers et al. (Springer, Berlin, 2009), pp. 107–127
22. M. Gleicher, Motion path editing, in *Proceeding of 2001 ACM Symposium on Interactive 3D Graphics* (2001), pp. 195–202
23. M. Gleicher, H.J. Shin, L. Kovar, A. Jepsen, Snap-together motion: assembling run-time animation. ACM Trans. Gr. SIGGRAPH '03 **22**(3), 702–702 (2003)
24. K. Grochow, S.L. Martin, A. Hertzmann, Z. Popovic, Style-based inverse kinematics. ACM Trans. Gr. **23**(3), 522–531 (2004)
25. D. Helbing, P. Molnár, I.J. Farkas, K. Bolay, Self-organizing pedestrian movement. Environ. Plan. B: Plan. Design **28**, 361–383 (2001)
26. A. Herdt, H. Diedam, P.-B. Wieber, D. Dimitrov, K. Mombaur, M. Diehl, Online walking motion generation with automatic foot step placement. Adv. Robot. **24**(5–6), 719–737 (2010)
27. E. Hsu, K. Pulli, J. Popovic, Style translation for human motion. ACM Trans. Gr. **24**(3), 1082–1089 (2005)
28. Y. Huang, M. Kallmann, Planning motions for virtual demonstrators, in *Intelligent Virtual Agents* (Springer, Berlin, 2014), pp. 190–203
29. A.J. Ijspeert, J. Nakanishi, H. Hoffmann, P. Pastor, S. Schaal, Dynamical movement primitives: learning attractor models for motor behaviors. Neural Comput. **25**(2), 328–373 (2013)
30. L. Ikemoto, O. Arikan, D.A. Forsyth, Generalizing motion edits with Gaussian processes. ACM Trans. Gr. **28**(1), 1–12 (2009)
31. Y. Ivanenko, R. Poppele, F. Lacquaniti, Five basic muscle activation patterns account for muscle activity during human locomotion. J. Physiol. **556**, 267–282 (2004)
32. S. Kajita, F. Kanehiro, K. Kaneko, K. Fujiwara, K. Harada, K. Yokoi, H. Hirukawa, Biped walking pattern generation by using preview control of zero-moment point, in *Proceedings of International Conference on Robotics and Automation* (2003), pp. 1620–1626
33. J. Koschorreck, K. Mombaur, Modeling and optimal control of human platform diving with somersaults and twists. Optim. Eng. **13**(1), 29–56 (2012)
34. L. Kovar, M. Gleicher, F. Pighin, Motion graphs. Proc. SIGGRAPH **2002**, 473–482 (2002)
35. S. Kuindersma, R. Deits, M. Fallon, A. Valenzuela, H. Dai, F. Permenter, T. Koolen, P. Marion, R. Tedrake, Optimization–based locomotion planning, estimation, and control design for the atlas humanoid robot. Auton. Robot. 1–27 (2015)
36. T. Kwon, K.H. Lee, J. Lee, S. Takahashi, Group motion editing. *ACM Trans. Gr. SIGGRAPH 200827*(3), 80–87 (2008)
37. W.M. Land, D.A. Rosenbaum, S. Seegelke, T. Schack, Whole-body posture planning in anticipation of a manual prehension task: prospective and retrospective effects. Acta Psychol. **114**, 298–307 (2013)
38. M. Lau, Z. Bar-Joseph, J. Kuffner, Modeling spatial and temporal variation in motion data. ACM Trans. Gr. *28*(5), Art.No.171 (2009)
39. N.D. Lawrence, Learning for larger datasets with the Gaussian process latent variable model. J. Mach. Learn. Res. - Proc. Track **2**, 243–250 (2007)
40. N.D. Lawrence, R. Court, Local distance preservation in the GP-LVM through back constraints, in *ICML* (2006), pp. 513–520
41. A. Lerner, E. Fitusi, Y. Chrysanthou, D. Cohen-Or, Fitting behaviors to pedestrian simulations, in *Proceedings of Eurographics/ACM SIGGRAPH Symposium on Computer Animation* (2009), pp. 199–208
42. S. Levine, J.M. Wang, A. Haraux, Z. Popović, V. Koltun, Continuous character control with low-dimensional embeddings. ACM Trans. Gr. ACM SIGGRAPH 2012 **31**(4), Art.No.28 (2012)
43. Y. Li, T. Wang, H.Y. Shum, Motion texture: a two level statistical model for character motion synthesis. Proc. SIGGRAPH **2002**, 465–472 (2002)

44. G. Liu, M. Xu, Z. Pan, A. El Rhalibi, Human motion generation with multifactor models. J. Comput. Anim. Virtual Worlds **22**(4), 351–359 (2011)
45. W. Lohmiller, J.J.E. Slotine, On contraction analysis for nonlinear systems. Automatica **34**(6), 683–696 (1998)
46. N. Mansard, O. Stasse, P. Evrard, A. Kheddar, A versatile generalized inverted kinematics implementation for collaborative working humanoid robots: the stack of tasks, in *Proceedings of International Conference on Advanced Robotics (ICAR)* (2009), p. art.119
47. R.G. Marteniuk, C.P. Bertram, Contributions of gait and trunk movement to prehension: perspectives from world- and body centered coordinates. Motor Control **5**, 151–164 (2001)
48. M. Mühlig, M. Gienger, J.J. Steil, Human-robot interaction for learning and adaptation of object movements, in *In Proceedings of IEEE/RSJ International Conference on Intelligent Robots and Systems (IROS 2010)* (2010), pp. 4901–4907
49. A. Mukovskiy, W. Land, T. Schack, M.A. Giese, Modeling of predictive human movement coordination patterns for applications in computer graphics. J. WSCG **23**(2), 139–146 (2015)
50. A. Mukovskiy, A.-N. Park, L. Omlor, J.-J. Slotine, M.A. Giese, Self-organization of character behavior by mixing of learned movement primitives, in *In Proceedings of the 13th Fall Workshop on Vision, Modeling, and Visualization (VMV)* (2008), pp. 121–130
51. A. Mukovskiy, J.J.E. Slotine, M.A. Giese, Analysis and design of the dynamical stability of collective behavior in crowds. J. WSCG **19**(1–3), 69–76 (2011)
52. A. Mukovskiy, J.J.E. Slotine, M.A. Giese, Dynamically stable control of articulated crowds. J. Comput. Sci. **4**(4), 304–310 (2013)
53. A. Mukovskiy, C. Vassallo, M. Naveau, O. Stasse, P. Souères, M.A. Giese, Adaptive synthesis of dynamically feasible full-body movements for the humanoid robot HRP-2 by flexible combination of learned dynamic movement primitives. Robot. Auton. Syst. J. Comput. Sci. (submitted to) (2016)
54. R. Narain, A. Golas, S. Curtis, M. Lin, Aggregate dynamics for dense crowd simulation. ACM Trans. Gr. Art.122 **28**(5), 1–8 (2009)
55. M. Naveau, M. Kudruss, O. Stasse, C. Kirches, K. Mombaur, P. Souères, A reactive walking pattern generator based on nonlinear model predictive control. IEEE Robot. Autom. Lett. (2016) (in press)
56. L. Omlor, M.A. Giese, Anechoic blind source separation using Wigner marginals. J. Mach. Learn. Res. **12**, 1111–1148 (2011)
57. D.A. Paley, N.E. Leonard, R. Sepulchre, D. Grunbaum, J.K. Parrish, Oscillator models and collective motion: spatial patterns in the dynamics of engineered and biological networks. IEEE Control Syst. Mag. **27**, 89–105 (2007)
58. S. Paris, J. Pettré, S. Donikian, Pedestrian reactive navigation for crowd simulation: a predictive approach. Proc. Eurographics 2007 **26**(3), 665–674 (2007)
59. A. Park, A. Mukovskiy, L. Omlor, M.A. Giese, Self organized character animation based on learned synergies from full-body motion capture data, in *Proceedings of International Conference on Cognitive Systems, (CogSys, 2008)* (2008)
60. A. Park, A. Mukovskiy, L. Omlor, M.A. Giese, Synthesis of character behaviour by dynamic interaction of synergies learned from motion capture data, in *The 16-th International Conference in Central Europe on Computer Graphics, Visualization and Computer Vision'2008, WSCG'08* (2008), pp. 9–16
61. A. Park, A. Mukovskiy, J.J.E. Slotine, M.A. Giese, Design of dynamical stability properties in character animation. Proc. VRIPHYS **09**, 85–94 (2009)
62. S.I. Park, H.J. Shin, S.Y. Shin, On-line locomotion generation based on motion blending, in *Proceedings of the 2002 ACM SIGGRAPH/Eurographics Symposium on Computer Animation* (2002), pp. 105–111
63. N. Pelechano, J.M. Allbeck, N.I. Badler, Controlling individual agents in high-density crowd simulation, in *Proceedings of Eurographics/ACM SIGGRAPH Symposium on Computer Animation* (2007), pp. 99–108
64. Q.C. Pham, J.J.E. Slotine, Stable concurrent synchronization in dynamic system networks. Neural Netw. **20**(3), 62–77 (2007)

65. C.L. Roether, L. Omlor, A. Christensen, M.A. Giese, Critical features for the perception of emotion from gait. J. Vis. **9**(6), 15 (2009)
66. C. Rose, M. Cohen, B. Bodenheimer, Verbs and adverbs: multidimensional motion interpolation using radial basis functions. IEEE Comput. Gr. Appl. **18**(5), 32–40 (1998)
67. C. Rose, B. Guenter, B. Bodenheimer, M. Cohen, Efficient generation of motion transitions using spacetime constraints, in *Proceedings of ACM SIGGRAPH'96 International Conference on Computer Graphics and Interactive Techniques***30**, 147–154 (1996)
68. D.A. Rosenbaum, Reaching while walking: reaching distance costs more than walking distance. Psychon. Bull. Rev. **15**, 1100–1104 (2008)
69. D.A. Rosenbaum, R.G. Cohen, S.A. Jax, D.J. Weiss, R. van der Wel, The problem of serial order in behavior: Lashley's legacy. Hum. Mov. Sci. **26**(4), 525–554 (2007) (Europ, Workshop on Mov, Sci., 2007)
70. A. Safonova, J. Hodgins, N. Pollard, Synthesizing physically realistic human motion in low-dimensional, behavior-specific spaces. ACM Trans. Gr. **23**(3), 514–521 (2004)
71. M. Santello, M. Flanders, J.F. Soechting, Postural hand synergies for tool use. J. Neurosci. **18**(23), 10105–10115 (1998)
72. L. Scardovi, R. Sepulchre, Collective optimization over average quantities, in *Proceedings of the 45th IEEE Conference on Decision and Control, San Diego, California* (2006), pp. 3369–3374
73. S. Schaal, S. Kotosaka, D. Sternad, Nonlinear dynamical systems as movement primitives, in *Proceedings of 1st IEEE-RAS International Conference on Humanoid Robots, Humanoids* (Springer, Berlin, 2000), pp. 117–124
74. G. Schöner, M. Dose, C. Engels, Dynamics of behavior: theory and applications for autonomous robot architectures. Robot. Auton. Syst. **16**(2–4), 213–245 (1995)
75. A. Shoulson, N. Marshak, M. Kapadia, N.I. Badler, ADAPT: the agent development and prototyping testbed. IEEE Trans. Vis. Comput. Gr. (TVCG) **99**, 1–14 (2014)
76. M. Sreenivasa, P. Souères, J.-P. Laumond, Walking to grasp: modeling of human movements as invariants and an application to humanoid robotics. IEEE Trans. Syst. Man Cybern. Part A: Syst. Hum. **42**(4), 880–893 (2012)
77. O. Stasse, *Habilitation Thesis*. Paul Sabatier University, CNRS, Toulouse (2013)
78. O. Stasse, B. Verelst, A. Davison, N. Mansard, F. Saidi, B. Vanderborght, C. Esteves, K. Yokoi, Integrating walking and vision to increase humanoid autonomy. Int. J. Humanoid Robot. Spec. Issue Cogn. Humanoid Robot. **5**, 287–310 (2008)
79. M. Taïx, M.T. Tran, E. Souères, P. Guigon, Generating human-like reaching movements with a humanoid robot: a computational approach. J. Comput. Sci. **4**, 269–284 (2013)
80. N. Taubert, A. Christensen, D. Endres, M.A. Giese, Online simulation of emotional interactive behaviors with hierarchical Gaussian Process Dynamical Models, in *Proceedings of SAP'12* (ACM Press, New York, 2012), pp. 25–32
81. N. Taubert, D. Endres, A. Christensen, M.A. Giese, Shaking hands in latent space: modeling emotional interactions with Gaussian process latent variable models, in *Proceedings of KI 2011: Advances in Artificial Intelligence, LNAI*, ed. by S. Edelkamp, J. Bachpages (Springer, Berlin, 2011), pp. 330–334
82. N. Taubert, D. Endres, M.A. Giese, Reactive virtual reality avatar with controllable emotional style based on hierarchical Gaussian process dynamical models, in *Proceedings of ICANN 2014* (2014), p. Art.No.25
83. N. Taubert, M. Löffler, N. Ludolph, A. Christensen, D. Endres, M.A. Giese, A virtual reality setup for controllable, stylized real-time interactions between humans and avatars with sparse Gaussian process dynamical models, in *Proceedings of SAP'13* (2013), p. 41–44
84. D. Velychko, D. Endres, The variational Gaussian process dynamical model, in *Proceedings of the Workshop on Advances in Approximate Bayesian Inference* (NIPS, Montreal, Canada, 2015), pp. 1–6
85. D. Velychko, D. Endres, N. Taubert, M.A. Giese, Coupling Gaussian process dynamical models with product-of-experts kernels, in *Proceedings of the 24th International Conference on Artificial Neural Networks*. LNCS, vol. 8681 (Springer, Berlin, 2014), pp. 603–610

86. M. Vukobratović, Yu. Stepanenko, On the stability of anthropomorphic systems. Math. Biosci. **15**, 1–37 (1972)
87. J.M. Wang, D.J. Fleet, A. Hertzmann, Multifactor Gaussian process models for style-content separation, in *Proceedings of ICML* (2007)
88. J.M. Wang, D.J. Fleet, A. Hertzmann, Gaussian process dynamical models for human motion. IEEE Trans. Pattern Anal. Mach. Intell. **30**(2), 283–298 (2008)
89. W. Wang, J.J.E. Slotine, On partial contraction analysis for coupled nonlinear oscillators. Biol. Cybern. **92**(1), 38–53 (2005)
90. Y. Wang, Z.-Q. Liu, L.-Z. Zhou, Learning style-directed dynamics of human motion for automatic motion synthesis, in *Proceedings of IEEE International Conference on Systems, Man, and Cybernetics* (2006), pp. 4428–4433
91. W.H. Warren, The dynamics of perception and action. Psychol. Rev. **113**(2), 358–389 (2006)
92. M. Weigelt, T. Schack, The development of end-state comfort planning in preschool children. Exp. Psychol. **57**(6), 476–782 (2010)
93. A.P. Witkin, Z. Popović, Motion warping. Proc. ACM SIGGRAPH'95 **29**, 105–108 (1995)
94. K. Yamane, Y. Nakamura, Dynamics filter - concept and implementation of on-line motion generator for human figures, in *Proceedings of IEEE International Conference on Robotics and Automation* (2000), pp. 688–695
95. Y. Ye, C.K. Liu, Synthesis of responsive motion using a dynamic model. Comput. Gr. Forum (Proc. Eurographics) **29**(2), 555–562 (2010)
96. E. Yoshida, A. Mallet, F. Lamiraux, O. Kanoun, O. Stasse, M. Poirier, P-F. Dominey, J.-P. Laumond, K. Yokoi, 'Give me the Purple Ball' – he said to HRP-2 N.14, in *Proceedings of IEEE-RAS International Conference on Humanoid Robots (Humanoids'07)* (2007)

Physical Interaction via Dynamic Primitives

Neville Hogan

Abstract Humans out-perform contemporary robots despite vastly slower 'wetware' (e.g. neurons) and 'hardware' (e.g. muscles). The basis of human sensory-motor performance appears to be quite different from that of robots. Human haptic perception is not compatible with Riemannian geometry, the foundation of classical mechanics and robot control. Instead, evidence suggests that human control is based on dynamic primitives, which enable highly dynamic behavior with minimal high-level supervision and intervention. Motion primitives include submovements (discrete actions) and oscillations (rhythmic behavior). Adding mechanical impedance as a class of dynamic primitives facilitates controlling physical interaction. Both motion and interaction primitives may be combined by re-purposing the classical equivalent electric circuit and extending it to a nonlinear equivalent network. It highlights the contrast between the dynamics of physical systems and the dynamics of computation and information processing. Choosing appropriate task-specific impedance may be cast as a stochastic optimization problem, though its solution remains challenging. The composability of dynamic primitives, including mechanical impedances, enables complex tasks, including multi-limb coordination, to be treated as a composite of simpler tasks, each represented by an equivalent network. The most useful form of nonlinear equivalent network requires the interactive dynamics to respond to deviations from the motion that would occur without interaction. That suggests some form of underlying geometric structure but which geometry is induced by a composition of motion and interactive dynamic primitives? Answering that question might pave the way to achieve superior robot control and seamless human-robot collaboration.

Keywords Tool use · Physical interaction · Dynamic primitives · Equivalent networks · Mechanical impedance

---

Submitted to: Laumond, J.-P. Geometric and Numerical Foundations of Movement.

N. Hogan (✉)
Department of Mechanical Engineering, Department of Brain and Cognitive Sciences, Massachusetts Institute of Technology, 77 Massachusetts Avenue, Room 3-146, Cambridge, MA 02139, USA
e-mail: neville@mit.edu

J.-P. Laumond et al. (eds.), *Geometric and Numerical Foundations of Movements*, Springer Tracts in Advanced Robotics 117, DOI 10.1007/978-3-319-51547-2_12

1 The Paradox of Human Performance

Using tools is a hallmark of human behavior. While some animals have been shown capable of making and using tools, this ability remains the distinctive signature that has given humans an evolutionary advantage [10, 51, 53, 60]. Tool use requires dexterous control of physical interaction, and we excel at it. Yet one of the most critical features of the human neuromuscular system is that it is agonizingly slow. The fastest neural transmission speed in humans is no more than 120 m/s [56]. That compares very poorly with information transmission in electro-mechanical systems such as robots, which can conservatively be estimated at 10^8 m/s, about a million times faster. Moreover, muscles are slow. The typical isometric twitch contraction time[1] for the human biceps brachii is about 50 ms [56]. Assuming a linearized model to approximate this behavior implies a bandwidth of about 3 Hz. In comparison, electro-mechanical actuator technology routinely achieves bandwidths from tens to hundreds of Hz [9, 76, 84] and can achieve motion up to 1 KHz, albeit in specialized applications [8, 77]. Furthermore, our brains are slow. A now-classical study of human mental rotations to assess congruency of visually-presented objects demonstrated a reaction time of about 1 second plus 1 additional second per 60° of rotation (i.e. ~4 s for a 180° rotation) [96].

Despite slow neurons, muscles and brains, humans achieve astonishing agility and dexterity manipulating objects—and especially using tools—far superior to anything yet achieved in robotic systems. Slow neuro-mechanical response implies that *prediction* using some form of internal representation is a key aspect of human motor control, yet the nature of that representation remains unclear [18, 27, 57, 111]. Consider fly-casting or cracking a whip: These objects comprise flexible materials that interact with complex compressible fluid dynamics and, in the case of whip-cracking, operate into the hypersonic regime. Models of their behavior based on mechanical physics tax even modern super-computers. The likelihood that anything resembling a physics-based model underlies real-time human control of these objects seems slim, yet some humans can manipulate them with astonishing skill. If humans use internal models for planning and predictive control of dynamic objects—which seems likely—what form might their internal models take?

2 Human Performance Is Not Consistent with Riemannian Geometry

To address this question we studied *haptic illusions* [26]. 'Haptic' refers to the combination of motor and sensory information used when we feel objects, sometimes also called 'active touch'. 'Illusion' in this context refers to the fact that, like other perceptual modalities (e.g. vision), haptic perception is distorted. Experimentally, it

[1] Twitch contraction time is the time from an impulsive stimulus (e.g. electrical) to peak isometric tension.

is observed that the perceived length of a line segment depends on its orientation with respect to the subject; line segments oriented radially from the shoulder are perceived as being longer than line segments oriented tangentially to circles centered at the shoulder [63, 73]. Moreover, the amount of distortion is configuration dependent; distortion becomes more pronounced as the center of the object moves away from the shoulder [45].

During contact and physical interaction, afferent and efferent information is acquired. Afferent information comes from mechanoreceptors such as cutaneous and deep tissue sensors, muscle spindles, Golgi tendon organs and joint capsule receptors [56]. Efferent information is available from so-called corollary discharge, information available from motor areas of the central nervous system (CNS) that project onto sensory areas [56]. Percepts of external objects are formed based on afferent and efferent information acquired during interaction. Perception of objects can be viewed as an integrative, computational process in which geometric properties are inferred from acquired efferent and/or afferent information, combined with prior knowledge. Geometric properties of objects, such as lengths of segments, continuity of paths, angles between edges, etc., may be determined based on the spatial stimulus alone. It therefore seems reasonable to describe the perceptual processes as implementing an underlying, abstract geometrical reasoning system.

What geometric structure underlies human sensory-motor performance? A distorted haptic perception might reflect a Riemannian geometry, consistent with classical mechanics. Riemannian geometry is a mathematically simple extension of Euclidean geometry based on an inner product of vectors v and w denoted $\langle v, w\rangle = v^T Gw$, where the metric G is characterized by a symmetric, positive-definite matrix. This metric can vary from location to location and, in general, haptic perceptual distortion is known to be location dependent [45]. In our study we were concerned only with haptic perceptual distortion in a small region, hence we assumed the metric was effectively constant.

Inner products provide measures of length and measures of angle. The length of a vector v is the square root of the inner product of that vector with itself, $\|v\| = \langle v, v\rangle^{1/2}$. The angle α between two vectors v and w may be determined from $\langle v, w\rangle = \|v\|\,\|w\| \cos\alpha$. To be metrically consistent, the perception of length and angle must be related. If the metric is constant in a given locality (as we assumed) the Riemannian geometry corresponds to a linear stretch of Euclidean geometry (Fig. 1). If, due to perceptual distortion, a certain rectangle is perceived to be square, then if that rectangle is cut in half along the diagonal, a metrically consistent observer would perceive the acute angles of the resulting right triangle to be equal.

To test the metric consistency of human haptic perception, we measured subjects' judgment of length and angle at the same workspace location; details are in [26]. In a length-judgment experiment, subjects felt rectangular holes oriented at 0° and 45° (the latter shown schematically in Fig. 2, Panel (a) and judged which pair of sides was longer. The rectangular holes were simulated by a planar robotic manipulandum with two degrees of freedom. Rectangles of the same area but with 15 different aspect ratios were presented, allowing accurate assessment of the rectangle which

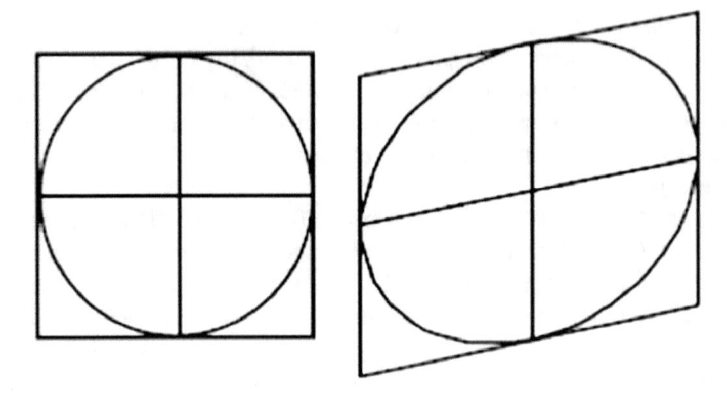

Fig. 1 With a constant metric, a Riemannian geometry is a linear stretch of a Euclidean geometry. Reproduced from [26]

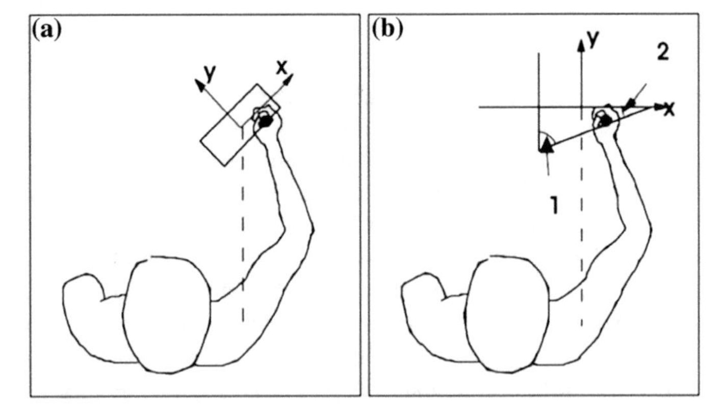

Fig. 2 Panel **a** Subjects felt rectangular holes and were asked to judge which sides were longer. Panel **b** Subjects felt triangular holes and were asked to judge which acute angle was larger. Reproduced from [26]

was perceived to be square. That information for the two orientations was sufficient to identify a metric underlying haptic length perception.

A metric can be used to generate geometrical shapes similar to Euclidean shapes. For example, a Riemannian circle of radius r can be identified with the set of displacement vectors of length r from its center, $\{v|v^T Gv = r\}$. This is the equation of an ellipse which can be used to depict the 'subjective circle' corresponding to haptic length perception. The 'subjective circles' for 8 subjects were remarkably similar, with an eccentricity of $\epsilon = 1.29$ and a major axis oriented at $\theta = 17°$ counter-clockwise from the line joining the shoulders (Fig. 3, Panel a).

In an angle-judgment experiment at the same location, subjects felt triangular holes oriented at 0° and 45° (the former shown schematically in Fig. 2, Panel b) and judged which acute angle was larger. To prevent inference based on judging

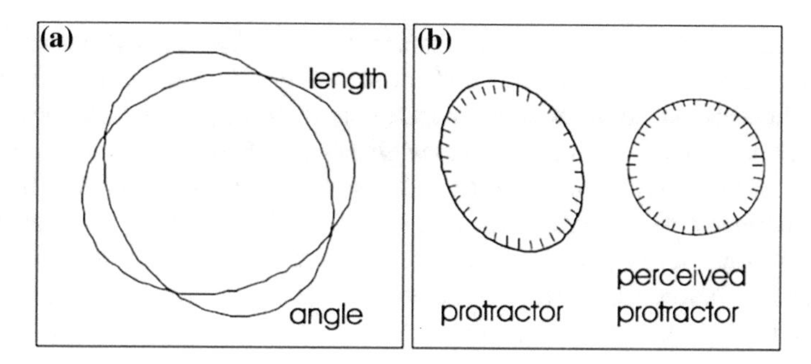

Fig. 3 **Panel a** Average subjective circles determined from the length experiment ($\epsilon = 1.29, \theta = 17°$) and the angle experiment ($\epsilon = 1.28, \theta = -62°$). **Panel b** The angle judgment experiment implies the observer uses the distorted protractor shown on the *left*, which is perceived as the Euclidean protractor on the *right*. Reproduced from [26]

the lengths of the perpendicular sides, the right-angled corner was inaccessible (see [26] for details). Triangles with a constant base length and 19 different aspect ratios were presented, allowing accurate assessment of the angles that were perceived to be equal. That information for the two orientations was sufficient to identify a metric underlying haptic angle perception. The 'subjective circles' corresponding to haptic angle perception were more variable between subjects but had an average eccentricity of 1.28 and a major axis oriented at –62° counter-clockwise from the line joining the shoulders (Fig. 3, Panel a). This difference was highly significant ($p < 0.01$).

This is remarkable. Riemannian geometry is a foundation of mechanical physics. Yet, at least in the context of haptic perception, the brain's 'internal representation' is not consistent with Riemannian geometry.

3 Dynamic Primitives

Combined with the slow response of muscles, the long communication delays due to slow neural transmission impair reactive control. It therefore appears that a major component of human motor control requires planning and 'pre-computation' using some internal representation of the relevant dynamics. Neural evidence has been presented to support this hypothesis, and suggests that the cerebellum is one of the major structures instantiating this 'internal model' [7, 111]. Prediction based on the mathematical models of mechanical physics figures prominently in the control of modern robots yet it appears that our 'internal models' are incompatible with those of mechanical physics—even the underlying geometry is incompatible. Nevertheless, humans substantially out-perform robots. What might be alternative bases for our internal models?

One possibility is that human motor performance is based on dynamic primitives [47, 48, 52]. A dynamic primitive is conceived as an attractor that emerges from the nonlinear dynamics of a neuro-mechanical system—a network of neurons and/or their interaction with the musculo-skeletal periphery [20, 52, 97, 102]. Examples include the familiar point attractor (which may underlie the maintenance of posture) and limit-cycle attractor (which may give rise to rhythmic behavior). Attractor dynamics confers important stability and robustness. It also accounts for nonlinear interference between primitives [19, 94, 100, 101]. Evoking or 'launching' a dynamic primitive may require minimal central control and reduce the need for continuous intervention. At the same time, because each primitive is a highly dynamic behavior, highly dynamic performance may be achieved.

4 Evidence of Dynamic Primitives

Biological evidence supports this account. The most compelling comes from observations of persons recovering their ability to move after having survived a stroke (cerebral vascular accident) that left them partially paralyzed. In the course of studying the feasibility and effectiveness of using physically-interactive robots to aid neuro-recovery, kinematic records were obtained of the earliest movements made by patients as they recovered [61]. These first recovered movements were conspicuously fragmented. Even simple point-to-point reaching movements exhibited a highly irregular speed profile, with large speed fluctuations and frequent stops. This is quite unlike unimpaired movements, which tend to be smooth [29].

Remarkably, each of the movement fragments exhibited a highly stereotyped speed profile—even for patients with brain lesions of widely differing location and extent [61]. This suggests that human movements are composed of primitive submovements. A submovement may be defined as an attractor that describes a smooth sigmoidal transition of a variable from one value to another with a stereotyped time profile [47]. For limb position, the variable is a vector in some coordinate frame, e.g., hand position in visually-relevant coordinates $\mathbf{x} = [x_1, x_2 \ldots x_n]^T$. Each coordinate's speed profile has the same unimodal shape which has finite support: $\dot{x}_j(t) = \hat{v}_j \sigma(t), \; j = 1 \ldots n$ where $\hat{v}_j$ is the peak speed of the submovement; $\sigma(t) > 0$ iff $b < t < e$ where b is the time when the submovement begins and e is the time it ends, otherwise $\sigma(t) = 0$; and the speed profile has only one peak: there is only one point $t_p \in (b, e)$ at which $\dot{\sigma}(t_p) = 0$ and at that point $\sigma(t_p) = 1$. This definition was used to identify sequences of submovements underlying continuous movements.

Reliably extracting overlapping submovements from a continuous kinematic record is a notoriously hard problem. The common practice of examining zero-crossings of progressively higher derivatives (acceleration, jerk, etc.) is fundamentally misleading. Even aside from the practical difficulty of obtaining reliable higher-order derivatives from kinematic data, a composition of two single-peaked speed profiles may yield a composite speed profile with one, two or three speed peaks, hence one to five zero-crossings in the acceleration profile, etc. [92]. Instead, submovement identification is better approached as an optimization problem, minimizing

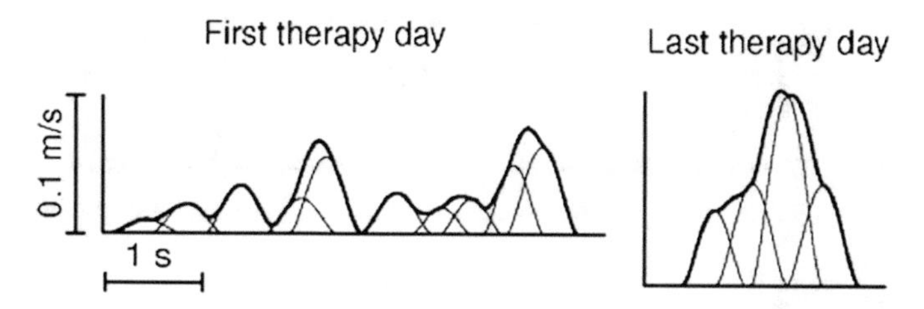

Fig. 4 Typical movements of one representative patient on the first and last therapy days. *Bold lines* indicate tangential speed measured during movement. The later movement is briefer with a single speed peak, while the earlier movement has an irregular speed profile with multiple peaks. Fine lines indicate underlying submovements. The later movement shows fewer submovements which have greater peak speed, duration and overlap than the earlier movement. Reproduced from [91]

mean-squared error between kinematic data and its reconstruction as a sequence of submovements. That avoids the problems mentioned above and yields robust identification even in the presence of substantial measurement noise [92, 93].

This approach was applied to identify submovement sequences in the movements of a cohort of 41 sub-acute and chronic phase stroke survivors as they progressed through robot-aided therapy [91]. Although there was substantial variability across patients, who had widely differing brain lesions, as they recovered they made fewer submovements, which had higher peak speed, longer duration and greater temporal overlap; these changes were statistically significant ($p < 0.05$). Figure 4 shows typical submovements of one patient observed on the first and last days of therapy. These observations indicate that the ability to generate stereotyped submovements appears to be preserved after injury to the CNS and that a major part of the recovery process manifests as re-learning how to combine and blend these dynamic primitives to produce desired behavior.

5 Consequences of Control via Dynamic Primitives

Motor control based on dynamic primitives may facilitate performance of highly dynamic behavior without the need for continuous intervention by the higher levels of the CNS. However, it may also lead to limitations of motor performance that cannot be ascribed to biomechanics. In particular, the parameters of submovements appear to be limited. Their maximum duration is typically on the order of a second. There appears to be a 'refractory period', a minimum interval (on the order of 100 ms) between the onsets of adjacent submovements. Together, these limitations imply that humans would have difficulty generating slow, smooth movements. Observations of unimpaired human subjects have confirmed this prediction. In one experiment, unimpaired subjects made horizontal planar discrete reaching movements between two targets 14 cm apart in the mid-sagittal plane. Subjects were instructed to move smoothly at three different self-paced speeds: 'comfortable', 'fast' (instructed to be

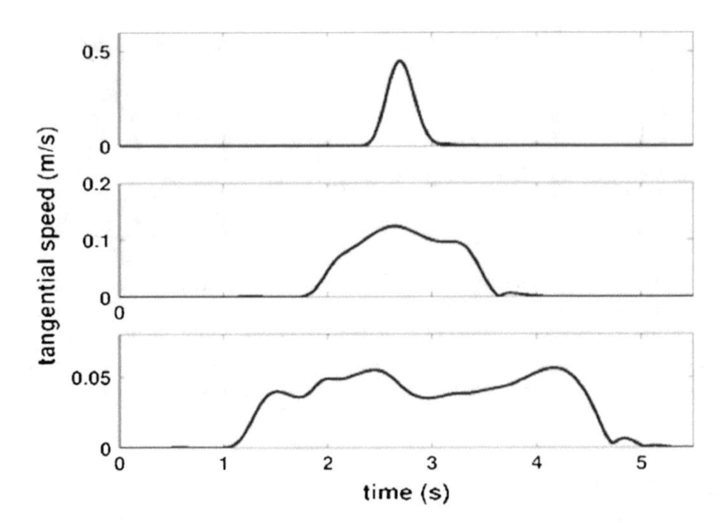

Fig. 5 Tangential speed profiles of discrete reaching movements made by an unimpaired subject at three self-paced speeds. *Top* 'fast'; middle: 'comfortable'; *Bottom* 'slow'. Note the different vertical scales. Slower movements were progressively more irregular

twice as fast as 'comfortable') and 'slow' (instructed to be twice as slow as 'comfortable'). Averaged across subjects, peak speeds were 0.28 ± 0.04 m/s (mean $\pm$ standard deviation) for 'fast' movements; 0.10 ± 0.03m/s for 'comfortable' movements; and 0.05 ± 0.01m/s for 'slow' movements, demonstrating that subjects could successfully follow task instructions.

Figure 5 shows typical speed profiles for the three cases. The 'fast' movement has a single speed peak with a 'bell-shaped' profile similar to that of a maximally-smooth movement [29]. The speed profile of the 'comfortable' movement also has a single speed peak, but is noticeably more irregular. Irregularity of the speed profile is most pronounced in the 'slow' movement, which has multiple peaks. Figure 6 shows a different set of movements for the three cases, and includes the minimal sequences of overlapping submovements that fit the speed profiles with residual error less than 3%. Each submovement has a support-bounded lognormal speed profile, which may be lepto- or platy-kurtic and positively or negatively skewed [85]. With a statistical significance $p < 0.01$ the number of submovements increased with movement duration: $n_{slow} > n_{comfortable} > n_{fast}$.

Taken together, these data provide strong evidence that discrete reaching movements are composed of submovements. The observations of submovements in stroke patients as they recovered was serendipitous (they were not the focus of the experiments) but could not be overlooked. The observation that speed fluctuations increased as unimpaired subjects moved slowly cannot be attributed to mechanics or biomechanics. Factors that might contribute to movement irregularity, such as the torques required to compensate for nonlinear kinematic and inertial coupling between joints or the nonlinear and noisy behavior of muscles, all decline as movements slow. The

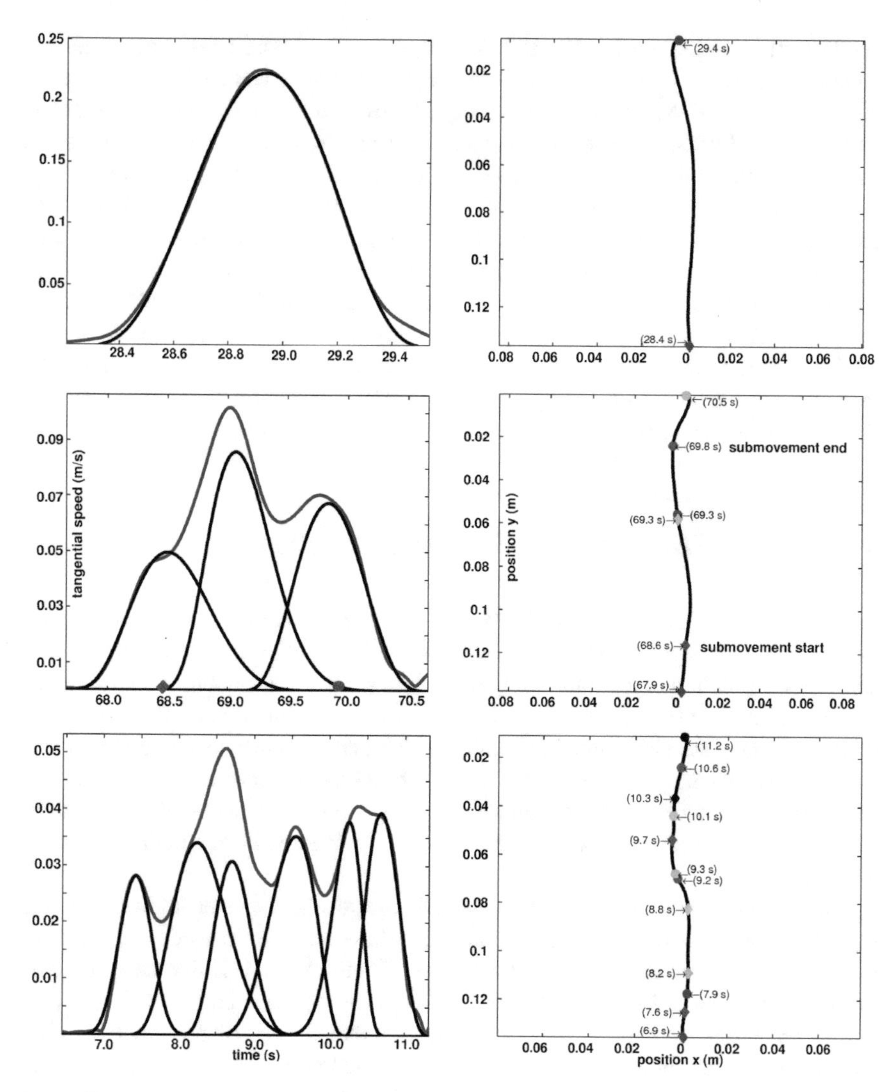

Fig. 6 Tangential speed profiles (*left*) and paths (*right*) of discrete reaching movements made by an unimpaired subject at three self-paced speeds: 'fast' (*top*); 'comfortable' (*middle*); 'slow' (*bottom*). The underlying sequences of submovements are superimposed. *Colored dots* denote the beginning and end of each submovement

fact that, instead, they increased strongly implicates the 'software' underlying motor control. Moving slowly and smoothly is hard for humans.

Similar results have been reported for rhythmic movements: unimpaired subjects were unable to sustain smoothly rhythmic performance as period increased; instead, kinematic irregularity increased as movement slowed, consistent with composition as a sequence of submovements [22, 23]. A complementary result was reported in another study: unimpaired subjects made sequential back-and-forth discrete movements, instructed to dwell at rest at the end of each movement for a duration equal to the movement time. Movements were paced by a metronome which slowly decreased its period. As frequency increased, subjects were progressively less able to sustain the dwell at the end of movement, eventually producing smoothly rhythmic movements. Importantly, changing the duration of the metronome sound (to 50% of the metronome period) significantly reduced the frequency at which dwell time disappeared. In that case, the passage to zero dwell time cannot be attributed to biomechanical limitations, because with different sensory conditions subjects were demonstrably capable of faster discrete movements with non-zero dwell time. Instead, subjects switched to using an oscillatory dynamic primitive. This was confirmed by a 'discreteness index' which changed abruptly from values corresponding to discrete movements to those corresponding to smoothly rhythmic movements [103].

6 Dynamic Primitives for Physical Interaction

Submovements and oscillations may provide a basis for unconstrained movements, but contact and physical interaction are essential for that quintessentially human ability, manipulating objects and using tools. It may seem reasonable to control force when in contact with objects, but that is not sufficient. Simple hand tools illustrate this point; many are elaborated versions of a stick that you push on. A woodworker's chisel is a stick with a sharpened tip; axial compression is required to cut with it. A screwdriver is a stick with a specialized tip designed to mate with a corresponding shape in the head of a screw; to use it effectively, it must be maintained in axial compression.

Unfortunately, pushing on a stick destabilizes its posture. Consider a stick of length R pushed against a surface. To keep matters simple, assume its tip cannot slip but may pivot about the point of contact on the surface.[2] If the stick is initially perpendicular to the surface, the compressive force f_c exerted by the hand on the stick must be strictly axial, also perpendicular to the surface. Small angular displacements $\Delta\theta$ of the stick's orientation from the perpendicular (which might arise from fluctuations or "noise" in the neuromuscular system) displace the point of action of the force laterally (i.e. parallel to the surface) by an amount $\Delta x \cong R\Delta\theta$. If the force exerted by the hand is maintained constant in magnitude and direction, displacement evokes a torque about the tip of $\tau_{tip} = f_c\Delta x \cong (f_c R)\,\Delta\theta$ which acts to increase the deviation. Force control in this situation is statically unstable [89].

To counteract this effect the hand must generate the equivalent of a lateral translational stiffness k_{xx} at the point of contact with the tool so as to produce a lateral restoring force $f_x = -k_{xx}\Delta x$. This generates a rotational restoring torque $\tau = -\left(R^2 k_{xx}\right)\Delta\theta$ about the tip. The minimum stiffness required to maintain static

[2]This is one advantage of a Phillips (cross-head) screwdriver, invented by John P. Thompson, U.S. Patent 1,908,080 May 9, 1933, assigned to Henry F. Phillips.

stability is $k_{xx} > f_c/R$. This highly simplified analysis demonstrates a point that may easily be overlooked: even in this idealized *static* task (nothing varies with time) to exert a *force independent of motion would be unworkable*. Because the act of exerting force may destabilize posture, stiffness must also be present to ensure stability, and greater stiffness is required to stabilize greater forces. Humans generate the required stiffness via the grip of the fingers on the handle, supported by the wrist, the shoulder, and so forth. Remarkably, because the minimum required stiffness increases with applied force, the maximum force a human can exert in this task is determined by the limits of muscle-generated stiffness rather than by the limits of muscle-generated force [90].

Generating stiffness is a minimum requirement for controlling interaction. More generally, other effects equivalent to viscous damping and/or higher-order phenomena will also be required, collectively termed mechanical impedance. Loosely speaking, mechanical impedance is a generalization of stiffness to encompass nonlinear dynamic behavior [39, 44]. Mathematically, it is a dynamic operator that determines the force (time-history) evoked by an imposed displacement (time-history). Physical interaction requires including mechanical impedances as an additional class of dynamic primitives to describe force evoked by motion [47, 48]. Biologically, mechanical impedance at the hand may be modulated by adjusting neural reflex gains [37, 80], co-activating opposing muscles [38, 50], selecting the pose or configuration of the limbs [42] or combinations of these approaches.

The practicality of mechanical impedance as a dynamic primitive underlying human dexterity was demonstrated by implementing it in a 'bio-mimetic' controller for a motorized trans-humeral amputation prosthesis [2]. Control inputs were derived from surface myoelectric activity (EMG) obtained from antagonist muscles in the limb residuum. The difference of their amplitudes determined a 'zero-force' trajectory along which the prosthesis actuator torque was zero. Displacement from that trajectory evoked torque determined by mechanical impedance that was implemented as position and velocity feedback to a highly-back-drivable electro-mechanical actuator. The sum of EMG amplitudes determined the prosthesis mechanical impedance, mimicking the action of natural muscles [1]. Comparison of this controller with conventional velocity control (difference of EMG amplitudes determined prosthesis angular velocity with high mechanical impedance) showed a marked superiority, especially in tasks requiring coordination of natural and artificial joints (e.g. accommodation of a kinematic constraint as in turning a crank), bi-manual coordination and production of mechanical work [3, 86].

7 Combining Motion and Interaction Primitives

An important question is how interactive and motion primitives work together to enable and potentially simplify dexterous manipulation. Two distinct domains are involved. Motion planning belongs in the domain of 'signals' or information-processing, which permeates conventional computation and control theory.

Information-processing operations are *uni-directional* (input affects output but not vice-versa) and the only constraints appear to be temporal causality (no output before input) and boundedness (no infinite quantities). In contrast, interactions due to physical contact are fundamentally *bi-directional*—each system affects the other with mutual causality, as expressed in Newton's 3rd law [79]. They are subject to the numerous additional constraints that arise from the storage and transmission of energy, e.g. conservation of energy, production of entropy, etc.

A combination of dynamic behavior arising from computational information-processing with that arising from physical systems may be described in a unified framework by re-purposing and extending a remarkably effective tool of engineering analysis, the *equivalent electric circuit*. A comprehensive history of this concept is presented by Johnson [54, 55]. Originally Helmholtz and later (independently) Thévenin showed that any electric circuit containing electromotive forces (voltage sources) and resistors could be replaced at any pair of terminals by a single voltage source in series with a single resistor [34, 105]. Subsequently Mayer and Norton simultaneously (and independently) formulated an equivalent electric circuit composed of a current source in parallel with a resistor [75, 81]. The concept of impedance introduced by Heaviside and its dual, admittance, allowed equivalent electric circuits to be extended to include dynamic behavior (e.g. capacitance and inductance) [32, 33].

An electric circuit comprising *arbitrarily complicated* networks of voltage sources, current sources, and linear resistors, capacitors and inductors may be represented by a Thévenin or Norton equivalent circuit. At a terminal pair where the circuit interacts with its environment—an *interaction port*—it behaves as though composed of only two parts with a simple connection. Moreover, each of those parts may be identified unambiguously by simple experiments performed at the interaction port. This prodigious simplification is one reason why equivalent circuits remain a core conceptual tool of engineering analysis.

An equivalent circuit describes an interface between the domain of signals and the domain of energy. In an audio amplifier, signals with negligible power (e.g. retrieved from a storage medium or synthesized by a computer) control some of the amplifier's internal voltage and/or current sources which act to deliver substantial power to a loudspeaker, thereby generating sound energy. An equivalent circuit 'parses' the dynamics of the entire audio amplifier into two pieces. The Thévenin or Norton equivalent source describes the 'forward-path' dynamics relating the low-power input signals to the high-power electrical excitation delivered to the loudspeaker—independent of interaction with the loudspeaker. The equivalent admittance or impedance describes the dynamics of the high-power interaction between the amplifier and the loudspeaker—independent of the forward-path dynamics.

The equivalent circuit concept may be re-purposed to describe the relation between the dynamics of computational information processing and the dynamics of physical systems. The 'equivalent source' (Thévenin or Norton) describes unidirectional forward path dynamics through which computation may influence physical events. The 'equivalent resistance' (admittance or impedance) describes bidirectional dynamic interactions through which physical events evoke a physical response.

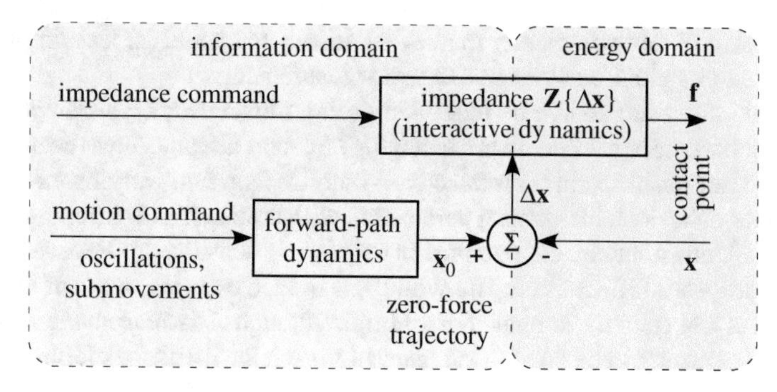

Fig. 7 A nonlinear equivalent network relating the information and energy domains of dynamic behavior. Reproduced from [43]

Equivalent circuits were originally applied to electrical systems with linear dynamics. The concept is readily extended to non-electrical systems though, as general physical systems do not necessarily form closed circuits, the term 'equivalent network' is more appropriate. The concept may further be extended to important classes of nonlinear systems, especially actuators (including mammalian muscles) which occupy the interface between physical and informational dynamics [43].

Extending classical circuit theory to nonlinear systems combining informational and physical dynamics provides a *unified* description of how central commands and peripheral mechanics cooperate to produce observable behavior (Fig. 7). It specifies how three classes of dynamic primitives may be related. Independent of interaction with the environment, the 'equivalent source' describes the nominal unidirectional forward-path dynamic response to central commands, which may consist of submovements and oscillations. Bidirectional interactive dynamics (also modifiable by central commands) are characterized by mechanical impedances. These two parts, *unidirectional* and *bidirectional*, can be identified unambiguously by simple experiments [43]. This disambiguation teases apart the contributions of mechanical dynamics and the problems solved by computation.

8 Identifying the Equivalent Source Without 'Opening the Box'

The challenge of describing and detailing human interactive dynamics is particularly acute. First, the biological actuator (muscle) has notoriously complicated and highly nonlinear dynamics [112]. Second, the neural control system is prodigiously complicated and largely uncharted [56]. Third, there is as yet no ethical way to 'open the box' and reliably observe relevant variables internal to the human neural control system. To date, imaging technologies provide only a coarse-grained measure of limited parts of this system. If applicable, an equivalent network representation could summarize all of that complexity in a few elements that could (at least in principle) be identified unambiguously from external measurements.

Can the equivalent source be identified during movement, i.e. when commands from the central nervous system are changing? Several attempts have been made by assuming a reasonable form for interactive dynamics (e.g. time-varying mass-spring-and-damper behavior), identifying parameters of that model, and extrapolating from the results. Unfortunately, the outcome is exquisitely sensitive to the assumed form of the model—see [30] but compare with [31]. In fact, even the *order* of interactive dynamics is not reliably known. For example, though it is reasonable to assume that high-frequency behavior is dominated by skeletal inertia, yielding 2nd order dynamics, in fact there is evidence of anti-resonance due to muscle mass moving relative to skeletal inertia, and that requires higher-order dynamics [108, 109].

However, an equivalent network motion source can, in principle, be identified without *any* knowledge of the neuro-muscular actuator impedance. A workable method to do so during arm movements was presented in [36]. The essence of the method was to estimate multi-variable skeletal inertia, then generate exogenous forces with a robotic manipulandum so that the neuro-muscular forces were nominally zero throughout a discrete reaching movement. The resulting trajectory was the motion source of an equivalent network model of the neuro-muscular actuator. Iteration over several nominally-identical movements was required to improve the estimate and the exogenous forces were presented only on randomly selected movements to preclude human adaptation to the stimulus. Passing over the details, which are presented in [35, 36], a significant result was that, despite the intrinsic variability of human motor control, the method converged rapidly. The result was a reliable estimate of the motion source output, independent of any assumptions about neuro-muscular mechanical impedance.

9 The Preferred Form of Equivalent Network Models

A linear equivalent circuit may be expressed in four different and fully interchangeable ways: there are two choices for the equivalent source (Thévenin or Norton) and two choices for the operational form of the interactive dynamics (admittance or impedance—force in, motion out or *vice versa*). A nonlinear equivalent network is more restricted. Because a nonlinear dynamic operator may not have a well-defined inverse, the interactive dynamics may be expressible in only one of the two operational forms (admittance or impedance). For example, mammalian muscle is always well-defined in impedance operational form (motion in, force out) but not necessarily in admittance operational form.

The type of source may also be restricted. Key properties of physical system dynamics manifest as symmetries or invariances—features that do not change when other factors do. Noether's theorem famously identifies conservation principles with symmetries (e.g. energy conservation with time-shift invariance, etc.) [106]. One desirable geometric symmetry for interactive dynamics is *translation invariance*. If the reference frame origin is translated, interactive behavior should not change. That is, the commands required to perform a contact task should ideally be identical at all locations. While this may be challenging in some cases—a fixed-based robot's limited workspace obviously limits translation invariance to within its reach—it is highly desirable, at least within a region near the center of the robot's workspace. Translation invariance restricts the choice of equivalent source as follows.

For many contact and interaction tasks it seems natural for the forward path dynamics to specify a nominal force (or torque) $\mathbf{f}_0(t)$. Examples include the nominal force that must be exerted by the feet on the ground during normal locomotion; averaged over a gait cycle, the net vertical foot-ground force must equal the vehicle's (or animal's) weight. Interactive dynamics (in impedance operational form) $\mathbf{Z}\{\cdot\}$ modify that nominal force based on actual motion $\mathbf{x}(t), \mathbf{f}(t) = \mathbf{f}_0(t) - \mathbf{Z}\{\mathbf{x}(t)\}$.

To clarify the following argument, consider that impedance is like a dynamic, nonlinear version of a linear spring of stiffness k. In one dimension $f = f_0 - kx$. Further consider translating the coordinate frame to a new origin so that $x' = x + c$ where c is a constant. In the new coordinate frame $f' = (f_0 + kc) - kx'$. In this case, the translated 'force source' is $f_0' = f_0 + kc$. This is *not* translation invariant; it depends both on the origin of the coordinate frame c and on the stiffness k. The latter is especially troubling as it compromises the separation of forward path dynamics from interactive dynamics—yet that is one of the particular advantages of an equivalent network representation.

Instead, consider an equivalent network with forward path dynamics that specifies a nominal or 'zero-force' motion $\mathbf{x}_0(t)$. Interactive dynamics (in impedance operational form) generate forces in response to deviations of actual motion from nominal motion $\mathbf{f}(t) = \mathbf{Z}\{\Delta\mathbf{x}(t)\}$ where $\Delta\mathbf{x}(t) = \mathbf{x}_0(t) - \mathbf{x}(t)$. As above, impedance is like a dynamic, nonlinear version of a linear spring of stiffness k. In one dimension $f = k(x_0 - x)$. Again consider translating the coordinate frame to a new origin so that $x' = x + c$. With $x_0' = x_0 + c$, $\Delta x' = \Delta x$, hence $f' = f$ and $k' = k$ or, more generally, $\mathbf{Z}'\{\cdot\} = \mathbf{Z}\{\cdot\}$. The advantages of an equivalent network representation have been preserved.

Maxwell identified a correspondence between electrical and mechanical systems, with electrical voltage analogous to mechanical force and current to velocity [74]. Using that analogy, a Norton equivalent network (i.e. with a motion source) is translation invariant whereas a Thévenin equivalent network (i.e. with a force source) is not. Both the structure of the mathematical representation and (more important) the separation of forward path dynamics from interactive dynamics are independent of the choice of coordinate frame origin. It is interesting (and probably no coincidence) that a Norton equivalent network also permits unambiguous identification of the source term from observations made at the interaction port, while a Thévenin equivalent network does not [43].

Summarizing, actuators such as muscles are at the interface between the computational dynamics of the information domain and the physical dynamics of the energy domain. A nonlinear equivalent network provides a competent 'canonical model' of these dynamic objects that may be used to compare alternatives. An equivalent network of the Norton (motion source) type appears to be superior, both unambiguously identifiable and invariant under coordinate frame translation. This may seem counter-intuitive as some tasks seem naturally to require specification of nominal forces. Nevertheless, the more robust description is in terms of nominal motion.

Remarkably, the human motor control system exhibits a strong preference to plan motions rather than forces. Point-to-point reaching movements are generally executed by an approximately straight, smooth hand path [29]. Exposed to mechanical perturbations, subjects spontaneously adapt their muscle forces to restore an approximately straight, smooth hand path [64, 95]. Exposed to a distorted mapping between motion of the hand and motion of a cursor on a screen, subjects spontaneously adapt their hand path to restore an approximately straight, smooth path of the screen cursor [28]. This emphasis on motions is the basis of a successful robot-aided approach to neuro-recovery [46].

10 Choosing Task-Specific Impedance

The most appropriate physical interaction dynamics varies with the task to be accomplished. One effective way to choose that dynamic behavior is to describe the task as an optimization problem. Optimization is a powerful and general approach to robot motion planning which has become more practical with advances in both computational speed and algorithmic sophistication [21, 62].

For optimization to yield specifications for interactive dynamics, the objective function to be optimized should include terms involving both motion and 'exertion' at the interaction port. The term motion is here intended as an 'umbrella' label for velocity and its integrals and derivatives (e.g. displacement, acceleration, etc.) The term exertion is here intended as an 'umbrella' label for force and its integrals and derivatives (momentum, force rate, etc.) The essential distinction between motion and exertion is articulated in [43]. An interaction port is defined by any set of motion variables and their energetic conjugates such that energy and its integrals or derivatives are well-defined. Thus mechanical work is defined by $W = \int \mathbf{f}^T d\mathbf{x}$ where $\mathbf{f}$ is a vector of forces or torques and $d\mathbf{x}$ is a vector of translational or angular displacements. The displacements need not refer to the same physical location (e.g. they may be displacements of a robot's several degrees of freedom) provided the corresponding forces are energetic conjugates such that work is correctly defined.

A simple 'toy' example may illustrate the point. Assume a manipulator and its control system are modeled as a mass m_m moving in 1 degree of freedom, retarded by linear damping b and driven by a linear spring referenced to a 'zero-force' point x_0. With no interaction force, $m_m\ddot{x} + b\dot{x} = k(x_0 - x)$. Assume the manipulator interacts with an object modeled as mass m_o such that both move with common motion x. The connection between manipulator and object is an interaction port, and the force exerted by the manipulator on the object is $f_o = m_o\ddot{x}$. Further assume the object is subject to stochastic perturbation forces w_e, modeled as zero-mean Gaussian white noise of strength S. Defining $m = m_m + m_o$, state-determined equations for the coupled system are

$$\frac{d}{dt}\begin{bmatrix} x \\ v \end{bmatrix} = \begin{bmatrix} 0 & 1 \\ -k/m & -b/m \end{bmatrix}\begin{bmatrix} x \\ v \end{bmatrix} + \begin{bmatrix} 0 \\ k/m \end{bmatrix} x_0 + \begin{bmatrix} 0 \\ 1/m \end{bmatrix} w_e$$

$$f_o = \begin{bmatrix} -\frac{m_o}{m}k - \frac{m_o}{m}b \end{bmatrix}\begin{bmatrix} x \\ v \end{bmatrix} + \begin{bmatrix} \frac{m_o}{m}k \end{bmatrix} x_0$$

The objective function should include force and displacement at the interaction port. Define displacement $\Delta x = x_0 - x$ and the objective function

$$Q = E\left\{\frac{1}{t_{final}}\int_0^{t_{final}}\left(\frac{f_o^2}{f_{tol}^2} + \frac{\Delta x^2}{\Delta x_{tol}^2}\right)dt\right\}$$

where f_{tol} and Δx_{tol} are tolerances on interface force and displacement. The state (x, v) and output f_o are random variables due to the presence of the stochastic input. The expectation operator $E\{\cdot\}$ makes the objective Q a deterministic scalar. The stochastic perturbation is included only to ensure the optimization yields non-trivial stable solutions. Once a solution with non-zero noise strength S has been identified, we may consider the limit as the noise strength approaches zero.

For simplicity, assume $x_0(t) = \text{constant} = 0$; i.e. the object is to be held at a constant position despite perturbations. A summary of subsequent analysis is presented in an appendix, based on a method presented in [38, 41]. A steady-state solution for the optimal stiffness and damping is

$$\lim_{S\to 0} k_{opt} = \frac{m_m + m_o}{m_o}\frac{f_{tol}}{\Delta x_{tol}}$$

$$\lim_{S\to 0} b_{opt} = \sqrt{2k_{opt}(m_m + m_o)}$$

The optimal damping coefficient b_{opt} is such that the 2nd order coupled system (manipulator plus object) has a dimensionless damping ratio of $\zeta = b_{opt}/2\sqrt{k_{opt}(m_m + m_o)} = \sqrt{2}/2$. As a result the 2nd order frequency response function relating object motion to perturbations is 'optimally flat' up to a break frequency defined by the optimal stiffness and the total mass, manipulator plus object.

The optimal stiffness k_{opt} is proportional to the ratio of force tolerance to displacement tolerance. That is physically reasonable; if an object is delicate and cannot tolerate large applied forces, the manipulator should be compliant in proportion. Perhaps less obvious is that the optimal stiffness is also proportional to the ratio $(m_m + m_o)/m_o$ of total mass (manipulator plus object) to object mass. As a result the 2nd order frequency response function relating object motion to perturbations has a break frequency Ω_{break} independent of manipulator mass.

$$\Omega_{break} = \sqrt{\frac{k_{opt}}{m_m + m_o}} = \sqrt{\frac{1}{m_o}\frac{f_{tol}}{\Delta x_{tol}}}$$

A minimal stiffness $k_{opt} = f_{tol}/\Delta x_{tol}$ of would be obtained if the object mass was much larger than the manipulator mass apparent at the interaction port. That may be achieved with some of the newer designs of back-drivable or compliant robot 'hands' or end-effectors [83]. Unfortunately, the converse is usually true. For a typical robot, the mass apparent at its gripper dwarfs the mass of the objects it can manipulate. Greater manipulator apparent mass implies greater optimal stiffness for the same ratio of force and displacement tolerances.

Despite its simplicity, this 'toy' example may provide insight. Apparent mass matters, even if the emphasis is on choosing the optimal stiffness, e.g. to be implemented via one of the recent variable-stiffness actuator designs [11, 107]. Apparent mass includes actuator inertia, 'reflected' or transformed through the transmission system relating end-effector motion to actuator motion. Electro-mechanical robot transmissions commonly include high-ratio gear trains to amplify motor torques. That dramatically increases the motor's contribution to end-effector apparent mass, which is proportional to the square of the gear ratio. A recent study of a commercially-available robot showed that the contribution of its motors to end-effector apparent mass was more than 2.5 times the contribution of its link segments [49].

This 'toy' example also hints at one of the challenges of choosing impedance. Constructing an objective function to include terms involving both motion and 'exertion' at an interaction port is straightforward but solving the resulting optimization problem is not. Even this linear 1 degree of freedom example with 2nd order dynamics required solving 6 simultaneous nonlinear differential equations, and only a steady-state solution was presented (see Appendix). The complexity of the computational problem may be expected to grow exponentially with the number of degrees of freedom and the order of the dynamics associated with each degrees of freedom. Some tasks—especially if they require active vibration absorption—will require higher-order interaction dynamics. Furthermore, a general task will require a time-varying 'trajectory' of impedances rather than a steady-state time-invariant solution. Powerful methods for numerical optimization are now available but unfortunately, variable impedance makes the optimization non-convex, and this appears to be fundamental. Nevertheless, despite the formidable challenges, solutions have been presented [12, 13, 72]. Advances in computational algorithms and processing power may be expected to yield further progress.

11 Using Composability to Meet Multiple Task Objectives

Unlike the 'toy' example above, realistic tasks may have multiple objectives which may present conflicting requirements. In addition, realistic tasks are commonly performed against the backdrop of other ongoing activities, which may interfere. For example, fly-casting is usually performed from a standing position. The required arm motions generate inertial and gravitational perturbations to balance and posture; in turn, changing posture influences those arm motions. In principle, multiple conflicting goals may be incorporated into a single optimization problem; indeed, a well-formulated objective function must quantify some compromise between conflicting requirements if a non-trivial solution is to be identified. In principle, all of the human body's roughly 200 degrees of freedom could be included in the dynamics of the system to be optimized. Unfortunately, the exponential growth of computational complexity with degrees of freedom renders all but the simplest problems infeasible; this is Richard Bellman's notorious 'curse of dimensionality'.

Even if continuing advances in algorithms and processor speed may push back the boundaries of what can be computed in practice, it seems doubtful that global optimization is the best description of processes underlying human motor control. The *composability* of dynamic primitives provides an alternative. 'Composability' refers to the fact that dynamic primitives may be combined to produce more complex behavior. Experimental evidence indicates that some human movements are composed of a sequence of overlapping submovements (Figs. 4, 5 and 6). Oscillatory movements may also be combined, though there appears to be a strong preference for a limited set of phase relations between component oscillations [58, 59, 98, 99, 104]. Mechanical impedances are also composable. Remarkably, multiple impedances may be combined by *linear* superposition, even if the interactive dynamic relations they embody are *nonlinear* [39, 41]. This is due to a fact of Newtonian mechanics: an inertial object such as the skeleton or a tool determines acceleration in response to the linear sum of forces to which it is subjected.

Taking advantage of composability can dramatically simplify control. A simple example illustrates this point. Exerting force on a tool requires producing a concomitant minimum stiffness [90]. Expressed as an equivalent network, the required static behavior may be written as $f = k\,(x_0 - x)$ where f, k and x are end-point force, stiffness and position and x_0 is the zero-force position. This may be transformed to joint coordinates as $\tau = J^T(\theta)\,k\,(x_0 - L(\theta))$ where τ and θ are joint torque and angle, $x = L(\theta)$ describes the forward kinematics and $J(\theta)$ its derivative, the Jacobian matrix. With a high-level controller that specifies k and x_0 and controllable-torque actuators, that expression may be implemented as a nonlinear joint-space controller to achieve the specified end-point equivalent network behavior.

With kinematic redundancy—more joint than end-point degrees of freedom ($\dim\theta > \dim x$)—that equivalent network alone is insufficient to control configuration. Many joint configurations θ yield the same stiffness $f = k\,(x_0 - x)$ and 'null-space' motions that leave x unchanged are unaffected by this controller. However, configuration may be managed by a controller that implements a joint-space equivalent network behavior $\tau_j = K\Delta\theta = K\,(\theta_r - \theta)$ where K is a non-singular joint-space stiffness and θ_r is a zero-torque configuration. Even though one of these controllers is nonlinear, they may be superimposed by simple addition to achieve desirable behavior, $\tau_{net} = \tau + \tau_j = J^T(\theta)\,k\,(x_0 - L(\theta)) + K\,(\theta_r - \theta)$. This composite controller readily manages redundant degrees of freedom. Importantly, *inversion of the kinematic equations is not required*. Not only is a difficult computational problem avoided but, unlike controllers that fundamentally require inversion of the kinematic equations, this approach is indifferent to kinematic singularities. It can operate *at and into* kinematic singularities (e.g. at maximum reach).

12 Modulating Inertia via Multi-limb Coordination

Managing redundant degrees of freedom is especially important for modulating inertial behavior, which dominates interactive dynamics at the transitions between free and constrained motion. Modulating a robot's inertial behavior using feedback control is challenging. It usually requires expensive and delicate force/torque sensors. Moreover, the extent of feedback modulation of inertial behavior is severely constrained if contact stability is to be guaranteed [14, 16]. However, choosing the configuration of the joints has a profound influence on the inertial behavior apparent at a point of contact such as the hand [40]. Importantly, it allows inertial dynamics, which determine the magnitude of impulsive forces, to be *pre-tuned prior to contact* thereby avoiding possible problems with time delays due to reactive control.

The advantages of dynamic primitives and the composability of impedance extend to multi-limb coordination. Controlling inertia with a single limb is challenging due to the distribution of physical inertia along the limb segments. In particular, translational force impulses almost always induce undesirable rotational motion. In contrast, two-handed control of interaction with a tool affords advantages. In particular, with two hands, the inertial terms that couple translational impulses to rotational motion can be made to cancel, making the response to the collision that occurs on contact fundamentally more predictable.

Unfortunately, wielding a tool with two hands 'closes the kinematic chain' relating joint motions to end-point motions. Closed-chain kinematic equations are notoriously challenging. However, *this challenge may be avoided entirely* by taking advantage of the composability of impedance. If the motion of the two hands at the point of contact with the tool is common (more generally, if they are kinematically related through the tool) their interactive behaviors superimpose linearly. Each limb may be endowed with stiffness as described above based on the *open-chain* kinematics of each limb separately (and without inverting its kinematic equations). The net stiffness of both arms interacting with the tool is simply the sum of their individual stiffnesses. The net inertia of both arms interacting with the tool is the sum of their individual inertias. The undesirable coupling terms of the 'left' arm are generally equal and opposite to those of the 'right' arm; combined, they cancel.

The above considered only stiffness and inertia. Dissipative behavior (e.g. damping) is also important, indeed essential to ensure stability. Once again, the composability of impedance allows damping terms to be implemented independent of stiffness or inertia, separately for each limb (or even for each joint), then combined by simple superposition. Higher-order dynamic terms, should they be relevant, may be treated in an exactly analogous manner. Moreover, if each of these terms, singly or in combination, is configured to exhibit attractor dynamics, then it becomes a dynamic primitive in the sense that we have defined [47, 48]. This, in turn, confers an important robustness to the behavior implemented.

13 Advantages and Consequences of Composability

The composability of dynamic primitives provides one way to 'work around' the curse of dimensionality, allowing the challenge of coordinating many degrees of freedom in a multi-objective task to be broken down into a set of much smaller problems. Each sub-task may be expressed in the form of an equivalent network (Fig. 7) which combines uni-lateral forward-path dynamic behavior, which outputs a zero-force trajectory $x_0(t)$, with bi-lateral interactive dynamic behavior, the impedance $Z\{\cdot\}$. Each equivalent network responds to the motion of the inertial object with which it interacts. Interactions may be at different locations; for example, one equivalent network may specify a desired behavior of the hand, another a desired behavior of the elbow, and so forth. In this way, different body parts may be used to 'manipulate' the world, even simultaneously; humans do this frequently. Each equivalent network determines an output force or torque which adds to the net force or torque applied to the inertial object (e.g. the skeleton) and produces force or motion, depending on the totality of all interacting equivalent networks and physical objects (e.g. tools). If conflicting goals are expressed by different equivalent networks, their respective impedances determine the resolution. Coordination emerges from the combined action of all equivalent networks. Because of its generality, this approach has been extended to non-contact tasks such as avoiding obstacles while acquiring targets, even obstacles which may move [6, 41, 78].

This '*divide et impera*' approach may have interesting consequences. Using it as outlined above to manage kinematic redundancy, to ensure control of configuration the joint-space stiffness matrix K must be positive-definite. Its inverse exists, defining a joint-space compliance $\Delta\theta = K^{-1}\tau_j$. For small $\Delta\theta$, the corresponding end-point compliance is $\Delta x \cong J(\theta)K^{-1}J^T(\theta)f$, where $\Delta x = x - x_r = x - L(\theta_r)$.[3] The end-point compliance $c_j(\theta) = J(\theta)K^{-1}J^T(\theta)$ provides a 'default' interactive behavior which renders force control difficult. Its inverse[4] determines a minimum end-point stiffness. Even with extremely back-drivable actuators (e.g. current-controlled electric motors) 'perfect' force control, corresponding to infinite compliance or zero stiffness, cannot be achieved. Of course, this is also a limitation of human motor control. Whether this is a disadvantage (a 'bug') or an advantage (a 'feature') depends on context. As outlined above, in most tool-using tasks, simultaneous modulation of stiffness and force is essential.

---

[3]This result may be extended to large $\Delta\theta$ and Δx: for any Δx imposed within the workspace, at equilibrium the linkage assumes a pose that minimizes total potential energy; the analysis is omitted for brevity.

[4]When it can be computed; in some configurations and some directions, e.g., arm fully outstretched, compliance approaches zero and stiffness approaches infinity.

14 Integrating Tool Use with Posture, Balance and Locomotion

The composability of dynamic primitives may simplify the control of complex behavior, and using mechanical impedance to manage physical interaction may facilitate integration of arm motion with posture, balance and locomotion. Indeed, consideration of posture is essential to understand contact tasks and force control. Many tools are used from a standing position. In that case the dynamics of force production depends critically on the posture of the feet. Pushing hard by leaning with the feet together introduces an unstable dynamic zero—force must transiently increase before it can decrease and vice versa. Spreading the feet apart eliminates this behavior—foot pose affects hand dynamics [88].

However, the precise nature of dynamic motion primitives underlying posture and locomotion is unclear. At first blush it might seem that a point attractor is the appropriate dynamic primitive for posture and balance, but time-series analysis of center of pressure variation during standing indicates that multiple limit cycles are present [17, 25]. Rhythmic walking might seem to require a limit-cycle attractor and, consistent with this model, human walking exhibits entrainment to periodic perturbations, both on a treadmill and overground [4, 5, 82]. However, rhythmic walking might alternatively emerge as a consequence of a 'capture point' foot-placement strategy: the swing foot is placed at a location where present momentum would bring the body to rest over it; observations of human walking appear consistent with this model [24, 87, 110].

Unimpaired human walking is highly dynamic, to the extent that it may be regarded as 'controlled falling'; during single-leg stance the system is unstable (like an inverted pendulum). From that perspective, one important function of the foot—and especially the ankle—is analogous to the function of an automobile shock-absorber, acting to 'catch' the descending body. The essential 'shock-absorbing' behavior is characterized by mechanical impedance. While neural feedback of motion and force contributes to net mechanical impedance, the delays due to neural transmission render feedback modulation of mechanical impedance ineffective during the rapid events associated with heel-strike. Consequently, we may expect ankle/foot mechanical impedance to be pre-tuned prior to heel-strike. Observations of multi-variable human ankle mechanical impedance show that it is reliably increased by simultaneous co-activation of opposing muscles [65, 66, 69, 71]. Furthermore, observations of the time-varying 'trajectory' of ankle mechanical impedance during treadmill walking show that it is elevated by co-contracting opposing muscles prior to heel-strike [67, 70]. Remarkably, these measurements also show that ankle mechanical impedance is *energetically passive*, even when muscles are active (up to 30% of maximum voluntary contraction) [68]. The significance of this observation is that, in general, physical interaction (e.g. due to foot-ground contact) might compromise stability. Energetic passivity guarantees that physical contact cannot induce instability [15].

Summarizing, while it is clear that dynamic primitives such as oscillations and mechanical impedances likely play an important role in posture and locomotion, many of the details of how this is accomplished remain to be uncovered.

15 A Geometry of Dynamic Primitives?

The paradox of human performance—how do we out-perform modern robots despite inferior 'wetware' and 'hardware'?—is both a challenge and opportunity. The challenge is to understand how it is done; the opportunity is to identify bio-inspired approaches to improve robot performance. This may require substantial re-thinking of robot control, even down to the fundamentals of the underlying geometry. Model-based robot control implicitly assumes a Riemannian geometry, yet human haptic perception appears to be incompatible with Riemannian geometry.

A growing body of evidence suggests that human motor performance is based on dynamic primitives. Combinations of motion primitives (submovements and oscillations) account for recovery after neurological injury as well as some counter-intuitive limitations of human motor control (moving slowly is hard for humans). Controlling physical interaction may also be based on interactive dynamic primitives (impedance or admittance). Forward-path dynamics and interactive dynamics may be combined by re-purposing and extending the classical linear equivalent electric circuit to define a nonlinear equivalent network.

Interestingly, the most useful form of nonlinear equivalent network requires the forward-path dynamics to prescribe motions, not forces. That is consistent with unimpaired human motor behavior and recovery after neural injury. It suggests some form of underlying geometric structure but prompts an open question: Which geometry is induced by a composition of motion and interactive dynamic primitives? What are its properties? Answering those questions might pave the way to achieve superior robot control and seamless human-robot collaboration.

Acknowledgements I would especially like to acknowledge Professor Dagmar Sternad's seminal contribution to the concepts of dynamic primitives presented herein. This work was supported in part by the Eric P. and Evelyn E. Newman Fund and by NIH Grant R01-HD087089 and NSF EAGER Grant 1548514.

Appendix: Choosing Impedance via Stochastic Optimization

Assume a manipulator and its control system are modeled as a mass m_m moving in 1 degree of freedom, retarded by linear damping b and driven by a linear spring referenced to a 'zero-force' point x_0. It interacts with an object modeled as mass m_o such that both move with common motion x. State-determined equations for the coupled system are

$$\frac{d}{dt}\begin{bmatrix} x \\ v \end{bmatrix} = \begin{bmatrix} 0 & 1 \\ -k/m & -b/m \end{bmatrix}\begin{bmatrix} x \\ v \end{bmatrix} + \begin{bmatrix} 0 \\ k/m \end{bmatrix} x_0 + \begin{bmatrix} 0 \\ 1/m \end{bmatrix} w_e$$

$$f_o = \begin{bmatrix} -\frac{m_o}{m}k - \frac{m_o}{m}b \end{bmatrix}\begin{bmatrix} x \\ v \end{bmatrix} + \begin{bmatrix} \frac{m_o}{m}k \end{bmatrix} x_0$$

where $m = m_m + m_o$, f_o is the force exerted by the manipulator on the object, and w_e denotes stochastic perturbation forces, modeled as zero-mean Gaussian white noise of strength S, i.e. $E\{w_e(t)\} = 0$, $E\{w_e(t)\, w_e(t+\tau)\} = S\delta(\tau)$, where $E\{\cdot\}$ is the expectation operator and $\delta(\cdot)$ denotes the unit impulse function. The objective function to be minimized is

$$Q = E\left\{ \frac{1}{t_{final}} \int_0^{t_{final}} \left(\frac{f_o^2}{f_{tol}^2} + \frac{\Delta x^2}{\Delta x_{tol}^2} \right) dt \right\}$$

where $\Delta x = x_0 - x$ and f_{tol} and Δx_{tol} are tolerances on interface force and displacement. Due to the stochastic input, the state $(x(t), v(t))$ and output $f_o(t)$ are random variables. Assume $x_0(t) = \text{constant} = 0$ and find conditions for a steady-state solution (i.e. consider the limit as $t_{final} \to \infty$). The mean state and output variables propagate deterministically, i.e. $E\{x(t)\} = E\{v(t)\} = E\{f_o(t)\} = 0$. Define the input covariance $W_e(t) = E\{(w_e^2(t))\} = S$ and the state covariance matrix

$$\Sigma(t) = E\left\{ \begin{bmatrix} x(t) \\ v(t) \end{bmatrix} \begin{bmatrix} x(t) & v(t) \end{bmatrix} \right\} = E\left\{ \begin{bmatrix} x^2(t) & x(t)\,v(t) \\ x(t)\,v(t) & v^2(t) \end{bmatrix} \right\}$$

For notational convenience, omit the explicit time dependence and use overbar notation $\Sigma = \begin{bmatrix} \overline{x^2} & \overline{xv} \\ \overline{xv} & \overline{v^2} \end{bmatrix}$. Covariance propagation through a linear time-invariant system is described by the dynamic equation $\dot{\Sigma} = A\Sigma + \Sigma A^T + BSB^T$ where A and B are system and input weighting matrices.

$$\frac{d}{dt}\begin{bmatrix} \overline{x^2} & \overline{xv} \\ \overline{xv} & \overline{v^2} \end{bmatrix} = \begin{bmatrix} 2\overline{xv} & \overline{v^2} - \overline{x^2}k/m - \overline{xv}b/m \\ \overline{v^2} - \overline{x^2}k/m - \overline{xv}b/m & -2\overline{xv}k/m - 2\overline{v^2}b/m \end{bmatrix} + \begin{bmatrix} 0 & 0 \\ 0 & 1/m^2 \end{bmatrix} S$$

Re-write as 3 coupled scalar differential equations

$$\frac{d}{dt}\overline{x^2} = 2\overline{xv}$$
$$\frac{d}{dt}\overline{xv} = \overline{v^2} - \overline{x^2}k/m - \overline{xv}b/m$$
$$\frac{d}{dt}\overline{v^2} = S/m^2 - 2\overline{xv}k/m - 2\overline{v^2}b/m$$

The scalar to be minimized is

$$Q = \frac{1}{t_{final}} \int_{0}^{t_{final}} \left(\frac{\overline{f_o^2}}{f_{tol}^2} + \frac{\overline{\Delta x^2}}{\Delta x_{tol}^2} \right) dt$$

$$f_o^2 = \left(\frac{m_o^2}{m^2} \right) \left(k^2 x^2 + 2kbxv + b^2 v^2 \right)$$

Defining $q = (f_{tol}/\Delta x_{tol})(m/m_o)$

$$Q = \frac{1}{f_{tol}^2} \frac{m_o^2}{m^2} \frac{1}{t_{final}} \int_{0}^{t_{final}} \left(k^2 \overline{x^2} + 2kb\overline{xv} + b^2 \overline{v^2} + q^2 \overline{x^2} \right) dt$$

Construct the 'control Hamiltonian'

$$\begin{aligned} H = \left(k^2 + q^2 \right) \overline{x^2} + 2kb\overline{xv} + b^2 \overline{v^2} \\ +\lambda_1 2\overline{xv} \\ +\lambda_2 \left(\overline{v^2} - \overline{x^2} k/m - \overline{xv} b/m \right) \\ +\lambda_3 \left(S/m^2 - 2\overline{xv} k/m - 2\overline{v^2} b/m \right) \end{aligned}$$

where λ_i denote Lagrange multipliers. Minimize with respect to k and b.

$$\frac{\partial H}{\partial k} = 2k\overline{x^2} + 2b\overline{xv} - \lambda_2 \overline{x^2}/m - \lambda_3 2\overline{xv}/m = 0$$
$$\frac{\partial H}{\partial b} = 2k\overline{xv} + 2b\overline{v^2} - \lambda_2 \overline{xv}/m - \lambda_3 2\overline{v^2}/m = 0$$

The Lagrange multipliers are defined by 'co-state' equations.

$$\frac{\partial H}{\partial \overline{x^2}} = -\dot{\lambda}_1 = k^2 + q^2 - \lambda_2 k/m$$
$$\frac{\partial H}{\partial \overline{xv}} = -\dot{\lambda}_2 = 2kb + 2\lambda_1 - \lambda_2 b/m - \lambda_3 2k/m$$
$$\frac{\partial H}{\partial \overline{v^2}} = -\dot{\lambda}_3 = b^2 + \lambda_2 - \lambda_3 2b/m$$

Assume steady state exists and set all rates of change to zero.

$$
\begin{gathered}
2\overline{xv} = 0 \\
\overline{v^2} - \overline{x^2}k/m - \overline{xv}b/m = 0 \\
S/m^2 - 2\overline{xv}k/m - 2\overline{v^2}b/m = 0 \\
k^2 + q^2 - \lambda_2 k/m = 0 \\
2kb + 2\lambda_1 - \lambda_2 b/m - \lambda_3 2k/m = 0 \\
b^2 + \lambda_2 - \lambda_3 2b/m = 0
\end{gathered}
$$

A little manipulation shows that

$$
\begin{aligned}
\overline{xv} &= 0 \\
\overline{v^2} &= \overline{x^2}k/m \\
\overline{x^2} &= S/2kb \\
\overline{v^2} &= S/2bm
\end{aligned}
$$

The first co-state equation yields

$$k^2 + q^2 = \lambda_2 k/m$$

$$\lambda_2 = km + q^2 m/k$$

The optimal stiffness is defined by

$$2k_{opt}\overline{x^2} = \lambda_2 \overline{x^2}/m$$

$$k_{opt} = q = \frac{f_{tol}}{\Delta x_{tol}} \frac{m}{m_o}$$

Note that this manipulation requires $\overline{x^2} \neq 0$ and hence $S \neq 0$. However, k_{opt} is independent of noise strength S.

The third co-state equation yields

$$b^2 + \lambda_2 = \lambda_3 2b/m$$

$$\lambda_3 = \left(b^2 + \lambda_2\right) m/2b = \frac{bm}{2} + \frac{km}{2b} + \frac{q^2 m^2}{2kb}$$

The optimal damping is defined by

$$2b_{opt}\overline{v^2} = \lambda_3 2\overline{v^2}/m$$

$$b_{opt} = \sqrt{2k_{opt} m}$$

This manipulation requires $\overline{v^2} \neq 0$ and hence $S \neq 0$. However, b_{opt} is independent of noise strength S.

References

1. C. Abul-Haj, N. Hogan, An emulator system for developing improved elbow-prosthesis designs. IEEE Trans. Biomed. Eng. **34**, 724–737 (1987)
2. C.J. Abul-Haj, N. Hogan, Functional assessment of control-systems for cybernetic elbow prostheses. 1. Description of the technique. IEEE Trans. Biomed. Eng. **37**, 1025–1036 (1990a)
3. C.J. Abul-Haj, N. Hogan, Functional assessment of control-systems for cybernetic elbow prostheses. 2. Application of the technique. IEEE Trans. Biomed. Eng. **37**, 1037–1047 (1990b)
4. J. Ahn, N. Hogan, A simple state-determined model reproduces entrainment and phase-locking of human walking dynamics. PLoS ONE **7**, e47963 (2012a)
5. J. Ahn, N. Hogan, Walking is not like reaching: evidence from periodic mechanical perturbations. PLoS ONE **7**, e31767 (2012b)
6. J.R. Andrews, N. Hogan, Impedance Control as a Framework for Implementing Obstacle Avoidance in a Manipulator, in BOOK, D. E. H. A. W. J. (ed.) *Control of Manufacturing Processes and Robotic Systems* (ASME, 1983)
7. A.J. Bastian, T.A. Martin, J.G. Keating, W.T. Thach, Cerebellar ataxia: abnormal control of interaction torques across multiple joints. J. Neurophysiol. **76**, 492–509 (1996)
8. A. Bissal, J. Magnusson, E. Salinas, G. Engdahl, A. Eriksson, On the design of ultra-fast electromechanical actuators: a comprehensive multi-physical simulation model, in *Sixth International Conference on Electromagnetic Field Problems and Applications (ICEF)* (2012)
9. T. Boaventura, C. Semini, J. Buchli, M. Frigerio, M. Focchi, D.G. Caldwell, Dynamic torque control of a hydraulic quadruped robot, in *IEEE International Conference on Robotics and Automation* (IEEE, Saint Paul, Minnesota, USA, 2012)
10. C. Boesch, H. Boesch, Tool use and tool making in wild chimpanzees. Folia Primatologica **54**, 86–99 (1990)
11. D.J. Braun, S. Apte, O. Adiyatov, A. Dahiya, N. Hogan, Compliant actuation for energy efficient impedance modulation, in *IEEE International Conference on Robotics and Automation* (2016)
12. J. Buchli, F. Stulp, E. Theodorou, S. Schaal, Learning variable impedance control. Int. J. Robot. Res. **30**, 820–833 (2011)
13. M. Cohen, T. Flash, Learning impedance parameters for robot control using an associative search network. IEEE Trans. Robot. Autom. **7**, 382–390 (1991)
14. E. Colgate, On the intrinsic limitations of force feedback compliance controllers, in *Robotics Research - 1989*, eds. by K. Youcef-Toumi, H. Kazerooni (ASME, 1989)
15. J.E. Colgate, N. Hogan, Robust control of dynamically interacting systems. Int. J. Control **48**, 65–88 (1988)
16. J.E. Colgate, N. Hogan, The interaction of robots with passive environments: application to force feedback control, in *Fourth International Conference on Advanced Robotics*, June 13–15 (Columbus, Ohio, 1989)
17. J.J. Collins, C.J. de Luca, Open-loop and closed-loop control of posture: a random-walk analysis of center-of-pressure trajectories. Exp. Brain Res. **95**, 308–318 (1993)
18. F. Crevecoeur, J. McIntyre, J.L. Thonnard, P. Lefèvre, Movement stability under uncertain internal models of dynamics. J. Neurophysiol. **104**, 1301–1313 (2010)
19. A. de Rugy, D. Sternad, Interaction between discrete and rhythmic movements: reaction time and phase of discrete movement initiation against oscillatory movements. Brain Res. **994**, 160–174 (2003)
20. S. Degallier, A. Ijspeert, Modeling discrete and rhythmic movements through motor primitives: a review. Biol. Cybern. **103**, 319–338 (2010)

21. R. Deits, R. Tedrake, Efficient mixed-integer planning for UAVs in cluttered environments, in *IEEE International Conference on Robotics and Automation (ICRA)* (IEEE, Seattle, WA, 2015)
22. J.A. Doeringer, N. Hogan, Intermittency in preplanned elbow movements persists in the absence of visual feedback. J. Neurophysiol. **80**, 1787–1799 (1998a)
23. J.A. Doeringer, N. Hogan, Serial processing in human movement production. Neural Netw. **11**, 1345–1356 (1998b)
24. J. Englsberger, C. Ott, A. Albu-Schaffer, Three-dimensional bipedal walking control based on divergent component of motion. IEEE Trans. Robot. **31**, 355–368 (2015)
25. C.W. Eurich, J.G. Milton, Noise-induced transitions in human postural sway. Phys. Rev. E **54**, 6681–6684 (1996)
26. E.D. Fasse, N. Hogan, B.A. Kay, F.A. Mussa-Ivaldi, Haptic interaction with virtual objects - spatial perception and motor control. Biol. Cybern. **82**, 69–83 (2000)
27. J. Flanagan, P. Vetter, R. Johansson, D. Wolpert, Prediction precedes control in motor learning. Curr. Biol. **13**, 146–150 (2003)
28. J.R. Flanagan, A.K. Rao, Trajectory adaptation to a nonlinear visuomotor transformation: evidence of motion planning in visually perceived space. J. Neurophysiol. **74**, 2174–2178 (1995)
29. T. Flash, N. Hogan, The coordination of arm movements - an experimentally confirmed mathematical model. J. Neurosci. **5**, 1688–1703 (1985)
30. H. Gomi, M. Kawato, Equilibrium-point control hypothesis examined by measured arm stiffness during multijoint movement. Science **272**, 117–120 (1996)
31. P. Gribble, D.J. Ostry, V. Sanguinetti, R. Laboissiere, Are complex control signals required for human arm movement? J. Neurophysiol. **79**, 1409–1424 (1998)
32. O. Heaviside, *Electrical Papers* (Massachusetts, Boston, 1925a)
33. O. Heaviside, *Electrical Papers* (Massachusetts, Boston, 1925b)
34. H.V. Helmholtz, II. Uber einige Gesetze der Vertheilung elektrischer Ströme in körperlichen Leitern mit Anwendung auf die thierisch-elektrischen Versuche [Some laws concerning the distribution of electrical currents in conductors with applications to experiments on animal electricity]. Annalen der Physik und Chemie **89**, 211–233 (1853)
35. A.J. Hodgson, Inferring Central Motor Plans from Attractor Trajectory Measurements, Ph.D, Institute of Technology, Massachusetts, 1994
36. A.J. Hodgson, N. Hogan, A model-independent definition of attractor behavior applicable to interactive tasks. IEEE Trans. Syst. Man Cybern. Part C- Appl. Rev. **30**, 105–118 (2000)
37. J.A. Hoffer, S. Andreassen, Regulation of soleus muscle stiffness in premammillary cats: intrinsic and reflex components. J. Neurophysiol. **45**, 267–285 (1981)
38. N. Hogan, Adaptive control of mechanical impedance by coactivation of antagonist muscles. IEEE Trans. Autom. Control **29**, 681–690 (1984)
39. N. Hogan, Impedance control - an approach to manipulation. 1. Theory. J. Dyn. Syst. Meas. Control Trans. Asme **107**, 1–7 (1985a)
40. N. Hogan, Impedance control - an approach to manipulation. 2. Implementation. J. Dyn. Syst. Meas. Control Trans. Asme **107**, 8–16 (1985b)
41. N. Hogan, Impedance control - an approach to manipulation. 3. Applications. J. Dyn. Syst. Meas. Control Trans. Asme **107**, 17–24 (1985c)
42. N. Hogan, Mechanical impedance of single-and multi-articular systems, in *Multiple Muscle Systems: Biomechanics and Movement Organization*, eds. by J. Winters, S. Woo (Springer, New York, 1990)
43. N. Hogan, A general actuator model based on nonlinear equivalent networks. IEEE/ASME Trans. Mechatron. **19**, 1929–1939 (2014)
44. N. Hogan, S.P. Buerger, Impedance and interaction control, in *Robotics and Automation Handbook*, ed by T.R. Kurfess (CRC Press, Boca Raton, FL, 2005)
45. N. Hogan, B.A. Kay, E.D. Fasse, F.A. Mussaivaldi, Haptic illusions - experiments on human manipulation and perception of virtual objects. Cold Spring Harbor Symp. Quant. Biol. **55**, 925–931 (1990)

46. N. Hogan, H.I. Krebs, B. Rohrer, J.J. Palazzolo, L. Dipietro, S.E. Fasoli, J. Stein, R. Hughes, W.R. Frontera, D. Lynch, B.T. Volpe, Motions or muscles? Some behavioral factors underlying robotic assistance of motor recovery. J. Rehab. Res. Dev. **43**, 605–618 (2006)
47. N. Hogan, D. Sternad, Dynamic primitives of motor behavior. Biol. Cybern. **106**, 727–739 (2012)
48. N. Hogan, D. Sternad, Dynamic primitives in the control of locomotion. Front. Comput. Neurosci. **7**, 1–16 (2013)
49. L.A. Hosford, *Development and Testing of an Impedance Controller on an Anthropomorphic Robot for Extreme Environment Operations*, Master of Science, Massachusetts Institute of Technology (2016)
50. D.R. Humphrey, D.J. Reed, Separate cortical systems for control of joint movement and joint stiffness: reciprocal activation and coactivation of antagonist muscles, in *Motor Control Mechanisms in Health and Disease*, ed. by J.E. Desmedt (Raven Press, New York, 1983)
51. G.R. Hunt, Manufacture and use of hook-tools by New Caledonian crows. Nature **379**, 259–251 (1996)
52. A.J. Ijspeert, J. Nakanishi, H. Hoffmann, P. Pastor, S. Schaal, Dynamical movement primitives: learning attractor models for motor behaviors. Neural Comput. **25**, 328–373 (2013)
53. S.H. Johnson-Frey, The neural basis of complex tool use in humans. Trends Cogn. Sci. **8**, 71–78 (2004)
54. D.H. Johnson, Origins of the equivalent circuit concept: the current-source equivalent. Proc. IEEE **91**, 817–821 (2003a)
55. D.H. Johnson, Origins of the equivalent circuit concept: the voltage-source equivalent. Proc. IEEE **91**, 636–640 (2003b)
56. E.R. Kandel, J.H. Schwartz, T.M. Jessell (eds.), *Principles of Neural Science* (McGraw-Hill, New York, 2000)
57. M. Kawato, Internal models for motor control and trajectory planning. Curr. Opin. Neurobiol. **9**, 718–727 (1999)
58. J.A. Kelso, Phase transitions and critical behavior in human bimanual coordination. Am. J. Physiol. Regul. Integr. Comp. Physiol. **246**, R1000–R1004 (1984)
59. J.A.S. Kelso, On the oscillatory basis of movement. Bull. Psychon. Soc. **18**, 49–70 (1981)
60. B. Kenward, A.A.S. Weir, C. Rutz, A. Kacelnik, Behavioral ecology: Tool manufacture by naïve juvenile crows. Nature **433** (2005)
61. H.I. Krebs, M.L. Aisen, B.T. Volpe, N. Hogan, Quantization of continuous arm movements in humans with brain injury. Proc. Natl. Acad. Sci. U.S.A. **96**, 4645–4649 (1999)
62. S. Kuindersma, R. Deits, M. Fallon, A. Valenzuela, H. Dai, F. Permenter, T. Koolen, P. Marion, R. Tedrake, Optimization-based locomotion planning, estimation, and control design for the Atlas humanoid robot. Auton. Robot. **40**, 429–455 (2016)
63. T.M. Kunnapas, An analysis of the "vertical-horizontal illusion". J. Exp. Psychol. **49**, 134–140 (1955)
64. J.R. Lackner, P. Dizio, Rapid adaptation to coriolis force perturbations of arm trajectory. J. Neurophysiol. **72**, 299–313 (1994)
65. H. Lee, P. Ho, M.A. Rastgaar, H.I. Krebs, N. Hogan, Multivariable static ankle mechanical impedance with relaxed muscles. J. Biomech. **44**, 1901–1908 (2011)
66. H. Lee, P. Ho, M.A. Rastgaar, H.I. Krebs, N. Hogan, Multivariable static ankle mechanical impedance with active muscles. IEEE Trans. Neural Syst. Rehab. Eng. **22**, 44–52 (2013)
67. H. Lee, N. Hogan, Time-varying ankle mechanical impedance during human locomotion. IEEE Trans. Neural Syst. Rehab. Eng. (2014)
68. H. Lee, N. Hogan, Energetic passivity of the human ankle joint. IEEE Trans. Neural Syst. Rehab. Eng. (2016)
69. H. Lee, H. Krebs, N. Hogan, Multivariable dynamic ankle mechanical impedance with active muscles. IEEE Trans. Neural Syst. Rehab. Eng. **22**, 971–981 (2014a)
70. H. Lee, H.I. Krebs, N. Hogan, Linear time-varying identification of ankle mechanical impedance during human walking, in *5th Annual Dynamic Systems and Control Conference* (ASME, Fort Lauderdale, Florida, USA, 2012)

71. H. Lee, H.I. Krebs, N. Hogan, Multivariable dynamic ankle mechanical impedance with relaxed muscles. IEEE Trans. Neural Syst. Rehab. Eng. **22**, 1104–1114 (2014b)
72. Y. Li, S.S. GE, Impedance learning for robots interacting with unknown environments. IEEE Trans. Control Syst. Technol. **22** (2014)
73. F.M. Marchetti, S.J. Lederman, The haptic radial-tangential effect: two tests of Wong's 'moments-of-inertia' hypothesis. Bull. Psychon. Soc. **21**, 43–46 (1983)
74. J.C. Maxwell, *A Treatise on Electricity and Magnetism* (1873)
75. H.F. Mayer, Ueber das Ersatzschema der Verstärkerröhre [On equivalent circuits for electronic amplifiers]. Telegraphen- und Fernsprech-Technik **15**, 335–337 (1926)
76. Moog, Moog G761/761 Series Flow Control Servovalves (Moog Inc, 2014)
77. J.M. Morgan, W.W. Milligan, A 1 kHz servohydraulic fatigue testing system, in *Conference on High Cycle Fatigue of Structural Materials*, ed. by Srivatsan, W. O. S. A. T. S. (Warrendale, PA, 1997)
78. W.S. Newman, N. Hogan, High speed robot control and obstacle avoidance using dynamic potential functions, in *IEEE International Conference on Robotics and Automation* (IEEE, New Jersey, 1987)
79. I. Newton, Philosophiæ Naturalis Principia Mathematica (1687)
80. T.R. Nichols, J.C. Houk, Improvement in linearity and regulation of stiffness that results from actions of stretch reflex. J. Neurophysiol. **39**, 119–142 (1976)
81. E.L. Norton, *Design of Finite Networks for Uniform Frequency Characteristic* (Western Electric Company Inc, New York, 1926)
82. J. Ochoa, D. Sternad, N. Hogan, Entrainment of overground human walking to mechanical perturbations at the ankle joint, in *International Conference on Biomedical Robotics and Biomechatronics (BioRob)* (IEEE, Singapore, 2016)
83. L.U. Odhner, L.P. Jentoft, M.R. Claffee, N. Corson, Y. Tenzer, R.R. Ma, M. Buehler, R. Kohout, R.D. Howe, A.M. Dollar, A compliant, underactuated hand for robust manipulation. Int. J. Robot. Res. **33**, 736–752 (2014)
84. N. Paine, S. Oh, L. Sentis, Design and control considerations for high-performance series elastic actuators. IEEE/ASME Trans. Mechatron. **19**, 1080–1091 (2014)
85. R. Plamondon, A.M. Alimi, P. Yergeau, F. Leclerc, Modelling velocity profiles of rapid movements: a comparative study. Biol. Cybern. **69**, 119–128 (1993)
86. R.A. Popat, D.E. Krebs, J. Mansfield, D. Russell, E. Clancy, K.M. Gillbody, N. Hogan, Quantitative assessment of 4 men using above-elbow prosthetic control. Arch. Phys. Med. Rehab. **74**, 720–729 (1993)
87. J. Pratt, J. Carff, S. Drakunov, A. Goswami, Capture point: a step toward humanoid push recovery, in *Humanoids 2006* (IEEE, New Jersey, 2006)
88. D. Rancourt, N. Hogan, Dynamics of pushing. J. Mot. Behav. **33**, 351–362 (2001a)
89. D. Rancourt, N. Hogan, Stability in force-production tasks. J. Mot. Behav. **33**, 193–204 (2001b)
90. D. Rancourt, N. Hogan, The biomechanics of force production, in *Progress in Motor Control: A Multidisciplinary Perspective*, ed by D. Sternad (Springer, Heidelberg, 2009)
91. B. Rohrer, S. Fasoli, H.I. Krebs, B. Volpe, W.R. Frontera, J. Stein, N. Hogan, Submovements grow larger, fewer, and more blended during stroke recovery. Mot. Control **8**, 472–483 (2004)
92. B. Rohrer, N. Hogan, Avoiding spurious submovement decompositions: a globally optimal algorithm. Biol. Cybern. **89**, 190–199 (2003)
93. B. Rohrer, N. Hogan, Avoiding spurious submovement decompositions II: a scattershot algorithm. Biol. Cybern. **94**, 409–414 (2006)
94. R. Ronsse, D. Sternad, P. Lefevre, A computational model for rhythmic and discrete movements in uni- and bimanual coordination. Neural Comput. **21**, 1335–1370 (2009)
95. R. Shadmehr, F.A. Mussa-Ivaldi, Adaptive representation of dynamics during learning of a motor task. J. Neurosci. **14**, 3208–3224 (1994)
96. R.N. Shepard, J. Metzler, Mental rotation of three-dimensional objects. Science **171**, 701–703 (1971)

97. D. Sternad, Towards a unified framework for rhythmic and discrete movements: behavioral, modeling and imaging results, in *Coordination: Neural, Behavioral and Social Dynamics*, eds. by A. Fuchs, V. Jirsa (Springer, New York, 2008)
98. D. Sternad, E.L. Amazeen, M.T. Turvey, Diffusive, synaptic, and synergetic coupling: an evaluation through inphase and antiphase rhythmic movements. J. Mot. Behav. **28**, 255–269 (1996)
99. D. Sternad, D. Collins, M.T. Turvey, The detuning factor in the dynamics of interlimb rhythmic coordination. Biol. Cybern. **73**, 27–35 (1995)
100. D. Sternad, A. de Rugy, T. Pataky, W.J. Dean, Interactions of discrete and rhythmic movements over a wide range of periods. Exp. Brain Res. **147**, 162–174 (2002)
101. D. Sternad, W.J. Dean, Rhythmic and discrete elements in multi-joint coordination. Brain Res. **989**, 152–171 (2003)
102. D. Sternad, W.J. Dean, S. Schaal, Interaction of rhythmic and discrete pattern generators in single-joint movements. Hum. Mov. Sci. **19**, 627–664 (2000)
103. D. Sternad, H. Marino, S.K. Charles, M. Duarte, L. Dipietro, N. Hogan, Transitions between discrete and rhythmic primitives in a unimanual task. Front. Comput. Neurosci. **7** (2013)
104. D. Sternad, M.T. Turvey, R.C. Schmidt, Average phase difference theory and 1:1 phase entrainment in interlimb coordination. Biol. Cybern. **67**, 223–231 (1992)
105. L.C. Thévenin, Sur un nouveau théorème d'électricité dynamique [On a new theorem of dynamic electricity]. Comptes Rendus des Séances de l'Académie des Sciences **97**, 159–161 (1883)
106. W.J. Thompson, *Angular Momentum: An Illustrated Guide to Rotational Symmetries for Physical Systems* (Wiley-Interscience, 1994)
107. B. Vanderborght, A. Albu-Schaeffer, A. Bicchi, E. Burdet, D.G. Caldwell, R. Carloni, M. Catalano, O. Eiberger, W. Friedl, G. Ganesh, M. Garabini, M. Grebenstein, G. Grioli, S. Haddadin, H. Hoppner, A. Jafari, M. Laffranchi, D. Lefeber, F. Petit, S. Stramigioli, N. Tsagarakis, M.V. Damme, R.V. Ham, L.C. Visser, S. Wolf, Variable impedance actuators: a review. Robot. Autonom. Syst. **61**, 1601–1614 (2013)
108. J.M. Wakeling, A.-M. Liphardt, B.M. Nigg, Muscle activity reduces soft-tissue resonance at heel-strike during walking. J. Biomech. **36**, 1761–1769 (2003)
109. J.M. Wakeling, B.M. Nigg, Modification of soft tissue vibrations in the leg by muscular activity. J. Appl. Physiol. **90**, 412–420 (2001)
110. Y. Wang, M. Srinivasan, Stepping in the direction of the fall: the next foot placement can be predicted from current upper body state in steady-state walking. Biol. Lett. **10** (2014)
111. D.M. Wolpert, R.C. Miall, M. Kawato, Internal models in the cerebellum. Trends Cogn. Sci. **2**, 338–347 (1998)
112. G.I. Zahalak, Modeling muscle mechanics (and energetics), in *Multiple Muscle Systems: Biomechanics and Movement Organization*, eds. by J.M. Winters, S.L.-Y. Woo (Springer, New York, 1990)

Human Control of Interactions with Objects – Variability, Stability and Predictability

Dagmar Sternad

Abstract How do humans control their actions and interactions with the physical world? How do we learn to throw a ball or drink a glass of wine without spilling? Compared to robots human dexterity remains astonishing, especially as slow neural transmission and high levels of noise seem to plague the biological system. What are human control strategies that skillfully navigate, overcome, and even exploit these disadvantages? To gain insight we propose an approach that centers on how task dynamics constrain and enable (inter-)actions. Agnostic about details of the controller, we start with a physical model of the task that permits full understanding of the solution space. Rendering the task in a virtual environment, we examine how humans learn solutions that meet complex task demands. Central to numerous skills is redundancy that allows exploration and exploitation of subsets of solutions. We hypothesize that humans seek solutions that are stable to perturbations to make their intrinsic noise matter less. With fewer corrections necessary, the system is less afflicted by long delays in the feedback loop. Three experimental paradigms exemplify our approach: throwing a ball to a target, rhythmic bouncing of a ball, and carrying a complex object. For the throwing task, results show that actors are sensitive to the error-tolerance afforded by the task. In rhythmic ball bouncing, subjects exploit the dynamic stability of the paddle-ball system. When manipulating a "glass of wine", subjects learn strategies that make the hand-object interactions more predictable. These findings set the stage for developing propositions about the controller: We posit that complex actions are generated with dynamic primitives, modules with attractor stability that are less sensitive to delays and noise in the neuro-mechanical system.

Keywords Human motor control · Motor learning · Neuromotor noise · Variability · Stability · Predictability · Tool use

---

Submitted to: Laumond, J.-P. Geometric and Numerical Foundations of Movement.

D. Sternad (✉)
Departments of Biology, Electrical and Computer Engineering, and Physics, Center for the Interdisciplinary Research on Complex Systems, Northeastern University, 134 Mugar Life Science Building, 360 Huntington Avenue, Boston, MA 02115, USA
e-mail: dagmar@northeastern.edu

J.-P. Laumond et al. (eds.), *Geometric and Numerical Foundations of Movements*,
Springer Tracts in Advanced Robotics 117, DOI 10.1007/978-3-319-51547-2_13

1 Introduction

Imagine a dancer, perhaps Rudolf Nureyev or Margaret Fonteyn, both legends in classical ballet: we can only marvel at how they are in complete control of their body, combining extraordinary flexibility and strength with technical difficulty and elegance. And yet, I submit that Evgenia Kanaeva, two-times all-around Olympic champion in rhythmic gymnastics, equals, if not surpasses their level of skill: Not only does she move her lithe body with perfection and grace, she also plays with numerous objects: she throws, catches and bounces a ball, she rolls and swivels a hoop, and sets a 6 m-long ribbon into smooth spirals with the most exquisite movements of her hands and fingers – and yes, sometimes also using her arms, shoulders, or her legs and feet. Her magical actions and interactions with objects arguably represent the pinnacle of human motor control.

How do humans act and interact with objects and tools? After all, tool use is what gave humans their evolutionary advantage over other animals. In robotics, manipulation of tools has clearly been one of the primary motivations to develop robots, going back to the first industrial robots designed to automate repetitive tasks such as placing parts or tightening screws. However, these actions lack the dexterity that not only elite performers, but all healthy humans display. Opening a bottle of wine with a corkscrew or eating escargot with a fork and tongs are skills that require subtle interactions with complex tools and objects. How do humans control these actions and interactions?

Research in motor neuroscience has only arrived at limited answers. To assure experimental control and rigor, computational research has confined itself to simple laboratory tasks, most commonly reaching to a point target, while restricting arm movements to two joints moving in the horizontal plane [57, 58]. Research on sequence learning has typically been limited to finger presses evaluated with simple discrete metrics of timing and serial errors [43, 81]. Grasping has been reduced to isometric finger presses with predetermined contact points to analyze contact forces [37, 82]. The obvious benefit of such simplifications is that the data are accessible and tractable for testing theory-derived hypotheses. Over the past 20 years, numerous studies in computational neuroscience have embraced control-theoretical concepts, such as Kalman filters [39], Bayesian multi-sensory integration [2, 80], and optimal feedback control [74] to account for such experimentally controlled human data. While advances have been made, nobody would deny that this approach encounters challenges when the actions become more complex and realistic. This is particularly problematic when actions are no longer free, as in reaching, but involve contact with objects, ranging from pouring a glass of wine to moving the ribbon in gymnastics. Needless to say, current state-of-the-art movements of robots are still a far cry from those of Elena Kanaeva. Why do humans perform so much better, at least to date? What can robotic control learn from human neuromotor control?

1.1 *The Paradox: Delays and Noise in the Human Neuromotor System*

A first look into the biological neuromotor control system reveals some puzzling facts: information transmission in the human central nervous system is extremely slow and also very noisy. Action potentials, the basic unit of information coding, travel at a speed of approximately 100 m/s [32]; the shortest feedback loop is around 50 ms and reserved for startle reactions [35, 47]. When feedback is integral to more meaningful responses, loop times of 200 ms and longer are a more realistic estimate. In addition to such long delays, the biological neuromotor system displays noise and fluctuations at all levels [13]. The biological system is an extremely complex non-linear system with multiple levels of spatiotemporal scales, ranging from molecular and cellular processes to motor units and muscle contractions, and to overt behavior. Noise and fluctuations from all these levels manifest themselves at the behavioral level as ubiquitous variability. For example, in simple rhythmic finger tapping even trained musicians exhibit at least 5% variance of the period [19, 72]. In a discrete throwing action, humans display a limit in timing resolution of 9 ms [8]. Such large delays and high levels of noise pose extreme challenges for any control model. And yet, humans are amazingly agile and dexterous.

While the human controller appears clearly inferior to robotic systems, the biological "hardware" with its compliant muscles and soft tissues defy any comparison with the heavy actuators of robots. It seems highly likely that the dexterous human controller exploits these features. More recent developments in robotics have produced actuators with variable compliance, such as hands or grippers made of soft material [12] or actuators with mechanically adjustable series compliance [78]. However, the flexibility that comes with variable stiffness may also incur costs, such as loss in precision or higher energy demands. How do humans combine their software limitations and use their compliant and high-dimensional actuators to solve complex task demands?

1.2 *Intrinsic and Extrinsic Redundancy*

The biological sensorimotor system has a large number of hierarchical levels with high dimensionality on each level. One important consequence of this high dimensionality is that it affords redundancy and thereby an infinite variety of ways a given action can be performed. At the behavioral level, hammering a nail into a wooden block can be achieved with multiple different arm trajectories and muscle activation patterns. The adage "repetitions without repetition" conveys that the ubiquitous and ever-present fluctuations prevent any action to be the same as another one. Importantly, this *intrinsic redundancy* faces an additional *extrinsic redundancy* that is inherent to the task. Imagine dart throwing: the bull's eye or the rings on the dartboard allow a set of hits that achieve a given score. Further, orientation angle of the dart stuck on the board does not change the score. Hence, the task has extrinsic redundancy that permits a *manifold of solutions* [68]. However, not all solutions are equally suitable: some may not be biomechanically optimal, others may be risky, yet others may have a lot of tolerance to error and noise. Examining human performance may reveal how humans navigate the task's redundancy and preferences may give insight into the controller. Hence, a suitably constructed extrinsic redundancy presents an important entry point into examining human control, strategies, or objective functions.

1.3 An Agnostic Approach to Human Motor Control

Recognizing these challenges, our research has adopted an approach with minimal assumptions about human neuromotor control. Instead of starting with a hypothesized controller and the plant, i.e., the brain and the musculo-skeletal system, connected by forward and feedback loops transmitting motor and sensory signals, we take an agnostic stance. We begin with what is known and can be analyzed: the physical task that the actor performs. Under simplified conditions, very few assumptions need to be made about the human controller.

This chapter will review this task-dynamic approach as it was developed in three experimental paradigms that examine human interactive skills. These three skills progress from the simple action of throwing a ball, to rhythmic intermittent bouncing of a ball, to the continuous manipulation of a complex object, a cup with a rolling ball inside, mimicking a cup of coffee – or a glass of wine. Mathematical analyses and exemplary results will show that variability, stability and predictability matters in human motor control. I will close with a still largely speculative hypothesis on how the human control system generates such actions, a perspective that may be less hampered by long delays and noise: control via dynamic primitives.

2 A Task-Dynamic Approach to Understanding Control of Interactions

Using mathematical modeling and virtual technology we developed a task-dynamic approach to study the acquisition and control of simple and more complex interactive skills. Following a brief outline of the methodological steps, three exemplary lines of research will be reviewed with some selected results.

2.1 Identifying a Motor Task

The important initial step is choosing a motor task that satisfies several desiderata: First, it should represent a core aspect germane to many other tasks that is "interesting" from a control perspective. Second, the motor task should have redundancy: the well-defined goal should allow for a variety of solutions to achieve the task goal. Third, the task should be novel and sufficiently challenging to require practice to achieve success. The changes over practice provide an important lens to reveal how humans navigate through the space of solutions. (Note this differs from studying everyday behaviors, such as reaching or grasping, where only adaptations to novel scenarios produce change.) Fourth, improvement should happen within one or few experimental session(s), but should also allow for fine-tuning over a longer time scale. These stages are likely to reveal processes underlying motor learning.

We selected and designed three tasks: The arguably simplest (inter-)active task is to throw a ball to a target. While the ball only needs to be released, the size and location of the target imposes constraints on the release that fully determine the projectile's trajectory and thereby the hitting accuracy. A next step in interaction is to repeatedly contact the ball – such as in bouncing a ball rhythmically in the air. This intermittent interaction extends the control demands, as the propelled object needs to be intercepted again. Any error at one contact influences the subsequent contact – these repeated interactions render the task a dynamic system. The third task takes interactions one significant step further: motivated by the seemingly mundane action of carrying a cup of coffee or glass of wine, we designed a simplified task that exemplifies the continuous interaction with a complex object.

2.2 Mathematical Model of the Task

Once the core control challenge is identified, the task is modeled mathematically to formalize and prune away irrelevant aspects of the real-life task. A simple physical model also facilitates subsequent analyses of both model and human data. What system captures the essential demands of ball release and permits a full analysis of the solution space? What is the simplest intermittent dynamical system that lends itself to mathematical analysis? What is the simplest physical system that captures the continuous interaction between the human and a dynamically complex object? One core element in our mathematical modeling and analysis is the distinction between the execution variables $\mathbf{x}$ and the result variables $\mathbf{r}$: The result variable(s) are defined by the task goal and the instruction to the subject and quantify the quality of performance. This is typically an error measure, although this error measure can take many forms. Execution variables are under control of the performer and determine the task result. For the analysis it is important to identify all execution variables that fully determine the result, in order to have an analytic or numerical understanding of the space of solutions. The functional relation between execution and result is the essence of the model and analysis: $\mathbf{r} = f(\mathbf{x})$.

2.3 Mathematical Analysis and Derivation of Hypotheses

Based on the physical model, the space of all possible solutions to the task can be derived. As the model system is typically nonlinear, the space of solutions may be complex and subsets of solutions may have additional properties, such as dynamic stability, risk, or predictability, as elaborated below. The model structure determines the mathematical tools that can be used to derive predictions. Core to our task-dynamic approach are analyses of stability, error sensitivity, or robustness to perturbations and noise. Importantly, exact quantitative hypotheses can be formulated that define those solutions with the greatest probability of success.

2.4 Implementation in a Virtual Environment

Based on the explicit mathematical model, the task is rendered in a virtual environment that permits precise measurement of human execution and errors, i.e., the execution and result variables. The execution variables are those that the subject controls via interfacing with the virtual system. For example, while the subject performs a throwing task, the real arm trajectory controls the ball release, but the ball and the target are virtual. The virtual rendering has the advantage that it confines the task to exactly the model variables and its known parameters. There are no uncontrolled aspects as would occur in a real experiment. Further, the parameters and result variables can be freely manipulated to test hypotheses about human control strategies.

2.5 Measurement, Analysis, and Hypothesis Testing of Human Performance

Subjects interact with the virtual physics of the task via manipulanda that simultaneously render the task dynamics and measure human performance strategies. The measured execution variables and the task result are then evaluated against the mathematical analysis of the solution space. The virtual environment affords easy manipulation of the model, its parameters, and specific task goals. Hypotheses about preferred solutions are derived from model analysis and can be evaluated based on the human data. As shown below, the task can be parameterized to create interesting task variations to contrast alternative control hypotheses.

2.6 Interventions

Based on the findings, the controlled virtual environment can also be used to create interventions that guide or shape behavior. This is significant for clinical applications, where scientifically-grounded quantitative assessments and interventions are still rare. While this review will focus on the basic science issues, some applications to questions on motor control in children with dystonia or on interventions for the elderly can be found in Sternad [60], Chu et al. [5], Hasson and Sternad [24].

3 Throwing a Ball to Hit a Skittle – Variability, Noise, and Error-Tolerance

3.1 The Motor Task

This experimental paradigm was motivated by a ball game found in many pubs and playgrounds around the world: The actor throws a ball that is tethered to a virtual post by a string like a pendulum; the goal is to hit a target skittle (or skittles) on the opposite side of the pole (Fig. 1a). Accurate throwing requires a controlled hand/ball trajectory that prepares the ball release at exactly the right position with the right velocity to send the ball onto a trajectory that knocks over the target skittle. The practical advantage of this game is that the tethered ball cannot be lost and the game can be played in a small space; the theoretical advantage is that the pendular motions of the ball introduce "interesting" dynamics with a nonlinear solution space including discontinuities that present challenges to trivial learning strategies such as gradient descent. Importantly, the task has redundancy and thereby offers a manifold of solutions with different properties.

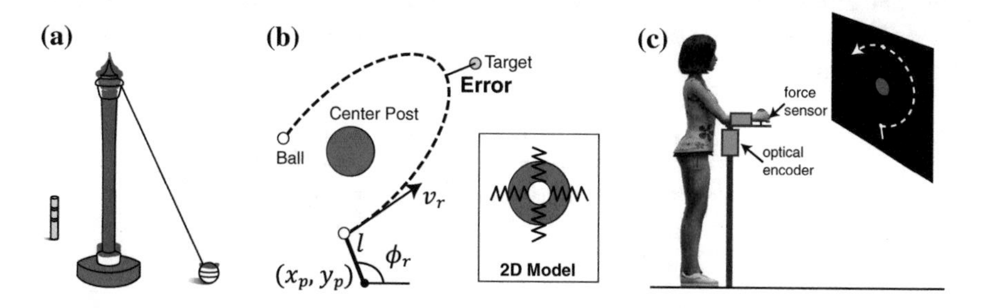

Fig. 1 The virtual throwing task. **a** Schematic of the real task. **b** The 2D model from a top-down view. **c** The experimental set-up with force and position sensors for recording of human movement. Measured movements are shown in real time on the screen (Reproduced from [68])

3.2 The Model and Its Virtual Implementation

To simplify the three-dimensional task, the ball was confined to the horizontal plane, eliminating the pendular elevation during excursion (Fig. 1b). In the model, the ball is attached to two orthogonal, massless springs with its rest position at the center post. In the virtual implementation, the actor views the workspace from above on a backprojection screen (Fig. 1c). S/he throws the virtual ball by moving his/her real arm in a manipulandum that measures the forearm rotations with an optical encoder; these measured movements are shown online by a virtual lever arm (Fig. 1b). Deflecting the ball from the rest position and throwing the ball with a given release angle and velocity, the ball traverses an elliptic path generated by the restoring forces of the two springs. The following equations describe the ball motion in the $x - y$ coordinates of the workspace:

$$\begin{pmatrix} x(t) \\ y(t) \end{pmatrix} = \begin{pmatrix} x_p \\ y_p \end{pmatrix} \cos \omega t + \begin{pmatrix} \cos \phi_r & -\sin \phi_r \\ -\sin \phi_r & \cos \phi_r \end{pmatrix} \begin{pmatrix} l \cos \omega t \\ v_r/\omega t \end{pmatrix} \tag{1}$$

ω denotes the natural frequency of the springs, (x_p, y_p) denotes the lever's pivot point, and l the length of the arm (Fig. 1b). Damping of the springs can be added; asymmetric damping and also stiffness may be used to introduce a more complex force field in the workspace. For a given throw, the two execution variables angle ϕ_r and velocity v_r of the virtual hand at ball release fully determine the ball trajectory in the workspace $x(t)$, $y(t)$ (for more details see [7]).

The actor's goal is to throw the ball to hit the target skittle, without hitting the center post. The latter restriction eliminates simple ball releases with zero velocity. Post hits are therefore penalized with a large fixed error. Otherwise, error is defined as the minimum distance between the ball trajectory and the target center (Fig. 1b). Thus, the result variable is the scalar error that is fully determined by ϕ_r and v_r. Importantly, there is more than one combination of ϕ_r and v_r that leads to zero error, i.e. the task has the simplest kind of redundancy: two variables map onto one. While this low dimensionality permits easy visualization in 3D to develop intuitions, the manifold of zero-error solutions can also be analytically derived and expressed in implicit form:

$$\frac{v_r}{\omega} = \frac{\left|\left(-l \sin \phi_r - y_p\right) x_t + \left(l \cos \phi_r + x_p\right) y_t\right|}{\sqrt{\left(l + \cos \phi_r x_p + \sin \phi_r y_p\right)^2 - \left(\cos \phi_r x_t + \sin \phi_r y_t\right)^2}} \tag{2}$$

3.3 Geometry of the Solution Space

Figure 2 illustrates two different target constellations that generate two different topologies of the result space [61]. Figure 2a, b show the top-down view of the

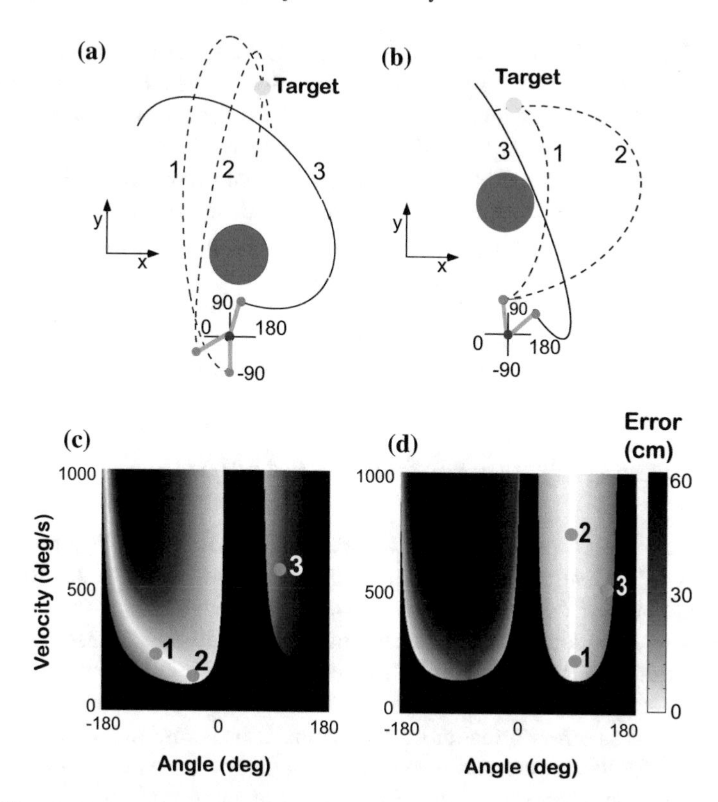

Fig. 2 Two target constellations (**a**, **b**) and their corresponding result spaces (**c**, **d**). For each task, three exemplary ball trajectories are shown which correspond to the three release points plotted in the result spaces (*green dots*). *White* denotes zero-error solutions, increasing error is shown by increasingly *darker grey shades*, *black* denotes a post hit. In both constellations, two ball trajectories exemplify how different release variables can lead to the same result with zero error (*1, 2, dashed lines*). Trajectory 3 shows a trajectory that does not intersect the target (Modified from [61])

workspace with the red circle representing the center post and the yellow circle the target. The manipulandum is shown at the bottom with its angular coordinates. The three elliptic trajectories are three exemplary ball trajectories with different release angles and velocities. In both work spaces two ball trajectories (1, 2) go through the target and have zero error, while one (3) has a non-zero error. Figure 2c, d show the respective result spaces, spanned by release angle and velocity; error is depicted by shades of gray, with lighter shades indicating smaller errors. White denotes the zero-error solutions, or the solution manifold. Black signifies those releases that hit the center post, which incur a penalty in the experiment. The three points are the ball releases pertaining to the three ball trajectories above.

The two result spaces present several interesting features: In target constellation (a) the solution manifold has a nonlinear J-shape that represents solutions over a wide range of release velocities and angles. As indicated by the grey shades, the regions adjacent to the solution manifold have different gradients and the sensitivity of the zero-error solution changes along the solution manifold. Further, the region on the J-shaped manifold with the lowest sensitivity is directly adjacent to the black penalty region. Hence, strategies with the lowest velocity were adjacent to penalized post hits; this poses risk and a simple gradient descent may run into problems. In target constellation (b) the zero-error solutions are independent of velocity and fully specified by the release angle, as the solution manifold runs parallel to velocity. As visible from color shading, low-velocity solutions have slightly less error tolerance compared to high-velocity solutions and again transition directly into the penalty region. Note that other target locations have yet different geometries of the solution manifold creating different challenges to the performer [68].

3.4 Generating Hypotheses from Task Analysis

One study created two result spaces with different topologies to generate specific predictions [61]. Given that humans have limited control accuracy due to the pervasive noise in their neuromotor system, we hypothesized that in such redundant tasks humans seek solutions that are tolerant to their intrinsic noise and also to extrinsic perturbations (*Hypothesis 1*). Such error-tolerant solutions have higher likelihood to be accurate and would therefore also obviate some error corrections. This is advantageous as error corrections incur computational cost and, importantly, the sensorimotor feedback loop suffers from the long delays in the human system. Note that our definition of error tolerance differs from standard sensitivity analyses that assess local sensitivity in a linearized neighborhood. As humans make relatively large errors and the topology is highly nonlinear, we calculated error tolerance as the average error over an extended neighborhood around a chosen solution; this neighborhood is defined by the individual's variability. An alternative hypothesis was that humans select strategies that minimize velocity at release to avoid costs associated with higher effort or signal-dependent noise (*Hypothesis* 2). There is much evidence that movements at slow velocities are preferred, as higher speed tends to decrease accuracy (speed-accuracy trade-off) [16, 17, 42]. This observation concurs with the information-theoretical expectation that noise increases with signal strength. In motor control, signal strength is typically equated with firing rate of action potentials, i.e. force magnitude in the isometric case or, in the dynamic case, movement velocity or acceleration. A third hypothesis discussed in the human motor control literature is that risk is avoided, and participants stay at a distance from the penalty area (*Hypothesis* 3) [6, 40, 48].

3.5 Error Tolerance Over Minimizing Velocity and Risk

Nine participants practiced 540 and 900 throws with Task (a) and (b), respectively. Figure 3 illustrates the predictions as computed for *Hypothesis* 1 and 2 in the top two

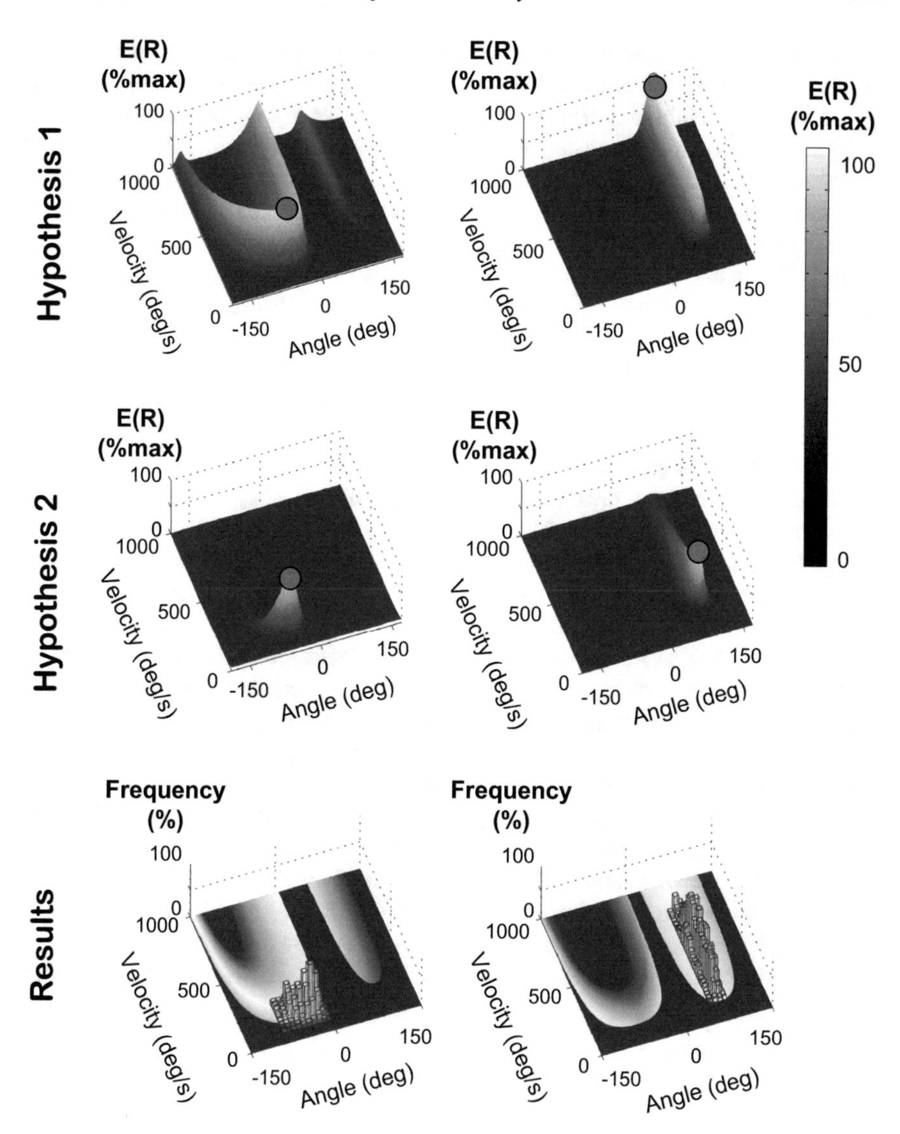

Fig. 3 Hypotheses and experimental results for two task **a** (*left column*) and task **b** (*right column*). The *top row* shows the expected results, E(R) for *Hypothesis* 1: Maximizing error tolerance; the *second row* shows simulated predictions for *Hypothesis* 2: Minimizing velocity and signal-dependent noise. The expected result E(R) was computed as average error over a neighborhood scaled by a softmax function (for details see [61]). The peaks highlighted by the *red circles* denote the expected solutions. The *third row* shows the data as histograms plotted over the result spaces to compare against the predicted solutions (Modified from [61])

rows. Error tolerance was quantified as the expected error or result E(R) over a neighborhood around each strategy, simulating that human strategies are noisy: it was then by a softmax function. For *Hypothesis* 2, expected velocity was computed over the same neighborhood, again scaled by a softmax function. The solutions that are most error-tolerant and those with lowest velocity are indicated by red circles in the middle panels. Examining all throws after removing the initial transients, the bottom panels show the histograms of all subjects' releases in both result spaces (from Fig. 2c, d). In Task (a) the data distribution clustered along the solution manifold at low velocities and close to the discontinuity. The mode at angle 236 deg and velocity 136 deg/s was close to the maximally error-tolerant point as predicted by *Hypothesis* 1. However, the solutions also had relatively low velocity, which was consistent with *Hypothesis* 2. These two benefits seemed to outweigh that these solutions were close to the high-penalty area, i.e. risky strategies were not avoided, counter to *Hypothesis* 3. Task (b) was designed to dissociate *Hypotheses* 1 and 2. The histograms on the right panel illustrate that the data were distributed across a large range of velocities between 140 and 880 deg/s, with the mode of the data distribution at 544 deg/s, although individual preferences were more clustered on the velocity axes. The fact that individuals chose solutions over a wide range of velocities, without a specific preference for low-velocity or the high-tolerance point was at first sight inconsistent with both *Hypotheses* 1 and 2. However, in further analyses the observed variability of each individual was regressed against release velocity, which revealed that variability did not increase at higher velocities, as would be expected from *Hypothesis* 2. Instead, these analyses showed that strategies were better explained by error-tolerance, consistent with *Hypothesis* 1 (for details see [61]).

Taken together, this first study showed how a task analysis can generate predictions that permit direct tests based on human data. The conclusion from this study is that humans seek out error-tolerant strategies, i.e., those where variability at the execution level has minimal detrimental effect on the result. As these strategies attenuate noise effects on the result, fewer errors occur that in turn require fewer corrections to stay on target. This not only reduces computations but also diminishes the negative effect that delays may cause.

3.6 Tolerance, Covariation, and Noise

Increasing error-tolerance is only one of three avenues to deal with unavoidable variability in execution. Two more, conceptually different avenues exist for how variability can be transformed to lessen its effect on the task result. Figure 4 illustrates this notion with data from a representative subject who practiced the same throwing task for 15 days, 240 throws per day [7]. The geometry of the result space shows a U-shaped solution manifold due to a different target constellation. The broad scatter of the data on Day 1 reflects initial exploratory attempts with inferior results compared to those after some practice. Most visibly, on Day 6 the data not only *translated* to a location on the solution manifold with more error-tolerance (shown as a wider band

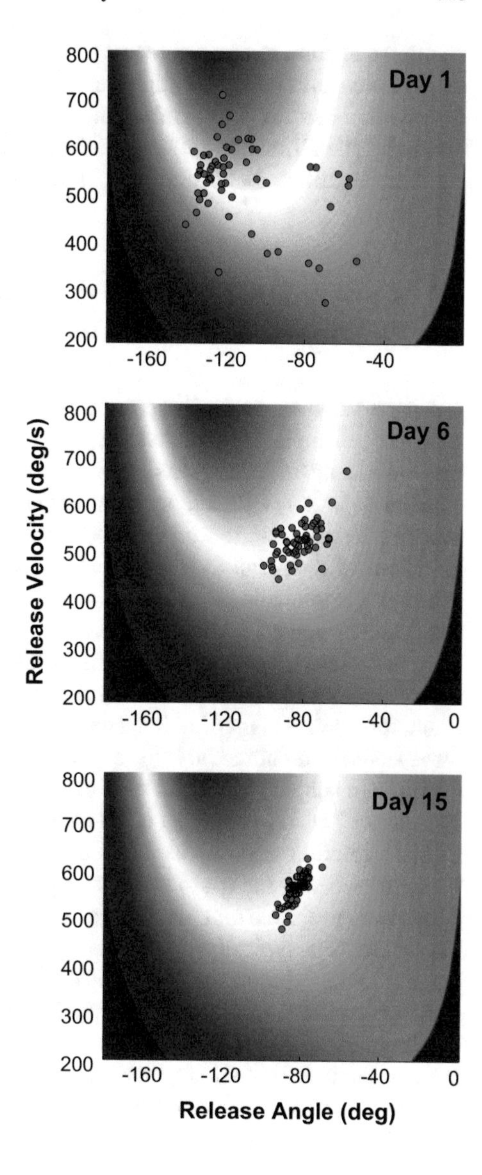

Fig. 4 Data from an exemplary subject who practiced the throwing task for 15 days. The initially broad scatter translated to a more error-tolerant strategy, rotated to covary with the solution manifold (*white*) and scaled of reduced the amplitude of dispersion over the course of practice (Modified from Cohen and Sternad [7])

of white), but the observed variability also started to *covary* with the direction of the solution manifold, while overall variability was only moderately reduced. The distribution on Day 15 clearly reveals a third transformation: the overall dispersion was significantly reduced or *scaled*, over and above the further enhanced covariation. These three data transformations, corresponding to the matrix transformations of translation, rotation, and scaling, were numerically quantified from individual data distributions as costs: The average result of a given data set could be improved by 1.2 cm on Day 1, if it were translated to the optimal location. The difference in average result from actual to optimal renders *Tolerance-cost*. If the actual data were rotated or permuted optimally, the difference in result with the real data would quantify *Covariation-cost*. If the real data distribution was scaled or its noise was reduced optimally, the difference between initial and optimal data quantifies *Noise-cost*. The parallel, but differential evolution of the three costs was shown in Cohen and Sternad [7].

3.7 Covariation, Sensitivity to Geometry of Result Space in Trial-by-Trial Learning

A separate study specifically focused on covariation and examined not only the distributions of the data, but also their temporal evolution to assess whether subjects' trial-by-trial updates were sensitive to the direction of the solution manifold [1]. Three detailed hypotheses guided our experimental evaluation: *Hypothesis* 1: Humans are sensitive to the direction of the solution manifold, which is reflected in preferred directions of their trial-to-trial updates. *Hypothesis* 2: This direction-sensitivity becomes more pronounced with practice. *Hypothesis* 3: The distributional and temporal structure is oriented in directions orthogonal and parallel to the solution manifold. Note that sensitivity to the directions of the null space is also core to several other approaches, which employ covariance-based analyses that linearize around the point of interest using standard null space analysis [10, 55]. In contrast to our approach, those analyses do not exploit the entire nonlinear geometry of the result space.

Thirteen subjects practiced for 6 days throwing to the same target as above, with 240 throws per day (4 blocks of 60 trials). To assess the distribution and also trial-to-trial evolution, each block of 60 throws was examined as illustrated in Fig. 5a. To assess whether the trial-to-trial changes had a directional preference, the 60 data points were projected onto lines through the center of the data set (red lines in Fig. 5a). The center was typically on or was close to the solution manifold. The direction parallel to the solution manifold was defined as θ_{par}, the direction orthogonal to the solution manifold was defined as θ_{ort}. The black horizontal line in Fig. 5a defines the direction of $\theta = 0$ deg. The time series of the projected data was then analyzed using autocorrelation and Detrended Fluctuation Analysis (DFA).

This line was then rotated through $0 < \theta < \pi$ rad, in 100 steps, with its pivot at the center of the data. At each rotation angle θ, the data were projected onto the line and time series analyses conducted. We expected that in directions orthogonal to the solution manifold θ_{ort} successive trials show negative lag-1 autocorrelation, reflecting error corrections; in the parallel direction θ_{par} correction was not necessary, as deviations have no effect on the task result. Note that the result space is spanned by angle and velocity, i.e. with different units; hence, both axes had to be normalized to each individual's variance to ensure orthogonality and a metric.

Figure 5b shows two time series of projected data from those directions that rendered maximum and minimum autocorrelation. Note the visible difference in temporal structure, reflecting that direction in the result space does matter. Plotting

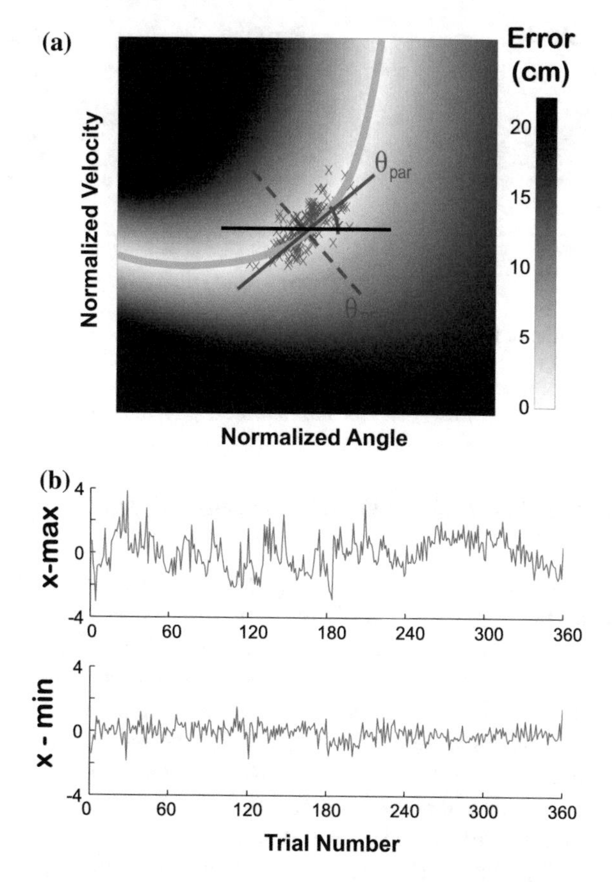

Fig. 5 **a** Result space with solution manifold (*green*), with angle and velocity normalized to variability of each individual. *Red lines* denote directions parallel and orthogonal to the solution manifold. The *black line* denotes = 0 rad. Data are projected onto lines between $0 < \theta < \pi$ rad and autocorrelations are computed for each projection. **b** Time series of projected data where autocorrelation was at a minimum and a maximum. Note that these directions do not necessarily correspond to parallel and orthogonal directions (Reproduced from [1])

the results of the lag-1 autocorrelations across angle of the projection in Fig. 6 reveals a marked modulation: The red lines (with variance across subjects shown by shaded bands) show autocorrelation values for each rotation angle. The modulation supports *Hypothesis* 1 that trial-by-trial updates are sensitive to the angle, and implicitly, the direction of the solution manifold. The green vertical lines denote the orthogonal and parallel directions of the solution manifold. The minima and maxima of the autocorrelation values are indicated by triangles. Consistent with *Hypothesis* 2, the modulation gets more pronounced across the three practice blocks, expressing that after the initial stage, trial-to-trial dynamics became more directionally sensitive. The structure in the orthogonal direction changed from initially positive autocorrelations to white noise and eventually very small negative values [1].

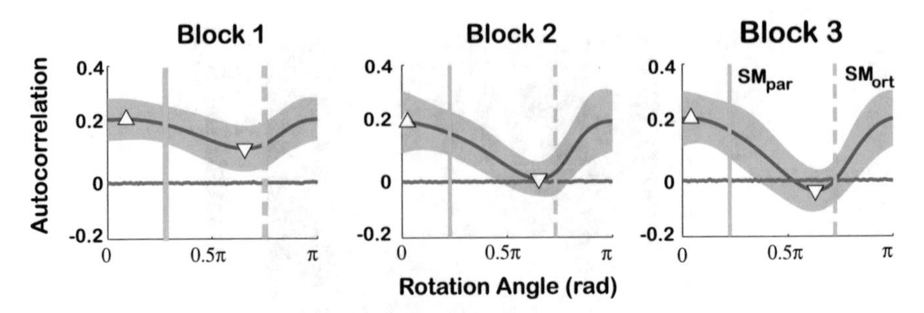

Fig. 6 Autocorrelation of time series of projected data in all directions in result space. The modulation across directions becomes more pronounced with practice, expressing increased sensitivity to the geometry of the result space. Note that while the extrema are close to the directions of the solution manifold (SM_{par} and SM_{ort}) they are not coincident (Modified from [1])

3.8 Orthogonality and Sensitivity to Coordinates

This analysis also revealed important discrepancies to *Hypothesis* 3. The directions of minimum and maximum autocorrelation were near, but not coincident with the orthogonal and parallel directions, as hypothesized. This finding alerts to an important issue: orthogonality is sensitively dependent on the chosen variables. In the present case, the original physical variables, angle and velocity, had different units and required normalization. While technically correct, it raises the question whether these units accurately reflect the units of the central nervous system. One important caveat for this and related approaches is that the structure of variability is fundamentally sensitive to the chosen coordinates.

This fact was highlighted in a separate study, which showed that this sensitivity is particularly pertinent for covariance-based analyses [69]. Even simple linear transformations can critically alter the results, as demonstrated by a simulation that examined 2-joint pointing movements to a target line in the horizontal plane. Given the univariate error measure, distance to the line, the mapping between error and bivariate joint angles was redundant. Analysis of variability of error as a function of joint angles, revealed that the anisotropy of the data distribution depends on the definition of joint angles: relative angles or absolute angles with reference to the shoulder. While covariance-based analysis of anisotropy of data is dependent on the coordinates, we also demonstrated that our analysis of error tolerance, covariation and noise is significantly less sensitive, as it projects the execution variables into the result space. Nevertheless, these critical questions open an interesting avenue for conceptually deeper questions: What are the coordinates of the nervous system? What is the appropriate metric? What is the best or most suitable representation of the problem? While data may be dependent on the coordinates, can data be used to reversely shed light on the coordinates that the nervous system uses?

To pursue these questions, the study by Abe and Sternad further examined how a rescaling of the execution variables in a simple model of task performance with similar redundancy may reproduce these deviations [1]. While this revealed possible sources for these observations, much more work is needed. For example, scaling noise in different execution variables or sensory signals might also give rise to such "deviations". These are clearly important issues for understanding biological movement control, and possibly also worth reflection when designing control in robotic systems.

3.9 Interim Summary

The throwing skill illustrated our model-based approach and its opportunities to shed light on human control. The findings showed that humans choose strategies that obviated the potentially detrimental effects of intrinsic noise. With less noise and variability, less error corrections are needed. Error corrections are not only computationally costly, they are also hampered by the slow transmission speed in biological systems. Are similar strategies also possible in different tasks, especially when interacting with an object?

4 Rhythmic Bouncing of a Ball – Dynamic Stability in Intermittent Interactions

4.1 The Motor Task

Rhythmically bouncing a ball on a racket is a playful and seemingly simple task. Yet, it requires a high degree of visually-guided coordination to intercept the ball at the right position and with the right velocity to reach a target amplitude and perform in a rhythmic fashion (Fig. 7a–c). As in the throwing task, success is determined at one critical moment when the racket intercepts the ball, as this impact fully determines its amplitude. Hence, the core challenge of this task is the control of collisions, a feature germane to numerous other behaviors, ranging from controlling foot-ground impact in running to playing the drums. One key difference to throwing is that these impacts are performed in a repeated fashion, and errors from one contact propagate to the next. Hence, the actor becomes part of a hybrid dynamical system combining discrete and continuous dynamics [11, 44, 46, 53].

4.2 The Model

The physical model for this task is again an extremely simple dynamical system, originally developed for a particle bouncing on a vibrating surface [21, 75]. The model consists of a planar surface moving sinusoidally in the vertical direction; a point mass moving in the gravitational field impacts the surface with instantaneous contact (Fig. 7b). The vertical position of the ball x_b between the kth and the $k+1$th racket-ball impact follows ballistic flight:

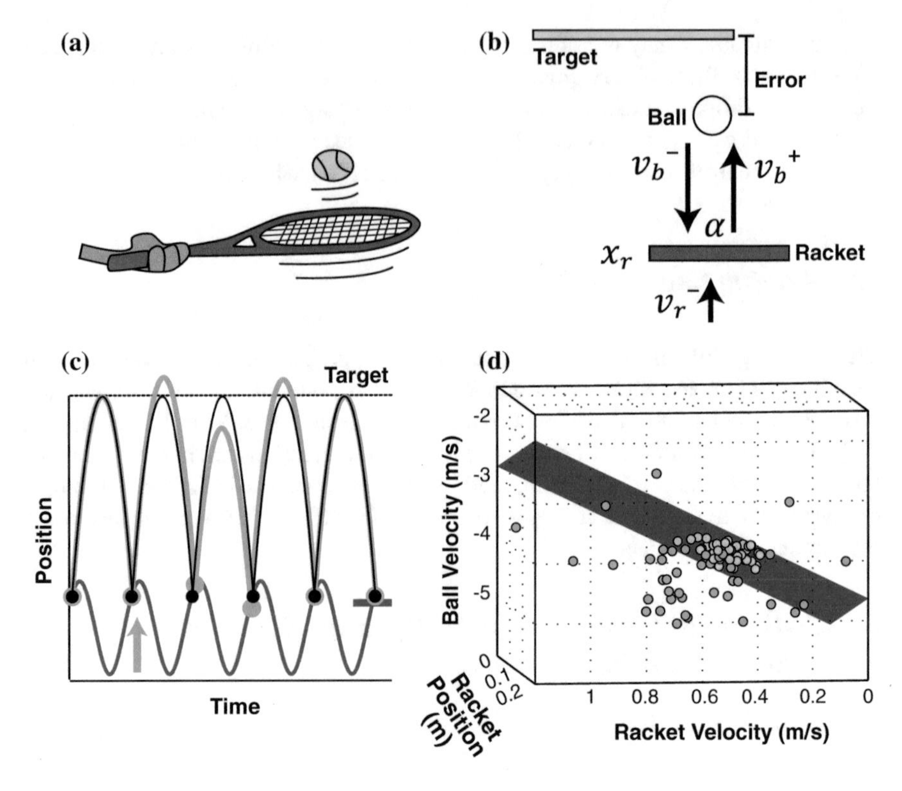

Fig. 7 Bouncing a ball with a racket. **a** The real task. **b** The physical and mathematical model. **c** Simulated time series assuming invariant sine waves of the racket. **d** Redundancy of the result space: Racket position and velocity and ball velocity determine ball amplitude. *Blue* data points are from early practice, *yellow* data points are from late practice (Reproduced from [68])

$$x_b(t) \;=\; x_r(t_k) \;+ v_b^+(t - t_k) - g/2(t - t_k)^2$$

where x_r is racket position, v_b^+ is the ball velocity just after impact, t_k is the time of the kth ball-racket impact, and g is the acceleration due to gravity. With the assumption of instantaneous impact, the ball velocity just after impact v_b^+ is determined by:

$$v_b^+ \;= ((1+\alpha)v_r^- - \alpha v_b^-)$$

where v_b^- and v_r^- are the ball and racket velocities just before impact, and the energy loss at the collision is expressed in the coefficient of restitution α. The maximum height of the ball between t_k and t_{k+1} depends on v_b^- and v_r^- and the position at impact x_r:

$$max_{t_k \le t \le t_{k+1}} x_b(t) = \;x_r(t_k) + (((1+\alpha)v_r^- - \alpha v_b^-)(t - t_k))^2/2g \qquad (3)$$

4.3 Redundancy

The task goal is to bounce the ball to a target height, and the error is defined as the deviation from the target height (Fig. 7c). Even in this simplified form, the task has redundancy, as the result variable error is determined by three execution variables: v_b^-, v_r^- and x_r. Figure 7d shows the execution space with the solution manifold, i.e. the planar surface that represents all solutions leading to zero error. The blue and yellow data points are two exemplary data sets from early and late practice, respectively; each data point corresponds to one ball-racket contact. As to be expected, the early (blue) data show a lot of scatter, while the late practice data (yellow) cluster around the solution manifold.

4.4 Dynamic Stability

While the redundancy analysis is performed on separate collisions, the racket and ball model also lends itself to dynamic stability analysis. To facilitate analysis, the racket movements are assumed to be sinusoidal, such that racket position and velocity at impact collapse into a single state variable, racket phase θ_k. Applying a Poincare section at the ball-racket contact, where x_r and x_b are identical, a discrete map can be derived with v_k^+ and θ_k as state variables:

$$\begin{aligned} v_{k+1}^+ &= (1+\alpha)A\omega \, cos\,\theta_{k+1} - \alpha v_k^+ + g\alpha(\theta_{k+1}-\theta_k)/\omega \\ 0 &= A\omega^2(\,sin\,\theta_k - \,sin\,\theta_{k+1}) + v_k^+\omega(\theta_{k+1}-\theta_k) - g/2(\theta_{k+1}-\theta_k)^2 \end{aligned} \tag{4}$$

A and ω are the amplitude and frequency of the sinusoidal racket movements [11, 53, 65]. This nonlinear system displays dynamic stability and, despite its simplicity, shows the complex dynamics of a period-doubling route to chaos [21, 75]. For present purposes, only stable fixed-point solutions are considered as they correspond to rhythmic bouncing. Local linear stability analysis of this discrete map identifies a stable fixed point, if racket acceleration at impact a_r satisfies the inequality:

$$-2g\frac{(1+\alpha^2)}{(1+\alpha)^2} < a_r < 0 \tag{5}$$

4.5 Hypotheses

In this dynamically stable state, small perturbations of the racket or ball die out without requiring corrections. Hence, if subjects establish such dynamically stable regime, they need not correct for small perturbations that may arise from the persistent neuromotor noise. Thus, we hypothesized that subjects learn these "smart" solution and exploit dynamic stability by hitting the ball with negative racket acceleration (*Hypothesis* 1). Further, due to the system's redundancy infinitely many stable solutions can be adopted. Hence, we administered perturbations to test if subjects established and re-established such stable states (*Hypothesis* 2).

4.6 Virtual Implementation

In the experiments, the participant stood in front of a projection screen and rhythmically bounced the virtual ball to a target line using a real table tennis racket. Similar to the throwing task, the projected racket movements were shown on the screen in real time impacting the ball. The display was minimal and only showed the modeled and measured elements, a horizontal racket and a ball, both moving vertically to a target height (Fig. 7b). A light rigid rod was attached to the racket and ran through a wheel, whose rotations were registered by an optical encoder, which measured the vertical displacement of the racket, in analogy with the model, and shown on the screen. Racket velocity was continuously calculated. The vertical position of the virtual ball was calculated using the ballistic flight equation initialized with values at contact. To simulate the haptic sensation of a real ball-racket contact, a mechanical brake, attached to the rod, was activated at each bounce and decelerated the upward motions. Racket acceleration at or just before the impact was analyzed after the experiment and served as the primary measure of dynamic stability to test *Hypothesis* 1 [79]. Ball position and velocity and racket velocity at contact were measured and analyzed to evaluate the data with respect to the solution manifold (*Hypothesis* 2).

4.7 Learning and Adaptation to Perturbations

Did human subjects seek and exploit dynamic stability of the racket-ball system? How robust is this system if the actor has to change and adapt to new situations? An experiment tested these questions in two stages: On Day 1, 8 subjects performed a sequence of 48 trials of rhythmic bouncing to a target height, each trial lasting 60 s. With the target height at 0.8 m from the lowest racket position, and $\alpha = 0.6$, the average period between repeated contacts was 0.6 s, leading to approximately 100 contacts per trial. On Day 2, subjects performed 10 trials under the same conditions as on Day 1, but then performed another 48 trials after a perturbation was implemented.

Stage 1: Figure 8a shows the ball amplitude errors averaged of all subjects across 48 trials. As expected, the error decreased with practice with a close-to exponential decline. Concomitantly, the acceleration of the racket at contact decreased from an initially positive to a negative value, indicative of performance attaining dynamic stability (Fig. 8b). Importantly, it took approximately 11 trials for subjects to "discover" this strategy, showing that it was not trivial and required practice to learn it. The parallel evolution of both error and racket acceleration with practice provide strong support for *Hypothesis* 1 that subjects seek dynamic stability.

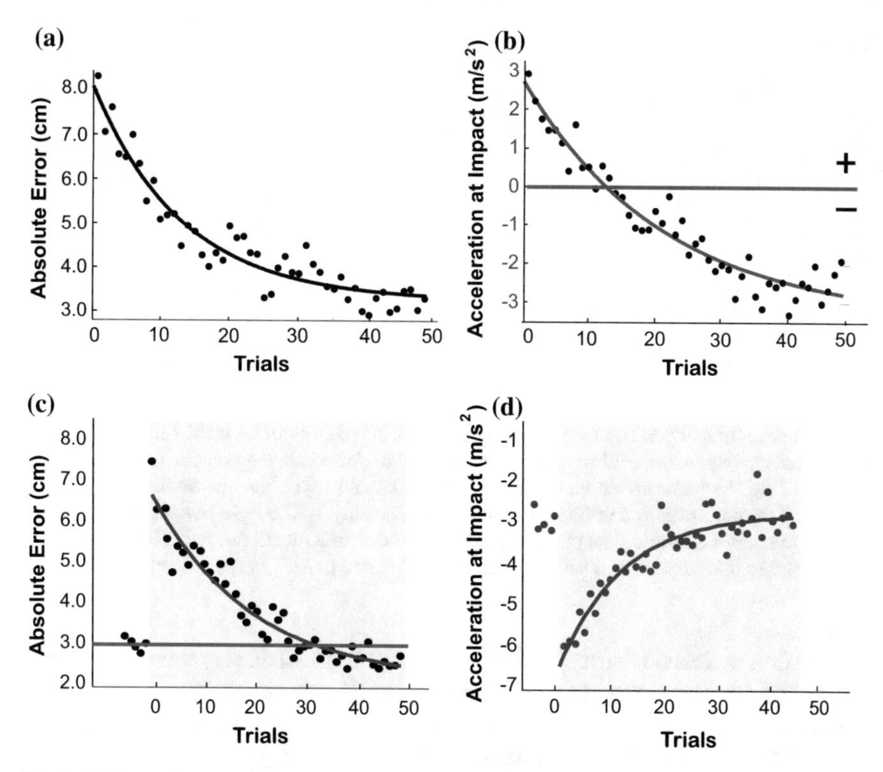

Fig. 8 Ball amplitude errors and racket accelerations over 48 trials. All data points are averages over 8 subjects. **a**, **b** Stage 1 of the experiment. **c**, **d** Stage 2 of the experiment. The shading denotes the perturbed trials

Stage 2: The second experimental session presented an even stronger test. Starting with 10 regular trials as on Day 1, subjects were exposed to a perturbation over the subsequent 48 trials (yellow shading in Fig. 8c, d). This perturbation was calculated using the redundancy of the execution: three execution variables, v_b^-, v_r^- and x_r, determined the one result variable, absolute error of ball peak amplitude to the target height. Following Day 1, the average and standard deviations of v_b^- and v_r^- and x_r of the first 10 baseline trials were calculated for each individual to render an ellipsoid in result space representing the individually preferred solution (9). In the subsequent perturbed trials this preferred strategy was penalized with an error in ball amplitude. This error was delivered by replacing the veridical ball release velocity with one calculated based on the execution ellipsoid. This new ball velocity over- or undershot the target height as calculated. By simply replacing the ball velocity at the contact discontinuity, subjects did not explicitly perceive the perturbation. Within the ellipsoid, the penalty was maximal at its centroid and it linearly decreased to zero towards the boundaries (defined by one standard deviation around its centroid). Hence, assuming sensitivity to the error gradient in result space and the redundancy of the task, subjects were expected to search for a new un-penalized solution. This

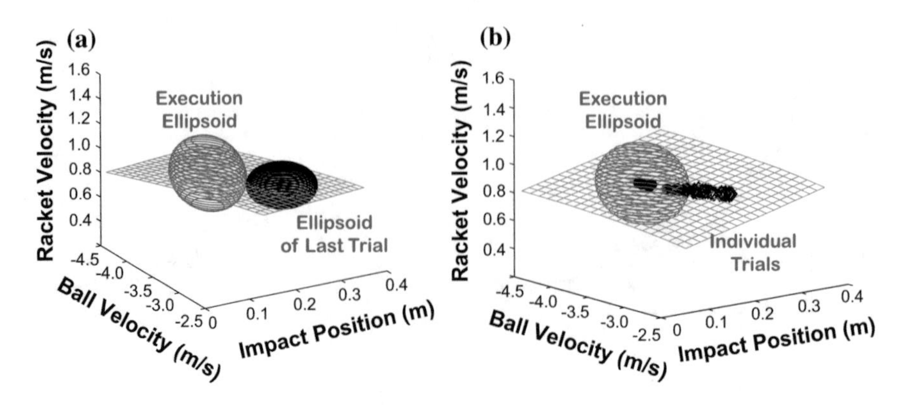

Fig. 9 Presentation of performance in execution space; the planar surface is the solution manifold. **a** The large execution ellipsoid represents the initially preferred strategy that is subsequently penalized during the perturbation phase. The smaller ellipsoid represents the final strategy that is established during the perturbation phase to avoid the penalty. **b** The *right panel* shows the same data and execution ellipsoid. The points are the sequence of trial means following the perturbation onset. It shows that subjects stay on the manifold but migrate outside the penalty ellipsoid

perturbation was calculated and delivered only in the virtual display such that subjects saw their drop in performance, but did not notice its cause explicitly.

Figure 9 illustrates the performance of one representative subject. Starting with the (larger) execution ellipsoid from the initial 10 trials (Fig. 9a), upon onset of the perturbation the subject gradually translated her execution along the planar solution manifold to a new location. The smaller and darker ellipsoid on the right depicts the average execution of the last trial: The strategy shifted and the variability decreased even further; importantly, there was no overlap with the initial ellipsoid (*Hypothesis* 2). This illustrates that the subject not only found a new successful solution without penalty, but the non-overlap also suggested that the subject was aware of her variability.

Returning to the measures or error and racket acceleration at impact for these same data, shown in Fig. 8c, d, reveals that upon perturbation onset, both errors and racket acceleration changed significantly as expected. However, over the course of the 48 perturbed trials, subjects incrementally decreased their errors and reestablished the previously preferred racket acceleration of -3 m/s^2. In fact, this acceleration value was determined to be optimal for the given parameters in additional Lyapunov analyses of the model system [53]. This result shows that subjects successfully established dynamic stability in multiple different ways.

Experimental evidence that subjects learn to hit the ball with a decelerating racket has been replicated in several different scenarios. The different experimental set-ups included a pantograph linkage with precise control of the haptic contact, a real tennis racket to bounce a real ball attached to a boom, and freely bouncing a real ball in 3D [65, 66]. The findings were robust: with experiences, performers learn to hit the ball with negative racket acceleration; based on stability analyses of the model we concluded that they learn to tune into the dynamic stability of the racket-ball system. Based on these findings, we also designed an intervention to guide subjects towards this dynamically stable solution. Manipulating the contact parameters via a state-based shift indeed successfully accelerated subjects' learning the dynamically stable solution, which correlated with faster performance improvement [30].

4.8 Interim Summary

These studies provided strong evidence that humans seek dynamic stability in a task, a solution that is computationally efficient as small errors and noise converge without necessitating explicit error correction. In the face of perturbations, subjects successfully navigated the result space and established new solutions available due to the redundancy. There was also evidence that they were aware of their own variability. As in skittles, subjects seek solutions where noise matters less.

5 Chaos in a Coffee Cup – Predictability in Continuous Object Control

5.1 The Motor Task

Leading a cup of coffee to one's mouth to drink is a seemingly straightforward action. However, transporting a cup filled with sloshing fluid to safely contact the mouth without spilling remains a challenge not to be underestimated for both humans and robots. Carrying a cup of coffee (or a glass of wine) exemplifies a class of tasks that require continuous control of an object that has internal degrees of freedom. How do humans control interactions with such an object, where the sloshing fluid creates time-varying, state-dependent forces that have to be preempted and compensated to avoid spill? Can humans or robots really have a sufficiently accurate internal model of the complex fluid dynamics to online predict and react to the complex interaction forces? In search of human strategies that apparently deal with this problem easily, we started again with the analysis of the task dynamics, following the steps outlined above.

5.2 The Model

In principle, the task presents a problem in fluid dynamics [38, 49]. To make this complex infinitely-dimensional system more tractable, several simplifications were made [23]: (1) the 3D cup was reduced to 2D, (2) the sloshing coffee was reduced to

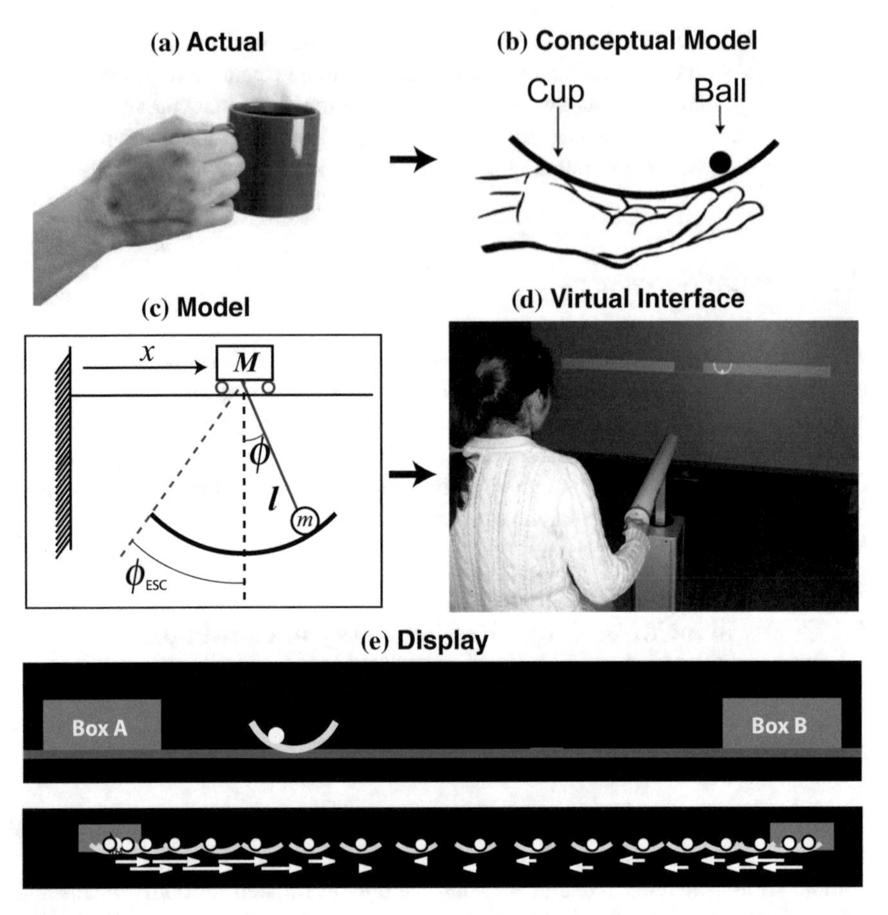

Fig. 10 Carrying a cup of coffee. **a** The model task. **b** The conceptual model: a 2D arc with a ball rolling inside. **c** Control model of the cart-and-pendulum. **d** Virtual implementation with the HapticMaster robot to control the cup in the horizontal direction. **e** The interactive screen display; the *green* rectangles specify the amplitude of the cup movement. The *lower panel* shows a sequence of moving cups with the *arrows* depicting the respective forces of cup and ball (Reproduced from [60])

a ball with point mass rolling in a cup, (3) the hand contact with the cup was reduced to a single point of interaction, (4) the cup transport was limited to a horizontal line (Fig. 10a–c). More precisely, the moving liquid is represented by a pendulum suspended to a cart that is translated in the horizontal x-direction. The pendulum is a point mass m (the ball) with a mass-less rod of length l with one angular degree of freedom θ. Subjects control the ball indirectly by applying forces to the cup, and the ball can escape if its angle exceeds the rim of the cup. The cup is a point mass M that moves horizontally. The hand moving the cup is represented by a horizontal force $F(t)$. Despite these simplifications, the model system retained essential elements of complexity: it is nonlinear and creates complex interaction forces between hand and object. The equations of the system dynamics are:

$$(m+M)\ddot{x} = ml(-\ddot{\theta}cos\phi + \dot{\theta}^2 sin\phi) + F(t)$$
$$l\ddot{\theta} = -\ddot{x}cos\theta - gsin\theta \quad (6)$$

where θ, $\dot{\theta}$, and $\ddot{\theta}$ are angular position, velocity, and acceleration of the ball/pendulum; x, $\dot{x}$, and $\ddot{x}$ and are the cart/cup position, velocity, and acceleration, respectively; F is the force applied to the cup by the subject; g is gravitational acceleration. The model has four state variables $x, \dot{x}, \theta, \dot{\theta}$ and the externally applied force $F(t)$ that determines the behavior of the ball and cup system. Hence, only one variable *F(t)* is under direct control of the subject, but this is co-determined by the ball/pendulum interacting with the cart. These instantaneous interaction forces make the distinction into execution and result variables significantly more complicated than in the previous two examples.

5.3 Virtual Implementation

The ball-and-cup system was implemented in a virtual environment. The cart and the pendulum rod was hidden, leaving only the ball visible. In addition, a semicircular arc with radius equal to pendulum length l was drawn on the screen so that the ball appeared to roll in a cup (Fig. 10d, e). Subjects manipulate the virtual cup-and-ball system via a robotic arm, which measures hand forces $F_{External}$ applied to the cup but also exerts forces from the virtual object onto the hand (HapticMaster, Motek [76]). θ and $\dot{\theta}$ were computed online and the ball force F_{Ball} was computed based on system equations such that the force that accelerated the virtual mass $(m+M)$ was $F_{applied} = M\ddot{x} = F_{External} + F_{Ball}$. Two rectangular target boxes set the required movement distance and spatial accuracy (for more details see [23]).

5.4 Model Analysis and Hypothesis

The cup of coffee can be moved as a relatively short discrete placement to a target, or in a more continuous fashion, as for example carrying the cup while walking. A previous study examined a single placement onto a target focusing on the discontinuous aspect of the task: the coffee can be spilled [23, 24]. Given the noise intrinsic to the neuromotor system and the fluctuations created by the extrinsic cart-and-pendulum system, avoiding spilling coffee, or losing the ball, became the core challenge when the task was to move as fast as possible. The "distance" from losing the ball was quantified by an energy margin, defined as the difference between the current energy state and the one where the ball angle would exceed the rim angle. Results showed that this continuous metric sensitively captured performance quality and learning in healthy and also older subjects.

Here, we review another study that examined more prolonged interaction, where the nonlinear dynamics manifests its full complexity and, technically, displays chaos [41, 67]. To this end, the task instruction was to move the cup rhythmically between two very large targets leaving amplitude under-specified; the task-specified frequency defined the *result variable*. Movement strategies were fully described by the *execution variables* cup amplitude, frequency, and initial angle and velocity of the ball, $A, f, \theta_0, \dot{\theta}_0$. To derive hypotheses about the space of solutions, inverse dynamics analysis was conducted to calculate the force $F(t)$ required to satisfy the task. Numerical simulations were run for combinations of the scalar execution variables $A, f, \theta_0, \dot{\theta}_0$. To keep the number of simulations manageable, frequency f was fixed to the task-required frequency, and $\dot{\theta}_0$ was set to zero.

Figure 11 shows two example profiles generated by inverse dynamics calculations with two different initial ball states $\theta_0(\dot{\theta}_0 = 0)$ that both result in a sinusoidal cup trajectory $x(t)$. The left profile $F(t)$ shows irregular unpredictable fluctuations for $\theta_0 = 0.4$ rad, while the right profile initialized at $\theta_0 = 1.0$ rad shows a periodic waveform with high regularity. To characterize the pattern of force profiles with respect to the cup dynamics, $F(t)$ was strobed at every peak of cup position $x(t)$. The marginal distributions of the strobed force values are plotted as a function of initial ball phase θ_0 in the bottom panel. This input-output relation reveals a bifurcation diagram with a pattern similar to the period-doubling behavior of chaotic systems, indicating chaos in the cup-and-ball system.

5.5 Hypotheses for Human Control Strategies

It seems uncontested that controlling physical interaction requires “knowledge” and prediction of object dynamics. On the other hand, it is reasonable to doubt that the complex details of object dynamics are known or faithfully represented in an internal model. In chaotic dynamics, small changes in initial states can dramatically change the long-term behavior and, technically, lead to unpredictable solutions. Can or should internal models be able to represent this complex dynamics? To make this challenge more tractable for the neural control system we hypothesized that subjects seek solutions that render the object behavior more predictable to reduce computational effort and facilitate at least some prediction.

To quantify the concept of predictability of the object dynamics based on the human’s applied force, we computed mutual information *MI* between the applied force and the kinematics of the cup, i.e. long-term predictability of the object’s dynamics [9]. *MI* is a nonlinear correlation measure defined between two probability density distributions that quantifies the information shared by two random variables, $F(t)$ and the kinematics of the cup $x(t)$:

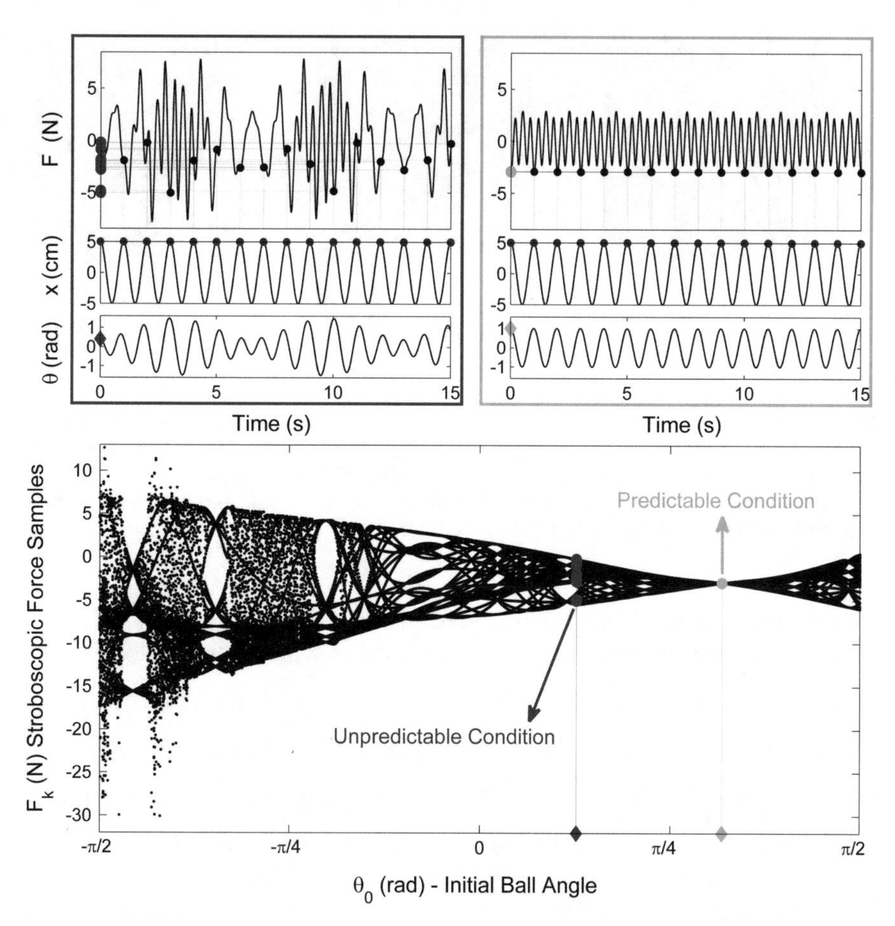

Fig. 11 Inverse dynamics simulations of the cart-and-pendulum model. *Top panels* show two different simulation runs with different initial ball angles θ_0, requiring a complex and a relatively simple input force (*top row*). Strobing force values at maxima of the cup profile x and plotting the marginal distributions against all initial ball angles renders the bifurcation-like diagram (Reproduced from [41])

$$MI\,(x, F) = \iint p\,(x, F)\,log_e \frac{p(x, F)}{p\,(x)\,p(F)} dxdF \tag{7}$$

MI presents a scalar measure of the performer's strategy calculated at each point of the 4D execution space spanned by $A, f, \theta_0, \dot{\theta}_0$. The higher *MI*, the more predictable the relation between force and object dynamics. Hence, we expected that subjects would seek strategies with high *MI* (*Hypothesis* 1, Fig. 12a). Predictability as a control priority had to be tested against alternative hypothesis. The experiments permitted testing two alternative control priorities: minimizing effort (*Hypothesis* 2, Fig. 12b) and maximizing smoothness (*Hypothesis* 3, Fig. 12c); both

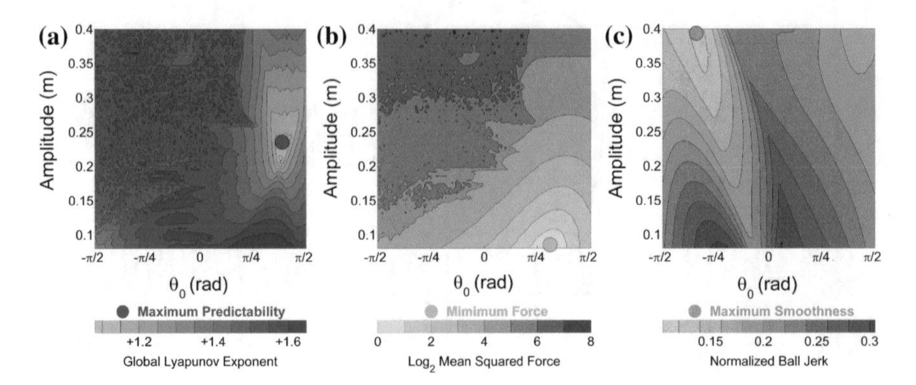

Fig. 12 Result space computed for three different hypothesized control priorities. The space is computed for different initial ball angles and cup amplitudes; frequency is set to 1 Hz, and ball velocity is set to zero. **a** Mutual information. **b** Effort defined as mean squared force over a given trial. **c** Smoothness or mean squared jerk defined over a given trial. The optimal strategy for each hypothesis is noted by the large dot (Reproduced from [41])

are commonly accepted and widely supported criteria in free unconstrained movements. To calculate the effort required for each strategy, the Mean Squared Force of the force profile was calculated: $MSF = \frac{1}{nT}\int_0^{nT} F(t)^2 dt$, where n denoted the number of cycles and $T = 1/f$ the period of each cycle. Mean Square Jerk was calculated as $MSJ = \frac{1}{T(\dddot{\theta}_{max} - \dddot{\theta}_{min})}\int_0^T |\dddot{\theta}|^2 dt$, where the value was normalized with respect to ball jerk amplitude to make it dimensionless [27]. Similar to *MI*, *MSF*-values were calculated for all strategies in 4D execution space. To constrain the calculations, the initial value of the angular velocity $\dot{\theta}_0$ was set to zero, consistent with the experimental data. Figure 12 compares the corresponding predictions for *MI*, *MSF*, and *MSJ*. Color shades express the degree as explained in the legend. The large dots denote the points of maximum *MI*, minimum *MSF* and *MSJ*. Importantly, these predicted strategies are at very different locations in result space.

To test these hypotheses, equivalent measures had to be calculated from the experimental data to evaluate observed human strategies against the simulated result space. In contrast to the simulations, the experimental trajectories were not fully determined by initial values as online corrections were likely. Therefore, to attain better estimates of the execution variables from the experimental trajectories, estimates were extracted at each cycle k of the cup displacement x during each 40 sec trial (see Fig. 11); trial averages $\bar{A}, \bar{f}, \bar{\theta}_0, \bar{\dot{\theta}}_0$ served as correlates for the variables in the simulations. *MI*, *MSF*, and *MSJ* were calculated for each measured strategy $\bar{A}_k, \bar{f}_k, \bar{\theta}_k, \bar{\dot{\theta}}_k$.

5.6 Predictable Interactions

An experimental study provided first evidence that subjects indeed favored predictable solutions over those that minimized the expended force and smoothness [41]. Subjects performed rhythmic cup movements paced at the natural frequency of the pendulum, which corresponded to the anti-resonance of the coupled system. This facilitated the emergence of the system's nonlinear characteristics with chaotic solutions that maximized the challenge. Amplitude was free to choose and relative phase between ball and cup was also unspecified. Each subject performed 50 trials (40 s each). By choosing the cup amplitude and phase, subjects could manipulate interaction forces of different complexity and predictability.

The main experimental results are summarized in Fig. 13; the plot shows *MI* in shades of purple (lighter shades denote higher *MI*) and contours of selected values of *MSF* (green) from the simulations overlaid with the results from human subjects; each data point represents one trial (red). The data clearly show how subjects gravitated towards areas with higher *MI*, i.e. strategies with more predictable interactions, consistent with *Hypothesis* 1. The left panel shows individual trials pooled over all subjects; darker red indicates early practice and lighter red indicates late practice.

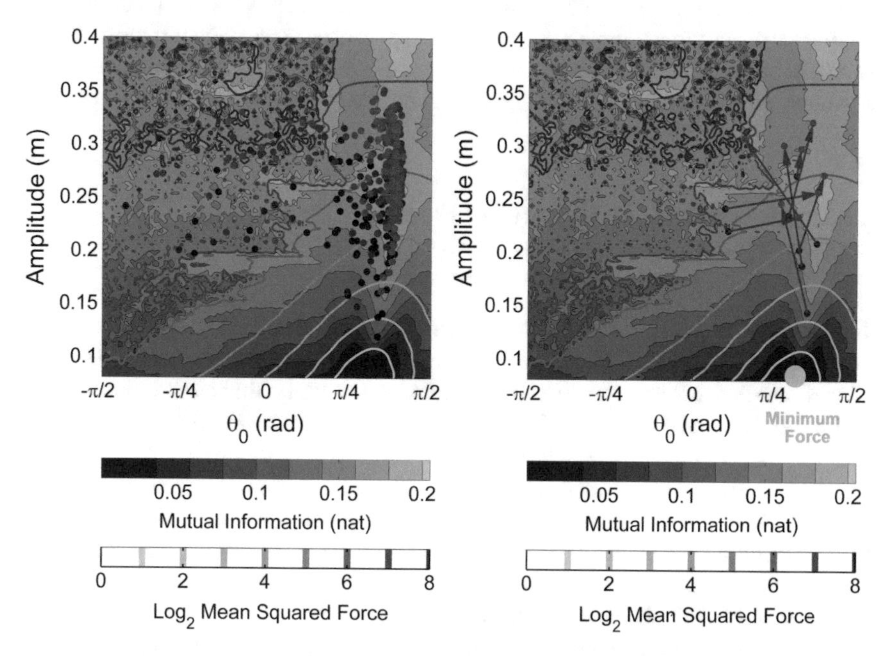

Fig. 13 Result space with Mutual Information as the result variable, shown by shades of *purple*. The *left panel* plots trial data from all 9 subjects showing that they converge to the area with highest MI. Each data point is one trial; *darker color* shades denote later in practice. The *arrows* in the *right panel* show each subject with initial trial values the start of the *arrow* and the final practice trial the tip of the *arrow* (Reproduced from [41])

The right panel shows the same data separated by subject: the red arrows mark how each subject's average strategy changed from early practice (mean of first 5 trials) to late practice (mean of last 5 trials). The majority of subjects switched from low- to high-predictability regions in the result space. Both figures also show that all subjects increased their movement amplitude, associated with an increase in overall exerted force. None of the subjects moved toward the minimum force strategy, nor towards a strategy with maximum smoothness (counter to *Hypotheses* 2 and 3). In fact, overall force exerted, or *MSF*, rather increased with practice.

5.7 Interim Summary

These results highlight that humans are sensitive to object dynamics and favor strategies that make interactions predictable. In the case shown, these predictable solutions were even favored over those with less effort. This is plausible because unpredictable interaction forces are experienced as disturbances that continuously require reactions and corrections. Knowing that in real life we carry a glass of wine without paying much attention to the carrying, more predictable strategies appear plausible. Analogous to the dynamically stable solutions in ball bouncing, predictable solutions may require fewer computations as they obviate error corrections. Given that in chaotic solutions small changes due to external or internal perturbations lead to unpredictable behavior, noise matters less in predictable solutions.

6 From Analysis to Synthesis: Dynamic Primitives for Movement Generation

This brief overview of our research revealed potential control priorities or cost functions that humans may use to coordinate simple and complex interactions. Humans favor strategies that are sensitive to dynamics and stability, that exploit redundancy of the solution space to channel their intrinsic noise into task-irrelevant dimensions, and that exploit predictable solutions of potentially very complex task dynamics. The review also demonstrated what can be learnt from analysis of human data in conjunction with mathematical understanding of the task and its solution space. The only assumption is that the dynamics and stability properties of the task are fundamental and determine "opportunities" and "costs". The known solution space serves as reference to evaluate human movement.

The task-dynamic approach as outlined is analytic and largely agnostic about details of the controller. This contrasts with other research in computational motor neuroscience that starts with a hypothesized controller and then compares the predicted with the experimentally observed behavior. One recent prominent example for this direction is work that has sought evidence that the brain operates like an optimal feedback controller [56, 73, 74]. Other control models include internal models with Kalman-filters or tapped-delay lines, to mention just a few [39]. Our approach refrains from such assumptions directly borrowed from control theory; rather, we aim to extract principles from human data with as few assumptions as possible. Nevertheless, the question of synthesis remains: what controller or control policy would generate these strategies? While still largely speculative, our task-dynamic perspective presents a sound foundation for a generative hypothesis.

To begin, let's return to the initial pointer to the seemingly inferior features of the human neuromotor system - the high degree of noise and the slow information transmission. These features seem puzzling given the extraordinary dexterity of humans that by far surpasses that of robots, at least to date. Therefore, the direct translation of control policies that heavily rely on central control and feedback loops may remain inadequate to achieve human dexterity. As mentioned earlier, the human wetware with its compliant actuators and high dimensionality appears to provide an advantage. Hence, lower levels of the hierarchical neuromotor system should be given more responsibility. Consistent with our task-dynamic perspective, we have therefore suggested that the biological system generates movements via dynamic primitives, defined over the high-dimensional nonlinear neuromotor system [26, 28, 45, 50, 51, 59, 64]. We propose that the human neuromotor system exploits attractors states, defined over both the neural and mechanical nonlinear system. If the neuromotor system is parameterized to settle into such stable states, central control may only need to occasionally intervene. In principle, nonlinear autonomous systems have three possible stable attractor states: fixed point, limit cycle, and chaotic attractors. Putting chaotic attractors aside for now, we proposed fixed-point and limit cycle attractors for primitives.

The two main stable attractors fixed points and limit cycles directly map onto discrete and rhythmic movements. To understand discrete movements such as reaching to a target as convergence to a stable end state is not completely new. Equilibrium-point control was first posited by Feldman for simple position control [14, 15]. Numerous subsequent studies, both behavioral and neurophysiological, have given evidence for attractive properties in reaching behavior [4, 20, 25, 36]. This work has widened to include a virtual trajectory, even though details are still much contested. For rhythmic behavior a similar host of experimental and modeling studies have presented support for stable limit cycle dynamics. For example, bimanual rhythmic finger movements showed transitions from anti-phase to in-phase coordination that bear the hallmarks of nonlinear phase transitions in coupled nonlinear oscillators [22, 33]. Our own work has shown how extremely simple oscillator models can account for synchronization in bimanual rhythmic coordination, including subtle phase differences between oscillators with different natural frequencies [62, 70, 71]. Several different oscillator models have been developed that produce autonomous oscillations to represent central pattern generators in the spinal cord of invertebrates [31, 45]. Support for the distinction between rhythmic and discrete movements also came from a neuroimaging study [54]. Brain activation revealed that in rhythmic movements only primary motor areas were activated, while significantly more areas were needed to control discrete movements.

In an attempt to synthesize this evidence from largely disparate research groups, our own research made first forays into combining the two types of building blocks. Playing piano is after all a combination of complex rhythmic finger movements combined with reaches across the keyboard. Note that in principle, optimal feedback control could also achieve such movements, including those with dynamic stability. In fact, there is no inherent limit to what optimal feedback control may achieve. It is this omnipotence that contrasts with the well-known coordinative limitations that may reveal features of the human controller. Beyond "patting your head while rubbing your stomach", research has revealed that rhythmic bimanual actions tends to settle into in-phase and anti-phase coordination [34, 71], humans avoid moving very slowly [3, 77], and the 2/3 power law in handwriting and drawing may reveal intrinsic geometry or other limitations [18, 52]. Several modeling and experimental studies showed the possibilities and limitations of combining two dynamic primitives. Wiping a table rhythmically, while translating the hand across the table revealed that rhythmic and discrete elements cannot be combined arbitrarily [63, 64].

However, research is still far from having generated conclusive evidence that dynamic motion primitives underlie observed behavior. More specifically, interactions with objects cannot be addressed with the two primitives alone. Therefore, recently Hogan and myself argued that impedance is needed as a third dynamic primitive to enable the system to interact with objects and the environment [28, 29]. Combining discrete and rhythmic primitives with impedance in an equivalent network is a first proposal on how humans may interact with objects in the environment. More details and first theoretical developments can be found in the chapter of Hogan in the same volume. With these theoretical efforts under way, also further complementary empirical work is needed. The challenge for the future is to combine analysis and synthesis. How can dynamic primitives be employed to pour—and enjoy—a glass of wine?

Acknowledgements This work was supported by the National Institute of Health, R01-HD045639, R01-HD081346, and R01HD087089, and the National Science Foundation DMS-0928587 and EAGER-1548514. I would like to acknowledge all my graduate and postdoctoral students who worked hard to generate this body of research. I would also like to thank Neville Hogan for many inspiring discussions and contributions.

References

1. M. Abe, D. Sternad, Directionality in distribution and temporal structure of variability in skill acquisition. Front. Hum. Neurosci. **7** (2013). doi:10.3389/fnhum.2013.002
2. D. Angelaki, Y. Gu, G. Deangelis, Multisensory integration: psychophysics, neurophysiology, and computation. Current Opin. Neurobiol. **19**, 452–458 (2009)
3. B. Berret, F. Jean, Why don't we move slower? The value of time in the neural control of action. J. Neurosci. **36**, 1056–1070 (2016)
4. E. Bizzi, N. Accornero, W. Chapple, N. Hogan, Posture control and trajectory formation during arm movements. J. Neurosci. **4**, 2738–2744 (1984)

5. W. Chu, S.-W. Park, T. Sanger, D. Sternad, Dystonic children can learn a novel motor skill: strategies that are tolerant to high variability. IEEE Trans. Neural Syst. Rehabil. Eng. **24**(8), 847–858 (2016)
6. W. Chu, D. Sternad, T. Sanger, Healthy and dystonic children compensate for changes in motor variability. J. Neurophysiol. **109**, 2169–2178 (2013)
7. R.G. Cohen, D. Sternad, Variability in motor learning: relocating, channeling and reducing noise. Exp. Brain Res. **193**, 69–83 (2009)
8. R.G. Cohen, D. Sternad, State space analysis of intrinsic timing: exploiting task redundancy to reduce sensitivity to timing. J. Neurophysiol. **107**, 618–627 (2012)
9. T.M. Cover, J.A. Thomas, *Elements of Information Theory* (Wiley, Hoboken, 2006)
10. J.P. Cusumano, P. Cesari, Body-goal variability mapping in an aiming task. Biol. Cybern. **94**, 367–379 (2006)
11. T.M.H. Dijkstra, H. Katsumata, A. de Rugy, D. Sternad, The dialogue between data and model: passive stability and relaxation behavior in a ball bouncing task. Nonlinear Stud. **11**, 319–345 (2004)
12. A.M. Dollar, R.D. Howe, Towards grasping in unstructured environments: grasper compliance and configuration optimization. Adv. Robot. **19**, 523–543 (2005)
13. A.A. Faisal, L.P. Selen, D.M. Wolpert, Noise in the nervous system. Nat. Rev. Neurosci. **9**, 292–303 (2008)
14. A.G. Feldman, Functional tuning of the nervous system with control of movement or maintenance of a steady posture: II) Controllable parameters of the muscle. Biophysics **11**, 565–578 (1966a)
15. A.G. Feldman, Functional tuning of the nervous system with control of movement or maintenance of a steady posture: III) Mechanographic analysis of execution by man of the simplest motor task. Biophysics **11**, 667–675 (1966b)
16. P.M. Fitts, The information capacity of the human motor system in controlling the amplitude of movement. J. Exp. Psychol. **47**, 381–391 (1954)
17. P.M. Fitts, J.R. Peterson, Information capacity of discrete motor responses. J. Exp. Psychol. **67**, 103–112 (1964)
18. T. Flash, A.A. Handzel, Affine differential geometry analysis of human arm movements. Biol. Cybern. **96**, 577–601 (2007)
19. M. Franek, J. Mates, T. Radil, K. Beck, E. Pöppel, Finger tappping in musicians and nonmusicians. Int. J. Psychophysiol. **11**, 277–279 (1991)
20. H. Gomi, M. Kawato, Equilibrium-point control hypothesis examined by measured arm stiffness during multijoint movement. Science **272**, 117–220 (1996)
21. J. Guckenheimer, P. Holmes, *Nonlinear Oscillations, Dynamical Systems, and Bifurcations of Vector Fields* (Springer, New York, 1983)
22. H. Haken, J.A.S. Kelso, H. Bunz, A theoretical model of phase transition in human hand movements. Biol. Cybern. **51**, 347–356 (1985)
23. C. Hasson, T. Shen, D. Sternad, Energy margins in dynamic object manipulation. J. Neurophysiol. **108**, 1349–1365 (2012)
24. C. Hasson, D. Sternad, Safety margins in older adults increase with improved control of a dynamic object. Fronti. Aging Neurosci. **6** (2014). doi:10.3389/fnagi.2014.00158
25. N. Hogan, An organizing principle for a class of voluntary movements. J. Neurosci. **4**, 2745–2754 (1984)
26. N. Hogan, D. Sternad, On rhythmic and discrete movements: reflections, definitions and implications for motor control. Exp. Brain Res. **18**, 13–30 (2007)
27. N. Hogan, D. Sternad, Sensitivity of smoothness measures to movement duration, amplitude, and arrests. J. Motor Behavior **41**, 529–534 (2009)
28. N. Hogan, D. Sternad, Dynamic primitives of motor behavior. Biol. Cybern. **106**, 727–739 (2012)
29. N. Hogan, D. Sternad, Dynamic primitives in the control of locomotion. Front. Comput. Neurosci. **7** (2013). doi:10.3389/fncom.2013.00071

30. M. Huber, D. Sternad, Implicit guidance to stable performance in a rhythmic perceptual-motor skill. Exp. Brain Res. **233**, 1783–1799 (2015)
31. A. Ijspeert, Central pattern generators for locomotion control in animals and robots: a review. Neural Netw. **21**, 642–653 (2008)
32. E.R. Kandel, T.M.J. Schwartz, T.M. Jessel, *Principles of Neural Sciences* (Elsevier, New York, 1991)
33. J.A.S. Kelso, Phase transitions and critical behavior in human bimanual coordination. Am. J. Physiol.: Regul. Integr. Comp. Physiol. **15**, R1000–R1004 (1984)
34. J.A.S. Kelso, Elementary coordination dynamics, in *Interlimb coordination: Neural, dynamical, and cognitive constraints*, ed. by S. Swinnen, H. Heuer, J. Massion P. Casaer (Academic Press, New York, 1994)
35. I. Kurtzer, J. Pruszynski, S. Scott, Long-latency reflexes of the human arm reflect an internal model of limb dynamics. Current Biol. **18**, 449–453 (2008)
36. M.L. Latash, *Control of human movements*. Human Kinetics (Urbana, Champaign, IL, 1993)
37. Z. Li, M. Latash, V. Zatsiorsky, Force sharing among fingers as a model of the redundancy problem. Exp. Brain Res. **119**, 276–286 (1998)
38. H.C. Mayer, R. Krechetnikov, Walking with coffee: why does it spill? Phys. Rev. E **85**, 046117 (2012)
39. B. Mehta, S. Schaal, Forward models in visuomotor control. J. Neurophysiol. **88**, 942–953 (2002)
40. A. Nagengast, D. Braun, D. Wolpert, Optimal control predicts human performance on objects with internal degrees of freedom. PLoS Comput. Biol. **5**, e1000419 (2009)
41. B. Nasseroleslami, C. Hasson, D. Sternad, Rhythmic manipulation of objects with complex dynamics: predictability over chaos. PLoS Comput. Biol. **10**, e1003900 (2014). doi:10.1371/journal.pcbi.1003900
42. R. Plamondon, A.M. Alimi, Speed/accuracy trade-offs in target-directed movements. Behavior Brain Sci. **20**, 1–31 (1997)
43. E. Robertson, The serial reaction time task: implicit motor skill learning? J. Neurosci. **27**, 10073–10075 (2007)
44. R. Ronsse, D. Sternad, Bouncing between model and data: stability, passivity, and optimality in hybrid dynamics. J. Motor Behavior **6**, 387–397 (2010)
45. R. Ronsse, D. Sternad, P. Lefevre, A computational model for rhythmic and discrete movements in uni- and bimanual coordination. Neural Comput. **21**, 1335–1370 (2009)
46. R. Ronsse, K. Wei, D. Sternad, Optimal control of cyclical movements: the bouncing ball revisited. J. Neurophysiol. **103**, 2482–2493 (2010)
47. J. Rothwell, *Control of Human Voluntary Movement* (Springer, New York, 2012)
48. T. Sanger, Risk-aware control. Neural Comput. **26**, 2669–2691 (2014)
49. A. Sauret, F. Boulogne, J. Cappello, E. Dressaire, H. Stone, Damping of liquid sloshing by foams: from everyday observations to liquid transport. Phys. Fluids **27**, 022103 (2015)
50. S. Schaal, S. Kotosaka, D. Sternad, Nonlinear dynamical systems as movement primitives, in *Proceedings of the 1st IEEE-RAS International Conference on Humanoid Robotics (Humanoids 2000), Cambridge, MA, September 7–9 2000*
51. S. Schaal, D. Sternad, Programmable pattern generators, in *International Conference on Computational Intelligence in Neuroscience (ICCIN '98), Research Triangle Park, NC, Oct 24–26 1998*
52. S. Schaal, D. Sternad, Origins and violations of the 2/3 power law. Exp. Brain Res. **136**, 60–72 (2001)
53. S. Schaal, D. Sternad, C.G. Atkeson, One-handed juggling: a dynamical approach to a rhythmic movement task. J. Motor Behavior **28**, 165–183 (1996)
54. S. Schaal, D. Sternad, R. Osu, M. Kawato, Rhythmic arm movement is not discrete. Nature Neurosci. **7**, 1136–1143 (2004)
55. J. Scholz, G. Schöner, The uncontrolled manifold concept: identifying control variables for a functional task. Exp. Brain Res. **126**, 289–306 (1999)

56. S.H. Scott, Optimal feedback control and the neural basis of volitional motor control. Nature Rev. Neurosci. **5**, 532–546 (2004)
57. R. Shadmehr, F.A. Mussa-Ivaldi, Adaptive representation of dynamics during learning of a motor task. J. Neurosci. **14**, 3208–3224 (1994)
58. R. Shadmehr, S.P. Wise, *Computational Neurobiology of Reaching and Pointing: A Foundation for Motor Learning* (MIT Press, Cambridge, 2005)
59. D. Sternad, Towards a unified framework for rhythmic and discrete movements: behavioral, modeling and imaging results, in *Coordination: Neural, Behavioral and Social Dynamics*, ed. by A. Fuchs, V. Jirsa (Springer, New York, 2008)
60. D. Sternad, From theoretical analysis to clinical assessment and intervention: three interactive motor skills in a virtual environment, in Proceedings of the IEEE International Conference on (ICVR) Virtual Rehabilitation, June 9–12 2015, Valencia, Spain (2015), pp. 265–272
61. D. Sternad, M.O. Abe, X. Hu, H. Müller, Neuromotor noise, sensitivity to error and signal-dependent noise in trial-to-trial learning. PLoS Comput. Biol. **7**, e1002159 (2011)
62. D. Sternad, D. Collins, M.T. Turvey, The detuning factor in the dynamics of interlimb rhythmic coordination. Biol. Cybern. **73**, 27–35 (1995)
63. D. Sternad, W.J. Dean, Rhythmic and discrete elements in multijoint coordination. Brain Res **989**, 151–172 (2003)
64. D. Sternad, W.J. Dean, S. Schaal, Interaction of rhythmic and discrete pattern generators in single-joint movements. Hum. Mov. Sci. **19**, 627–665 (2000a)
65. D. Sternad, M. Duarte, H. Katsumata, S. Schaal, Dynamics of a bouncing ball in human performance. Phys. Rev. E **63**, 011902-1–011902-8 (2000)
66. D. Sternad, M. Duarte, H. Katsumata, S. Schaal, Bouncing a ball: tuning into dynamic stability. J. Exp. Psychol.: Hum. Percept. Perform. **27**, 1163–1184 (2001)
67. D. Sternad, C. Hasson, Predictability and robustness in the manipulation of dynamically complex objects, Adv. Exp. Med. Biol. **957**, 55–77 (2016)
68. D. Sternad, M.E. Huber, N. Kuznetsov, Acquisition of novel and complex motor skills: stable solutions where intrinsic noise matters less. Adv. Exp. Med. Biol. **826**, 101–124 (2014)
69. D. Sternad, S. Park, H. Müller, N. Hogan, Coordinate dependency of variability analysis. PLoS Comput. Biol. **6**, e1000751 (2010)
70. D. Sternad, M.T. Turvey, E.L. Saltzman, Dynamics of 1:2 coordination in rhythmic interlimb movement: I. Generalizing relative phase. J. Motor Behavior **31**, 207–223 (1999)
71. D. Sternad, M.T. Turvey, R.C. Schmidt, Average phase difference theory and 1:1 phase entrainment in interlimb coordination. Biol. Cybern. **67**, 223–231 (1992)
72. S. Sternberg, R. Knoll, P. Zukovsky, Timing by skilled musicians, in *The Psychology of Music* (Academic Press, New York, 1982), pp. 181–239
73. E. Todorov, Optimality principles in sensorimotor control. Nature Neurosci. **7**, 907–915 (2004)
74. E. Todorov, M.I. Jordan, Optimal feedback control as a theory of motor coordination. Nature Neurosci. **5**, 1226–1235 (2002)
75. N.B. Tufillaro, T. Abbott, J. Reilly, *An Experimental Approach to Nonlinear Dynamics and Chaos* (Redwood City, Addison-Wesley, 1992)
76. R. van der Linde, P. Lammertse, HapticMaster - a generic force controlled robot for human interaction. Ind. Robot - An Int. J. **30**, 515–524 (2003)
77. R.P.R.D. van der Wel, D. Sternad, D.A. Rosenbaum, Moving the arm at different rates: Slow movements are avoided. J. Motor Behavior **1**, 29–36 (2010)
78. R. van Ham, T. Sugar, B. Vanderborght, K. Hollander, D. Lefeber, Compliant actuator design. IEEE Robtoics Autom. Mag. **9**, 81–94 (2009)
79. K. Wei, T.M.H. Dijkstra, D. Sternad, Passive stability and active control in a rhythmic task. J. Neurophysiol. **98**, 2633–2646 (2007)
80. K. Wei, K.Körding, Uncertainty of feedback and state estimation determines the speed of motor adaptation. Front. Comput. Neurosci. **4**, 11 (2010)
81. A.M. Wing, A.B. Kristofferson, The timing of interresponse intervals. Percept. Psychophys. **1**, 455–460 (1973)
82. V. Zatsiorsky, R. Gregory, M. Latash, Force and torque production in static multifinger prehension: biomechanics and control. I. Biomech. Biol. Cybern. **87**, 50–57 (2002)

Part IV
Robot Motion Generation

Momentum-Centered Control of Contact Interactions

Ludovic Righetti and Alexander Herzog

Abstract The control and planning of interaction forces is fundamental for locomotion and manipulation tasks since it is through the interaction with the environment that a robot can walk forward or manipulate objects. In this chapter we present a control and planning strategy focused on the control of interaction forces to generate multi-contact whole-body behaviors. Centered around the robot momentum dynamics, our approach consists of a hierarchical inverse dynamics controller that treats the control of the robot's momentum as a contact force task and a trajectory optimization algorithm that can generate desired whole-body motions, momentum and desired contact forces for multiple contacts. Experimental results demonstrate the capabilities of the approach on a humanoid robot.

1 Introduction

The planning and control of interaction forces is fundamental for both locomotion and manipulation tasks. Indeed, it is only through the action of external forces (gravity and contact with the environment) that a robot can move its center of mass or that an object can be manipulated. While external (and internal) forces can explain the motion of humans, robots or objects by applying Newton's laws of motion, the converse is not always true. It is not always possible to deduct all the important forces in action during a manipulation or locomotion task by solely observing movements. Indeed, it is possible to create contact forces with the environment that do not create motion and yet are important for behavior. For example, squeezing an object in one's hand changes the forces exerted on the object without creating motion. Since the acceleration of the center of mass of a body is proportional to the sum of all external forces, it is therefore not always possible to infer each individual force uniquely from

L. Righetti (✉) · A. Herzog
Max-Planck Institute for Intelligent Systems, Paul Ehrlich Str. 15,
72076 Tuebingen, Germany
e-mail: ludovic.righetti@tuebingen.mpg.de

A. Herzog
e-mail: alexander.herzog@tuebingen.mpg.de

J.-P. Laumond et al. (eds.), *Geometric and Numerical Foundations of Movements*,
Springer Tracts in Advanced Robotics 117, DOI 10.1007/978-3-319-51547-2_14

the observed motion without making additional assumptions (i.e. only the sum of the forces can be deducted).

Each time a robot (or a human) creates multiple contacts with the environment, there is an infinite number of possible contact forces that will explain a certain motion. Contact forces can therefore be controlled to achieve additional objectives. For example, they can be chosen to ensure that all contact forces reside within their friction cone to prevent the robot from slipping. We demonstrated in Righetti et al. [30] how the optimization of contact forces in the nullspace of the movement could help a robot traverse difficult terrains. The controller was reorienting contact forces to stay inside the friction cone without changing the actual motion of the robot.

The importance of controlling interaction with the environment has been recognized for quite some time, for example by developing impedance control approaches [14] or more recently for balancing biped robots [15, 27]. However, most motion planning or optimal control methods for locomotion or multi-contact behaviors have either ignored the importance of contact forces to simplify the problem, or made additional assumptions to make the contact model invertible by pre-defining an actuation redundancy resolution rule [34], such as minimizing torque effort, therefore significantly reducing the opportunities to exploit interaction forces.

In this chapter, we present our recent results on the control and planning of multi-contact behaviors for legged robots. Based on the robot dynamics model and more particularly on the (linear and angular) momentum dynamics, we develop a method to plan and control interaction forces and robot motion concurrently. In a first step, we use a hierarchical inverse dynamics controller that makes a special use of interaction forces to control the momentum of the robot. In a second step, optimal control is used to plan momentum trajectories and contact forces together with whole-body motion. The central aspect of robot momentum in this approach allows to explicitly consider the role of interaction forces for robot control while providing a natural problem decomposition that simplifies our optimization problems.

1.1 Structure of the Robot Dynamics Equations

The approach strongly relies on the structure of the robot dynamics, in particular the way the interaction forces enter the equations related to the robot's momentum. It leads to a task decomposition for feedback control that we use in our hierarchical inverse dynamics where momentum control is central and to a natural decomposition of the whole-body multi-contact planning problem.

The rigid-body dynamics model of a legged robot is usually written as

$$\mathbf{M}(\mathbf{q})\ddot{\mathbf{q}} + \mathbf{h}(\mathbf{q}, \dot{\mathbf{q}}) = \mathbf{S}^T \boldsymbol{\tau} + \mathbf{J}^T \boldsymbol{\lambda} \tag{1}$$

where $\mathbf{q} \in \mathbb{SE}(3) \times \mathbb{R}^n$ is a vector of position and orientation of the robot in space and its joint configuration, $\mathbf{M}(\mathbf{q})$ is the inertia matrix, $\mathbf{h}(\mathbf{q}, \dot{\mathbf{q}})$ is the vector of generalized forces including Coriolis, centrifugal and gravitational forces, $\boldsymbol{\tau}$ is the vector of actuation torques, $\boldsymbol{\lambda}$ the contact forces, $\mathbf{S}^T = [\mathbf{0}_{n\times 6}\ \mathbf{I}_{n\times n}]^T$ is the actuation matrix reflecting the unactuated pose of the robot in space and $\mathbf{J}$ is the contact Jacobian.

This equation has a notable structure that we will exploit throughout the chapter. We can separate the robot dynamics into actuated and unactuated parts

$$\mathbf{M}_u\ddot{\mathbf{q}} + \mathbf{h}_u = \mathbf{J}_u^T\boldsymbol{\lambda} \tag{2}$$

$$\mathbf{M}_a\ddot{\mathbf{q}} + \mathbf{h}_a = \boldsymbol{\tau} + \mathbf{J}_a^T\boldsymbol{\lambda} \tag{3}$$

Equation (2) corresponds to the 6-dimensional unactuated dynamics ($\mathbf{M}_u \in \mathbb{R}^{6\times(n+6)}$). It is in fact equivalent to the Newton–Euler equations of the robot that describe the change of (angular and linear) momentum of the robot through external forces as explained in Wieber [37]. Equation (3), which is n dimensional, is equivalent to the equations of a manipulator in contact with no under-actuation. For any combination of $\ddot{\mathbf{q}}$ and $\boldsymbol{\lambda}$ there exists a unique vector of actuation torques $\boldsymbol{\tau}$ that satisfy this equation. It means that for any combination of desired motion and contact forces it is in principle possible to find a control command that will achieve both, if we neglect actuation limits.

Equation (2) will play a key role in associating motion and contact forces. It represents the under-actuated dynamics of any legged robots: i.e. the part of the dynamics that cannot be directly influenced by our choice of control $\boldsymbol{\tau}$. It therefore constitutes the main dynamic constraint of any control and planning algorithm. Equation (3) will be mainly useful to express constraints on the actuation torques (or equivalently by enforcing bounds on $\mathbf{M}_a\ddot{\mathbf{q}} + \mathbf{h}_a - \mathbf{J}_a^T\boldsymbol{\lambda}$) or to gain information about the kinematics of the robot: this equation tells us how to create joint accelerations from actuation torques.

Since Eq. (2) relates to the momentum of the robot, the robot dynamics can equivalently be written

$$\mathbf{H}(\mathbf{q})\ddot{\mathbf{q}} + \dot{\mathbf{H}}(\mathbf{q})\dot{\mathbf{q}} = \begin{bmatrix} M\mathbf{g} + \sum_e \mathbf{f}_e \\ \sum_e(\boldsymbol{\kappa}_e + (\mathbf{x}_e(\mathbf{q}) - \mathbf{x}_{CoM}(\mathbf{q})) \times \mathbf{f}_e) \end{bmatrix} \tag{4}$$

$$\mathbf{M}_a(\mathbf{q})\ddot{\mathbf{q}} + \mathbf{h}_a(\mathbf{q}, \dot{\mathbf{q}}) = \boldsymbol{\tau} + \mathbf{J}_a(\mathbf{q})^T\boldsymbol{\lambda} \tag{5}$$

where $\mathbf{H}$ is the centroidal momentum matrix that maps the generalized velocities $\dot{\mathbf{q}}$ to the linear and angular momentum of the robot as described by Orin and Goswami [26]. M is the mass of the robot and $\mathbf{g}$ the vector of accelerations due to gravity. In Eq. (4), the contact forces λ were decomposed into the sum of wrenches applied at each contact point e, where $\mathbf{f}_e$ is the force component and $\boldsymbol{\kappa}_e$ the torque component of the wrench. $\mathbf{x}_e(\mathbf{q})$ is the position of the contact e and $\mathbf{x}_{CoM}(\mathbf{q})$ is the position of the robot CoM. Equation (4) is nothing else than the Newton–Euler equations of a rigid multi-body, i.e. the (linear and angular) momentum rate of change is equal to the sum of external forces (for the linear part) and moments (for the angular part).

1.2 Momentum-Centric Hierarchical Optimal Control

This decomposition will play a key role in the rest of the chapter to plan and control whole-body motion and interaction forces. Our approach is decomposed in two levels, depicted in Fig. 1, and is centered around the control and optimization of contact forces and momentum.

The first level, described in Sect. 2, consists of a hierarchical inverse dynamics controller. This controller is used to compute actuation torques that will allow the concurrent execution of several tasks while enforcing certain constraints. The hierarchical aspect of the controller allows to enforce execution priorities between tasks when it is not possible to execute all the tasks at once. Each task is written as a desired closed-loop behavior. One of the important task in this controller is the control of the momentum of the robot.

In the second level, described in Sect. 3, trajectory optimization is used to find optimal momentum and whole-body motion together with contact forces over a certain time-horizon. The problem is split into two sub-problems: the optimization of contact forces and momentum (as described by Eq. (4)) which creates a dynamically consistent momentum profile and the optimization of the whole-body motion of the robot, which is a kinematic problem. Moreover, along the momentum trajectory a time-varying feedback control law is computed that finds the locally optimal change

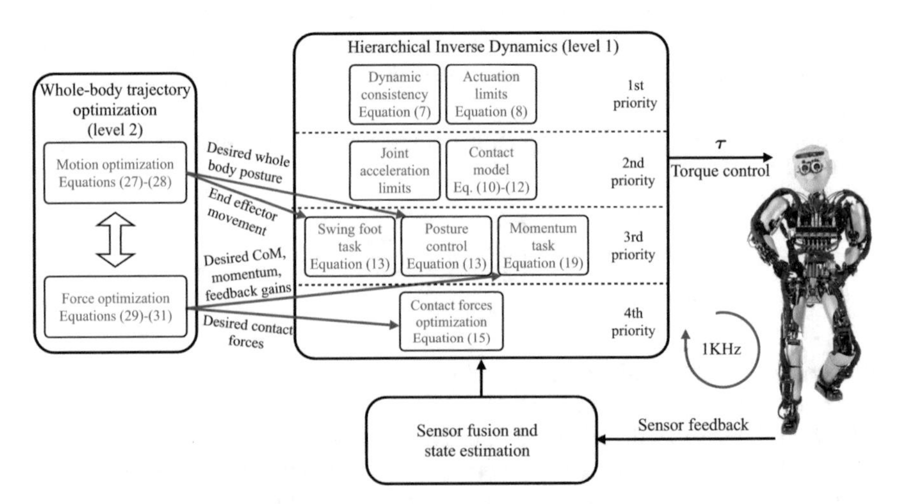

Fig. 1 Schematic representation of the system presented in the chapter. The trajectory optimization algorithm computes desired motions for each end-effector, the entire body and the center of mass. It also generates optimal momentum trajectories, desired contact forces and optimal force feedback gains to stabilize the momentum. The hierarchical inverse dynamics computes the required actuation torques to achieve these optimized motion and force trajectories on the robot. The state estimation algorithm is used to compute the position, orientation and velocity of the robot in space from inertial and kinematic measurements as in Rotella et al. [31]. The equations mentioned in the diagram are developed in Sects. 2 and 3

of contact forces necessary to correct tracking errors in momentum trajectories. The computed force, momentum and desired robot motion (e.g. the motion of each end-effector) directly define all the objectives of the tasks controlled in the first level.

With this approach, the control of whole-body multi-contact behaviors is decomposed into simpler optimization problems that can be run at different time scales. The idea is to use only the model information that is relevant for reasoning at a specific time scale. For example, at the highest control frequency, it is necessary to take into account the detailed dynamic constraints to ensure, for example, that joint limits are not violated and that the commands stay within actuation limits. However, it is not necessarily important to replan foot steps or momentum motions at 1 kHz and therefore this can be decided on a slower time scale using reduced dynamic models.

2 Feedback Control with Hierarchical Inverse Dynamics

The first level of our control approach consists of the hierarchical inverse dynamics controller presented in Herzog et al. [11]. In a complex locomotion and manipulation scenario, it is desirable to achieve several tasks concurrently using all the motion capabilities of robots with a large number of degrees of freedom. For example, a robot tasked with drilling a hole in a wall would need to: (1) keep its CoM above the support polygon to ensure it does not fall, (2) move the drill at a specific position on the wall, (3) apply a force on the wall with the drill bit to drill the hole and (4) move its head to keep the drill visible to the cameras to allow visual servoing. While ideally all these tasks should be achieved concurrently, they do not have the same importance. In case of a disturbance, it might be better to ensure that balance is preserved while keeping the drill bit in sight might not be so important. Keeping the drill at a proper location might be more important than the vision task but not as crucial as balancing. We see that there exists natural priorities between tasks (i.e. a hierarchy of tasks). It is therefore desirable to design feedback controllers that can enforce the task hierarchy when needed.

This example is interesting because it also shows that the control tasks can be of very different nature. Keeping the drill at a specific location on the wall is a position task, it means that the controller should regulate the position of the drill bit as well as possible. Drilling inside the wall is a task that requires to regulate the amount of force applied on the wall, it is a force task.[1] Keeping the CoM above the support polygon, especially when trying to reject disturbances from drilling, can be seen as an impedance task with hard constraints on the allowed location of the CoM. Tasks can consist in the regulation or tracking of positions, forces or impedance in task space. They can also consist in limits on these quantities (e.g. allowed CoM position or maximum allowed torque).

---

[1] One could argue that it could be seen as an impedance task, which is also valid and does not change the argument.

Hierarchical inverse dynamics algorithms have recently been developed by several groups to allow the execution of concurrent tasks and the satisfaction of inequality constraints while preserving priorities [8, 20, 22]. These algorithms have also been used to achieve complex whole-body tasks, as in Jarquin et al. [16] or Saab et al. [32]. While it is possible to control interaction forces with these algorithms (e.g. in Sherikov et al. [33] or Herzog et al. [12]), focus is generally put on kinematic tasks.

In this section, we detail the inverse dynamics algorithm used in this chapter. In particular we explain how it can be used to control interaction forces and how a task that controls the linear and angular momentum of the robot can be written as a contact force task. All tasks are written as desired closed loop behaviors.

2.1 Tasks Formulation

The hierarchical inverse dynamics controller computes at each instant of time the accelerations $\ddot{\mathbf{q}}$, contact forces $\boldsymbol{\lambda}$ and actuation torques $\boldsymbol{\tau}$ that will best satisfy all the tasks and constraints and their hierarchy. For this controller, all tasks are written as equalities or inequalities that are linear in the decision variables $\ddot{\mathbf{q}}$, $\boldsymbol{\lambda}$ and $\boldsymbol{\tau}$, i.e. they are written in the form

$$\mathbf{A}\begin{bmatrix}\ddot{\mathbf{q}}\\ \boldsymbol{\lambda}\\ \boldsymbol{\tau}\end{bmatrix} = \mathbf{a} \quad \text{or} \quad \mathbf{B}\begin{bmatrix}\ddot{\mathbf{q}}\\ \boldsymbol{\lambda}\\ \boldsymbol{\tau}\end{bmatrix} \leq \mathbf{b} \tag{6}$$

In the following we detail the type of tasks and constraints that we will use in this chapter.

Dynamic Model of the Robot

It is the controller's highest priority to ensure that the control law is compatible with the dynamics of the robot. As explained in Sect. 1, to ensure dynamic consistency of the control law, it is only necessary to consider the underactuated dynamics of the robot

$$\mathbf{M}_u\ddot{\mathbf{q}} + \mathbf{h}_u = \mathbf{J}_u^T\boldsymbol{\lambda} \tag{7}$$

The actuated part of the dynamic equations of the robot is only necessary to express tasks related to actuation torques. For example, we will write actuation limits

$$\boldsymbol{\tau}_{min} \leq \boldsymbol{\tau} \leq \boldsymbol{\tau}_{max} \tag{8}$$

by substituting $\boldsymbol{\tau}$ with Eq. (3)

$$\boldsymbol{\tau}_{min} \leq \mathbf{M}_a\ddot{\mathbf{q}} + \mathbf{h}_a - \mathbf{J}_a^T\boldsymbol{\lambda} \leq \boldsymbol{\tau}_{max} \tag{9}$$

Therefore, in our control problem $\boldsymbol{\tau}$ will never explicitly appear as an optimization variable. Only accelerations and contact forces will be optimized and the resulting actuation torques will be computed using Eq. (3). Joint limits can also be enforced by setting limits on the admissible accelerations $\ddot{\mathbf{q}}$.

Contact Model

A contact model is necessary to predict contact forces. We use the following contact model

$$\mathbf{J}\ddot{\mathbf{q}} = -\dot{\mathbf{J}}\dot{\mathbf{q}} \tag{10}$$

$$\mathbf{A}_{friction}\boldsymbol{\lambda} < \mathbf{0} \tag{11}$$

$$\mathbf{A}_{cop}\boldsymbol{\lambda} < \mathbf{b}_{cop} \tag{12}$$

The first equality imposes as a constraint that the robot parts in contact with the environment should not move. The first inequality is a linear approximation of the friction cones. The second inequality is a limit on the individual centers of pressure on each contact surface to ensure that these center of pressure stay inside the boundary of the contact surface. These constraints ensure that the robot can create physically consistent forces and avoid slippage (cf. Herzog et al. [11] for more details).

Motion and Force Control Tasks

Motion tasks are written using the Jacobian of the task as

$$\ddot{\mathbf{x}}_{des} = \mathbf{J}_x\ddot{\mathbf{q}} + \dot{\mathbf{J}}_x\dot{\mathbf{q}} \tag{13}$$

where $\mathbf{x}$ represents the task vector, $\mathbf{J}_x(\mathbf{q})$ is the Jacobian of the task and $\ddot{\mathbf{x}}_{des}$ is the desired acceleration for the task. The desired acceleration is computed using a feedback controller, for example a linear feedback controller for the task would be written as

$$\ddot{\mathbf{x}}_{des} = \mathbf{P}(\mathbf{x}_{ref} - \mathbf{x}) + \mathbf{D}(\dot{\mathbf{x}}_{ref} - \dot{\mathbf{x}}) + \ddot{\mathbf{x}}_{ref} \tag{14}$$

where $\mathbf{P}$ and $\mathbf{D}$ are positive definite gain matrices and $\mathbf{x}_{ref}$ is a reference trajectory computed by a motion planner. In our case, these reference trajectories will be computed using the whole-body planner presented in Sect. 3.

Force tasks are written as

$$\boldsymbol{\lambda} = \boldsymbol{\lambda}_{des} \tag{15}$$

where $\boldsymbol{\lambda}_{des}$ is a desired contact force. Force tasks, specially used to distribute the weight of the robot during multi-contact turned out to be very important to get more stable behavior in real robot experiments. They also have a central role in the execution of the optimal plan computed in Sect. 3.

Momentum Control Task

While most methods (e.g. [18, 23, 36]) regulate momentum and CoM dynamics using kinematics information, i.e. through $\ddot{\mathbf{q}}$, it is also possible to explicitly use the contact forces to derive a desired closed loop behavior [9, 11]. While both approaches are equivalent in principle, we prefer to use the formulation involving contact forces because it will allow us to use optimal feedback gains to track a desired momentum and CoM trajectory through desired contact forces (Sect. 3.2).

Using Eq. (4), we can write the CoM and momentum dynamics in terms of generalized contact forces $\boldsymbol{\lambda}$

$$\dot{\mathbf{x}}_{CoM} = \frac{1}{M}\mathbf{h}_{lin} \tag{16}$$

$$\dot{\mathbf{h}} = \begin{bmatrix} \mathbf{I}_{3\times 3} & \mathbf{0}_{3\times 3} & \cdots \\ (\mathbf{x}_e(\mathbf{q}) - \mathbf{x}_{CoM}(\mathbf{q}))_\times & \mathbf{I}_{3\times 3} & \cdots \end{bmatrix} \boldsymbol{\lambda} + \begin{bmatrix} M\mathbf{g} \\ \mathbf{0} \end{bmatrix} \tag{17}$$

where $\mathbf{h} = [\mathbf{h}_{lin}^T \mathbf{h}_{ang}^T]^T$ is the vector of linear and angular momentum and $\mathbf{a}_\times$ is the matrix such that $\mathbf{a}_\times \cdot \mathbf{b} = \mathbf{a} \times \mathbf{b}$.

With this formulation, we can consider the contact forces $\boldsymbol{\lambda}$ to be the control input in Eq. (17). Therefore, we can write a control law for momentum control as

$$\boldsymbol{\lambda}_{des} = -\mathbf{K}\begin{bmatrix} \mathbf{x}_{CoM} \\ \mathbf{h} \end{bmatrix} + \mathbf{k}(\mathbf{x}_{CoM_{ref}}, \mathbf{h}_{ref}) \tag{18}$$

where $\mathbf{K}$ is a gain matrix and $\mathbf{k}$ is a feedforward control term depending on the reference CoM and momentum trajectories. Both the gain matrix and feedforward terms will be computed by the optimal control approach presented in Sect. 3.2.

The desired momentum corresponding to the desired contact forces is found by using Eq. (18) inside Eq. (17)

$$\dot{\mathbf{h}}_{des} = \begin{bmatrix} \mathbf{I}_{3\times 3} & \mathbf{0}_{3\times 3} & \cdots \\ (\mathbf{x}_e(\mathbf{q}) - \mathbf{x}_{CoM}(\mathbf{q}))_\times & \mathbf{I}_{3\times 3} & \cdots \end{bmatrix} \left(-\mathbf{K}\begin{bmatrix} \mathbf{x}_{CoM} \\ \mathbf{h} \end{bmatrix} + \mathbf{k}(\mathbf{x}_{CoM_{ref}}, \mathbf{h}_{ref})\right) + \begin{bmatrix} M\mathbf{g} \\ \mathbf{0} \end{bmatrix}$$

And therefore the desired closed-loop behavior $\dot{\mathbf{h}} = \dot{\mathbf{h}}_{des}$ can be written in terms of contact forces as

$$\begin{bmatrix} \mathbf{I}_{3\times 3} & \mathbf{0}_{3\times 3} & \cdots \\ (\mathbf{x}_e(\mathbf{q}) - \mathbf{x}_{CoM}(\mathbf{q}))_\times & \mathbf{I}_{3\times 3} & \cdots \end{bmatrix} \left(\boldsymbol{\lambda} + \mathbf{K}\begin{bmatrix} \mathbf{x}_{CoM} \\ \mathbf{h} \end{bmatrix} - \mathbf{k}(\mathbf{x}_{CoM_{ref}}, \mathbf{h}_{ref})\right) = \mathbf{0} \tag{19}$$

This equation is linear in the contact forces $\boldsymbol{\lambda}$. While it can seem a bit complicated, this equation merely asks the contact forces $\boldsymbol{\lambda}$ to have the same effect on the momentum of the robot than the desired contact forces $\boldsymbol{\lambda}_{des}$ would have. The desired contact forces have been projected in the range space of the momentum. This formulation will allow to exploit the contact forces acting in the nullspace of the momentum (i.e. the contact forces that do not create movement) to achieve other force tasks. For example, to ensure that forces are within the friction cones or to redistribute the weight of the robot between the different contacts.

2.2 Hierarchical Inverse Dynamics Using a Cascade of Quadratic Programs

All the tasks and constraints described above can be written as equality or inequality constraints. Therefore, for each priority level the optimization problem consists of finding decision variables that will best satisfy these constraints and ensure that the solution is still optimal with respect to higher priority levels. As mentioned above, it is not necessary to explicitly incorporate the torque command in the optimization. For each priority level i, the problem can be written as

$$[\ddot{\mathbf{q}}_i^*,\ \boldsymbol{\lambda}_i^*,\ \mathbf{v}_i^*,\ \mathbf{w}_i^*] = \underset{\ddot{\mathbf{q}},\ \boldsymbol{\lambda},\ \mathbf{v}_i,\ \mathbf{w}_i}{\operatorname{argmin}} \ ||\mathbf{v}_i||^2 + ||\mathbf{w}_i||^2$$

$$\text{s.t. } \mathbf{A}_i \begin{bmatrix} \ddot{\mathbf{q}} \\ \boldsymbol{\lambda} \end{bmatrix} - \mathbf{a}_i = \mathbf{v}_i\ ,\ \ \mathbf{B}_i \begin{bmatrix} \ddot{\mathbf{q}} \\ \boldsymbol{\lambda} \end{bmatrix} - \mathbf{b}_i \leq \mathbf{w}_i \qquad \Big\}\text{Tasks of priority } i$$

$$\mathbf{A}_j \begin{bmatrix} \ddot{\mathbf{q}} \\ \boldsymbol{\lambda} \end{bmatrix} - \mathbf{a}_j = \mathbf{v}_j^*\ ,\ \ \mathbf{B}_j \begin{bmatrix} \ddot{\mathbf{q}} \\ \boldsymbol{\lambda} \end{bmatrix} - \mathbf{b}_j \leq \mathbf{w}_j^*,\ \ \forall j < i \qquad \Big\}\begin{array}{l}\text{Ensures optimality w.r.t.} \\ \text{higher priority tasks}\end{array}$$

This quadratic program tries to achieve all the inequality and equality tasks by minimizing the slack variables $\mathbf{v}_i$ and $\mathbf{w}_i$. The tasks are described in the first line of constraints where a slack variable of 0 means that all tasks can be achieved. It also ensures that the solution remains optimal with respect to the previous priorities, this is the second line of constraints. A series of quadratic programs can then be solved (one for each priority). The control command is then simply computed using the result from the last quadratic program

$$\boldsymbol{\tau} = \mathbf{M}_a \ddot{\mathbf{q}}_n^* + \mathbf{h}_a - \mathbf{J}_a^T \boldsymbol{\lambda}_n^* \tag{20}$$

There are several ways to efficiently compute the solution for this hierarchy of quadratic programs. In this chapter we numerically solve the problem using the algorithm developed in Herzog et al. [11]. It proved to be fast enough to be run in a 1 kHz feedback control loop. However, it is worth mentioning that other numerical algorithms might be faster, such as the ones recently presented in Escande et al. [8], Dimitrov et al. [6], at the cost of a potentially reduced ability to regularize the solutions of the optimizer.

3 Optimal Control of Contact Forces and Momentum

When a robot has to perform tasks in complex contact environments such as crossing a difficult terrain or getting up from a seat, it becomes necessary to take into account limb motion and contact forces over a complete execution horizon. Consider for instance a human sitting in a chair with both feet on the ground. It is almost impossible to move up in slow motion, because no feasible forces can be applied on the ground and surface of the chair that support the body throughout the motion. On the other hand, by building up momentum, forces can be applied that will swing up the body in a more dynamic manner. It is therefore important to consider the combination of joint motions and contact forces applied throughout the execution horizon to generate dynamically feasible motions to achieve a larger range of tasks.

Traditionally, simplified models such as the linear inverted pendulum model (LIPM) are used to plan a center of mass motion for legged locomotion, as in Kajita et al. [17], Englsberger et al. [7]. However, for more complex behaviors simplified models do not capture anymore the relevant parts of the dynamics or they keep parts of the state uncontrolled [1] and more complex models become necessary. Planning and optimal control algorithms using more complex models have also been used since the seminal work of Bretl [2] and Hauser et al. [10]. Complete dynamics models have been used in Mordatch et al. [25], Lengagne et al. [24] to plan complex whole-body behaviors but they usually come at a high computational cost and usually contact forces are not explicitly controlled during the execution of the task.

In this section, we discuss a whole-body motion and force generation approach that decouples the problem into a motion trajectory generator and optimal control of contact forces [12, 13]. As we will show, the two sub-problems are coupled only through the momentum that is defined by either the contact forces or the whole-body motion respectively leading to better structured and therefore more efficient optimization algorithms. Further, we describe an optimal control based feedback law to track resulting force and momentum profiles inside the hierarchical inverse dynamics framework presented in Sect. 2.

3.1 Trajectory Optimization

In Sect. 2 we discussed an approach to compute torque commands for fast feedback control. The input to the controller can be a desired execution plan in the form of end-effector or joint trajectories $\mathbf{q}(t)$, a desired momentum trajectory as well as contact wrenches $\boldsymbol{\lambda}(t)$. In this section, we discuss an optimization algorithm for computation of motion plans $\mathbf{q}(t), \boldsymbol{\lambda}(t)$ that minimize a task specific cost function $J = J_{\mathbf{q}}(\mathbf{q}) + J_{\lambda}(\boldsymbol{\lambda})$ and at the same time remain dynamically feasible (i.e. they satisfy Eq. (4)). The motion plans of interest are specified by the following mathematical program

$$\min_{\mathbf{q}(t),\boldsymbol{\lambda}(t)} \quad J_{\mathbf{q}}(\mathbf{q}) + J_{\lambda}(\boldsymbol{\lambda}) \tag{21}$$

$$\text{s.t.} \quad \mathbf{H}(\mathbf{q})\ddot{\mathbf{q}} + \dot{\mathbf{H}}(\mathbf{q})\dot{\mathbf{q}} = \begin{bmatrix} M\mathbf{g} + \sum_e \mathbf{f}_e \\ \sum_e (\boldsymbol{\kappa}_e + (\mathbf{x}_e(\mathbf{q}) - \mathbf{x}_{CoM}(\mathbf{q})) \times \mathbf{f}_e) \end{bmatrix} \tag{22}$$

$$\boldsymbol{\lambda} \in \mathscr{S}_{\lambda}, \mathbf{q} \in \mathscr{S}_{\mathbf{q}} \tag{23}$$

The scalar cost function $J_{\mathbf{q}} + J_{\boldsymbol{\lambda}}$ is designed to express how well motion and force trajectories $\mathbf{q}$, $\boldsymbol{\lambda}$ achieve a certain task. For example, a robot that is crossing a difficult terrain should move its end-effectors from one contact location to the other, i.e. $J_{\mathbf{q}}$ could penalize the distance to a contact point or to the final position of the robot on the terrain. Similarly, $J_{\boldsymbol{\lambda}}$ could penalize certain types of contacts and encourage others (e.g. to minimize the strain put on the hands during climbing). The cost J is shaped such that its minimizer will solve the desired task requirement but it does not consider dynamic feasibility. It is the Newton–Euler Equation (22) that couples motion and force and guarantees solutions to be consistent with physics. Additionally, we need to guarantee that solutions remain in the constraint sets $\mathscr{S}_{\mathbf{q}}$, $\mathscr{S}_{\boldsymbol{\lambda}}$, which describe kinematic constraints (e.g. joint limits) and contact constraints (e.g. the contact model defined in Eqs. (11)–(12)).

Note that the control commands $\boldsymbol{\tau}$ do not appear in our optimal control problem. Indeed, as we have seen in the previous sections, they are only necessary when enforcing actuation limits which we ignore when computing the motion and contact forces trajectories. Therefore, the only "control" commands present in the optimal control problem are the contact forces. Despite the fact that we do not have direct access to contact forces $\boldsymbol{\lambda}$ (i.e. the environment exerts these forces on the robot), the whole-body controller developed in Sect. 2 allows to indirectly control them by finding appropriate joint actuation torques $\boldsymbol{\tau}$.

Solving the problem defined by Eqs. (21)–(23) is challenging in general. At first glance, we have many degrees of freedom in $\mathbf{q}, \boldsymbol{\lambda}$ coupled in a non-linear way. Naive solvers would have to iteratively update all variables and constraints together suffering from polynomial computational complexity. However, realizing that the variables $\mathbf{q}, \boldsymbol{\lambda}$ are only coupled through the Newton–Euler equations describing the evolution of the momentum of the system, we can decompose our mathematical program into two better structured, simpler, sub-problems: one that depends only on kinematics $\mathbf{q}(t)$ and another optimizing for contact forces $\boldsymbol{\lambda}(t)$. As we showed in Herzog et al. [13], the two sub-problems have a more beneficial structure than the full problem (Eqs. (21)–(23)) and can be solved efficiently using better informed solvers.

In order to decompose our movement generation problem, we rewrite the Newton–Euler equations (Eq. (22)) into

$$\mathbf{x}_{CoM}(\mathbf{q}) = \mathbf{r} \tag{24}$$

$$\mathbf{x}_e(\mathbf{q}) = \mathbf{c}_e \tag{25}$$

$$\mathbf{H}(\mathbf{q})\ddot{\mathbf{q}} + \dot{\mathbf{H}}(\mathbf{q})\dot{\mathbf{q}} = \dot{\mathbf{h}} = \begin{bmatrix} M\mathbf{g} + \sum_e \mathbf{f}_e \\ \sum_e (\boldsymbol{\kappa}_e + (\mathbf{c}_e - \mathbf{r}) \times \mathbf{f}_e) \end{bmatrix} \tag{26}$$

Here we introduced variables for the CoM $\mathbf{r}$, contact locations $\mathbf{c}_e$ and momentum $\mathbf{h}$. Note that the left-hand side of Eqs. (24)–(26) depends only on kinematics ($\mathbf{q}$), while the right-hand side is only dependent on momentum and contact forces and locations ($\mathbf{r}, \mathbf{c}, \mathbf{h}$ and $\boldsymbol{\lambda} = \left[\ldots\ \mathbf{f}_e^T\ \boldsymbol{\kappa}_e^T\ \ldots\right]^T$). This means that $\mathbf{r}, \mathbf{c}, \mathbf{h}$ can be computed either purely through the kinematic motion of the whole-body or by applying external forces to the robot. It is therefore sufficient to solve independently the kinematic trajectory optimization problem and the optimal control problem associated to the contact forces while forcing both problems to converge to the same values of $\mathbf{r}$, $\mathbf{c}$, $\mathbf{h}$. When both problems agree on these values, a solution to the original problem is found. In what follows, we solve the problem by iteratively holding one side of Eqs. (24)–(26) fixed to a constant reference $\bar{\mathbf{r}}$, $\bar{\mathbf{c}}$, $\bar{\mathbf{h}}$ and optimizing the other side to match this reference.

Motion Optimization

First we keep the dynamic plan fixed (right-hand side), express our problem as a function of $\mathbf{q}$ and add in a cost term to penalize deviation of $\mathbf{r}$, $\mathbf{c}$ and $\mathbf{h}$ from the values found with the contact force optimization

Motion Optimization

$$\min_{\mathbf{q}(t)} \quad \underbrace{J_{\mathbf{q}}(\mathbf{q})}_{\text{Task Cost}} + \underbrace{||\mathbf{H}(\mathbf{q})\dot{\mathbf{q}} - \bar{\mathbf{h}}||^2 + ||\mathbf{x}_{CoM}(\mathbf{q}) - \bar{\mathbf{r}}||^2 + ||\mathbf{x}_e(\mathbf{q}) - \bar{\mathbf{c}}_e||^2}_{\text{Consistency with Force Optimization}} \tag{27}$$

$$\text{s.t.} \quad \mathbf{q} \in \mathscr{S}_{\mathbf{q}} \tag{28}$$

This mathematical program optimizes only over joint trajectories $\mathbf{q}$ and as such reduces the problem to a kinematic trajectory optimization. In this sub-problem we do not have to consider contact forces therefore significantly reducing computational complexity. The cost is constructed from two objective terms. On the one hand, we minimize the task specific cost $J_{\mathbf{q}}$ and on the other hand we bias our solution to match momentum and contact location references $\bar{\mathbf{r}}$, $\bar{\mathbf{c}}$, $\bar{\mathbf{h}}$. Since our sub-problem is only optimized over $\mathbf{q}$, we can reuse wide-spread kinematic solvers that are suited for this type of problem structure. Our plan contains CoM, $\mathbf{x}_{CoM}(\mathbf{q})$, end-effector trajectories, $\mathbf{x}_e(\mathbf{q})$, and also the momentum profile, $\mathbf{H}\dot{\mathbf{q}}$, that is induced by the motion of the robot, e.g. our plan might find a swing leg motion generating a momentum around the hip.

Force Optimization

In the next step, we compute admissible contact forces that accomplish the required motion-induced momentum and contact location references $\bar{\mathbf{r}}$, $\bar{\mathbf{c}}_e$, $\bar{\mathbf{h}}$. Therefore, we solve the problem

Force Optimization

$$\min_{\mathbf{r},\mathbf{c},\mathbf{h},\boldsymbol{\lambda}} \quad \underbrace{J_{\lambda}(\mathbf{r}, \mathbf{c}, \mathbf{h}, \boldsymbol{\lambda})}_{\text{Task Cost}} + \underbrace{||\mathbf{h} - \bar{\mathbf{h}}||^2 + ||\mathbf{r} - \bar{\mathbf{r}}||^2 + ||\mathbf{c}_e - \bar{\mathbf{c}}_e||^2}_{\text{Consistency with Motion Optimization}} \tag{29}$$

$$\text{s.t.} \quad \dot{\mathbf{h}} = \begin{bmatrix} M\mathbf{g} + \sum_e \mathbf{f}_e \\ \sum_e (\boldsymbol{\kappa}_e + (\mathbf{c}_e - \mathbf{r}) \times \mathbf{f}_e) \end{bmatrix} \tag{30}$$

$$\boldsymbol{\lambda}, \mathbf{r}, \mathbf{c} \in \mathscr{S}_{\lambda} \tag{31}$$

Contrary to the kinematic trajectory optimizer, here we optimize only over contact forces $\boldsymbol{\lambda}$, contact locations $\mathbf{c}_e$ and resulting CoM $\mathbf{r}$ and momentum $\mathbf{h}$. Similar to the *Motion Optimization* problem we trade-off a task relevant cost J_λ with the tracking of reference trajectories $\bar{\mathbf{h}}, \bar{\mathbf{r}}, \bar{\mathbf{c}}_e$ that were found by the *Motion Optimization* problem. The solutions to the *Force Optimization* problem are dynamically feasible since they remain in the admissible set $\mathscr{S}_\lambda$ and satisfy the Newton–Euler Equation (30).

Note that the *Force Optimization* problem can be seen as an optimal control problem if we assume that the contact forces are the control inputs. In fact, this allowed us to identify interesting structure in the problem [13] that is beneficial for dedicated solvers.

Alternating Algorithm for Whole-Body Planning

The two sub-problems *Motion Optimization* and *Force Optimization* consider separate cost and constraint terms and at the same time they try to accomplish a coherent momentum profile. We can now solve the two problems iteratively passing on reference momentum profiles from one solver to the other until both agree on a common solution as summarized in Algorithm 1. It exploits dedicated solvers that are suited for kinematic (line 3) and optimal control problems (line 5). Both solvers communicate only the common set of variables in lines 4, 6 but remain independent otherwise. In our experience, only a few iterations (typically 2) are necessary before convergence.

Algorithm 1 momentum-centric motion generation

1: initialize $\bar{\mathbf{r}}, \bar{\mathbf{c}}, \bar{\mathbf{h}}$
2: **repeat**
3: solve the *Motion Optimization* problem
4: $\bar{\mathbf{r}}, \bar{\mathbf{c}}, \bar{\mathbf{h}} \leftarrow \mathbf{x}_{CoM}(\mathbf{q}), \mathbf{x}_e(\mathbf{q}), \mathbf{H}(\mathbf{q})\dot{\mathbf{q}}$
5: solve the *Force Optimization* problem
6: $\bar{\mathbf{r}}, \bar{\mathbf{c}}, \bar{\mathbf{h}} \leftarrow \mathbf{r}, \mathbf{c}, \mathbf{h}$
7: **until** convergence

A Note on the Structure of the Force Optimization problem

In the *Force Optimization* Problem we are facing non-linear dynamics that result in non-convex optimization in general. The non-convexity stems from the cross product in Eq. (30). We have shown in Herzog et al. [13] that this expression can be written as the difference of two positive semi-definite quadratic functions in the form $\mathbf{s}^+(\mathbf{r}, \mathbf{c}, \lambda) - \mathbf{s}^-(\mathbf{r}, \mathbf{c}, \lambda)$, where $\mathbf{s}^+, \mathbf{s}^- : \mathbb{R}^9 \rightarrow \mathbb{R}^3$ are both convex functions with quadratic components. The decomposition into a difference of convex functions can be performed analytically and does not require additional computation effort inside the solver. Decomposing the problem into its convex and concave parts and exploiting sparsity patterns specific to the problem has the advantage that better approximations can be made inside of dedicated optimizers leading to more efficient solvers [13, 29]. For a more detailed discussion on the structure of the optimization problem please refer to Herzog et al. [13].

3.2 Contact Forces Feedback Control

In the previous section we discussed a motion generation algorithm that computes dynamically consistent joint trajectories $\mathbf{q}(t)$ together with momentum $\mathbf{h}(t)$ and contact forces $\boldsymbol{\lambda}(t)$. These trajectories will then be controlled with hierarchical inverse dynamics as presented in Sect. 2. This controller converts $\mathbf{q}(t)$, $\boldsymbol{\lambda}$ into joint torques $\boldsymbol{\tau}$ and regulates disturbances through feedback in task space directly. In particular, we explained in details how momentum control tasks could be thought as a force control problem. In this section, we show how the results from the *Force Optimization* problem can be used to create a locally optimal feedback control law for stabilizing momentum trajectories through contact forces.

Given optimal CoM and momentum trajectories $\mathbf{x}^* = \left[\mathbf{r}^{*T}, \mathbf{h}^{*T}\right]^T$ and contact wrenches $\boldsymbol{\lambda}^*(t)$, we linearize Eq. (30) along the optimal trajectories and compute a quadratic approximation of the performance cost (Eq. (21)). We now have a time-varying linear quadratic control problem along a desired trajectory where the control inputs are the contact forces.

We can use time-varying LQR to obtain a locally optimal feedback control policy of the form

$$\boldsymbol{\lambda}_{des} = \mathbf{k}(\mathbf{x}^*, \boldsymbol{\lambda}^*) - \mathbf{K}(t)\mathbf{x} \tag{32}$$

where $\mathbf{K}(t)$ is a time-varying feedback gain, $\mathbf{k}$ a feedforward term and $\mathbf{x}$ the current CoM and momentum measurements. These terms can then readily be used in the momentum control task defined by Eq. (19).

In the hierarchical inverse dynamics, we can now track momentum at high priority and use the contact force references from the whole-body planner at a lower priority. The part of the forces that do not create momentum will only be tracked as long as they do not conflict with higher priority tasks. With this feedback design approach, we fully exploit the coupling between CoM and angular momentum (cf. Eq. (4)) and find controls that optimize our desired task. We also found in practice [11] that the control law in Eq. (19) not only performs better than a manually tuned PD controller for momentum tracking, but also allows to remove any tuning of parameters in the momentum task (e.g. different contact configurations would require different gain tuning).

4 Experiments

In this section we present a set of experiments illustrating the capabilities of both the hierarchical inverse dynamics and the trajectory optimization for multi-contact motions.

4.1 Hierarchical Inverse Dynamics on a Humanoid Robot

Using the controller described above, we performed experiments on the lower body of a Sarcos Humanoid robot which is a torque controlled robot offering high performance torque tracking capabilities. In order to estimate the position, orientation and velocity of the robot in space, we use the base state estimator developed in Rotella et al. [31]. The control diagram is similar to what is described in Fig. 1 except that the trajectory optimization block is not used yet (except to compute LQR gains as explained in Sect. 3.2).

Several experiments were conducted to evaluate the capabilities of the robot to balance by regulating its linear and angular momentum. During double support, the robot was placed on moving platforms such as a seesaw and a rolling platform (shown in Fig. 2). In another experiment, the task was to move from a two leg balance to standing on one leg. Then the robot started to move its swing leg in a periodic manner (following sine trajectories for the swing foot). During this motion, the robot was pushed with forces of different magnitudes (as shown in Fig. 2).

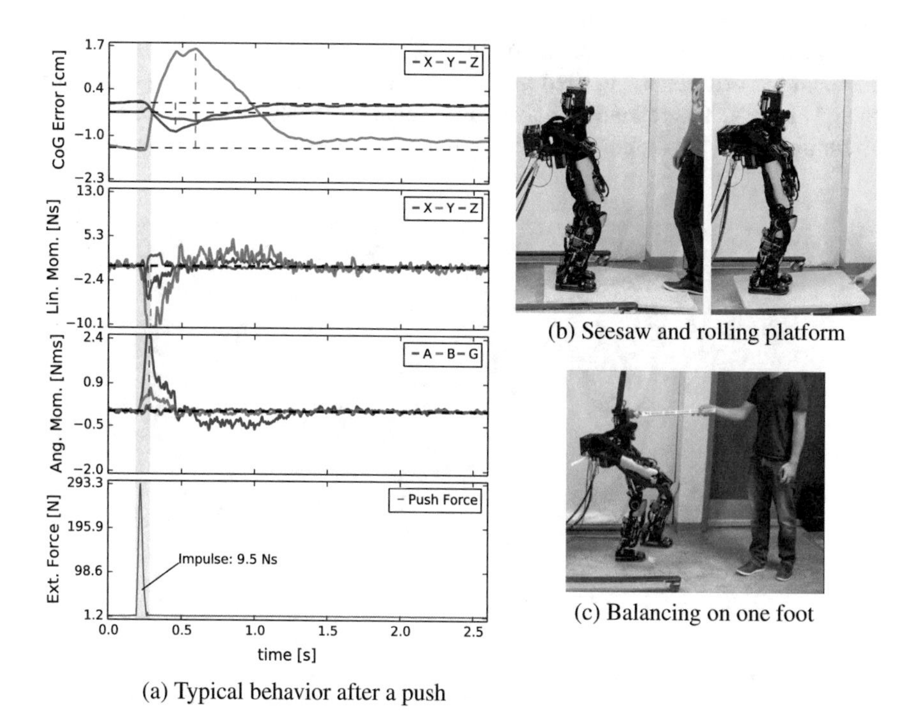

(a) Typical behavior after a push

(b) Seesaw and rolling platform

(c) Balancing on one foot

Fig. 2 On the *left*, typical disturbance rejection behavior of the center of mass and momentum when the robot is pushed while standing on two feet. We notice the fast damped dissipation of the push. On the *right*, pictures of the various experiments performed on the robot. The robot was balancing on a seesaw and a rolling platform with unknown disturbances. It also moves from double support to a single support balancing while executing a swing leg motion in the air under external disturbances (The figures are taken from Herzog et al. [11])

Experiments demonstrate a very high performance for balancing by using solely feedback control. We notice that regulating the angular momentum helped to create more stable balancing behaviors. Moreover, the use of inequality constraints for the center of pressure of each foot was very useful for balancing: the controller actively generates angular momentum to maintain the CoP within the constraints when necessary. During the experiments, the robot was pushed with a peak force up to 293 N and impulse of 9.5Ns which is rather high for balancing experiments (note that the robot weights 51 kg and each foot is 0.09 m wide and 0.25 m long). A systematic analysis of the results can also be found in Herzog et al. [11].

4.2 Control and Planning of Multi-contact Behaviors

We demonstrate how Algorithm 1 can generate locomotion on a terrain of stepping stones by jointly optimizing joint trajectories $\mathbf{q}(t)$ and contact forces $\boldsymbol{\lambda}(t)$. The stepping stones (cf. Fig. 3) are angled and vary in height, which breaks the LIPM assumptions and requires consideration interaction forces at individual contact locations. We initialize our algorithm with a reference CoM $\bar{\mathbf{r}}$ in form of a linear interpolation between the desired position of the robot at the beginning and at the end of the task. The reference momentum is simply initialized to $\bar{\mathbf{h}} = \mathbf{0}$. In line 3 of our algorithm the motion generator computes a constrained joint trajectory that

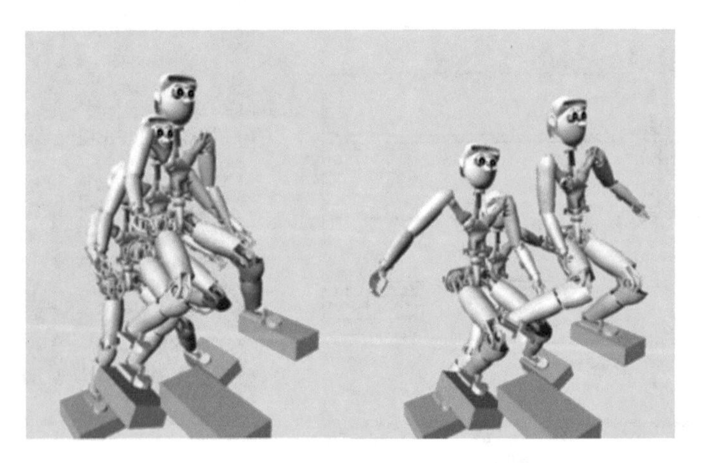

Fig. 3 A visualization of a kinematic plan generated by Algorithm 1. After the first iteration (*left-hand side*) the CoM remains above the center between the feet and the arms are at rest. Although, the joint trajectories remain inside of their limits and the task relevant cost is minimized, momenta are required that are not feasible. For instance, one might notice that the CoM moves forward, although the robot is in single support and can only push on a stone orthogonal to the motion. After the last iteration (*right-hand side*) the kinematic plan is in coherence with constrained force profiles. The CoM has a wider sway in lateral direction and the robot uses its arms to account for angular momentum (Figure taken from Herzog et al. [13])

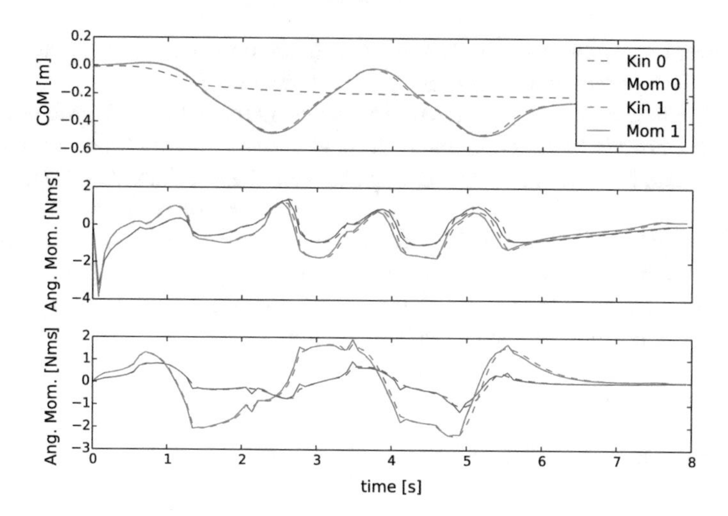

Fig. 4 Momentum plans generated by the motion optimization Algorithm 1. The *Motion Optimization* problem optimizes for an unfeasible CoM lateral motion (*top*), which is corrected after the first *Force Optimization* iterations. In the second iteration of the algorithm, the kinematic plan is adapted to realize a CoM sway, this time resulting in an admissible momentum profile. Horizontal angular momentum (*bottom* two plots) is also found to be consistent across kinematic as well as momentum optimization. It is interesting to note that non-trivial angular momentum trajectories are computed that would be difficult to design by hand (Figure taken from Herzog et al. [13])

Fig. 5 A visualization of a gain matrix as it is computed from LQR on the momentum equations. It is interesting to see that off-diagonal terms, which are not obvious to design by hand, contribute significantly to the feedback. For instance we can see that the *lower left* part of the matrix exploits coupling between linear and angular momentum and uses angular momentum in order to correct errors in CoM and linear momentum tracking (Figure taken from Herzog et al. [13])

minimizes the distance between end-effectors $\mathbf{c}_e$ and stepping stones as it tries to track the desired CoM $\bar{\mathbf{r}}$. As a result we obtain a motion that is kinematically feasible and crosses the terrain in a kinematically optimal way, however, ignoring limitations on contact wrenches. The desired motion is then optimized to ensure that contact wrenches are restricted to push on stepping stones without violating friction constraints (cf. Fig. 4). In practice we noted that 2 iterations were usually sufficient to converge, i.e. consecutive momentum profiles would not change.

Then the feedback policy around optimal momentum and wrench profiles is computed and embedded into the hierarchical inverse dynamics controller presented in Sect. 2. As can be seen in Fig. 5, coupling between linear and angular momentum is exploited. The resulting controller led to better stabilization behavior. The trajectories generated by the plan were difficult to stabilize with a hand-tuned PD controller while they were tracked without problem using our feedback law. Figure 3 shows an example of the motion before and after dynamic optimization, illustrating the effect of dynamic constraints on the whole-body motion.

5 Discussion: Strengths and Limitations

Algorithmic Considerations

In previous sections we have presented a method to plan and control dynamically consistent multi-contact behaviors. The planning and control methods are centered around the momentum equations and contact forces. While several recent work have considered the effect of (linear and angular) momentum for motion generation, they either do not plan it explicitly [36] or do not address the problem of controlling momentum and contact forces together [5], which is in contrast with the approach we proposed. We rely on a dynamic model of the robot and make no simplifying assumption to compute trajectories and control laws. Therefore, it should in principle be possible to generate arbitrary multi-contact movements for complex robots such as a humanoid. In our architecture we assumed that a sequence of desired contacts was given but recent algorithms [35] planning acyclic contact sequences could be used instead to find contact sequences automatically. One strong advantage with dynamic models and optimization methods is that the approach can generalize to arbitrary conditions, ensures locally optimized behaviors with guaranteed constraint satisfaction. However, using complex dynamic model can be problematic because they are usually synonymous to computational complexity. While recent results suggest that some of these computations can be done very efficiently (cf. Ponton et al. [29] for example), there remains significant research to be done to improve the numerical efficiency of such algorithms.

What About Multi-modal Sensing?

Methods based on models and optimal control usually make a very limited use of the available sensory information. Modern robots usually come equipped with a large number of sensor modalities (position and force sensors, inertial measurement units, artificial skin, etc.). Typically, these algorithms only use sensory information to reconstruct the state of the robot for feedback and then discard the information. It is also the case in the approach we proposed here and it constitutes a potential limitation of the approach. Indeed, it seems reasonable to think that much more information could be extracted from sensing, especially during contact interaction, to create more reactive and robust behaviors. Data-driven approaches have been successfully proposed to create more robust behaviors in manipulation research. For example, by learning impedance behaviors directly from demonstration [4] or using reinforcement learning [3]. Our previous work also demonstrated some results on learning force control for manipulation skills using reinforcement learning [19]. We have also shown how to create very reactive and robust manipulation skills by learning a model of the task directly in sensor space with little a-priori on the task dynamics [28]. This work was successfully extended to sequence more complex manipulation tasks in Kappler et al. [21] where the authors demonstrated very robust performance without the need for an explicit model of the task nor a computationally complex optimization algorithm. However, these data-driven approaches are still very limited and do not generalize well across tasks nor scale to more complex tasks such as multi-contact whole-body behavior. A more systematic use of multi-modal sensing, from a data-driven perspective, remains therefore an important challenge to complement the approach we presented in this chapter and optimal control approaches in general.

6 Conclusion

Centered around the recognition that interaction forces are fundamental to explain motion, we have proposed a method to control and plan multi-contact behaviors. The method is based on hierarchical inverse dynamics and optimal control for trajectory generation. The angular and linear momentum of the robot play a central role in the approach, first as a fundamental quantity that needs to be controlled through interaction forces and second as a natural way to decompose the optimization of dynamically consistent multi-contact trajectories. Future work will include the extension of the approach to more complex behaviors and a systematic inclusion of multi-modal sensing in the optimization approach to increase robustness.

Acknowledgements This work was supported by the Max-Planck Society, the European Research Council under the European Union's Horizon 2020 research and innovation programme (ERC StG CONT-ACT, grant agreement No 637935) and the Max Planck ETH Center for Learning Systems. We would also like to warmly thank two anonymous reviewers for their constructive comments that helped significantly improve the chapter.

References

1. H. Audren, J. Vaillant, A. Kheddar, A. Escande, K. Kaneko, E. Yoshida, Model preview control in multi-contact motion-application to a humanoid robot, in *2014 IEEE/RSJ International Conference on Intelligent Robots and Systems* (2014), pp. 4030–4035
2. T. Bretl, Motion planning of multi-limbed robots subject to equilibrium constraints: the free-climbing robot problem. Int. J. Robot. Res. **25**(4), 317–342 (2006)

3. J. Buchli, F. Stulp, E. Theodorou, S. Schaal, Learning variable impedance control. Int. J. Robot. Res. **30**(7), 820–833 (2011)
4. S. Calinon, I. Sardellitti, D.G. Caldwell, Learning-based control strategy for safe human-robot interaction exploiting task and robot redundancies, in *IEEE/RSJ International Conference on Intelligent Robots and Systems (IROS)* (2010), pp. 249–254
5. H. Dai, A. Valenzuela, R. Tedrake, Whole-body motion planning with centroidal dynamics and full kinematics, in *2014 IEEE-RAS International Conference on Humanoid Robots* (2014), pp. 295–302
6. D. Dimitrov, A. Sherikov, P.B. Wieber, Efficient resolution of potentially conflicting linear constraints in robotics, submitted (2015). https://hal.inria.fr/hal-01183003
7. J. Englsberger, C. Ott, A. Albu-Schäffer, Three-dimensional bipedal walking control based on divergent component of motion. IEEE Trans. Robot. **31**(2), 355–368 (2015)
8. A. Escande, N. Mansard, P.B. Wieber, Hierarchical quadratic programming: fast online humanoid-robot motion generation. Int. J. Robot. Res. **33**(7), 1006–1028 (2014)
9. S. Feng, X. Xinjilefu, C.G. Atkeson, J. Kim, Optimization based controller design and implementation for the atlas robot in the darpa robotics challenge finals, in *IEEE-RAS International Conference on Humanoid Robots (Humanoids)* (2015), pp. 1028–1035
10. K. Hauser, T. Bretl, J.C. Latombe, K. Harada, B. Wilcox, Motion planning for legged robots on varied terrain. Int. J. Robot. Res. **27**(11–12), 1325–1349 (2008)
11. A. Herzog, N. Rotella, S. Mason, F. Grimminger, S. Schaal, L. Righetti, Momentum control with hierarchical inverse dynamics on a torque-controlled humanoid. Auton. Robots **40**(3), 473–491 (2015)
12. A. Herzog, N. Rotella, S. Schaal, L. Righetti, Trajectory generation for multi-contact momentum control, in *IEEE-RAS International Conference on Humanoid Robots (Humanoids)* (2015), pp. 874–880
13. A. Herzog, S. Schaal, L. Righetti, Structured contact force optimization for kino-dynamic motion generation, in *IEEE/RSJ International Conference on Intelligent Robots and Systems (IROS)* (2016)
14. N. Hogan, Impedance control - an approach to manipulation. J. Dyn. Syst. Meas. Control (Trans. ASME) **107**(1), 1–24 (1985)
15. S.H. Hyon, Compliant terrain adaptation for biped humanoids without measuring ground surface and contact forces. IEEE Trans. Robot. **25**(1), 171–178 (2009)
16. G. Jarquin, A. Escande, G. Arechavaleta, T. Moulard, Y. Eoshida, V. Parra-Vega, Real-time smooth task transitions for hierarchical inverse kinematics, in *IEEE-RAS International Conference on Humanoid Robots (Humanoids)* (2013), pp. 528–533
17. S. Kajita, F. Kanehiro, K. Kaneko, K. Fujiwara, K. Harada, K. Yokoi, H. Hirukawa, Biped walking pattern generation by using preview control of zero-moment point, in *IEEE International Conference on Robotics and Automation (ICRA)* (2003), vol. 2, pp. 1620–1626
18. S. Kajita, F. Kanehiro, K. Kaneko, K. Fujiwara, K. Harada, K. Yokoi, H. Hirukawa, Resolved momentum control: humanoid motion planning based on the linear and angular momentum, in *IEEE/RSJ International Conference on Intelligent Robots and Systems (IROS)* (2003), pp. 1644–1650
19. M. Kalakrishnan, L. Righetti, P. Pastor, S. Schaal, Learning force control policies for compliant manipulation, in *IEEE International Conference on Intelligent Robots and Systems (IROS)* (2011), pp. 4639–4644
20. O. Kanoun, F. Lamiraux, P.B. Wieber, Kinematic control of redundant manipulators: generalizing the task-priority framework to inequality task. IEEE Trans. Robot. **27**(4), 785–792 (2011)
21. D. Kappler, P. Pastor, M. Kalakrishnan, M. Wuthrich, S. Schaal, Data-driven online decision making for autonomous manipulation, in *Proceedings of Robotics: Science and Systems* (Rome, Italy, 2015)
22. M. de Lasa, I. Mordatch, A. Hertzmann, Feature-based locomotion controllers. ACM Trans. Gr. **29**(4), 131:1–131:10 (2010)

23. S.H. Lee, A. Goswami, A momentum-based balance controller for humanoid robots on non-level and non-stationary ground. Auton. Robots **33**, 399–414 (2012)
24. S. Lengagne, J. Vaillant, E. Yoshida, A. Kheddar, Generation of whole-body optimal dynamic multi-contact motions. Int. J. Robot. Res. **32**(9–10), 1104–1119 (2013)
25. I. Mordatch, E. Todorov, Z. Popović, Discovery of complex behaviors through contact-invariant optimization. ACM Trans. Gr. **31**(4), 43:1–43:8 (2012)
26. D.E. Orin, A. Goswami, Centroidal momentum matrix of a humanoid robot: structure and properties, in *IEEE/RSJ International Conference on Intelligent Robots and Systems (IROS)* (2008), pp. 653–659
27. C. Ott, M.A. Roa, G. Hirzinger, Posture and balance control for biped robots based on contact force optimization, in *11th IEEE-RAS International Conference on Humanoid Robots (Humanoids)* (2011), pp. 26–33
28. P. Pastor, L. Righetti, M. Kalakrishnan, S. Schaal, Online movement adaptation based on previous sensor experiences, in *IEEE International Conference on Intelligent Robots and Systems (IROS)* (2011), pp. 365–371
29. B. Ponton, A. Herzog, S. Schaal, L. Righetti, A convex model of momentum dynamics for multi-contact motion generation, in *IEEE-RAS International Conference on Humanoid Robots (Humanoids)* (2016)
30. L. Righetti, J. Buchli, M. Mistry, M. Kalakrishnan, S. Schaal, Optimal distribution of contact forces with inverse-dynamics control. Int. J. Robot. Res. **32**(3), 280–298 (2013)
31. N. Rotella, M. Bloesch, L. Righetti, S. Schaal, State estimation for a humanoid robot, in *2014 IEEE/RSJ Conference on Intelligent Robots and Systems* (Chicago, 2014), pp. 952–958
32. L. Saab, O. Ramos, N. Mansard, P. Soueres, J.Y. Fourquet, Dynamic whole-body motion generation under rigid contacts and other unilateral constraints. IEEE Trans. Robot. **29**, 346–362 (2013)
33. A. Sherikov, D. Dimitrov, P.B. Wieber, Balancing a humanoid robot with a prioritized contact force distribution, in *2015 IEEE-RAS 15th International Conference on Humanoid Robots (Humanoids)* (2015), pp. 223–228
34. E. Todorov, A convex, smooth and invertible contact model for trajectory optimization, in *2011 IEEE International Conference on Robotics and Automation (ICRA)* (2011), pp. 1071–1076
35. S. Tonneau, N. Mansard, C. Park, D. Manocha, F. Multon, J. Pettre, A Reachability-based planner for sequences of acyclic contacts in cluttered environments, in *International Symposium on Robotics Research (ISSR 2015)* (2015)
36. P.M. Wensing, D.E. Orin, Generation of dynamic humanoid behaviors through task-space control with conic optimization, in *2013 IEEE International Conference on Robotics and Automation (ICRA)* (2013), pp. 3103–3109
37. P.B. Wieber, Holonomy and nonholonomy in the dynamics of articulated motion, in *Fast Motions in Biomechanics and Robotics: Optimization and Feedback Control*, ed. by M. Diehl, K. Mombaur (Springer, Berlin Heidelberg, 2006), pp. 411–425

A Tutorial on Newton Methods for Constrained Trajectory Optimization and Relations to SLAM, Gaussian Process Smoothing, Optimal Control, and Probabilistic Inference

Marc Toussaint

Abstract Many state-of-the-art approaches to trajectory optimization and optimal control are intimately related to standard Newton methods. For researchers that work in the intersections of machine learning, robotics, control, and optimization, such relations are highly relevant but sometimes hard to see across disciplines, due also to the different notations and conventions used in the disciplines. The aim of this tutorial is to introduce to constrained trajectory optimization in a manner that allows us to establish these relations. We consider a basic but general formalization of the problem and discuss the structure of Newton steps in this setting. The computation of Newton steps can then be related to dynamic programming, establishing relations to DDP, iLQG, and AICO. We can also clarify how inverting a banded symmetric matrix is related to dynamic programming as well as message passing in Markov chains and factor graphs. Further, for a machine learner, path optimization and Gaussian Processes seem intuitively related problems. We establish such a relation and show how to solve a Gaussian Process-regularized path optimization problem efficiently. Further topics include how to derive an optimal controller around the path, model predictive control in constrained k-order control processes, and the pullback metric interpretation of the Gauss–Newton approximation.

1 Introduction

It is hard to track down explicitly when Newton methods were first used for trajectory optimization. As the method is centuries old it seems fair to assume that they were used from the very beginning. More recent surveys, such as [3, 36], take Newton methods and standard non-linear constrained mathematical programming (NLP) methods as granted. Reference [3] for instance states that Newton methods were the standard in the 60's, often executed analytically by hand. Presumably the Apollo missions relied on Newton methods to compute paths. In the 70's, with rais-

M. Toussaint (✉)
Machine Learning and Robotics Lab, University Stuttgart, Stuttgart, Germany
e-mail: marc.toussaint@informatik.uni-stuttgart.de

J.-P. Laumond et al. (eds.), *Geometric and Numerical Foundations of Movements*, Springer Tracts in Advanced Robotics 117, DOI 10.1007/978-3-319-51547-2_15

ing computational powers and quasi-Newton methods (such as BFGS), they became prevalent for many kinds of control problems.

Why do we need, half a century later, a tutorial on Newton methods for trajectory optimization? Especially in the last decade the fields of machine learning, AI, robotics, optimization and control became more and more intertwined, with methods of one discipline fertilizing ideas or complementing methods in another. This often leads to great advances in the fields. However, the interrelations between methods in the different fields are sometimes hard to see and acknowledge because the languages differs, textbooks are not cross-disciplinary, and technical papers cannot focus on length on this.

Many interesting novel approaches to trajectory optimization have been proposed in the last decade. However, identifying and relating the actual state-of-the-art across disciplines is hard. An excellent and very necessary paper in the robotics community (TrajOpt; [27]), proposing non-linear mathematical programming (NLP) for trajectory optimization, might in other communities perhaps have been located decades earlier. That paper is in fact an important answer on previous papers within robotics, esp. (CHOMP; [24]), that have not compared to the NLP view on trajectory optimization. To comment also on own work, the Approximate Inference approach to Trajectory Optimization (AICO; [31]) establishes important relations between iterative message passing and trajectory optimization (see below) and still inspires great advances in the field [8]. But the optimization view on the same problem formulation leads to basic Newton methods that can more easily be extended to hard constraints and are more robust in practice. Similarly, it seems important to acknowledge the tight relations between the optimal control approaches DDP [20] and iLQG [30] and plain (Gauss–) Newton methods, as discussed in more detail below.

In this tutorial we take the stand that such methods and especially their relations are best understood by considering optimization as their common underlying foundation, in particular the Newton method. With this we hope to give a basis for fertilization and understanding across disciplines.

What is proposed in this tutorial is not fundamentally novel: We discuss basic Newton and NLP methods for a general problem formulation, including also control and model predictive control around the optimum. However, some specifics of the presentation are novel, for instance:

(i) The specific k-order path optimization formulation is in contrast to the more common phase-space formulation of path problems. This, and the banded problem Jacobians and Hessians were previously mentioned in [34].
(ii) The particular generalization of dynamic programming and model-predictive control to constrained k-order processes are, to our knowledge, novel. Also the related *approximate constrained* Linear-Quadratic Regulator (acLQR) around an optimal path has, to our knowledge, not been described in this form before. Related work is [28].
(iii) The intimate relations between Newton-based trajectory optimization and Graph-SLAM have only very recently been mentioned [8]; the recast of CHOMP as plain Newton that drops some terms seems novel.

(iv) Reference [8] introduced the interesting idea to consider global-scale Gaussian Process smoothness priors over paths and utilize GTSAM to optimize the resulting problem. Here we propose a simpler approach to account for "banded-support" covariance kernels in the path objective with leads to linear-in-T complexity of computing Newton steps.

(v) Throughout the paper we discuss complexities of computing the Newton steps, which has not been presented in this way before.

1.1 Structure of This Tutorial

Although the material presented is closely related to optimal control, we think it is insightful for this tutorial to first consider a pure trajectory optimization perspective. Controls and optimal control are not mentioned until Sect. 4. With this we aim to show how much we can learn about the structure of Newton-based path optimization that then relates intimately to optimal control methods.

Hence, in the first part, we formulate a path optimization problem of a particular k-order structure. Section 2.3 discusses the resulting banded structures of the problem Jacobian and Hessian and based on this derives the complexities of computing Newton steps. These basic properties of the Jacobian, Hessian and the computation of Newton steps seem technical, but they are the core to understand the relations discussed later. For instance, this allows us to understand relations to the pullback of Riemannian metrics in differential geometry, to Graph-SLAM methods, and to the CHOMP optimization method.

Section 3 asks how we can incorporate a more global smoothness objective in the optimization formulation. We briefly consider a B-spline representation of paths, which are intuitively very promising to enforce smoothness and speed up optimization. However, in practice they hardly reduce the number of Newton steps needed and the complexity of each Newton step is equal to non-spline representations. We then consider an alternative way to include more global smoothness objectives: with a covariance kernel function as in Gaussian Processes (GPs), efficiently optimizing the neg-log probability of a GP with a banded kernel function.

Section 4 then reconsiders the problem from an optimal control perspective. We first briefly introduce the basic optimal control framework and discuss direct versus indirect approaches. To tackle our specific k-order path optimization problem we then consider dynamic programming to compute cost-to-go functions under hard constraints and the respective approximate constrained Linear Quadratic Regulator, which, just for sanity, is shown to be equivalent to the Riccati equation in the unconstrained LQR case. We extend the dynamic programming formulation to a model predictive control (MPC) formulation (in fact, a constrained k-order version of MPC) that allow us to control around pre-optimized trajectories. Moving to the probabilistic setting the relations to DDP, iLQG and AICO become clear. On the conceptual level, this section establishes the relations between (i) inverting a banded Hessian (in a Newton step), (ii) dynamic programming and (iii) probabilistic message passing, all three of them making the linear-in-T complexity of computing Newton steps explicit.

2 *k*-Order Path Optimization and Its Structure

2.1 *Problem Formulation: k-Order Constrained Path Optimization (KOMO)*

Let $x \in \mathbb{R}^{T \times n}$ be the path[1] of T time steps in an n-dimensional *configuration* space X. That is, in the dynamic case, x_t does not include velocities and x is not a *state* space (or phase space) trajectory. Instead, x only represents a series of configurations.

A general non-linear program over a path x is of the form

$$\min_x f(x) \quad \text{s.t.} \quad g(x) \le 0 \,, \quad h(x) = 0 \,, \tag{1}$$

where $f : \mathbb{R}^{T \times n} \to \mathbb{R}$ is a scalar objective function, $g : \mathbb{R}^{T \times n} \to \mathbb{R}^{d_g}$ defines d_g inequality constraint functions, and $h : \mathbb{R}^{T \times n} \to \mathbb{R}^{d_h}$ defines d_h equality constraint functions. We generally assume f, g, and h to be smooth, but not necessarily convex or unimodal.

For the case of path optimization we make the following assumption:

Assumption 1 (*k-order Markov Assumption*) We assume

$$f(x) = \sum_{t=1}^{T} f_t(x_{t-k:t}) \,, \quad g(x) = \bigotimes_{t=1}^{T} g_t(x_{t-k:t}) \,, \quad h(x) = \bigotimes_{t=1}^{T} h_t(x_{t-k:t}) \,, \tag{2}$$

for a given *prefix* $x_{k-1:0}$, where each f_t is scalar, g_t is d_{gt}-dimensional, and h_t is d_{ht}-dimensional.

Here we use the tuple notation $x_{t-k:t} = (x_{t-k}, x_{t-k+1}, .., x_t)$. The prefix $x_{k-1:0}$ are the system configurations before the path; assuming this to be known simplifies the notation, without the need to introduce a special notation for the first k terms. The outer product $\bigotimes$ notation means that the constraint functions g_t of each time step are stacked to become the full ($d_g = \sum_t d_{gt}$)-dimensional constraint function g over the full path. Under this assumption, we define our problem as

[1] We use the words *path* and *trajectory* interchangeably: we always think of a path as a mapping $[0, T] \to \mathbb{R}^n$, including its temporal profile.

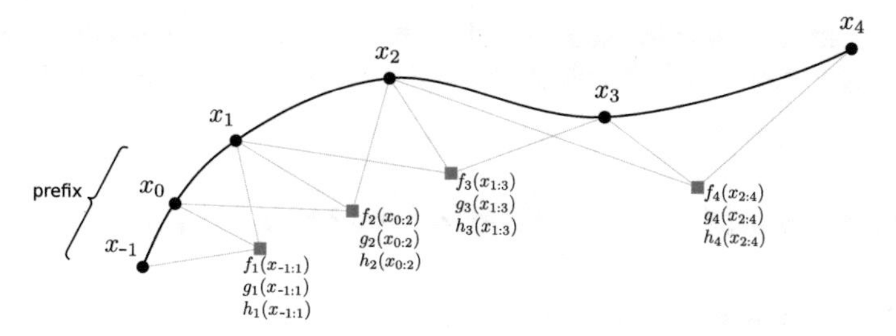

Fig. 1 Illustration of the structure implied by the k-order Markov Assumption (Eq. 2)

Definition 1 *(k-order Motion Optimization (KOMO; [34]))*

$$\min_x \sum_{t=1}^{T} f_t(x_{t-k:t}) \quad \text{s.t.} \quad \forall_{t=1}^{T}: \; g_t(x_{t-k:t}) \le 0 \,, \;\; h_t(x_{t-k:t}) = 0 \,. \tag{3}$$

Figure 1 illustrates the structure implied by the k-order Markov Assumption: Tuples $x_{t-k:t}$ of $k+1$ consecutive variables are coupled by the objectives and constraints

$$\phi_t(x_{t-k:t}) \triangleq \begin{pmatrix} f_t(x_{t-k:t}) \\ g_t(x_{t-k:t}) \\ h_t(x_{t-k:t}) \end{pmatrix} . \tag{4}$$

We call these $\phi_t(x_{t-k:t}) \in \mathbb{R}^{1+d_{gt}+d_{ht}}$ the **features** at time t, encompassing cost, inequality, and equality features. In Fig. 1, the coupling features $\phi_t(x_{t-k:t})$ are represented by the boxes. The graphical notation is used in analogy to factor graphs and conditional random fields (CRFs) [16, 18], helping us to discus these relations already on the level of the problem formulation.

The structure of CRFs is typically captured in the form

$$P(y|x) = \frac{1}{Z(x,\beta)} \exp\{\sum_i \tilde{\phi}_i(y_{\partial i}, x)^\top \beta_i\} \,, \tag{5}$$

where $\tilde{\phi}_i(y_{\partial i}, x)$ are *features* that couple the input x to a tuple $y_{\partial i}$ of output variables.[2] These features capture the structure of the output distribution $P(y|x)$. Going back to path optimization, in our case the features $\phi_t(x_{t-k:t})$ not only encompass costs, but also inequality and equality constraints. As plain path optimization is not a learning problem, we have no global model parameters β. However, as a side note, in the case of inverse optimal control it is exactly the case that we want to parameterize an

[2] ∂i denotes the neighborhood of feature i in the bipartite graph of features and variables; and thereby indexes the tuple of variables on which the ith feature depends.

unknown path cost function and learn it from data—which can be done exactly by introducing parameters β that weight potential cost features, as in CRFs [9].

2.2 Background on Basic Constrained Optimization

The field of optimization has developed a large amount of methods for non-linear programming—see [22] for an excellent introduction. These existing methods are highly relevant also in the context of path optimization. We cannot review in detail the material here. Instead we summarize, in a very subjective nutshell, a few essential insights from the field of optimization as follows:

(i) The core two issues in unconstrained optimization are *stepsize* and *step direction.*
(ii) Concerning stepsize, solid adaptation schemes with guarantees are line search (backtracking and Wolfe conditions) and trust regions.
(iii) Concerning step direction, the Newton direction is the golden standard. If Hessians are not readily available, try to approximate them (quasi-Newton methods, BFGS) or at least account for correlations of gradients or the search space metric (conjugate gradient, natural gradient). Never use plain gradients or even black-box sampling if there is a chance to be more informed towards Newton directions. The Hessian represents the structure of the problem, analogous to graphical models and factor graphs (see below)—and efficiency requires to exploit such structure.
(iv) There are various ways to address constrained programs by solving a series of unconstrained problems, in particular: log-barrier (interior point), primal-dual-Newton, augmented Lagrangian, and sequential quadratic programming (SQP). If done properly, each of these approaches might lead to comparable performance and the best choice depends on the specifics of the application. Arguably, this choice is less relevant than the previous two points.

As a consequence, in the case of path optimization we need to discuss especially the structure of the problem, that is, the structure of the Hessian. This will be a central topic of this tutorial, and we will discuss how this structure relates to factor graphs and graphical models, and how exploitation of this structure in terms of the respective linear algebra methods is analogous or equivalent to message passing or dynamic programming in such graphical models.

In the case of *un*constrained optimization ($d_g = d_h = 0$), we could directly consider the structure of Newton steps

$$-\nabla^2 f(x)^{-1} \nabla f(x) \tag{6}$$

under our assumptions. However, as we are concerned with a constrained problem we first want to recap standard approaches to constrained optimization and discuss what the implication of these approaches is w.r.t. the structure of the resulting Newton steps. We focus on sequential quadratic programming (SQP) and the augmented Lagrangian (AuLa) method, and only briefly mention standard log barrier and primal-dual Newton methods.

The Newton method steps, in every iteration, towards the optimum of a local 2nd-order Taylor approximation of $f(x)$. Sequential Quadratic Programming (SQP, see [22] for details) is a direct generalization of this: In every iteration we step towards the optimum of a local Taylor approximation of the original constrained problem (1). Concretely, we compute the local 2nd-order Taylor of the objective,

$$f(x+\delta) \approx f(x) + \nabla f(x)^\top \delta + \frac{1}{2}\delta^\top \nabla^2 f(x) \delta , \tag{7}$$

and the local 1st-order Taylor of the constraints,

$$g(x+\delta) \approx g(x) + \nabla g(x)^\top \delta , \quad h(x+\delta) \approx h(x) + \nabla h(x)^\top \delta . \tag{8}$$

This defines the sub-problem

$$\min_\delta \; f(x) + \nabla f(x)^\top \delta + \frac{1}{2}\delta^\top \nabla^2 f(x)\delta \;\text{ s.t. }\; g(x) + \nabla g(x)^\top \delta \le 0 , \quad h(x) + \nabla h(x)^\top \delta = 0 , \tag{9}$$

which can be solved with a standard Quadratic Programming solver. In a robotics context, the computation of the terms $\nabla f(x)$, $\nabla^2 f(x)$, $\nabla g(x)$, $\nabla h(x)$ is typically expensive, requiring to *query* kinematics, dynamics and collision models; but once these terms are computed locally at x, the sub-problem of computing δ^* considers these as constant and does not require further queries. The dimensionality of the sub-problem (9) is though still the same as that of (1). As in ordinary Newton methods, the optimal δ^* only defines a good search direction and we need to backtrack until we found a point that decreases f sufficiently (Wolfe condition) *and* that is feasible—these criteria again require the real kinematics, dynamics and collision models to be queried.

As a general conclusion, an optimizer should try to reduce the number of queries as much as possible by putting much effort in deciding on a good step direction and stepsize. SQP does so by solving the QP (9).

SQP became a standard in robotics. However, we want to also highlight another method that is not as frequently mentioned in the robotics context and not well documented for the inequality case: the augmented Lagrangian (AuLa) method [4, 33]. The method is simple and effective. First consider an imprecise but common practice to handle constraints, namely by adding squared penalty terms. Instead of solving (1) we address

$$F(x) = f(x) + \nu \sum_j h_j(x)^2 + \mu \sum_i [g_i(x) > 0] \, g_i(x)^2 , \tag{10}$$

which adds squared penalties if constraints are violated.[3] $F(x)$ can be efficiently minimized by a standard Gauss–Newton method, which approximates the Hessian of $F(x)$ by $\nabla^2 F(x) \approx \nabla^2 f(x) + \nu \sum_j \nabla h_j(x) \nabla h_j(x)^\top + \mu \sum_i [g_i(x) > 0] \nabla g_i(x) \nabla g_i(x)^\top$.

Because the squared penalties are flat at $h_j = 0$ and $g_i = 0$, minimizing $F(x)$ will lead to constraint violations for the critical (active) constraints. The amount of violation could be controlled by increasing ν and μ. However, there is a very elegant alternative: from the amount of violation we can guess Lagrange parameters that, in the next iteration, push out of constraint violations and "should" lead to satisfied constraints. Concretely, we define the augmented Lagrangian as

$$\hat{L}(x) = f(x) + \sum_j \kappa_j h_j(x) + \sum_i \lambda_i g_i(x) + \nu \sum_j h_j(x)^2 + \mu \sum_i [g_i(x) > 0]\, g_i(x)^2 , \quad (11)$$

which includes both, squared penalties and Lagrange terms.

In the first iteration, $\kappa = \lambda = 0$ and $\hat{L}(x) = F(x)$. We compute $x' = \min_x \hat{L}(x)$, and then reassign Lagrange parameters using the AuLa updates[4]

$$\kappa_j \leftarrow \kappa_j + 2\nu h_j(x') , \quad \lambda_i \leftarrow \max(\lambda_i + 2\mu g_i(x'), 0). \quad (12)$$

Note that $2\nu h_j(x')$ is the force (gradient) of the equality penalty at x', and $\max(\lambda_i + 2\mu g_i(x'), 0)$ is the force of the inequality penalty at x'. What this update does is it considers the forces exerted by the penalties, and translates them to forces exerted by the Lagrange terms in the next iteration. This tries to trade the penalizations for the Lagrange terms. It is straight-forward to prove that, if f, g and h are linear and the same constraints are active in two consecutive iterations, the AuLa update (12) assigns "correct" Lagrange parameters, all penalty terms are zero in the second iteration, and therefore the solution fulfills the first KKT condition after one iteration [33]. The convergence behavior and efficiency is, in practice, very similar to the simple and imprecise squared penalty approach, while it leads to precise constraint handling. Unlike SQP it does not need a QP solver for the sub-problems, but only a robust Gauss–Newton method on $\hat{L}(x)$. For reference, we include a basic robust Newton method in Table 1.

SQP and AuLa are excellent choices for constrained path optimization also because in practice they can be made rather robust against numerically imprecise and non-smooth objective and constraint functions. For instance, the distance between two convex 3D polyhedra is a continuous but only piece-wise smooth function; the gradients and Hessian discontinuously depend on what are the closest points on the polyhedra. Levenberg–Marquardt damping and the Wolfe conditions help to make standard Newton methods still lead to efficient monotone decrease. The log barrier method is an approach to constrained optimization that, in our experience, interferes

[3] $[expr]$ is the indicator function of a boolean expression.

[4] There is little literature on the AuLa updates to handle inequalities. The update rule described here is mentioned in by-passing in [22]; a more elaborate, any-time update that does not strictly require $x' = \min_x \hat{L}(x)$ is derived in [33], which also discusses more literature on AuLa.

Table 1 A basic robust Newton method. Line 3 computes the Newton step $d = -\nabla^2 f(x)^{-1}\nabla f(x)$; in practice, e.g., use the Lapack routine `dposv` to solve $Ax = b$ using Cholesky. The parameter λ controls the Levenberg–Marquardt damping, being dual to trust region methods, and makes the parabola steeper around current x

---

Input: initial $x \in \mathbb{R}^n$, functions $f(x)$, $\nabla f(x)$, $\nabla^2 f(x)$, tolerance θ, parameters (defaults: $\rho_\alpha^+ = 1.2$, $\rho_\alpha^- = 0.5$, $\rho_\lambda^+ = 1$, $\rho_\lambda^- = 0.5$, $\rho_{\text{ls}} = 0.01$)
Output: x
1: initialize stepsize $\alpha = 1$ and damping $\lambda = \lambda_0$
2: **repeat**
3: compute d to solve $[\nabla^2 f(x) + \lambda\mathbf{I}]\, d = -\nabla f(x)$
if $[\nabla^2 f(x) + \lambda\mathbf{I}]$ is not positive definite, increase $\lambda \leftarrow 2\lambda - \sigma_{\text{min}}$
4: **while** $f(x + \alpha d) > f(x) + \rho_{\text{ls}}\nabla f(x)^\top(\alpha d)$ **do** *// line search*
5: $\alpha \leftarrow \rho_\alpha^- \alpha$ *// decrease stepsize*
6: optionally: $\lambda \leftarrow \rho_\lambda^+ \lambda$ and recompute d *// increase damping*
7: **end while**
8: $x \leftarrow x + \alpha d$ *// step is accepted*
9: $\alpha \leftarrow \min\{\rho_\alpha^+ \alpha, 1\}$ *// increase stepsize*
10: optionally: $\lambda \leftarrow \rho_\lambda^- \lambda$ *// decrease damping*
11: **until** $\|\alpha d\|_\infty < \theta$

---

non-robustly with such imprecisions of constraint gradients—presumably because of the extreme conditioning of the barrier functions at convergence.

Primal-dual Newton methods are an equally strong candidate for path optimization as SQP and AuLa, as they share many structural aspects. The primal and dual variables are updated conjointly using Newton steps. Thereby we can equally exploit the structure of the Hessian as we will discuss it in the following. However, for the sake of brevity we do not go into more details of primal-dual Newton methods.

2.3 *The Structure of the Jacobian and Hessian*

We can summarize the previous section by observing that AuLa requires to compute Newton steps of $\hat{L}(x)$,

$$-\Big[\nabla^2 f(x) + \nu \sum_j \nabla h_j(x)\nabla h_j(x)^\top + \mu \sum_i [g_i(x) > 0]\, \nabla g_i(x)\nabla g_i(x)^\top\Big]^{-1}$$
$$(\nabla f(x) + (\kappa + 2\nu)^\top \nabla h(x) + (\lambda + 2\mu I_{[g_i(x)>0]})^\top \nabla g(x)) \;, \quad (13)$$

and SQP will apply Newton steps in one or another way to solve the sub-problem (9), which structurally will involve the same or similar terms as in (13). The efficiency

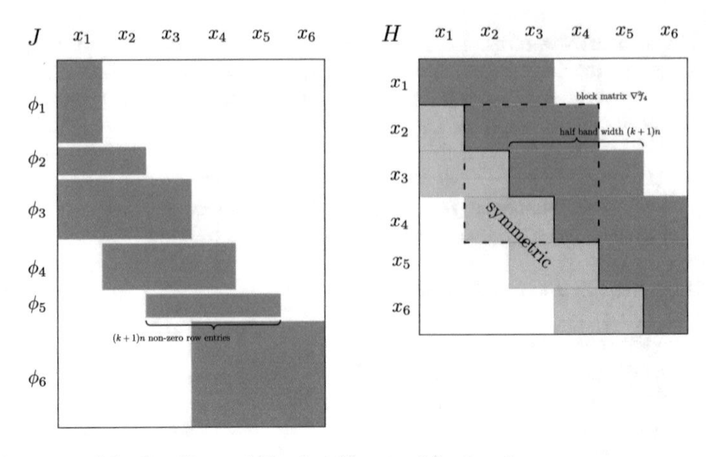

Fig. 2 Structure of the Jacobian and Hessian, illustrated for $k = 2$

of both approaches hinges on how efficiently we can compute such Newton steps, and this depends on the structure of the bracket term.

Going back to our k-order Markov Assumption (2), the Jacobian of the features

$$\phi(x) = \bigotimes_{t=1}^{T} \phi_t(x_{t-k:t}) \,, \quad J(x) = \frac{\partial \phi(x)}{\partial x} \tag{14}$$

reflects the factor graph structure illustrated in Fig. 1. Namely, Fig. 2 shows that the Jacobian is composed of blocks of rows, each one corresponding to a time t, which are non-zero only for those columns that correspond to the tuple $x_{t-k:t}$. Storing the dense Jacobian would require a $Tn \times (T + d_g + d_h)$-dimensional matrix with many zeros. A more natural storage of such a matrix is a *row-shifted packing*, which clips all the leading zeros of a row (shifting them to the left) and stores the number of zeros clipped. This leads to a matrix of at most $(k+1)n$ non-zero columns. Trivially we have:

Lemma 1 *If A is a row-shifted matrix of width l, the product $A^\top A$ is a banded symmetric matrix of band width $2l - 1$.*

Proof Let s_i be the shift (number of clipped zeros) of the ith row of A. Let $B = A^\top A$. We have

$$B_{ij} = \sum_{t=1}^{n} A_{ti} A_{tj} = \sum_{t=1}^{n} A_{ti} A_{tj} [s_t \le i < s_t + l][s_t \le j < s_t + l] \,. \tag{15}$$

If $|i - j| \ge l$, then i and j can never be in the same interval $[s_t, s_t + l)$, and $B_{ij} = 0$. Therefore B has a band width of $2l - 1$.

In (13) the constraints contribute to the approximate Hessian with terms $\nabla h_j(x) \nabla h_j(x)^\top$ and $\nabla g_j(x) \nabla g_j(x)^\top$. Therefore:

Corollary 1 *Under the k-order Markov Assumption, the matrix* $J(x)^\top J(x)$ *with* $J(x) = \frac{\partial \phi(x)}{\partial x}$ *is banded symmetric with width* $2(k+1)n - 1$.

The Hessian $\nabla^2 f(x)$ of the cost features has the structure

$$\nabla^2 f(x) = \sum_{t=1}^{T} \nabla^2 f_t(x_{t-k:t}) \,. \tag{16}$$

Each $\nabla^2 f_t(x_{t-k:t})$ is a $(k+1)n \times (k+1)n$ block matrix, as illustrated in Fig. 2. The sum of these block matrices is again banded symmetric and we have

Corollary 2 *Under the k-order Markov Assumption, the Hessian* $\nabla^2 f(x)$ *is banded symmetric with width* $2(k+1)n - 1$.

2.4 Computing Newton Steps for Banded Symmetric Hessians

In the previous section we established the banded symmetric structure of the Hessian of the augmented Lagrangian. Also when using SQP or other constrained optimization approaches, the Hessian for computing Newton steps in the sub-problems will have this structure, and the efficiency of path optimization will crucially hinge on the efficiency of computing these Newton steps. Specifically, we have:

Lemma 2 *The complexity of computing Newton steps* $-A^{-1}b$ *for a banded symmetric* A *of bandwidth* $2l - 1$ *and* $b \in \mathbb{R}^m$ *is* $O(ml^2)$.

Proof Reference [11] describes in Sect. 4.3.5 explicit Algorithms for computing the Cholesky decomposition of for banded symmetric matrices (Algorihtm 4.3.5) with complexity $O(ml^2)$. Solving the remaining banded triangular system (Algorithm 4.3.2) is $O(ml)$.

As a side note, these algorithms are accessible in LAPACK as `dpbsv`, which internally first computes the Cholesky decomposition using `dpbtrf` and then uses `dpbtrs` to solve the remaining linear equation system.

Corollary 3 *The complexity of computing Newton steps of the form* $[\nabla^2 f(x) + J(x)^\top J(x)]^{-1} b$ *(as for the KOMO problem* (3)*) is* $O(Tk^2n^3)$.

We emphasize that the complexity is only linear in the number T of time steps.

2.5 Sum-of-Square Costs, Gauss–Newton Methods, and the Pullback of Features Space Metrics

The path cost terms $f_t(x_{t-k:t})$ are, in practice, often sums-of-squares. For instance, to get smooth paths we might want to minimize squares of accelerations,

$$\|x_t + x_{t-2} - 2x_{t-1}\|^2 .$$

In optimal control, we typically want to minimize $\|u\|_H^2$ which, using a local approximation $u = M\ddot{q} + F$, implies cost terms

$$\|M(x_t + x_{t-2} - 2x_{t-1})/\tau^2 + F\|_H^2 .$$

If H is Cholesky decomposed as $H = A^\top A$, this is the sum-of-squares of the features $\hat{f}_t(x_{t-k:t}) = A[M(x_t + x_{t-2} - 2x_{t-1})/\tau^2 + F]$. Given a kinematic map $\psi : \mathbb{R}^n \to \mathbb{R}^d$ (e.g., mapping to an endeffector position), we often want to penalize a squared error $\|\psi(x_t) - y_t^*\|_{C_t}^2$ to a target y_t with precision C_t. Again, with a Cholesky decomposition $C_t = A_t^\top A_t$, defining $\hat{f}_t(x_{t-k:t}) = A_t[\psi(x_t) - y_t^*]$ renders this a sum-of-squares cost.

If all cost terms are sum-of-squares of features $\hat{f}_t(x_{t-k:t})$ we have

$$\hat{f}(x) \triangleq \bigotimes_{t=1}^{T} \hat{f}_t(x_{t-k:t}) \tag{17}$$

$$f(x) = \sum_{t=1}^{T} \hat{f}_t(x_{t-k:t})^\top \hat{f}_t(x_{t-k:t}) = \hat{f}(x)^\top \hat{f}(x) \tag{18}$$

$$\nabla f(x) = 2\nabla\hat{f}(x)^\top \hat{f}(x) \tag{19}$$

$$\nabla^2 f(x) = 2\nabla\hat{f}(x)^\top \nabla\hat{f}(x) + 2\hat{f}(x)^\top \nabla^2 \hat{f}(x) . \tag{20}$$

The Gauss–Newton method computes approximate Newton steps by replacing the full Hessian $\nabla^2 f(x)$ with the approximation $2\nabla\hat{f}(x)^\top \nabla\hat{f}(x)$, that is, approximating $\nabla^2 \hat{f}(x) \approx 0$. Note that the pseudo Hessian $2\nabla\hat{f}(x)^\top \nabla\hat{f}(x)$ is always semi-positive definite. Therefore, no problems arise with negative Hessian eigenvalues. The pseudo Hessian only requires the first-order derivatives of the cost features. There is no need for computationally expensive Hessians of features $\hat{f}_t$ or kinematic maps.

It is interesting to add another interpretation of the Gauss–Newton approximation, see also [25]: The mapping $\hat{f} : \mathbb{R}^{Tn} \to \mathbb{R}^{d_f}$ maps a path to a cost feature space. We may think of both spaces as Riemannian manifolds and $\hat{f}$ a differentiable map from one manifold to the other. In the feature space, the cost $f(x)$ is just the Euclidean norm $\hat{f}(x)^\top \hat{f}(x)$, which motivates us to think of the feature space as "flat" and define the Riemannian metric in feature space to be the Euclidean metric. Now, what is a reasonable metric to be defined on the path space? In differential geometry one defines the *pullback* of a metric w.r.t. a differentiable map $\hat{f}$ as

$$\langle x, x' \rangle_X = \left\langle d\hat{f}(x), d\hat{f}(x') \right\rangle_Y \tag{21}$$

where $d\hat{f}$ is the differential of $\hat{f}$ (a $\mathbb{R}^{d_f}$-valued 1-form) and $\langle \cdot, \cdot \rangle_Y$ is a metric in the output space of $\hat{f}$. In coordinates, and if $\langle \cdot, \cdot \rangle_Y$ is Euclidean as in our case, we have

$$\langle x, y \rangle_X = \nabla\hat{f}(x)^\top \nabla\hat{f}(x) \tag{22}$$

and therefore, *the pseudo Hessian* $2\nabla\hat{f}(x)^\top \nabla\hat{f}(x)$ *is the pullback of a Euclidean cost feature metric*. For instance, if some cost features $\hat{f}_t$ penalize velocities in feature space, finding paths x that minimize $f(x)$ corresponds to *computing geodesics* in the configuration space w.r.t. the pullback of a Euclidean feature space metric. If some cost features penalize accelerations (or control costs, as above) in some feature space, the result are geodesics in the system's phase space w.r.t. a pullback metric.

2.6 Relation to Graph-SLAM Methods

Simultaneous Localization and Mapping (SLAM) is closely related to path optimization. Essentially the problem is to find a path of the camera that is consistent with the sensor readings. Graph-SLAM [10, 29] explicitly formulates this problem as an optimization problem on a graph.

Following the conventions of G2O [17], the graph SLAM problem can be reduced to the form

$$\min_x \sum_{(i,j)\in\mathscr{C}} e(x_i, x_j, z_{ij})^\top \Omega_{ij} e(x_i, x_j, z_{ij}) , \tag{23}$$

where $e(x_i, x_j, z_{ij})$ is a "vector-valued error function" that indicates the consistency of states x_i and x_j with constraints z_{ij}. If we decompose the metric $\Omega_{ij} = A_{ij}^\top A_{ij}$ and define $f_{ij}(x) = A_{ij} e(x_i, x_j, z_{ij})$, this becomes a standard structured sum-of-squares problem. For $k = 1$, the KOMO problem (3) *without constraints* becomes a special case of (23), where the graph is just a chain. G2O is a highly-efficient solver for general graph least squares problems.

GTSAM [5] is another solver that allows for higher-order tuples of factors. It adopts a probabilistic interpretation of the problem (as also discussed below), but targets at computing the maximum-likelihood assignment of all random variables, which is equivalent to optimization on a factor graph. Again, unconstrained KOMO is the special k-order Markov case for such general least squares problems. Reference [8] exploit exactly these relations. They demonstrate the efficiency of using GTSAM for motion optimization, in addition to making the relation to Gaussian Processes (see below). As the approach fully exploits the structure of the problem's Hessian, the method is drastically more efficient as compared to other methods.

As a final note, none of the above consider hard constraints as we have them in KOMO. However, using, e.g., the AuLa methods it should not be hard to extend them to include hard constraints.

2.7 *Relation to CHOMP*

Let me briefly recap the notion of a *covariant* gradient of an objective function $f(x)$. The plain partial derivative $\nabla f(x)$ is, strictly speaking, not a vector, but a co-vector. The direction of $\nabla f(x)$ depends on the choice of coordinates. Related to this, $\nabla f(x)$ only describes the *steepest descent direction* w.r.t. a Euclidean metric. In general, the steepest descent direction should be defined depending on the metric as

$$\delta^* = \operatorname*{argmin}_{\delta} \nabla f(x)^\top \delta \quad \text{s.t.} \quad \langle \delta, \delta \rangle = 1 \; .$$

Here we take a step *of length one* and check how much $f(x)$ decreases in its linear approximation. The "*length one*" depends on the metric $\langle \cdot, \cdot \rangle$. If, in given coordinates, the metric is $\langle x, y \rangle = x^\top G y$, with metric tensor G, then one can show that

$$\delta^* \propto -G^{-1} \nabla f(x) \; . \tag{24}$$

It turns our that δ^* is a proper (covariant) vector that does not depend on the choice of coordinates. $\delta^*(x)$ is a *covariant* gradient of $f(x)$, more precisely, it is the covariant gradient w.r.t. the metric G. The Newton step is also a covariant vector: its direction is the covariant gradient of $f(x)$ w.r.t. the metric $H(x)$, that is, the Hessian acts as the local metric.

Covariant gradient descent therefore utilizes a metric in X to make the partial derivative become a covariant gradient. In the context of probability distributions, this metric is typically chosen to be the Fisher metric, also referred to as "natural gradient".

CHOMP [24] chooses the Hessian of smoothing costs as the path metric, and implements steepest descent (24) w.r.t. this metric. This is like a Newton step that drops the Hessian of the other, non-smoothing cost terms. More concretely, as smoothing cost terms CHOMP may, for instance, consider sum-of-squared accelerations $\sum_{t=1}^T f_t^2$ with cost features $\hat{f}_t = x_t + x_{t-2} - 2x_{t-1}$. The Hessian $H = 2\nabla \hat{f}^\top \nabla \hat{f}$ we established above is what CHOMP takes a path metric. In that sense, KOMO or any other classical Newton method generalize CHOMP to also include the Hessian of other cost terms in the Newton step.

However, this particular setting of CHOMP has the benefit that H (the Hessian of acceleration costs) is constant and sparse, making the linear algebra operations of computing quasi-Newton steps fast. Very fast kinematics and collision evaluations (using precomputed distance fields and a set-of-capsules approximation of the robot) further contributed to the performance and success of CHOMP.

3 Including More Global Smoothness Objectives

Smoothness is a basic objective we have about robot motion. Typically, smoothness is implied by minimizing accelerations, control costs, or jerk along a path. While these objectives are local and comply with out local k-order Markov assumption, they still imply a form global smoothness. E.g., it is well-known that B-splines minimize squared accelerations subject to the knot constraints.

However, it is interesting to consider objectives that directly imply a form of global smoothness. We have in particular Gaussian Processes in mind, where the kernel functions directly defines the correlatedness of distal points and thereby the desired form of smoothness. We will show below that such kind of smoothness objectives are not compliant with the k-order Markov assumption, but propose ways to handle them anyway.

Before discussing Gaussian Process smoothness objectives we first consider spline encodings of paths as a means to impose global smoothness.

3.1 Splines

Basis splines, or B-splines, are a simple way to reduce the dimensionality of the path representation. First assume we want to represent a continuous 1D path $x : [0, T] \to \mathbb{R}$ with $K+1$ knots $z_k \in \mathbb{R}, k = 0, .., K$. For a given degree p, let $t_k \in [0, T], k = 0, .., K + p + 1$ be a series of increasing time steps associated with the knots.[5] Then we can compute coefficients[6] $b(t) \in \mathbb{R}^{K+1}$ such that $x(t) = b(t)^\top z$. Therefore, $x(t)$ is linear in the spline parameters z.

We previously defined $x \in \mathbb{R}^{T \times n}$ to be a discrete time path in n-dimensional configuration space. In this case we can compute once the discrete time spline basis matrix $\bar{B} \in \mathbb{R}^{T+1 \times K+1}$ with $B_{t\cdot} = b(t/T)$ and then can represent

$$\bar{x} = \bar{B}\bar{z} , \tag{26}$$

with spline parameters $z \in \mathbb{R}^{K \times n}$. Here, $\bar{x} = x_{0:T}$ and $\bar{z} = z_{0:K}$ include the given start configuration $x_0 = z_0 \in \mathbb{R}^n$. To match better with the previous sections' notation we

---

[5]The time steps can, e.g., be chosen "uniformly" within $[0, T]$, $t_k = T \begin{cases} 0 & k \le p \\ 1 & k \ge K+1 , \\ \frac{k-p}{K+1-p} & \text{otherwise} \end{cases}$ which also assigns $t_{0:p} = 0$ and $t_{K+1:K+p+1} = T$, ensuring that $x_0 = z_0$ and $x_T = z_K$.

[6]The coefficients can be computed recursively. We initialize $b_k^0(t) = [t_k \le t < t_{k+1}]$ and then compute recursively for $d = 1, .., p$

$$b_k^d(t) = \frac{t - t_k}{t_{k+d} - t_k} b_k^{d-1}(t) + \frac{t_{k+d+1} - t}{t_{k+d+1} - t_{t+1}} b_{k-1}^{d-1}(t) , \tag{25}$$

up to the desired degree p, to get $b(t) \equiv b_{0:K}^p(t)$.

rewrite this as

$$x = Bz + bx_0^\top , \tag{27}$$

where $B = \bar{B}_{1:T,1:K}$ and $b = B_{1:T,0}$.

In conclusion, spline representations provide a simple linear re-representation of paths. In the spline representation, the feature Jacobian and Hessian are

$$J_z = J_x B \tag{28}$$
$$H_z = B^\top H_x B , \tag{29}$$

where J_x and H_x are the feature Jacobian and Hessian in the original path space.[7] Note that the spline basis matrix is also structured in a "banded", or row-shifted, manner, similar to the feature Jacobian. Namely,

$$b(t)_k \neq 0 \quad \Rightarrow \quad t_k \le t \le t_{k+p+1} \tag{30}$$
$$B_{tk} \neq 0 \quad \Rightarrow \quad \frac{T(k-p)}{K+1-p} \le t \le \frac{T(k+1)}{K+1-p} \tag{31}$$
$$\Leftrightarrow \quad (K+1-p)\, t/T - 1 \le k \le (K+1-p)\, t/T + p \,. \tag{32}$$

So the non-zero width of each row is $p+2$, and the non-zero height of each column is $T(p+2)/(K+1-p)$.

Corollary 4 *In a spline representation of degree p, the Hessian $B^\top H_x B$ has bandwidth $O(kpn)$.*

It is imperative to exploit this kind of sparsity of the spline basis matrix to ensure that the complexity of the matrix multiplication $J_x B$ in (28) is only $O(dknp)$ (recall, d is the number of features) instead of $O(dTnK)$. Equally, computing $H_x B$ in (29) is $O(Tn^2kp)$.

Now, does such a lower-dimensional spline representation of paths speed up Newton methods? We first note

Corollary 5 *The computational complexity of computing J_z is $O(dknp)$, of H_z is $O(Tn^2kp)$, of a Newton step $H_z^{-1}z$ is $O(Kn(kpn)^2)$.*

Overall, the complexity w.r.t. T is dominated by the computation of J_z and H_z and gives $O(T)$; and w.r.t. n, k it is dominated by the Newton step giving $O(k^2n^3)$. Both are exactly as for the original Newton step without spline representation. Note that also line search steps (e.g., checking the Wolfe condition) is $O(T)$ in both representations as the whole path needs to be evaluated.

If the complexity of computing Newton steps is not reduced in the spline representation, perhaps we need less iterations? We first note that Newton steps are

[7] As x is a matrix, J_x is, strictly speaking, a tensor and the above equations are tensor equations in which the t index of B binds to only one index of J_x and H_x.

covariant, that is, their direction is invariant under coordinate transforms. Therefore, if B *would* have full rank, the Newton steps are identical. Performing Newton steps on a lower-dimensional linear projection B is the same as projecting the high-dimensional Newton steps onto the low-dimensional hyperplane. There is no a priori reason for why this should lead to less iterations.

In conclusion, optimizing on a low-dimensional spline representation of the path does not necessarily lead to more efficient optimization. Empirically we often find that more Newton iterations are needed in a spline representation where the found path is less optimal.

Nevertheless, splines are a standard approach in path optimization. Perhaps the real motivation for using splines in practice is that they impose a large-scale smoothness on the solution which cannot efficiently be captured by cost features on $k+1$-tuples $x_{t:t+k}$. However, let us consider alternative approaches to large-scale smoothness in the following section.

3.2 Covariance Smoothness Objectives and Gaussian Process Priors

The k-order Markov structure allows us to express smoothness objectives in terms of cost features over the kth path derivatives. Such local smoothness objectives are different to global smoothness constraints as implied by spline projections, or the kind of smoothness implied by Gaussian Process (GP) priors.

Considering the latter, for discretized time, a GP is nothing but a joint Gaussian over all path points. For instance, a GP represents the prior

$$P(x) = \mathcal{N}(x|0, K) \propto \exp\{-\frac{1}{2}x^\top K^{-1}x\}\ , \quad K_{ts} = k(t, s)\ , \tag{33}$$

where the kernel function $k(t, s)$ is the correlation between the configurations x_t and x_s at two different times t and s. A typical kernel function used in GPs is the squared exponential kernel $k(t, s) = \exp\{-(t - s)^2/\sigma^2\}$ for some band width σ. Figure 3(left) illustrates such a covariance matrix K in gray shading.

In our optimization context, such a GP prior translates to neg-log-probability costs, namely

$$-\log P(x) \propto \frac{1}{2}x^\top K^{-1}x\ . \tag{34}$$

Note the matrix inversion here! Fig. 3(right) illustrates the matrix K^{-1}, which turns out to be in no way 'local' or banded. This precision matrix K^{-1} plays the role of a Hessian in the cost formulation. The checker board structure can vaguely be understood as penalizing *derivatives* of the path. The rather surprising non-local structure of K^{-1} clearly breaks our k-order Markov assumption. However, it turns

Fig. 3 *Left* The 20 × 20 covariance matrix $K_{ij} = \exp\{-((i-j)/3)^2\}$ (zero = *white*). *Right* its inverse (precision matrix) K^{-1} (zero = *gray*)

out that we can still compute Newton steps efficiently, in a manner that exploits the structure of K. To derive this, let us more formally define the generalized problem as

Definition 2 (*Covariance regularized KOMO (CoKOMO)*)

$$\min_x \sum_t f_t(x_{t-k:t}) + \frac{1}{2} x^\top K^{-1} x \quad \text{s.t.} \quad \forall_t : \; g_t(x_{t-k:t}) \le 0 \,, \quad h_t(x_{t-k:t}) = 0 \tag{35}$$

We define, as before, $H = \sum_t \nabla^2 f_t$ as the Hessian of the cost features, or $H = \nabla \hat{f}^\top \nabla \hat{f}$ in the Gauss–Newton case. The system's full Hessian is $H + K^{-1}$. Therefore

Corollary 6 *In CoKOMO, for a finite-support kernel, the total Hessian* $\bar{H} = H + K^{-1}$ *is a sum of a banded matrix* H *and the* inverse *of a banded matrix* K.

Computing a Newton step of the form $-\bar{H}^{-1} g$ for some g[8] can be tackled as follows

$$(H + K^{-1})^{-1} g = K(HK + I)^{-1} g \,. \tag{36}$$

Note that, if H and K are both banded, then $(HK + I)$ is banded and computing $(HK + I)^{-1} g$ is, exactly as before, $O(Tnb^2)$ if b is the bandwidth of HK. We have

Lemma 3 *If* H *is of semi-bandwidth* h *(that is, total bandwidth* $2h - 1$*) and* K *is of semi-bandwidth* c*, then* HK *is of semi-bandwidth* $h + c$.

[8] In the AuLa case, $g = \nabla \hat{L}(x)$, see Eq. (12). In the SQP case, the inner loop for solving the QP (9) would compute Newton steps w.r.t. the Hessian $\bar{H}$.

Proof

$$(HK)_{ij} = \sum_k H_{ik} K_{kj} = \sum_k [-h \le i-k \le h][-c \le j-k \le c]\, H_{ik} K_{kj} \tag{37}$$

$$= [-h-c \le i-j \le h+c] \sum_k H_{ik} K_{kj} \,. \tag{38}$$

Corollary 7 *Under the k-order Markov Assumption and including a banded covariance regularization of semi-bandwidth cn, the complexity of computing Newton steps of the form* $-(H+K^{-1})^{-1} g$ *is* $O(T(k+c)^2 n^3)$.

This is in comparison to the $O(Tk^2n^3)$ without the covariance regularization. We assumed a semi-bandwidth cn for K to account for the dimensionality of each $x_t \in \mathbb{R}^n$.

As a side note, the Woodbury identity and rank-one update (Sherman–Morrison formula) provide alternatives ideas to handle terms like $(H+K^{-1})^{-1}$, namely

$$(H+K^{-1})^{-1} = K - (I+KH)^{-1} KHK \tag{39}$$

$$(vv^\top + K^{-1})^{-1} = K - \frac{Kvv^\top K}{1+v^\top K v} \,. \tag{40}$$

The first line (Woodbury) involves only banded matrices, but seems less efficient than (36). The second line (Sherman–Morrison) provides a way to recursively compute $(H+K^{-1})^{-1}$ as a series of rank-one updates if $H = \sum_i v_i v_i^\top$—as is exactly the case in the Gauss–Newton approximation $H \approx 2\nabla\hat{f}^\top \nabla\hat{f}$. Again, all computations only rely on multiplication with banded matrices.

4 The Optimal Control Perspective

So far we have not mentioned controls at all. However, path optimization and KOMO are intimately related to standard optimal control methods. The aim of this section is two-fold, namely to clarify these relations as well as to derive algorithms for controlling a system around an optimal path.

Our starting point will be the discussion of an alternative solution approach to our optimization problem: a dynamic programming perspective on solving the general KOMO problem (3). This will be rather straight-forward, adapting Bellman's equation to the k-order constrained case, and leads to an optimal regulator around the path. This though leads to many insights:

(i) Using a 2nd-order approximation of all terms, the backward equation can be used to compute a Newton step—which now very explicitly shows the linear-in-T complexity of computing Newton steps and gives interesting insights in how the inversion of a banded symmetric matrix is related to dynamic programming on Markov chains.

(ii) Assuming a $k = 1$-order linear-quadratic control process, the 2nd-order approximate backward equation coincides with the Riccati equation. This gives insights in the tight interrelations between DDP, iLQG and Newton methods.

(iii) Moving to a probabilistic interpretation of the objective function we can connect to the recent work on using probabilistic inference methods for optimal control [26]. In particular, backward and forward dynamic programming in our KOMO problem become equivalent to backward and forward message passing in Markov chains. Based on this we can point to the relations with path integral control methods, AICO, Ψ-learning, Expectation Maximization and eNAC that are detailed in [26].

4.1 Background on Basic Optimal Control

Let us first recap the basic formulation of optimal control problems. In the discrete time setting, we consider a controlled system $x_{t+1} = f(x_t, u_t)$ and aim to minimize

$$\min_{x,u} \sum_{t=1}^T c_t(x_t, u_t) \quad \text{s.t.} \quad x_{t+1} = f(x_t, u_t) \ . \tag{41}$$

Here we optimize over both, the state path $x = x_{1:T}$ and the control path $u = u_{1:T}$. Both are of course related by the system dynamics. Given a control path u we can compute the state path $x = F(u)$ as a function of the start state and the controls by iterating the dynamics $f(x_t, u_t)$. The control problem can be recast as

$$\min_{u} \sum_{t=1}^T c_t(F(u)_t, u_t) \ , \tag{42}$$

and is typically solved by iteratively finding a better control path u (e.g. by a Newton step on u, or by dynamic programming, see below) and then recomputing the state path $x = F(u)$. This is called *indirect method* or multiple shooting. DDP and iLQG, which we discuss below, are such indirect methods.

This is in contrast to direct methods which instead consider x to be the optimization variable. Roughly, let $u(x_t, x_{t+1})$ be the control needed to transition from x_t to x_{t+1}. In non-holonomic systems, where not all transitions are feasible, let $h(x_t, x_{t+1}) = 0$ express an equality constraint that ensures the existence of a control signal $u(x_t, x_{t+1})$. Then the problem can be recast as

$$\min_{x} \sum_{t=1}^T c_t(x_t, u(x_t, x_{t+1})) \quad \text{s.t.} \quad h(x_t, x_{t+1}) = 0 \ . \tag{43}$$

Such *direct* methods eliminate the controls from the problem. Our KOMO formulation is therefore a direct method.

The *dynamic programming* approach to solving such problems is to define the optimal cost-to-go function (or value function). In the indirect view (see below for the Bellman equation in the direct view) we define

$$V_t(x) = \min_{u_{t:T}} \sum_{s=t}^T c_s(x_s, u_s) , \tag{44}$$

which, for every possible $x_t = x$, computes the optimal (minimal) cost for the remaining path. Bellman's optimality equation can be derived by separating out the optimization over the next control u_t,

$$V_t(x) = \min_{u_t} \Big[c_t(x, u_t) + \min_{u_{t+1:T}} \sum_{s=t+1}^T c_s(x_s, u_s) \Big] \tag{45}$$

$$= \min_{u_t} \Big[c_t(x, u_t) + V_{t+1}(f(x, u_t)) \Big] . \tag{46}$$

In a nutshell, the core implications of Bellman's equation are

(i) In principle we can compute all V_t recursively, iterating backward from $V_{T+1} \equiv 0$ to V_1 using Eq. (46). To retrieve the optimal control path u and state path x we then iterate forward

$$u_t = \operatorname*{argmin}_{u_t} \Big[c_t(x_t, u_t) + V_{t+1}(f(x, u_t)) \Big] , \quad x_{t+1} = f(x_t, u_t) , \tag{47}$$

starting from the start state x_1. This forward iteration is called *shooting*. Therefore, if we can compute all V_t exactly, we can solve the optimization problem.

(ii) In the LQ case, where $f(x, u) = Ax + Bu$ is linear and $c(x, u) = x^\top Qx + u^\top Hu$ is quadratic, all cost-to-go functions are quadratic of the form $V_t(x) = x^\top \hat{V}_t x$ and the minimization in the Bellman Eq. (46) is analytically given by the *Riccati equation*

$$\hat{V}_t = Q + A^\top [\hat{V}_{t+1} - \hat{V}_{t+1} B (H + B^\top \hat{V}_{t+1} B)^{-1} B^\top \hat{V}_{t+1}] A , \quad \hat{V}_{T+1} = 0 . \tag{48}$$

Given we computed all $\hat{V}_t$, the optimal controls for forward shooting (47) are

$$u_t = (H + B^\top \hat{V}_{t+1} B)^{-1} B^\top \hat{V}_{t+1} A x_t , \tag{49}$$

which is call the *Linear Quadratic Regulator*. The fact that we have this optimal regulator defined globally for all possible x_t adds a fully new perspective: We can not only use it for forward shooting to find the optimal path, but we can also use it during execution as a controller to react to perturbations of our system from the planned path.

(iii) The LQ case is the analogy to the 2nd-order Taylor approximation of a non-linear objective function: To solve a non-LQ control problem one typically starts with an initial path x, approximates the dynamics and costs to 1st- or

2nd-order around x, and then solves this locally approximate problem to yield a new, better path x. There are some alternatives on how to do this in detail:

- If we approximate all terms exactly up to 2nd-order, compute all V_t in (46) and also use this second order approximation for the forward shooting (47), then this is *exactly* equivalent to a Newton step on the path optimization problem: We approximated all terms up to 2nd order and found the minimum of that local approximation. However, this is not what typical control methods do:
- If we use 2nd-order approximations to compute all V_t in (46), but then the true non-linear dynamics for forward shooting in (47), this is referred to as Differential Dynamic Programming (usually formulated in continuous time) [20, DDP;].
- If we use an LQ-approximation (which neglects the dynamic's Hessian) to compute all V_t using the Riccati equation, but then the true non-linear dynamics for forward shooting in (47), this is referred to as iterative LQG [30, iLQG;].

Both, DDP and iLQG have additional details on how exactly to ensure convergence, analogous to Levenberg–Marquardt damping and backtracking in a robust Newton method. The fine difference to Newton's method has its origin in the fact that they are indirect methods, and therefore can use the exact non-linear dynamics in the forward shooting [19]. For very non-linear systems this may be beneficial [30].
In all three cases, the computed V_t in principle define a linear regulator around the path, which, however, does not guarantee to keep the system close to the state region where the local approximation is viable. This can be addressed using Model Predictive Control (MPC) as discussed below.

With this background, let us first discuss a (direct) dynamic programming approach to solve the KOMO problem, and then compare to standard LQG, DDP and iLQG methods.

4.2 k-Order Constrained Dynamic Programming and Constrained LQR Control

For easier reference we restate the general KOMO problem (3),

$$\min_x \sum_{t=1}^T f_t(x_{t-k:t}) \quad \text{s.t.} \quad \forall_{t=1}^T : \quad g_t(x_{t-k:t}) \le 0 \,, \quad h_t(x_{t-k:t}) = 0 \,. \tag{3}$$

Following the dynamic programming principle we define a value function over a *separator*[9] $x_{t-k:t-1}$,

Definition 3 (*k-order constrained Dynamic Programming (KODP)*)

$$J_t(x_{t-k:t-1}) \triangleq \min_{x_{t:T}} \sum_{s=t}^{T} f_s(x_{s-k:s}) \quad \text{s.t.} \quad \forall_{s=t}^{T}: \; g_s(x_{s-k:s}) \le 0\,, \; h_s(x_{s-k:s}) = 0\,, \tag{50}$$

$$= \min_{x_t} \Big[f_t(x_{t-k:t}) + J_{t+1}(x_{t-k+1:t}) \Big] \quad \text{s.t.} \quad g_t(x_{t-k:t}) \le 0\,, \; h_t(x_{t-k:t}) = 0\,, \tag{51}$$

$$J_{T+1} \triangleq 0\,. \tag{52}$$

Such k-order constrained Bellman equations are comparatively rare in the literature, but straight-forward and mentioned already by Bellman in the 50's [1]. See also [7]. Reference [28] presented a DP approach for the special case with constraints on the controls only. Solving the general non-linear constrained case, computing $J_t(x_{t-k:t-1})$ for all $x_{t-k:t-1}$, is infeasible.

If, as in DDP and SQP, we approximate all cost terms f_t as second order and constraints g_t, h_t in first order, [2] shows an explicit derivation of an optimal constrained LQR (C-LQR) controller. The computation is complex and the resulting C-LQR is piece-wise linear and continuous, where the pieces correspond to constrained activities of the underlying QP. Reference [2] emphasize the benefit of computing such optimal constrained regulators offline, for all $x_{t-k:t-1}$, rather than requiring a fast local MPC within the control loop to solve the resulting QP for the current $x_{t-k:t-1}$.

An alternative approximation to the problem (50) is to not only linearize around an optimal path, but also adopt the Lagrange parameters of the optimal path [1]. This clearly is not optimal, as the true path might hit constraints other than the optimal path and therefore require different Lagrange parameters. But it lends itself to a simple regulator that also, using a one-step-lookahead, is guaranteed to generate feasible paths.

For *fixed* Lagrange parameters κ_t, λ_t, the dynamic programming principle for the Lagrangian is

$$\tilde{J}_t(x_{t-k:t-1}) \triangleq \min_{x_{t:T}} \sum_{s=t}^{T} f_s(x_{s-k:s}) + \lambda_s^\top g_s(x_{s-k:s}) + \kappa_s^\top h_s(x_{s-k:s}) \tag{53}$$

$$= \min_{x_t} \Big[f_t(x_{t-k:t}) + \lambda_t^\top g_t(x_{t-k:t}) + \kappa_t^\top h_t(x_{t-k:t}) + \tilde{J}_{t+1}(x_{t-k+1:t}) \Big]\,. \tag{54}$$

This can efficiently be computed in the LQ approximation, see below. Given $J_t(x_{t-k:t-1})$ for all t, we define

[9] We use the word separator as in Junction Trees: a separator makes the sub-trees conditionally independent. In the Markov context, the future becomes independent from the past conditional to the separator.

Definition 4 *(Approximate (fixed Lagrangian) constrained LQR (acLQR))*

$$\pi_t : x_{t-k:t-1} \mapsto \operatorname*{argmin}_{x_t} \Big[f_t(x_{t-k:t}) + \tilde{J}_{t+1}(x_{t-k+1:t}) \Big]$$
$$\text{s.t.} \quad g_t(x_{t-k:t}) \le 0, \; h_t(x_{t-k:t}) = 0 \,. \tag{55}$$

Note that to determine the controls at time step t, we release the Lagrange parameters again and hard constrain w.r.t. g_t and h_t. Only the Lagrange-cost-to-go function $\tilde{J}_{t+1}(x_{t-k+1:t})$, computed via (53), employs the fixed Lagrange parameters. If for all t a feasible x_t is found, the whole path is guaranteed to be feasible.

To compute $\tilde{J}_t(x_{t-k:t-1})$ in the fixed Lagrange parameter case (53), the Lagrange terms can be absorbed in the cost terms f_s. To simplify the notation let us therefore focus only the unconstrained k-order dynamic programming case,

$$\tilde{J}_t(x_{t-k:t-1}) = \min_{x_t} \Big[\tilde{f}_t(x_{t-k:t}) + \tilde{J}_{t+1}(x_{t-k+1:t}) \Big] , \quad \tilde{J}_{T+1} = 0 \,, \tag{56}$$

where $\tilde{f} = f + \lambda^\top g + \kappa^\top h$, for all indices.

In the quadratic approximation we assume

$$\tilde{J}_t(x) = x^\top V_t x + 2 v_t^\top x + \bar{v}_t \tag{57}$$

$$\tilde{f}_t(x) \approx \nabla \tilde{f}_t(x^*)^\top (x - x^*) + \frac{1}{2}(x - x^*)^\top \nabla^2 \tilde{f}_t(x^*)(x - x^*) + \tilde{f}_t(x^*) \,. \tag{58}$$

To derive an explicit minimizer x_t in (56) we write the 2nd-order polynomial in block matrix form

$$\Big[\tilde{f}_t(x_{t-k:t}) + \tilde{J}_{t+1}(x_{t-k+1:t}) \Big] \triangleq \begin{pmatrix} x_{t-k:t-1} \\ x_t \end{pmatrix}^\top \begin{pmatrix} D_t & C_t \\ C_t^\top & E_t \end{pmatrix} \begin{pmatrix} x_{t-k:t-1} \\ x_t \end{pmatrix} + 2 \begin{pmatrix} d_t \\ e_t \end{pmatrix}^\top \begin{pmatrix} x_{t-k:t-1} \\ x_t \end{pmatrix} + c_t \,, \tag{59}$$

where the components $D_t, E_t, C_t, d_t, e_t, c_t$ are trivially defined in terms of $\nabla^2 \tilde{f}_t(x^*)$, $\nabla \tilde{f}_t(x^*)$, $\tilde{f}_t(x^*)$, V_{t+1}, v_{t+1} and $\bar{v}_{t+1}$. Then

$$x_t^* = \operatorname*{argmin}_{x_t} \Big[\tilde{f}_t(x_{t-k:t}) + \tilde{J}_{t+1}(x_{t-k+1:t}) \Big] = -E_t^{-1}(C_t^\top x_{t-k:t-1} + e_t) \tag{60}$$

$$V_t = D_t - C_t E_t^{-1} C_t^\top \,, \quad v_t = d_t - C_t E_t^{-1} e_t \,, \quad \bar{v}_t = c_t - e_t^\top E_t^{-1} e_t \,. \tag{61}$$

To get more intuition about this equation, let us first discuss the Riccati equation as special case.

4.3 Sanity Check in the LQG Case and Relation to DDP and iLQG

Let us assume $k = 1$ (a standard Markov chain) and standard linear control of a holonomic system,

$$x_t = Ax_{t-1} + Bu_{t-1} \,, \quad J_t(x_{t-1}) = \min_{x_{t:T}} \sum_{s=t}^{T} \|x_s - Ax_{s-1}\|^2_{\hat{H}} + \|x_s\|^2_Q \tag{62}$$

with $\hat{H} = B^{-\top} H B^{-1}$. Identifying $f_t(x_{t-1:t}) = \|x_t - Ax_{t-1}\|^2_{\hat{H}} + \|x_t\|^2_Q$ we have

$$\nabla f_t = 2\begin{pmatrix} -A^\top \\ 1 \end{pmatrix} \hat{H}(x_t - Ax_{t-1}) + 2\begin{pmatrix} 1 \\ 0 \end{pmatrix} Qx_t \,, \quad \nabla^2 f_t = 2\begin{pmatrix} A^\top \hat{H} A & -A^\top \hat{H} \\ -\hat{H} A & \hat{H} + Q \end{pmatrix} \tag{63}$$

$$D_t = A^\top \hat{H} A \,, \quad E_t = \hat{H} + Q + V_{t+1} \,, \quad C_t = -A^\top \hat{H} \tag{64}$$

$$V_t = A^\top \Big[\hat{H} - \hat{H}(\hat{H} + Q + V_{t+1})^{-1} \hat{H} \Big] A = A^\top \Big[(\hat{H}^{-1} + \hat{V}^{-1})^{-1}) \Big] A \tag{65}$$

$$= A^\top \Big[\hat{V} - \hat{V}(\hat{H} + \hat{V})^{-1} \hat{V}) \Big] A \,. \tag{66}$$

where $\hat{V} = Q + V_{t+1}$ and the last lines use the Woodbury identity $(A^{-1} + B^{-1})^{-1} = A - A(A + B)^{-1} A$ twice. The last line is the classical Riccati equation for $\hat{V}$.

This was just a sanity check, confirming that in the unconstrained LQ-case, the DP equation (53) reduces to the standard Riccati equation. Let us recap what we have found:

(i) We know that in the unconstrained LQ case, or KOMO problem is just an unconstrained quadratic program, where the first Newton step directly jumps to the optimum.
(ii) One way to compute this Newton step (or optimum) is via the methods we described in the first part of the paper where we emphasized the importance of the structure of the Hessian as a banded symmetric matrix, allowing for the complexity $O(Tk^2n^3)$ of computing Newton steps under the KOMO assumption. We derived this complexity by looking at the respective matrix operations, in particular the implicit Cholesky decomposition.
(iii) We have now seen a second way to compute the optimum, by recursing backward the explicit DP equation (53), or (61) in the LQ approximation, which equally has complexity $O(Tk^2n^3)$. This establishes an explicit relation between matrix inversion and the dynamic programming case.
(iv) If these methods are applied to local LQ approximations of a non-linear problem, the Newton step and Riccati equation lead to the same next iterate, that is, the same next path. In that view, the standard indirect multiple shooting methods DDP and iLQG can be viewed as Newton methods that use the Riccati equation (or DDP's equation) to compute Newton steps instead of banded

matrix inversion. Both algorithms also require step size controlling, such as Levenberg–Marquardt, to become robust.

(v) However, as mentioned already in Sect. 4.1, DDP and iLQG are different to Newton steps in one respect: Both use a Riccati sweep or 2nd-order Taylor approximations to compute the next *control path* u. However, the control path u is then used to compute the next *state path* $x = F(u)$ using the exact forward dynamics.
If we wanted to get equivalent iterates using Newton steps we would have to: (1) compute the next state path x using a Newton step, (2) compute the control path u for x, (3) use the exact non-linear dynamics $x' = F(u)$ to compute a corrected state path x'.

This clarifies the tight relations between classical DDP and iLQG and Newton steps in KOMO. A further technical difference is that in KOMO we can alternatively represent the problem as a $k = 2$-order process on the configuration variables, instead of as a $k = 1$-order process in phase space, which may be numerically more stable. Hard constraints have been considered in iLQG only for the special case with constraints on the controls [28]. The particular k-order constrained Dynamic Programming (50) has, to our knowledge, not been proposed before.

4.4 Constrained Model Predictive Control and Staying Close to the Reference

Stochasticity (or un-modelled additional aspects such as control delay or motor controller properties) will always lead us away from the optimal path. Depending how far we are off, the 2nd-order approximations we used to optimize the path and derive the acLQR around the path become imprecise and might lead to even more deviation from the optimal path. The standard approach to compensate for this is Model Predictive Control (MPC) (see, e.g., [6]).

In MPC we solve, in real time, at every time step t a finite horizon trajectory optimization problem given the concrete current state x_t. This finite horizon problem will also be non-linear and require local 2nd-order approximations, but these approximations are computed at the true x_t. When an optimal path x^* was precomputed, the finite-horizon MPC problem can be defined as finding controls that steer back to the reference path, e.g., $\min_{x_{t:t+H}} \|u\|^2_H$ s.t. $x_{t+H} = x^*_{t+H}$. However, MPC can also be viewed as an H-step lookahead variant of the optimal controller we get from the Bellman equation. In this view our acLQR (55) is a 1-step MPC. We can more generally define

Definition 5 (*Approximate (fixed Lagrangian) constrained MPC Regulator (acMPC)*)

$$\pi_t : x_{t-k:t-1} \mapsto \underset{x_{t:t+H-1}}{\operatorname{argmin}} \Big[\sum_{s=t}^{t+H-1} f_s(x_{s-k:s}) + \tilde{J}_{t+H}(x_{t+H-k:t+H-1}) + \rho \|x_{t+H-1} - x^*_{t+H-1}\|^2 \Big]$$
$$\text{s.t. } \forall_{s=t}^{t+H-1} : g_s(x_{s-k:s}) \le 0,\ h_s(x_{s-k:s}) = 0 \,. \tag{67}$$

Let's neglect the ρ-term first. For $H = 1$ this optimizes over only one configuration, x_t, and reduces to the acLQR (55) that relies on the (fixed Lagrangian) cost-to-go estimate $\tilde{J}_{t+1}$. For $H = T - t + 1$, acMPC becomes the full, long-horizon path optimization problem until termination T.

In typical applications, that is, for typical choices of the original KOMO problem (3) there is a caveat: Very often the objectives concern control costs and costs/constraints associated with the final state x_T only. The effect is that the value functions $\tilde{J}_{t+H}$ have, for $t \ll T$, very low eigenvalues. The resulting "gains" of the acLQR or acMPC will therefore also be very low. If the real system was linear, this would not be a problem—the Riccati equation tells us that this low-gain controller is optimal globally no matter how far we perturbed from the reference. However, in practice, for non-linear and imprecisely modeled systems, this would lead to a large and undesirable drift away from the reference, making the precomputed x^* and its local linearizations irrelevant, and be non-robust against small model errors.

The standard way to enforce staying closer to the reference during execution is to add the ρ-term to enforce steering back to the reference at horizon H. The second option is to introduce additional penalties $\tilde{f}_t \leftarrow \tilde{f}_t + \rho\|x_t - x^*_t\|^2$ for deviations in every time step[10] and use this $\tilde{f}_t$ in the backward dynamic programming (53) to compute value functions $\tilde{J}_t$ for the KOMO problem with cost terms $\tilde{f}_t$. Using such MPC we get robust trajectory tracking and can tune the stiffness of tracking by adjusting ρ and H.

4.5 Probabilistic Interpretation and the Laplace Approximation

Let us neglect constraints and consider problems of the form

$$\min_x f(x) \,, \quad f(x) = \sum_t f_t(x_{t-k:t}) \,. \tag{68}$$

There is a natural relation between cost (or "neg-energy", "error") and probabilities. Namely, if $f(x)$ denotes a cost for state x—or an error one assigns to choosing

[10]Note the relation to Levenberg–Marquardt regularization.

x—and $p(x)$ denotes a probability for every possible value of x, then a natural relation[11] is

$$P(x) \propto e^{-f(x)} , \quad f(x) = -\log p(x) . \tag{69}$$

Given a problem of the form (68) we may therefore define a distribution over paths as

$$P(x) \propto \prod_t \exp\{-f_i(x_{t-k:t})\} . \tag{70}$$

It is interesting to investigate how this probability distribution over paths is related to finding the optimal path, and to stochastic optimal control under the respective costs. Note that in the optimal control context ($k = 1$), $f_t(x_{t-1:t})$ subsumes control costs and state costs, e.g., $f_t(x_{t-1:t}) = \|u\|_H^2 + \|x_t\|_R^2$ where $u = u(x_{t-1}, x_t)$.

References [26, 31] discuss an approach to stochastic optimal control that considers the distribution

$$P(x_{0:T}, u_{0:T}) \propto P(x_0) \prod_{t=0}^{T} P(x_{t+1}|x_t, u_t)\, \pi(u_t|x_t)\, \exp\{-\eta c_t(x_t, u_t)\} . \tag{71}$$

Here, in contrast to (70), this is the joint distribution over controls and states. Reference [26] discuss in detail how inference, or more precisely, minimizing KL-divergences under such probabilistic models generalizes previous approaches such as path integral control methods [14], Approximate Inference Control [31], but also model-free Expectation Maximization and eNAC policy search methods [23, 35].

In all these contexts, a central problem of interest is to approximate the marginals of the path distribution (71). Above we already established the equivalence of DP programming and Newton steps in an LQ setting. Message passing in Gaussian factor graphs is generally equivalent to DP with squared cost-to-go functions. Typically one distinguishes between DP, which computes cost-to-go functions, and the forward unrolling of the optimal controller, to compute the optimal path. This can be viewed more symmetrically: computing optimal cost-to-go and cost-so-far functions forward and backward (or cost-to-go functions for all branches of a tree) equally yields the optimal path.

If the factor graph is not Gaussian, or the objective not 2nd-order polynomial, message passing as well as DP are approximated. Again, using Gaussian approximate message passing—e.g., as in extended Kalman filtering and smoothing—is equivalent to approximating the cost-to-go function locally as quadratic [31]. In

[11] Why is this a natural relation? Let us assume we have $p(x)$. We want to find a *cost* quantity $f(x)$ which is some function of $p(x)$. We require that if a certain value x_1 is more likely than another, $p(x_1) > p(x_2)$, then picking x_1 should imply less cost, $f(x_1) < f(x_2)$ (Axiom 1). Further, when we have two independent random variables x and y *probabilities are multiplicative*, $p(x, y) = p(x)p(y)$. We require that, for independent variables, *cost is additive*, $f(x, y) = f(x) + f(y)$ (Axiom 2). From both follows that f needs to be a logarithm of p.

conclusion, iterative Gaussian message passing to estimate marginals of (71) is very closely related to iterative DP using local LQG approximations and Newton methods to minimize (68).

So what are the motivations for the mentioned family of methods that build on the probabilistic interpretation? Essentially it is specific ideas on how exactly to do the approximation that arise from the probabilistic view, other than the Laplace approximation. For instance, in the probabilistic setting Gaussian messages can also be approximated using the Unscented Transform [12], or Expectation Propagation [21]. These are slightly different to local Laplace approximations. Importantly, if the path distribution cannot well be approximated as Gaussian, esp. because it is multi-modal, the probabilistic view provides alternative approaches to approximation, for instance, sampling from the path distribution [13]. Here we see that the goal of optimal control under a multi-modal path distribution really deviates from just computing an optimal path.

Incorporating hard constraints in approximate message passing is hard. In the context of Gaussian messages, truncated Gaussians could be used to approximate hard constraints [32]. However, in our experience this is far less precise and robust than using Lagrangian methods in optimization. Arguably, the handling of constraints, as well as the availability of robust optimization methods are the most important arguments in favor of the optimization view in comparison to the probabilistic interpretations. Multi-modality and true stochastic optimal control under such multi-modality are the strongest arguments for the probabilistic view.

As a side node on parallelizing message passing computations: KOMO, DDP, and iLQG all do full path updates in each iteration, that is, they compute a full new path $x_{0:T}$ in each Newton(-like) or line search step. This is a difference to AICO which allows to update individual states x_t in arbitrary order, not necessarily sweeping forward and backward. E.g. in AICO we can update a single x_t multiple times in sequence when the updates are large and therefore the local linearization changes significantly locally. This is possible because AICO computes backward *and* forward messages which define a local posterior belief for x_t that includes forward and backward messages. In the dynamic programming view this means that cost-to-go *and* cost-so-far functions are computed to define a local optimization problem over x_t only. In practice, however, these local path updates are harder to handle than global steps, especially because global monotonicity, as guaranteed by global Wolfe conditions, cannot easily be realized.

5 Conclusion

In this tutorial we chose the k-order cost and constraint feature convention to represent trajectory optimization problems as NLP. The implied structure of the Jacobians and Hessian is of central importance to understand the complexity of Newton steps in such settings.

Newton approaches are not just one alternative under many—they are at the core of efficient optimization as well as at the core of understanding the fundaments of the many related approaches mentioned in this tutorial. In particular, we've discussed the structure and complexity of computing Newton steps for banded symmetric Hessians and its relation to solving (tree- or Markov-) structured least squares problems using dynamic programming, both of which have a computational complexity linear in the number of variables. We have discussed control in the KOMO convention, especially constrained k-order dynamic programming to compute an approximate regulator around the optimal path with guaranteed feasibility, and its MPC extension. For the unconstrained LQ case we highlighted the relations to DDP, iLQG, and AICO.

An interesting line of future research based on this discussion is related to path optimization processes that are not strictly Markovian in the KOMO sense. One example is jointly optimizing over paths and model parameters, which equally implies non-banded terms in the Hessian [15]. Another example is sequential manipulation, in which costs that arise in some later part of the motion may directly depend on configuration decisions (grasps) made much earlier. The gradient of such costs then will always be non-zero w.r.t. the grasp configuration. These introduce "loops" in the dependencies that violate the k-order Markov assumption. However, Graph-SLAM has successfully addressed exactly this problem. The established relations between path optimization and Graph-SLAM may therefore be a promising candidate for an optimization-based approach to sequential manipulation.

Acknowledgements This work was supported by the DFG under grants TO 409/9-1 and the 3rdHand EU-Project FP7-ICT-2013-10610878.

References

1. R. Bellman, Dynamic programming and lagrange multipliers. Proc. National Acad. Sci. **42**(10), 767–769 (1956)
2. A. Bemporad, M. Morari, V. Dua, E.N. Pistikopoulos, The explicit linear quadratic regulator for constrained systems. Automatica **38**(1), 3–20 (2002)
3. J.T. Betts, Survey of numerical methods for trajectory optimization. J. Guid Control Dyn. **21**(2), 193–207 (1998)
4. A.R. Conn, N.I. Gould, P. Toint, A globally convergent augmented Lagrangian algorithm for optimization with general constraints and simple bounds. SIAM J. Numer. Anal. **28**(2), 545–572 (1991)
5. F. Dellaert, Factor graphs and GTSAM: A hands-on introduction. Technical Report Technical Report GT-RIM-CP&R-2012-002, Georgia Tech (2012)
6. M. Diehl, H.J. Ferreau, N. Haverbeke, Efficient numerical methods for nonlinear MPC and moving horizon estimation, in *Nonlinear Model Predictive Control* (Springer, 2009), pp. 391–417
7. C.R. Dohrmann, R.D. Robinett, Dynamic programming method for constrained discrete-time optimal control. J. Optim. Theory Appl. **101**(2), 259–283 (1999)
8. J. Dong, M. Mukadam, F. Dellaert, B. Boots, Motion planning as probabilistic inference using Gaussian processes and factor graphs, in *Proceedings of Robotics: Science and Systems (RSS-2016)* (2016)

9. P. Englert, M. Toussaint, Inverse KKT–learning cost functions of manipulation tasks from demonstrations, in *Proceedings of the International Symposium of Robotics Research* (2015)
10. J. Folkesson, H. Christensen, Graphical SLAM-a self-correcting map, in *2004 IEEE International Conference on Robotics and Automation, 2004. Proceedings. ICRA'04*, vol. 1 (IEEE, 2004), pp. 383–390
11. G.H. Golub, C.F. Van Loan, *Matrix Computations*, vol. 3 (JHU Press, Baltimore, 2012)
12. S.J. Julier, J.K. Uhlmann, New extension of the Kalman filter to nonlinear systems, in *AeroSense'97* (International Society for Optics and Photonics, 1997), pp. 182–193
13. M. Kalakrishnan, S. Chitta, E. Theodorou, P. Pastor, S. Schaal, STOMP: stochastic trajectory optimization for motion planning, in *2011 IEEE International Conference on Robotics and Automation (ICRA)* (IEEE, 2011), pp. 4569–4574
14. H.J. Kappen, V. Gómez, M. Opper, Optimal control as a graphical model inference problem. Mach. Learn. **87**(2), 159–182 (2012)
15. S. Kolev, E. Todorov, Physically consistent state estimation and system identification for contacts, in *2015 IEEE-RAS 15th International Conference on Humanoid Robots (Humanoids)* (IEEE, 2015), pp. 1036–1043
16. F.R. Kschischang, B.J. Frey, H.-A. Loeliger, Factor graphs and the sum-product algorithm. IEEE Trans. Inf. Theory **47**(2), 498–519 (2001)
17. R. Kümmerle, G. Grisetti, H. Strasdat, K. Konolige, W. Burgard, g2o: a general framework for graph optimization, in *2011 IEEE International Conference on Robotics and Automation (ICRA)* (IEEE, 2011), pp. 3607–3613
18. J. Lafferty, A. McCallum, F.C. Pereira, Conditional random fields: probabilistic models for segmenting and labeling sequence data, in *Proceedings of 18th International Conference on Machine Learning (ICML)* (2001), pp. 282–289
19. L.-z. Liao, C. A. Shoemaker, Advantages of differential dynamic programming over Newton's method for discrete-time optimal control problems. Technical report, Cornell University (1992)
20. D. Mayne, A second-order gradient method for determining optimal trajectories of non-linear discrete-time systems. Int. J. Control **3**(1), 85–95 (1966)
21. T.P. Minka, Expectation propagation for approximate Bayesian inference, in *Proceedings of the Seventeenth Conference on Uncertainty in Artificial Intelligence* (Morgan Kaufmann Publishers Inc., 2001), pp. 362–369
22. J. Nocedal, S. Wright, *Numerical Optimization* (Springer Science & Business Media, New York, 2006)
23. J. Peters, S. Schaal, Natural actor-critic. Neurocomputing **71**(7), 1180–1190 (2008)
24. N. Ratliff, M. Zucker, J.A. Bagnell, S. Srinivasa, CHOMP: gradient optimization techniques for efficient motion planning, in *IEEE International Conference on Robotics and Automation, 2009. ICRA'09* (IEEE, 2009), pp. 489–494
25. N. Ratliff, M. Toussaint, S. Schaal, Understanding the geometry of workspace obstacles in motion optimization, in *2015 IEEE International Conference on Robotics and Automation (ICRA)* (IEEE, 2015), pp. 4202–4209
26. K. Rawlik, M. Toussaint, S. Vijayakumar, On stochastic optimal control and reinforcement learning by approximate inference, in *Proceedings of Robotics: Science and Systems (R:SS 2012)* (2012). Runner Up Best Paper Award
27. J. Schulman, J. Ho, A.X. Lee, I. Awwal, H. Bradlow, P. Abbeel, Finding locally optimal, collision-free trajectories with sequential convex optimization, in *Robotics: Science and Systems*, vol. 9 (2013), pp. 1–10. Citeseer
28. Y. Tassa, N. Mansard, E. Todorov, Control-limited differential dynamic programming, in *2014 IEEE International Conference on Robotics and Automation (ICRA)* (IEEE, 2014), pp. 1168–1175
29. S. Thrun, M. Montemerlo, The graph SLAM algorithm with applications to large-scale mapping of urban structures. Int. J. Robot. Res. **25**(5–6), 403–429 (2006)
30. E. Todorov, W. Li, A generalized iterative LQG method for locally-optimal feedback control of constrained nonlinear stochastic systems, in *American Control Conference, 2005. Proceedings of the 2005* (IEEE, 2005), pp. 300–306

31. M. Toussaint, Robot trajectory optimization using approximate inference, in *Proceedings of the International Conference on Machine Learning (ICML 2009)* (ACM, 2009), pp. 1049–1056. ISBN 978-1-60558-516-1
32. M. Toussaint, Pros and cons of truncated Gaussian EP in the context of approximate inference control, in *NIPS Workshop on Probabilistic Approaches for Robotics and Control* (2009)
33. M. Toussaint, A novel augmented lagrangian approach for inequalities and convergent any-time non-central updates. e-Print arXiv:1412.4329 (2014)
34. M. Toussaint, KOMO: newton methods for k-order markov constrained motion problems. e-Print arXiv:1407.0414 (2014)
35. N. Vlassis, M. Toussaint, Model-free reinforcement learning as mixture learning, in *Proceedings of the International Conference on Machine Learning (ICML 2009)* (2009), pp. 1081–1088. ISBN 978-1-60558-516-1
36. O. Von Stryk, R. Bulirsch, Direct and indirect methods for trajectory optimization. Ann. Oper. Res. **37**(1), 357–373 (1992)

Optimal Control of Variable Stiffness Policies: Dealing with Switching Dynamics and Model Mismatch

Andreea Radulescu, Jun Nakanishi, David J. Braun and Sethu Vijayakumar

Abstract Controlling complex robotic platforms is a challenging task, especially in designs with high levels of kinematic redundancy. Novel variable stiffness actuators (VSAs) have recently demonstrated the possibility of achieving energetically more efficient and safer behaviour by allowing the ability to simultaneously modulate the output torque and stiffness while adding further levels of actuation redundancy. An optimal control approach has been demonstrated as an effective method for such a complex actuation mechanism in order to devise a control strategy that simultaneously provides optimal control commands and time-varying stiffness profiles. However, traditional optimal control formulations have typically focused on optimisation of the tasks over a predetermined time horizon with smooth, continuous plant dynamics. In this chapter, we address the optimal control problem of robotic systems with VSAs for the challenging domain of switching dynamics and discontinuous state transition arising from interactions with an environment. First, we present a systematic methodology to simultaneously optimise control commands, time-varying stiffness profiles as well as the optimal switching instances and total movement duration based on a time-based switching hybrid dynamics formulation. We demonstrate the effectiveness of our approach on the control of a brachiating robot with a VSA considering multi-phase swing-up and locomotion tasks as an illustrative application of our proposed method in order to exploit the benefits of the VSA and intrinsic dynamics of the system. Then, to address the issue of model

A. Radulescu · J. Nakanishi · D.J. Braun · S. Vijayakumar (✉)
School of Informatics, University of Edinburgh, Edinburgh, UK
e-mail: sethu.vijayakumar@ed.ac.uk

A. Radulescu
Department of Advanced Robotics, Istituto Italiano di Tecnologia, Genova, Italy
e-mail: andreea.radulescu@iit.it

J. Nakanishi
Department of Micro-Nano Systems Engineering, Nagoya University, Nagoya, Japan
e-mail: nakanishi@mein.nagoya-u.ac.jp

D.J. Braun
Engineering Product Development, Singapore University of Technology and Design, Tampines, Singapore
e-mail: david_braun@sutd.edu.sg

J.-P. Laumond et al. (eds.), *Geometric and Numerical Foundations of Movements*, Springer Tracts in Advanced Robotics 117, DOI 10.1007/978-3-319-51547-2_16

discrepancies in model-based optimal control, we extend the proposed framework by incorporating an adaptive learning algorithm. This performs continuous data-driven adjustments to the dynamics model while re-planning optimal policies that reflect this adaptation. We show that this augmented approach is able to handle a range of model discrepancies in both simulations and hardware experiments.

1 Introduction

Modern robotic systems are used in various fields and operate in environments highly dangerous to humans (e.g., space and deep sea exploration, search and rescue missions). Controlling these robotic platforms is a challenging task due to the design complexity and the discontinuity in the dynamics, e.g., introduced by mechanical contact with the environment. Inspired by the efficiency of biological systems in locomotion and manipulation tasks, the robotics community has recently developed a new generation of actuators equipped with an additional mechanically adjustable compliant mechanism [12, 24, 42]. These variable stiffness actuators (VSAs) can provide simultaneous modulation of stiffness and output torque with the purpose of achieving dynamic and flexible robotic movements. However, this adds further levels of actuation redundancy, making planning and control of such systems even more complicated, especially in the case of underactuated systems.

Several studies of stiffness modulation in the context of domains with contacts showed that VSAs provide a significant improvement in *energy efficiency* due to their energy storage capabilities and ability to modulate the system's dynamics [22, 40]. The use of stiffness modulation in scenarios involving interaction with the environment has been shown to provide several *safety* benefits [11, 38]. Other advantages of variable stiffness capabilities have been observed in terms of *robustness* and *adaptability*. These are often required by tasks involving unpredictable changes in the environment and noise [9, 47].

In this chapter, we first introduce a systematic methodology for movement optimisation with multiple phases and switching dynamics in robotic systems with VSAs arising from contacts and impacts with the environment [21]. By modelling such tasks as a hybrid dynamical system with time-based switching, our proposed method simultaneously optimises control commands, time-varying stiffness profiles and temporal aspect of the movement such as switching instances and total movement duration to exploit the benefits of the VSA and intrinsic dynamics of the system. We present numerical simulations and hardware experiments on a brachiating robot with a VSA to demonstrate the effectiveness of our proposed framework in achieving a highly challenging task on an underactuated system.

Then, we present an augmented method to improve the robustness of the proposed framework with respect to model uncertainty by incorporating an adaptive learning algorithm [29]. The performance of model-based control is dependent on the accuracy of the dynamics models, which are traditionally obtained by model-based parameter identification procedures. However, certain elements cannot be fully represented by simple analytical models (e.g., complex friction of the joints or dynamics resulting from cable movement [35]), while changes in the behaviour of the system can occur due to wear and tear, or due to the use of a tool [36]. Our proposed adaptive learning method performs continuous data-driven adjustments to the dynamics model while re-planning optimal policies that reflect this adaptation. We build on prior efforts to employ adaptive dynamics learning in improving the performance of robot control [15, 19, 23]. We present results showing that our augmented approach is able to handle a range of model discrepancies in both simulations and hardware experiments on a brachiating robot platform with a VSA.

2 Spatio-Temporal Optimisation of Multi-phase Movements in Domains with Contacts

Traditional optimal control approaches have focused on the formulation over a predetermined time horizon with smooth, continuous plant dynamics. In this section, we present our framework of optimal control problems for tasks with multiple phase movements including switching dynamics and discrete state transition and its application to the control of robotic systems with VSAs [21].

Dynamics with intermittent contacts and impacts such as locomotion and juggling are often modelled as hybrid dynamical systems [2, 6, 14]. From a control theoretic perspective, a significant effort has been made to address optimal control problems of various classes of hybrid systems [25, 34, 44]. However, only a few robotic applications of optimal control with hybrid dynamics formulation can be found on movement of optimisation over multiple phases [7, 14]. Instead of employing hybrid dynamics modelling formulation, different optimisation approaches to dealing with multiple contact events have been proposed such as model predictive control with smooth approximation of contact forces [37], a direct multiple-shooting based method [18], direct trajectory optimisation methods [26, 43], and a further extension of the direct collocation method with linear quadratic regulator and quadratic programming [27].

In this section, we present an approach to the hybrid optimal control problem proposed in [21] with an extension to the iterative linear quadratic regulator (iLQR[1]) algorithm [13] and generalisation of the time-based switched LQ control with state jumps [44]. We also incorporate temporal optimisation in order to simultaneously optimise control commands and temporal parameters (e.g., movement duration) [30]. iLQR/G is a practically effective method for iteratively solving optimal control problems and has been used in our previous work, e.g., [3, 4, 16]. Time-based switching approximation in hybrid dynamics is motivated due to the difficulties associated with the state-based switching formulation in dealing with the need of imposing constraints and finding the time for switching [5, 46]. Discussions on the benefits and

[1]iLQG is the stochastic extension to iLQR [13] and in the sequel, we may refer to these two interchangeably.

practical difficulties of alternative optimal control approaches such as indirect methods (e.g., multiple-shooting methods), direct methods, and successive approximation methods (e.g., iLQR/G and differential dynamic programming) can be found in [4]. Numerical simulations and hardware experiments on a brachiating robot driven by a VSA will be presented to demonstrate the effectiveness of the proposed approach.

2.1 Outline of Multi-phase Spatio-Temporal Hybrid Optimal Control Approach

We address spatio-temporal stiffness optimisation problems for tasks that consist of multiple phases of movements including switching dynamics and discrete state transitions (arising from interaction with the environment) in order to exploit the benefits of VSAs. In addition to optimising control commands and stiffness, we develop a systematic methodology to simultaneously optimise the temporal aspect of the movement (e.g., movement duration). Our proposed formulation also provides an optimal feedback control law while many trajectory optimisation algorithms typically compute only optimal feedforward controls.

The main ingredients of our proposed optimal control framework are as follows:

1. use of nonlinear time-based switching dynamics with continuous control input to model the dynamics of multi-phase movements;
2. use of nonlinear discrete state transition to model contacts and impacts;
3. use of realistic plant dynamics with a VSA model;
4. introduction of a composite cost function to describe task objectives with multi-phase movements;
5. simultaneous optimisation of joint torque and stiffness profiles across multiple phases;
6. optimisation of switching instances and total movement duration.

As presented below, we formulate this problem as time-based switching hybrid optimal control with temporal optimisation.

2.2 Problem Formulation

2.2.1 Time-Based Switching Hybrid Dynamics

In order to represent multi-phase movements having interactions with an environment, we consider the following time-based switching hybrid dynamics formulation composed of multiple sets of continuous dynamics (1) and discrete state transition (2) as in [8, 45]:

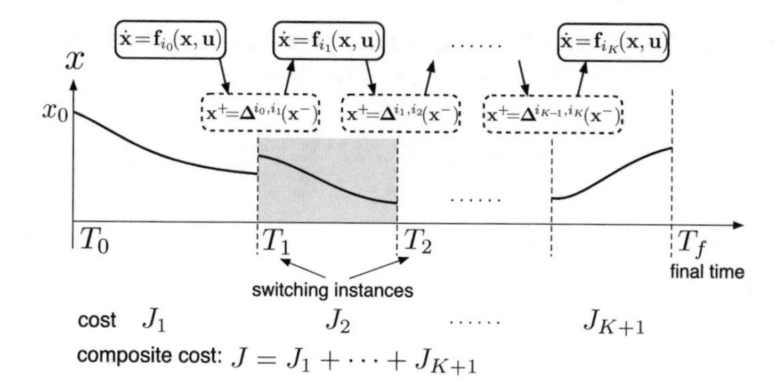

Fig. 1 A hybrid system with time-based switching dynamics and discrete state transition with a known sequence. The goal is finding an optimal control command $\mathbf{u}$, switching instances T_i and final time T_f which minimises the composite cost J

$$\dot{\mathbf{x}} = \mathbf{f}_{i_j}(\mathbf{x}, \mathbf{u}), \quad T_j \le t < T_{j+1} \text{ for } j = 0, \dots, K \tag{1}$$

$$\mathbf{x}(T_j^+) = \Delta^{i_{j-1}, i_j}(\mathbf{x}(T_j^-)) \text{ for } j = 1, \dots, K \tag{2}$$

where $\mathbf{f}_{i_j} : \mathbb{R}^n \times \mathbb{R}^m \to \mathbb{R}^n$ is the subsystem i_j, $\mathbf{x} \in \mathbb{R}^n$ is a state vector and $\mathbf{u} \in \mathbb{R}^m$ is a control input vector for subsystems. At the moment of dynamics switching from $\mathbf{f}_{i_{j-1}}$ to $\mathbf{f}_{i_j}$, we assume an instantaneous discrete (discontinuous) state transition according to the impact map in (2), where $\mathbf{x}(T_j^+)$ and $\mathbf{x}(T_j^-)$ denote the post- and pre-transition states, respectively (see Fig. 1).

2.2.2 Robot Dynamics with Variable Stiffness Actuation for Multi-phase Movement

Given the general form of the plant dynamics of our concern in a hybrid dynamics representation introduced in Sect. 2.2.1, we consider the following multiple set of robot dynamics with VSAs to describe multi-phase movements:

$$\mathbf{M}_i(\mathbf{q})\ddot{\mathbf{q}} + \mathbf{C}_i(\mathbf{q}, \dot{\mathbf{q}})\dot{\mathbf{q}} + \mathbf{g}_i(\mathbf{q}) + \mathbf{D}_i\dot{\mathbf{q}} = \tau_i(\mathbf{q}, \mathbf{q}_m) \tag{3}$$

where i denotes the i-th subsystem corresponding to its associated phase of the movement, $\mathbf{q} \in \mathbb{R}^n$ is the joint angle vector, $\mathbf{q}_m \in \mathbb{R}^m$ is the motor position vector of the VSA, $\mathbf{M}_i \in \mathbb{R}^{n\times n}$ is the inertia matrix, $\mathbf{C}_i \in \mathbb{R}^n$ is the Coriolis term, $\mathbf{g}_i \in \mathbb{R}^n$ is the gravity vector, $\mathbf{D}_i \in \mathbb{R}^{n\times n}$ is the viscous damping matrix, and $\tau_i \in \mathbb{R}^n$ is the joint torque vector from the VSA given in the form:

$$\tau_i(\mathbf{q}, \mathbf{q}_m) = \mathbf{N}_i^T(\mathbf{q}, \mathbf{q}_m)\mathbf{F}_i(\mathbf{q}, \mathbf{q}_m) \tag{4}$$

where $\mathbf{N}_i \in \mathbb{R}^{p\times n} (p \geq n)$ is the moment arm matrix and $\mathbf{F}_i \in \mathbb{R}^p$ is the forces generated by the elastic elements. The joint stiffness is defined by

$$\mathbf{K}_i(\mathbf{q}, \mathbf{q}_m) = -\frac{\partial \tau_i(\mathbf{q}, \mathbf{q}_m)}{\partial \mathbf{q}}. \tag{5}$$

We model the servo motor dynamics in the VSA as a critically damped second order dynamical system:

$$\ddot{\mathbf{q}}_m + 2\alpha_i \dot{\mathbf{q}}_m + \alpha_i^2 \mathbf{q}_m = \alpha_i^2 \mathbf{u} \tag{6}$$

where α_i determines the bandwidth of the servo motors[2] and $\mathbf{u} \in \mathbb{R}^m$ is the motor position command [3]. We assume that the range of the control command $\mathbf{u}$ is limited as $\mathbf{u}_{min} \preceq \mathbf{u} \preceq \mathbf{u}_{max}$.

In order to formulate an optimal control problem, we consider the following state space representation of the combined plant dynamics composed of the rigid body dynamics (3) and the servo motor dynamics (6):

$$\dot{\mathbf{x}} = \mathbf{f}_i(\mathbf{x}, \mathbf{u}) = \begin{bmatrix} \mathbf{x}_2 \\ \mathbf{M}_i^{-1}(\mathbf{x}_1)\left(-\mathbf{C}_i(\mathbf{x}_1, \mathbf{x}_2)\mathbf{x}_2 - \mathbf{g}_i(\mathbf{x}_1) - \mathbf{D}_i\mathbf{x}_2 + \tau_i(\mathbf{x}_1, \mathbf{x}_3)\right) \\ \mathbf{x}_4 \\ -\alpha_i^2\mathbf{x}_3 - 2\alpha_i\mathbf{x}_4 + \alpha_i^2\mathbf{u} \end{bmatrix} \tag{7}$$

where $\mathbf{x} = [\,\mathbf{x}_1^T, \mathbf{x}_2^T, \mathbf{x}_3^T, \mathbf{x}_4^T\,]^T = [\,\mathbf{q}^T, \dot{\mathbf{q}}^T, \mathbf{q}_m^T, \dot{\mathbf{q}}_m^T\,]^T \in \mathbb{R}^{2(n+m)}$ is the augmented state vector consisting of the robot state and the servo motor state.

2.2.3 Composite Cost Function for Multi-phase Movement Optimisation

For the given hybrid dynamics (1) and (2), in order to describe the full movement with multiple phases, we consider the following composite cost function

$$J = \phi(\mathbf{x}(T_f)) + \sum_{j=1}^{K} \psi^j(\mathbf{x}(T_j^-)) + \int_{T_0}^{T_f} h(\mathbf{x}, \mathbf{u})dt \tag{8}$$

where $\phi(\mathbf{x}(T_f))$ is the terminal cost, $\psi^j(\mathbf{x}(T_j^-))$ is the via-point cost at the j-th switching instance and $h(\mathbf{x}, \mathbf{u})$ is the running cost. When optimising multi-phase movements, it is possible to optimise each phase in a sequential manner. However, the total cost of such a sequential optimisation could result in a suboptimal solution [30].

For the given plant dynamics (1) and state transition (2), the optimisation problem we consider is to a) find an optimal feedback control law $\mathbf{u} = \mathbf{u}(\mathbf{x}, t)$ which minimises

[2] $\alpha = \text{diag}(a_1, \ldots, a_m)$ and $\alpha^2 = \text{diag}(a_1^2, \ldots, a_m^2)$ for notational convenience.

the composite cost (8) and b) simultaneously optimise switching instances $T_1, \ldots, T_k$ and the final time T_f. Note that in our formulation, we denote the final time separately from switching instances for notational consistency with the case of single phase optimisation. However, the final time can be absorbed as a part of switching instances, e.g., $T_f = T_{K+1}$.

2.2.4 Spatio-Temporal Multi-Phase Optimisation Algorithm

In this section, we present an overview of our framework for spatio-temporal optimisation for multi-phase movements. First, the iLQR method [13] is extended in order to incorporate timed switching dynamics with discrete and discontinuous state transitions. Then, we present a temporal optimisation algorithm to obtain the optimal switching instances and the total movement duration. For more detailed description, we refer the interested readers to [21].

In brief, the iLQR method solves an optimal control problem of the locally linear quadratic approximation of the nonlinear dynamics and the cost function around a nominal trajectory $\bar{\mathbf{x}}$ and control sequence $\bar{\mathbf{u}}$ in discrete time, and iteratively improves the solutions.

In order to incorporate switching dynamics and discrete state transition with a given switching sequence, the hybrid dynamics (1) and (2) are linearised in discrete time around the nominal trajectory and control sequence as

$$\delta \mathbf{x}_{k+1} = \mathbf{A}_k \delta \mathbf{x}_k + \mathbf{B}_k \delta \mathbf{u}_k \tag{9}$$

$$\delta \mathbf{x}_{k_j}^{+} = \Gamma_{k_j} \delta \mathbf{x}_{k_j}^{-} \tag{10}$$

$$\mathbf{A}_k = \mathbf{I} + \Delta t_j \left.\frac{\partial \mathbf{f}_{i_j}}{\partial \mathbf{x}}\right|_{\mathbf{x}=\mathbf{x}_k}, \quad \mathbf{B}_k = \Delta t_j \left.\frac{\partial \mathbf{f}_{i_j}}{\partial \mathbf{u}}\right|_{\mathbf{u}=\mathbf{u}_k} \tag{11}$$

$$\Gamma_{k_j} = \left.\frac{\partial \Delta^{i_{j-1}, i_j}}{\partial \mathbf{x}}\right|_{\mathbf{x}=\mathbf{x}_{k_j}^{-}} \tag{12}$$

where $\delta \mathbf{x}_k = \mathbf{x}_k - \bar{\mathbf{x}}_k$, $\delta \mathbf{u}_k = \mathbf{u}_k - \bar{\mathbf{u}}_k$, k is the discrete time step, Δt_j is the sampling time being optimised for the time interval $T_j \leq t < T_{j+1}$, and k_j is the j-th switching instance in the discretised time step. The composite cost function and the optimal cost-to-go function are locally approximated and the local optimal control problem is solved via modified Riccati-like equations as described in detail in [21]. Once we have a locally optimal control command $\delta \mathbf{u}$, the nominal control sequence is updated as $\bar{\mathbf{u}} \leftarrow \bar{\mathbf{u}} + \delta \mathbf{u}$. Then, the new nominal trajectory $\bar{\mathbf{x}}$ and the cost J are computed by running the obtained control $\bar{\mathbf{u}}$ on the system dynamics, and the above process is iterated until convergence (no further improvement in the cost within certain threshold).

In order to optimise the switching instance and the total movement duration, following our previous work in [30], we introduce a scaling parameter and sampling time for each duration of between switching as

$$\Delta t'_j = \frac{1}{\beta_j}\Delta t_j \quad \text{for} \quad T_j \le t < T_{j+1} \text{ where } \quad j = 0, \dots, K. \tag{13}$$

By optimising the vector of temporal scaling factors $\beta = [\ \beta_0,\ \dots,\ \beta_K\ ]^T$ via gradient descent, we can obtain each switching instance $T_{j+1} = (k_{j+1} - k_j)\Delta t'_j + T_j$ and the total movement duration $T_f = \sum_{j=0}^{K}(k_{j+1} - k_j)\Delta t'_j + T_0$, where $k_0 = 1$ and $k_{K+1} = N$.

In the complete optimisation, computation of the optimal feedback control law and update of the temporal scaling parameters are iteratively performed in an alternate manner until convergence. As a result, we obtain the optimal feedback control law

$$\mathbf{u}(\mathbf{x}, t) = \mathbf{u}_{opt}(t) + \mathbf{L}_{opt}(t)(\mathbf{x}(t) - \mathbf{x}_{opt}(t)) \tag{14}$$

and the optimal switching instances $T_1, \dots, T_K$ and the final time T_f, where $\mathbf{u}_{opt}$ is the feedforward optimal control sequence, $\mathbf{x}_{opt}$ is the optimal trajectory, and $\mathbf{L}_{opt}$ is the optimal feedback gain matrix.

2.3 Brachiating Robot Dynamics with VSA

The dynamics of a two-link brachiating robot with a VSA shown in Fig. 2 take the standard form of rigid body dynamics (3) where only the second joint has actuation (underactuation):

$$\mathbf{M}_i(\mathbf{q})\ddot{\mathbf{q}} + \mathbf{C}_i(\mathbf{q}, \dot{\mathbf{q}})\dot{\mathbf{q}} + \mathbf{g}_i(\mathbf{q}) + \mathbf{D}_i\dot{\mathbf{q}} = \begin{bmatrix} 0 \\ \tau(\mathbf{q}, \mathbf{q}_m) \end{bmatrix} \tag{15}$$

where $\mathbf{q} = [\ q_1,\ q_2\ ]^T$ is the joint angle vector. The same definitions for the elements in the rigid body dynamics are used as in (3). The index i in (15) is introduced to specify the configuration of the robot to indicate which hand is holding the bar. Since we assume that the robot has an asymmetric structure in the dynamics, we have two sets of subsystems denoted by the subscripts $i = 1$ (hand of link 1 is holding) and $i = 2$ (hand of link 2 is holding). In the multi-phase brachiation, the effective model switches between $i = 1$ and $i = 2$ interchangeably at the switching instance when the robot grasps the bar.

We use MACCEPA (Fig. 2) as our VSA implementation of choice [39], which has the desirable property that the joint can be passively compliant. This allows free swinging with a large range of movement by relaxing the spring, which is highly suitable for the brachiation task we consider. MACCEPA is equipped with two

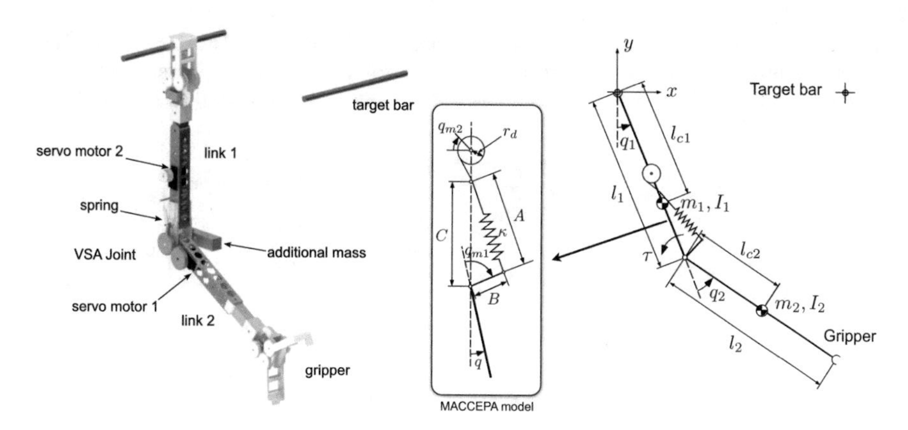

Fig. 2 Two-link brachiating robot model with the MACCEPA joint with the inertial and geometric parameters. The parameters of the robot are given in Table 1

position controlled servo motors, $\mathbf{q}_m = [\ q_{m1},\ q_{m2}\ ]^T$ which control the equilibrium position and the spring pre-tension, respectively.[3]

The joint torque for the MACCEPA model is given by:

$$\tau = \underbrace{\frac{BC\sin(q_{m1} - q)}{A}}_{\text{moment arm in (4)}} F \tag{16}$$

where $A = \sqrt{B^2 + C^2 - 2BC\cos(q_{m1} - q)}$, q is the joint angle.[4] F is the spring tension

$$F = \kappa_s(l - l_0) \tag{17}$$

where κ_s the spring constant, $l = A + r_d q_{m2}$ is the current spring length, $l_0 = C - B$ is the spring length at rest and r_d is the drum radius (see Fig. 2). The joint stiffness can be computed as $k = -\frac{\partial \tau}{\partial q}$. Note that the torque and stiffness relationships in MACCEPA are dependent on the current joint angle and two servo motor angles in a complicated manner and its control is not straightforward.

To formulate the multi-phase movement optimisation in brachiation, we use the state space representation in (7). At the transition at handhold, an affine discrete state transition $\mathbf{x}^+ = \Delta(\mathbf{x}^-) = \Gamma\mathbf{x}^- + \gamma$ is introduced to shift the coordinate system for the next handhold and reset the joint velocities of the robot to zero, which is defined as:

---

[3]We include position controlled servo motor dynamics as defined in (6). For the bandwidth parameters for the motors we use $\alpha = \mathrm{diag}(20, 25)$. The range of the commands of the servo motors are limited as $u_1 \in [-\pi/2, \pi/2]$ and $u_2 \in [0, \pi/2]$.

[4]In the brachiating robot model in Fig. 2, $q = q_2$.

Table 1 Model parameters of the two-link brachiating robot and the VSA. The index i in this table denotes the link number in Fig. 2. The final column (numbers in *in italic*) in the robot parameters table shows the change of parameters of the first link of the system under the changed mass distribution described in Sect. 3.2

Robot parameters		i=1	i=2	*i=1*
Mass	m_i (kg)	1.390	0.527	*1.240*
Moment of inertia	I_i (kgm^2)	0.0297	0.0104	*0.0278*
Link length	l_i (m)	0.46	0.46	*0.46*
COM location	l_{ci} (m)	0.362	0.233	*0.350*
Viscous friction	d_i (Nm/s)	0.03	0.035	*0.03*

MACCEPA parameters		value
Spring constant	κ_s (N/m)	771
Lever length	B (m)	0.03
Pin displacement	C (m)	0.125
Drum radius	r_d (m)	0.01

$$\Gamma = \text{diag}(\Gamma_1, \ldots, \Gamma_4), \tag{18}$$

where:

$$\Gamma_1 = \begin{bmatrix} 1 & 1 \\ 0 & -1 \end{bmatrix}, \Gamma_2 = \begin{bmatrix} 0 & 0 \\ 0 & 0 \end{bmatrix}, \Gamma_3 = \Gamma_4 = \begin{bmatrix} -1 & 0 \\ 0 & 1 \end{bmatrix} \tag{19}$$

and $\gamma = [\ -\pi,\ 0,\ \ldots,\ 0\ ]^T$. Note that in this example, we have $\Delta = \Delta^{1,2} = \Delta^{2,1}$.

2.4 *Exploitation of Passive Dynamics With Spatio-Temporal Optimisation of Stiffness*

In this section, we explore the benefits of simultaneous stiffness and temporal optimisation for tasks exploiting the intrinsic dynamics of the system. Brachiation is an example of a highly dynamic manoeuvre requiring the use of passive dynamics for successful task execution [10, 20, 31, 32].

2.4.1 Optimisation of a Single Phase Movement in Brachiation Task

In this section, we consider the brachiation task of swing locomotion from handhold to handhold on a ladder. A natural and desirable strategy for a swing movement in brachiation would be to make good use of gravity by making the joints passive and compliant. For a system with VSAs, our idea in exploiting passive dynamics is to frame the control problem in finding an appropriate (preferably small) stiffness profile to modulate the system dynamics only when necessary and compute the virtual equilibrium trajectory to fulfil the specified task requirement.

To implement this idea of passive control strategy, we consider the following cost function to encode the task:

$$J = (\mathbf{y}(T) - \mathbf{y}^*)^T \mathbf{Q}_T (\mathbf{y}(T) - \mathbf{y}^*) + \int_0^T \left(\mathbf{u}^T \mathbf{R}_1 \mathbf{u} + R_2 F^2 \right) dt \tag{20}$$

where $\mathbf{y} = [\, \mathbf{r},\ \dot{\mathbf{r}} \,]^T \in \mathbb{R}^4$ are the position and the velocity of the gripper in the Cartesian coordinates, $\mathbf{y}^*$ contains the target values when reaching the target $\mathbf{y}^* = [\, \mathbf{r}^*,\ \mathbf{0} \,]^T$ and F is the spring tension in the VSA given in (17). $\mathbf{Q}_T$ is a positive semi-definite matrix, $\mathbf{R}_1$ is a positive definite matrix and R_2 is a positive scalar. This objective function is designed in order to reach the target located at $\mathbf{r}^*$ at the time T, while minimising the spring tension F in the VSA. The term $\mathbf{u}^T \mathbf{R}_1 \mathbf{u}$ is added for regularisation with a small choice of the weights in $\mathbf{R}_1$, which is necessary in practice since iLQG requires a control cost in its formulation to compute the optimal control law.

2.4.2 Benefit of Temporal Optimisation

One of the issues in a conventional optimal control formulation is that the time horizon needs to be given in advance for a given task. While on fully actuated systems, control can be used to enforce a pre-specified timing, it is not possible to choose an arbitrary time horizon on underactuated systems. In a brachiation task, determination of an appropriate movement horizon is essential for successful task execution with reduced control effort.

Consider the swing locomotion task on a ladder with the intervals starting from the bar at $d_{start} = 0.42$ m to the target located at $d_{target} = 0.46$ m. We optimise both the control command $\mathbf{u}$ and the movement duration T. We use $\mathbf{Q}_T = \mathrm{diag}(10000, 10000, 10, 10)$, $\mathbf{R}_1 = \mathrm{diag}(0.0001, 0.0001)$ and $R_2 = 0.01$ for the cost function in (20). The optimised movement duration was $T = 0.806$ s.

Figure 3 shows the simulation result of (a) the optimised robot movement, (b) joint trajectories and servo motor positions, and (c) joint torque, spring tension and joint stiffness. In the plots, trajectories of the fixed time horizon ranging $T \in [0.7, 0.75, \ldots, 0.9]$ s are overlayed for comparison in addition to the case of the optimal movement duration $T = 0.806$ s. In the movement with temporal optimisation, the spring tension and the joint stiffness are kept small at the beginning and end of the movement resulting in nearly zero joint torque. This allows the joint to swing passively. The joint torque is exerted only during the middle of the swing by increasing the spring tension as necessary. In contrast, with non-optimal time horizon, larger joint torque and spring tension as well as higher joint stiffness can be observed resulting in the requirement of increased control effort. This result suggests that the natural plant dynamics are fully exploited for the desirable task execution with simultaneous stiffness and temporal optimisation.

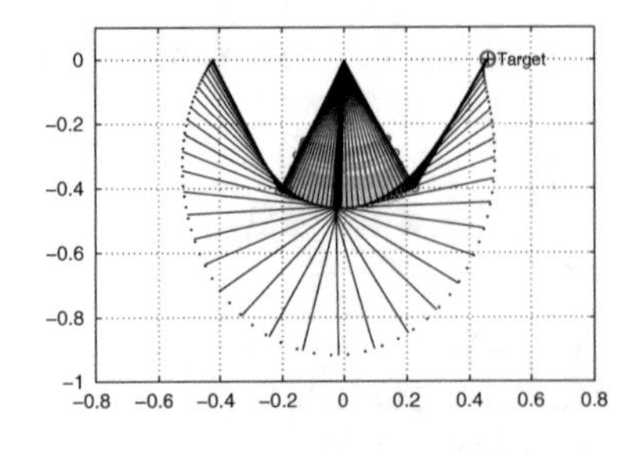

(a) Movement of the robot (simulation) with optimal variable stiffness control (optimised duration T=0.806 s).

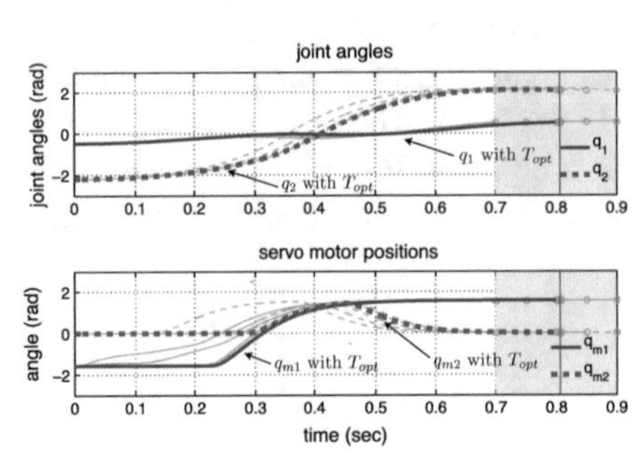

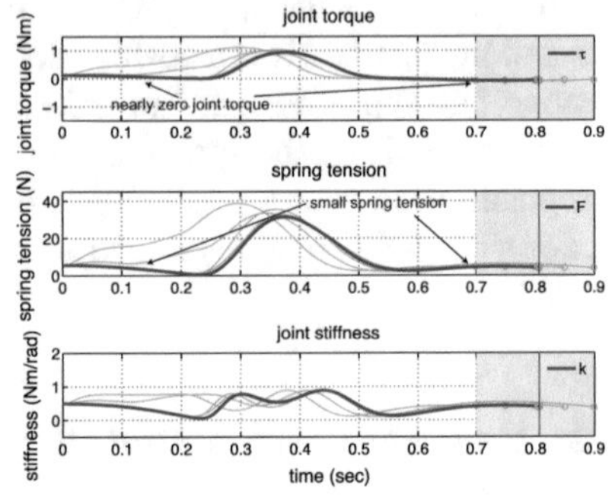

(b) Joint trajectories and servo motor positions (c) Joint torque, spring tension and joint stiffness

Fig. 3 Simulation result of the single phase brachiation task with temporal optimisation. In **b** and **c**, *grey thin lines* show the plots for non-optimised T in the range of $T = [0.7, 0.75, \ldots, 0.9]$ s and *blue thick lines* show the plots for optimised $T = 0.806$ s. With temporal optimisation, at the beginning and the end of the movement, joint torque, spring tension and joint stiffness are kept small allowing the joint to swing passively in comparison to the non-optimal time cases

2.5 Spatio-Temporal Optimisation of Multiple Swings in Robot Brachiation

To demonstrate the effectiveness of our proposed approach in multi-phase movement optimisation, we consider the following brachiation task with multiple phases of the movement: The robot initially swings up from the suspended posture to the target at $d_1 = 0.40$ m and subsequently moves to the target located at $d_2 = 0.42$ m and $d_3 = 0.46$ m. The composite cost function to encode this task is designed as:

$$J = (\mathbf{y}(T_f) - \mathbf{y}_f^*)^T \mathbf{Q}_{T_f} (\mathbf{y}(T_f) - \mathbf{y}_f^*) + \sum_{j=1}^{K} (\mathbf{y}(T_j^-) - \mathbf{y}_j^*)^T \mathbf{Q}_{T_j} (\mathbf{y}(T_j^-) - \mathbf{y}_j^*)$$

$$+ \int_0^{T_f} \left(\mathbf{u}^T \mathbf{R}_1 \mathbf{u} + R_2 F^2 \right) dt + w_T T_1 \tag{21}$$

where $K = 3$ is the number of phases, $\mathbf{y} = [\ \mathbf{r},\ \dot{\mathbf{r}}\ ]^T \in \mathbb{R}^4$ are the position and the velocity of the gripper in the Cartesian coordinates measured from the origin at

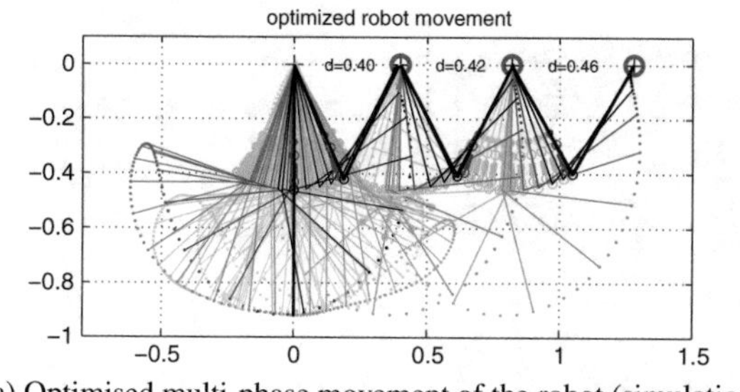

(a) Optimised multi-phase movement of the robot (simulation)

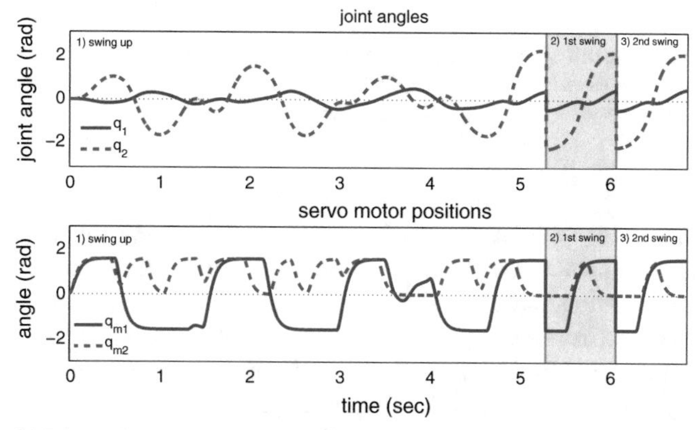

(b) Joint trajectories and servo motor positions with temporal optimisation.

Fig. 4 Simulation result of the multi-phase brachiation task with temporal optimisation

current handhold, $\mathbf{y}^*$ is the target values when reaching the target $\mathbf{y}^* = [\ \mathbf{r}^*,\ \mathbf{0}\ ]^T$ and F is the spring tension in the VSA. Note that this cost function includes the additional time cost $w_T T_1$ for the swing up manoeuvre in order to regulate the duration of the swing up movement. We use $\mathbf{Q}_{T_f} = \mathbf{Q}_{T_j} = \text{diag}(10000, 10000, 10, 10)$, $\mathbf{R}_1 = \text{diag}(0.0001, 0.0001)$ and $R_2 = 0.01$ and $w_T = 1$. In addition, we impose constraints on the range of the angle of the second joint during the course of the swing up manoeuvre as $q_{2_{min}} \le q_2 \le q_{2_{max}}$, where $[q_{2_{min}}, q_{2_{max}}] = [-1.745, 1.745]$ rad by adding a penalty term to the cost (21). This is empirically introduced and adjusted considering the physical joint limit of the hardware platform used in the experiments.

Figure 4a shows the sequence of the optimised multi-phase movement of the robot using the proposed algorithm including temporal optimisation in numerical simulations. The optimised switching instances and the total movement duration are $T_1 = 5.259$ s, $T_2 = 6.033$ s and $T_f = 6.835$ s, respectively, and the total cost is $J = 37.815$. Figure 4b shows the optimised joint trajectories and servo motor positions.

To illustrate the benefit of the proposed multi-phase formulation, we performed movement optimisation with a pre-specified nominal (fixed, non-optimal)

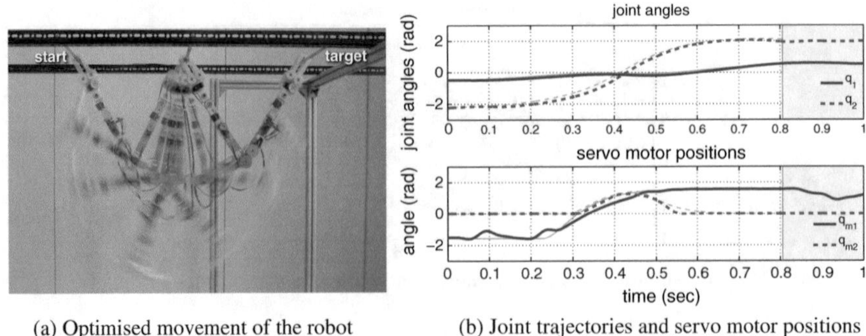

(a) Optimised movement of the robot (b) Joint trajectories and servo motor positions

Fig. 5 Experimental result of the single phase locomotion task on the brachiating robot hardware. In **b**, *red* and *blue think lines* show the experimental data, and *grey thin lines* show the corresponding simulation result with the optimised planned movement duration $T = 0.806$ s presented in Sect. 2.4.2

time horizon with $T_1 = 5.2$ s, $T_2 = 5.9$ s and $T_f = 6.7$ s using the same cost parameters both in sequential and multi-phase optimisation. With sequential individual movement optimisation, large overshoot was observed at the end of the final phase movement (distance from the target at $t = T_f$ was 0.0697 m) incurring a significantly large total cost of $J = 101.053$. In contrast, for the same pre-specified time horizon, by employing multi-phase movement optimisation, it was possible to find a feasible solution to reach the target bars. The error at the final swing at $t = T_f$ was 0.0020 m, which was significant improvement compared to the case of individual optimisation. The largest error observed in this sequence was 0.0109 m at the end of the first swing up phase. In this case, the total cost was $J = 50.228$. These results demonstrate the benefit of the multi-phase movement optimisation in finding optimal control commands and temporal aspect of the movement using the proposed method resulting in lower cost.

2.6 Evaluation on Hardware Platform

This section presents experimental implementation of our proposed algorithm on a two-link brachiating robot hardware developed in our laboratory [21]. The robot is equipped with a MACCEPA variable stiffness actuator and the parameters of the robot are given in Table 1.

Figure 5 shows the experimental result of a swing locomotion corresponding to the simulation in Sect. 2.4.2 with the optimal movement duration. In the experiments, we only use the open-loop optimal control command to the servo motors without state feedback as in [3].

Figure 6 shows the experimental result of multi-phase movements consisting of swing-up followed by two additional swings, which corresponds to the simulation

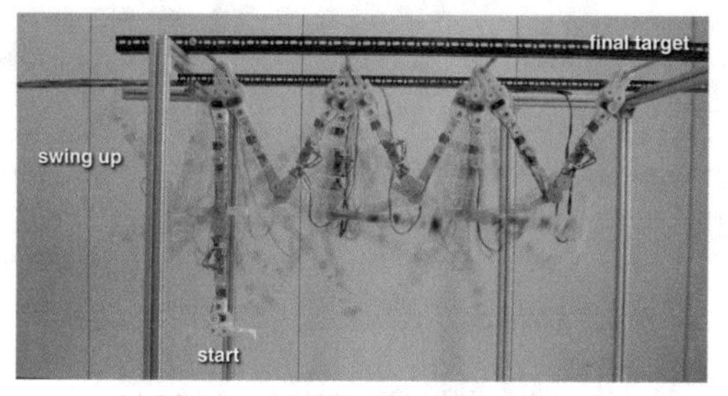

(a) Movement of the robot (experiment)

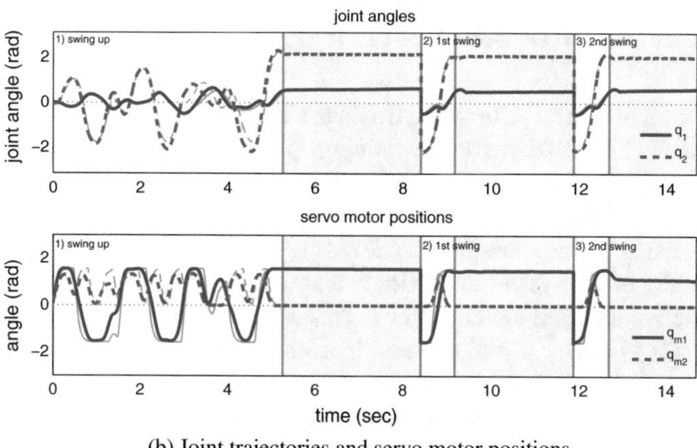

(b) Joint trajectories and servo motor positions

Fig. 6 **a** Experimental results of multi-phase locomotion task with the brachiating robot hardware. In **b**, *red* and *blue lines* show the actual hardware behaviour, and *grey lines* show the corresponding simulation results presented in Sect. 2.5

in Fig. 4 (Sect. 2.5). Note that at the end of each phase of the movement, switching to the next phase is manually done by confirming firm grasping of the bar in order to avoid falling off from the ladder at run-time.

The joint trajectories in the experiment closely match the planned movement in the simulation. The observed discrepancy is mainly due to the inevitable difference between the analytical nominal model and the hardware system. In the next section, we introduce an adaptive learning algorithm to improve the accuracy of the dynamics model used in optimisation.

These experimental results demonstrate the effectiveness and feasibility of the proposed framework in achieving highly dynamic tasks in compliantly actuated robots with variable stiffness capabilities under real conditions.

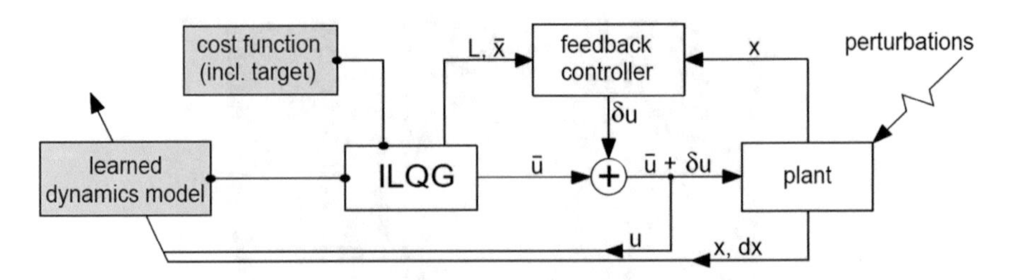

Fig. 7 The iLQG-LD learning and control scheme [16]

3 Optimal Control with Learned Dynamics

Classical optimal control is formulated using an analytical dynamics model, however, recent work [1, 17] has shown that combining optimal control with dynamics learning can produce an effective and principled control strategy for complex systems affected by model uncertainties.

In [17], using online (non-parametric) supervised learning methods, an adaptive internal model of the system dynamics is learned. The learned model is used afterwards to derive an optimal control law. This approach, named *iLQG with learned dynamics (iLQG-LD)*,[5] proved efficient in a variety of realistic scenarios including problems where the analytical dynamics model is difficult to estimate accurately or subject to changes and the system is affected by noise [16, 17]. The initial state and the cost function (which includes the desired final state) are provided to the iLQG planner alongside a preliminary model of the dynamics. An initial (locally optimal) command sequence $\bar{\mathbf{u}}$ is generated together with the corresponding state sequence $\bar{\mathbf{x}}$ and feedback correction gains $\mathbf{L}$. Applying the feedback controller scheme, at each time step the control command is corrected by $\delta\mathbf{u} = \mathbf{L}(\mathbf{x} - \bar{\mathbf{x}})$, where $\mathbf{x}$ is the true state of the plant. The model of the dynamics is updated using the information provided by the applied command $\mathbf{u} + \delta\mathbf{u}$ and observed state $\mathbf{x}$ (Fig. 7).

This iLQG-LD methodology employs a *Locally Weighted Learning (LWL)* method, or more specifically, the *Locally Weighted Projection Regression (LWPR)*, to train a model of the dynamics in an incremental fashion. LWL algorithms are non-parametric local learning methods that proved successful in the context of (online) motor learning scenarios [1]. Incremental LWL was proposed in [33] (*Receptive Field Weighted Regression (RFWR)* method) in order to achieve fast learning and computational efficiency. RFWR works by allocating resources in a data driven fashion, allowing online adaptation to changes in the behaviour. The LWPR [41] extends the RFWR method by projecting the input information into a lower dimensional manifold along selected directions before performing the fitting. Thus, it proves effective in high dimensionality scenarios where the data lies in a lower dimensional space. Consequently, iLQG-LD proved to be a robust and efficient technique for incremental learning of nonlinear models in high dimensions [17].

[5] Hereafter, we use the term iLQG for the optimisation algorithm of our concern.

We incorporate the iLQG-LD scheme into our approach involving learning the dynamics of a brachiating robot with a VSA and employing it in planning for multi-phase locomotion tasks [29]. The method proved capable of adapting to changes in the system's properties and provided a better accuracy performance than the optimisation without model adaptation. Based on these results, iLQG-LD could be a strong candidate for optimal control strategy for more complex hardware systems. We demonstrate the effectiveness of our adaptive learning approach in numerical simulations (Sect. 3.2) and hardware experiments (Sect. 3.3).

3.1 Multi-Phase Optimisation with Adaptive Dynamics Learning

In this section, we introduce the changes by the use of the LWPR method in the context of iLQG-LD for integration within the multi-phase optimisation approach described in Sect. 2.2. We assume that initially we have a nominal analytical dynamics model that takes the form presented in (3) and (6) which has inaccuracies. We use the LWPR method to model the error between the true behaviour of the system and the initial nominal model provided. For this purpose, we replace the dynamics $\mathbf{f}_i$ in (7) with a composite dynamics model $\mathbf{f}_{\mathbf{c}i}$:

$$\dot{\mathbf{x}} = \mathbf{f}_{\mathbf{c}i}(\mathbf{x}, \mathbf{u}) = \tilde{\mathbf{f}}_i(\mathbf{x}, \mathbf{u}) + \bar{\mathbf{f}}_i(\mathbf{x}, \mathbf{u}), \quad \mathbf{f}_{\mathbf{c}i} \in \mathbb{R}^{2(n+m)}, \tag{22}$$

where $\tilde{\mathbf{f}}_i$ is the initial nominal model and $\bar{\mathbf{f}}_i$ is the LWPR model to learn the discrepancy between $\tilde{\mathbf{f}}_i$ and the behaviour of the system.[6]

When using the composite model of the dynamics $\mathbf{f}_\mathbf{c}$ introduced in (22), the linearisation of the dynamics is provided in two parts. The linearisation of $\tilde{\mathbf{f}}$ is obtained by replacing $\mathbf{f}$ with $\tilde{\mathbf{f}}$ in (9) and (11). The derivatives of the learned model $\bar{\mathbf{f}}$ are obtained analytically by differentiating the LWPR model with respect to the inputs $\mathbf{z} = (\mathbf{x}; \mathbf{u})$ as suggested in [1]. With these modifications, the developed optimisation methodology is applied as described in Sect. 2.2 to obtain the locally optimal feedback control law.

3.2 Numerical Evaluations

In this section, we numerically demonstrate the effectiveness of the proposed model learning approach on a brachiating robot model with a VSA used in Sect. 2.3. In the nominal model, we introduce a mass (and implicitly mass distribution) discrepancy on one of the links (i.e., the mass of the true model is smaller by 150 g located

[6]Note that the changes introduced by iLQG-LD only affect the dynamics modelling in (1), while the instantaneous state transition map in (2) remains unchanged.

at the joint on link $i = 1$). The changed model parameters are shown in the right column of Table 1 (the numbers *in italic*). The introduced discrepancy affects the rigid body dynamics in the joint accelerations $(\ddot{q}_1, \ddot{q}_2)$. Thus, in the composite model $\mathbf{f}_c$, the information from the nominal model $\tilde{\mathbf{f}}$ requires correction only in those two dimensions (i.e., the required dimensionality of the LWPR model output $\bar{\mathbf{f}}$ is 2, the remaining dimensions can be filled with zeros).

We demonstrate the effectiveness of the proposed approach on a multi-phase swing-up and brachiation task with a VSA while incorporating *continuous, online model learning*. In the multi-phase task, we consider the same task of swing-up and multi-swing locomotion presented in Sect. 2.5. Since the system has an asymmetric configuration and the state space of the swing-up task is significantly different from that of a brachiation movement, we proceed by first learning a separate error model for each phase. This procedure contains two steps. The initial exploration phase is performed in order to pre-train the LWPR model $\bar{\mathbf{f}}_i$ (as an alternative to applying motor babbling), while the second phase is using iLQG-LD to refine the model in an online fashion. In our evaluations, the training data are obtained by using a simulated version of the true dynamics, which is an analytical model incorporating the discrepancy.

3.2.1 Individual Phase Learning

As presented in Sect. 2.5, using our multi-phase spatio-temporal optimisation framework with the correct dynamics model, we successfully achieved a multi-phase brachiation task with a position error of as small as 0.002 m at the target bar in numerical simulations. However, once the discrepancy is introduced to the nominal model as described in Sect. 3.2, the planned solution is no longer valid and the final position deviates from the desired target in each individual swing-up and locomotion movement as illustrated in simulations (Fig. 8, blue line). We demonstrate the effectiveness of the iLQG-LD framework in order to learn the characteristics of the system dynamics and recover the task performance.

As a measure of the model accuracy, we use the *normalised mean squared error* (nMSE) of the model prediction on the true optimal trajectory (if given access to the analytical form of the true dynamics of the system). The evolution of the nMSE at the stage of training for each movement phase is shown in Fig. 9.

In the first part (pre-training phase in Fig. 9), we generate ($p = 7$) random targets around the desired x_T. A movement is planned for these targets using the nominal model $\tilde{\mathbf{f}}$. The obtained optimal feedback controller for the nominal model is then applied to the simulated version of the true dynamics. We repeat this procedure for a set of 10 noise added versions of the commands. The collected data are used to train the model.

This pre-training phase seeds the model with information within the region of interest prior to its use for planning. The aim is to reduce the number of iterations required for convergence in iLQG-LD. For each movement, at the end of the procedure,

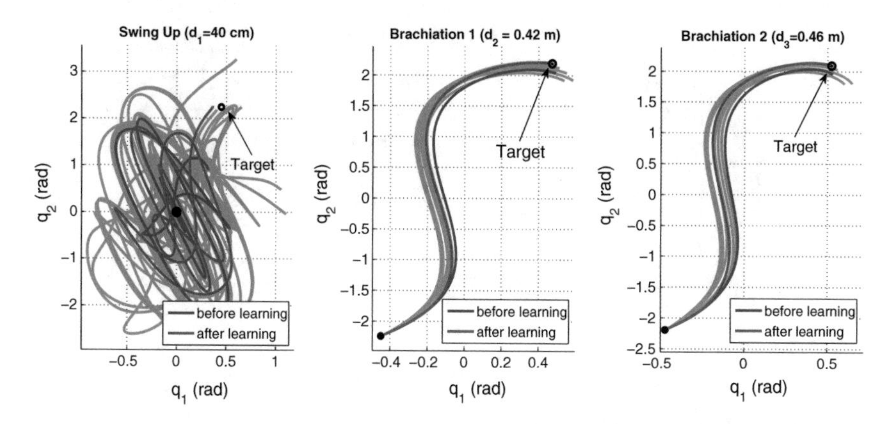

Fig. 8 Individual phase learning with model adaptation in simulation: Comparison of the final position achieved (for each individual phase) when using the initial planning (inaccurate nominal model in *blue*) and the final planning (composite model after learning in *red*). Intermediary solutions obtained at each step of the iLQG-LD run are depicted in *grey*

the planned trajectory matched the behaviour obtained from running the command solution on the real plant (the final nMSE has an order of magnitude of 10^{-4}).

Overall, the discrepancy is found to be small enough to reach the desired end-effector position within a tolerance of $\varepsilon_T = 0.040$ m. Figure 8 shows the effect of the learning by comparing the performance of the planning with the inaccurate nominal model and with the composite model obtained after training.

3.2.2 Multi-phase Performance

In order to evaluate the validity of the learned model, we optimise the multi-phase brachiation task with the composite cost function given in (21) using the obtained model from the individual phase learning procedure in Sect. 3.2.1. We use the optimal solutions obtained for each individual phase above as an initial command sequence for the multi-phase optimisation. The simulation result is shown in Fig. 10. The planner is able to successfully achieve the intermediate and final goals, while the expected behaviour provides a reliable match to the actual system's behaviour.[7] The cost of multi-phase optimised solution ($J = 35.13$) is significantly lower than the sum of the costs of the individual phase solutions ($J = 44.45$).

[7]We assume that if the position at the end of each phase is within a threshold $\varepsilon_T = 0.040$ m from the desired target, the system is able to start the next phase movement from the ideal location considering the effect of the gripper on the hardware.

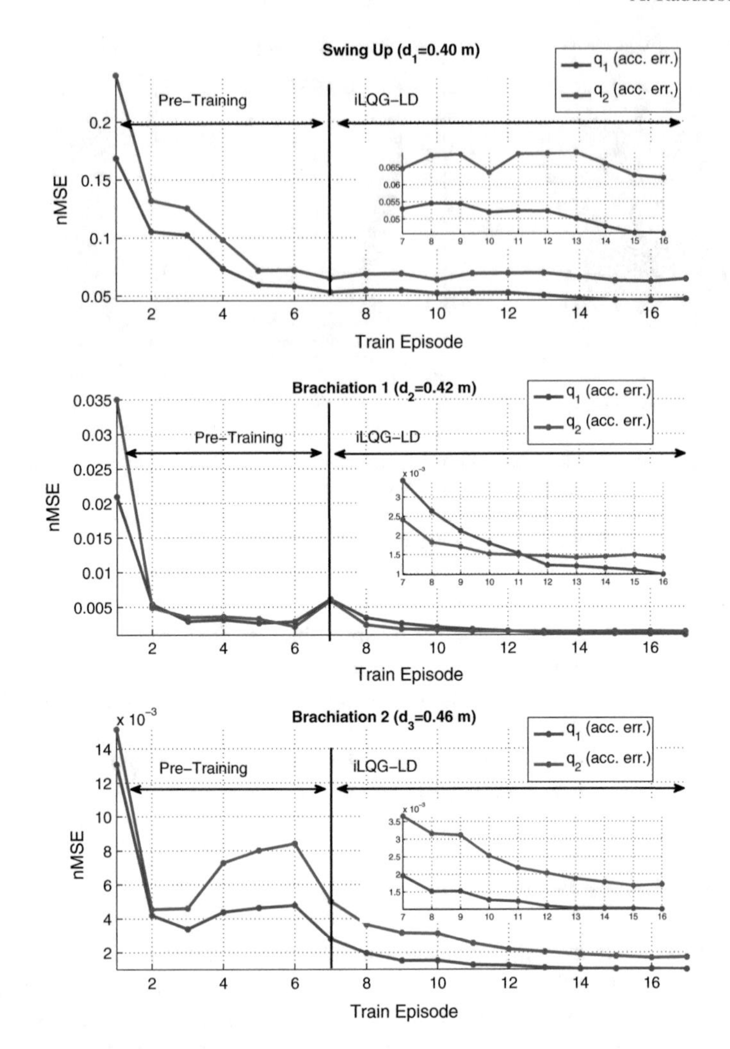

Fig. 9 Individual phase learning with model adaptation in simulation: Evolution of the nMSE for each phase of the movement at each episode

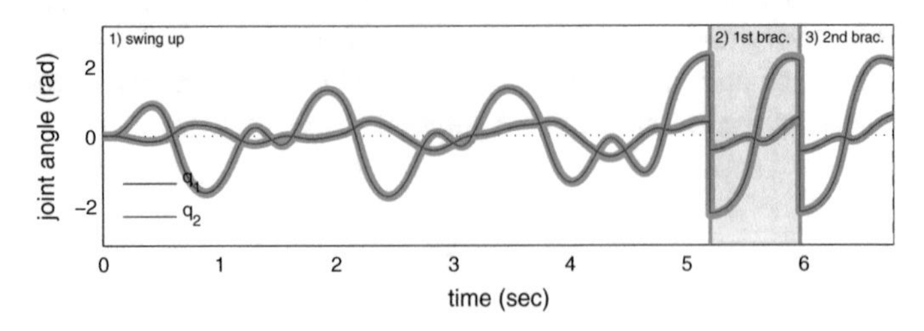

Fig. 10 Performance of the fully optimised multi-phase brachiation task using the composite learned model in simulation. *Thick grey lines*: planned movement. *Red* and *blue lines*: actual system movement

3.3 Hardware Experiments: Individual Phase Learning for a Brachiating Robot

In this section, we perform hardware experiments to evaluate the simulation results obtained in Sect. 3.2 on our two-link brachiating robot with a VSA. In the hardware experimental set-up, we only have access to sensory readings from two IMU units attached to each link, a potentiometer mounted on the main joint and the internal potentiometers of the servo motors in the VSA. The outputs of the IMU units are fairly accurate and adequate filtering provides reliable readings for estimating the positions and velocities of the robot links. However, the internal potentiometers of the servo motors suffer from significant amount of noise; with filtering, their readings can be used as an estimation of the motor position, but they are not reliable enough to derive the servo motor accelerations, which are needed for model learning. For this reason, we reduce the dimensionality of the input for the model approximation from 10 ($[\mathbf{q}^T, \dot{\mathbf{q}}^T, \mathbf{q}_m^T, \dot{\mathbf{q}}_m^T, \mathbf{u}^T]^T$) to 8 ($[\mathbf{q}^T, \dot{\mathbf{q}}^T, \mathbf{q}_m^T, \mathbf{u}^T]^T$).[8]

In the pre-training phase, we generate random targets around the desired x_T and plan movements for those targets using the nominal model $\tilde{\mathbf{f}}$ as in the numerical simulations described in Sect. 3.2.1 (with 150 g of mass discrepancy introduced). We apply the obtained solution to the hardware with a set of 10 noise added versions of the commands. The collected data are used to train the model. In the second phase, we apply iLQG-LD as described in Sect. 3.2. The evolution of the nMSE at the stage of training for each phase is shown in Fig. 11.

For each phase of the movement, at the end of the procedure, the planned trajectory matched the behaviour obtained from running the command solution on the real plant (the final nMSE has an order of magnitude of 10^{-4} and 10^{-2}, respectively).

In Fig. 12, we compare the performance of the system under the (i) solution obtained from the nominal (incorrect) analytical model (blue) and (ii) solution obtained after training the LWPR model (red). We can observe that the error in the position of the end-effector (i.e., open gripper) at the end of the allocated time improved significantly in both brachiation tasks from (i) 0.0867 m and 0.1605 m to (ii) 0.0161 m and 0.0233 m, respectively. The final positions are close enough to allow the gripper to compensate for the rest of the error by grasping it, thus resulting in the final error of 0.004 m. Note that the true positions of the gripper are actually at the target as it is securely locked on the target bar. The error comes from the variability of the sensor readings. The experimental conditions for the individual phase learning in hardware presented in this section correspond to the second and third phases of the movement from the multi-phase brachiation task considered in this chapter.

Besides the mass change, we perform an additional experiment in which we also modify the stiffness of the spring in the MACCEPA actuator from $\kappa_s = 771$ N/m to $\kappa_s = 650$ N/m in the nominal model. As before, at the end of the procedure,

[8] With the reduced input dimensionality, practically, there could be the case that it is not possible to predict the full state of the system particularly in the swing-up motion due to unobserved input dimensions. Thus, we only considered the swing locomotion task in the hardware experiment with model learning.

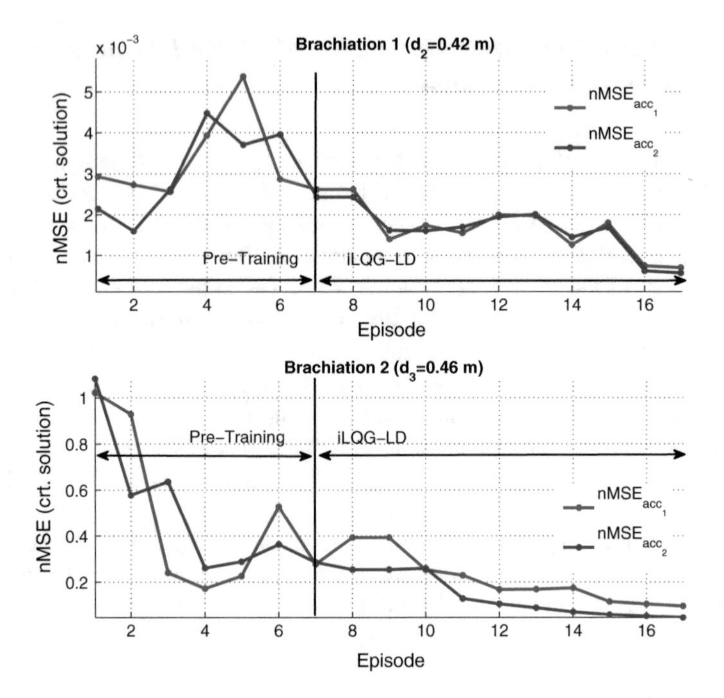

Fig. 11 Individual phase learning in hardware (mass discrepancy): Evolution of the nMSE for each phase of the movement at each episode

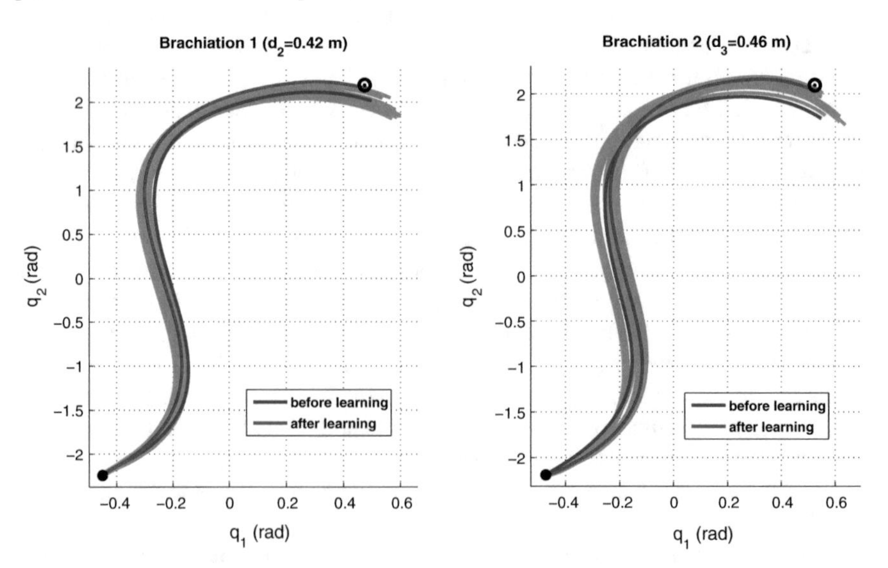

Fig. 12 Individual phase learning in hardware (mass discrepancy). Final position: initial planning (inaccurate nominal model in *blue*) and final planning (composite model after learning in *red*). Intermediary iLQG-LD solutions (*grey*)

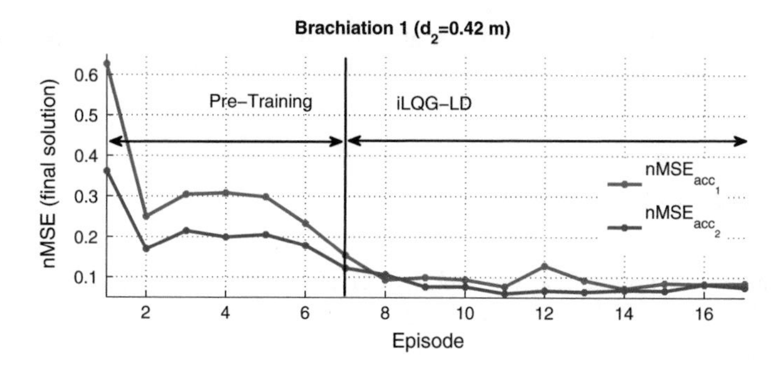

Fig. 13 Individual phase learning in hardware (mass and stiffness discrepancy): Evolution of the nMSE for each phase of the movement at each episode

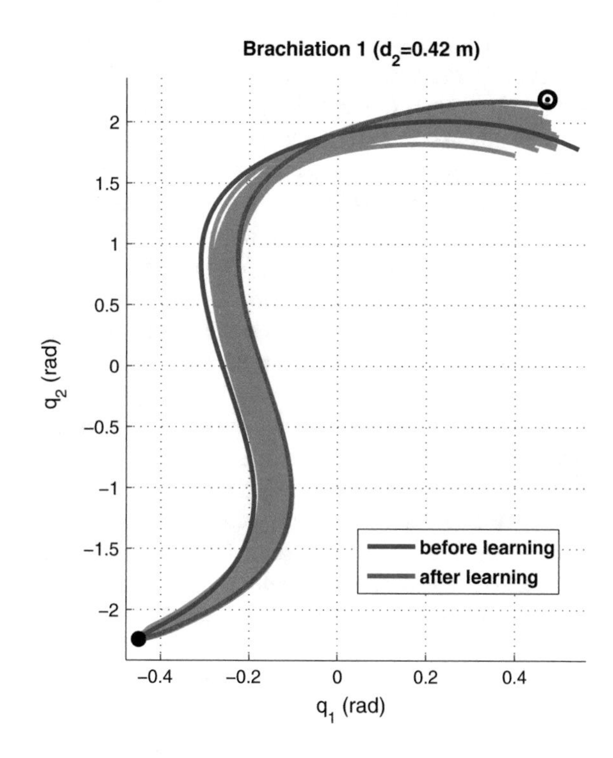

Fig. 14 Individual phase learning in hardware (mass and stiffness discrepancy): Final position achieved when using the initial planning (inaccurate nominal model in *blue*) and the final planning (composite model after learning in *red*). Intermediary solutions obtained at each step of the iLQG-LD run and exploration trajectories are depicted in *grey*

the planned trajectory matched the behaviour obtained from running the command solution on the real plant (the final nMSE $\approx 10^{-2}$, Fig. 13).

In Fig. 14, we observe the improvement in performance from an initial reaching error (at the end of the planned time) of (i) 0.1265 m (blue line) to (ii) 0.0167l m (red line). The robot was able to grab the bar located at the desired target with the optimised control command using the improved model. The experimental results demonstrate the feasibility of the developed adaptive learning framework for the application to real-world systems.

4 Summary

In this chapter, we addressed the optimal control problem of robotic systems with VSAs including switching dynamics and discontinuous state transitions.

First, we presented a systematic methodology for movement optimisation with multiple phases and switching dynamics in robotic systems with variable stiffness actuation with the focus on exploiting intrinsic dynamics of the system. Tasks including switching dynamics and interaction with an environment are approximately modelled as a hybrid dynamical system with time-based switching. We have demonstrated the benefit of simultaneous temporal and variable stiffness optimisation leading to reduction in control effort and improved performance. With an appropriate choice of the composite cost function to encode the task, we have demonstrated the effectiveness of the proposed approach in various brachiation tasks in numerical simulations and hardware implementation in a brachiating robot with VSA. In [21], we have presented additional numerical evaluations of the proposed approach on the control of a hopping robot model with a VSA having different mode of dynamics (flight and stance) and impact with the environment. Simulation results on the hopping robot control in [21] illustrated the feasibility of our approach and the robustness of the obtained optimal feedback controller against external perturbations.

Next, we extended our approach by incorporating adaptive learning, which allows for adjustments to the dynamics model, based on changes occurred to the system's behaviour, or when the behaviour cannot be fully represented by a rigid body dynamics formulation. The method employed (in the form of the LWPR algorithm) is particularly suited for certain regression situations such as non-linear function learning with the requirement of incremental learning. We demonstrated that the augmented developed methodology was successfully applied in the case of underactuated systems such as a brachiating robot. We provided results for a range of model discrepancies in both numerical simulations and real hardware experiments.

In our previous work, we have addressed movement optimisation of variable impedance actuators including damping [28], and our framework presented in this chapter can be generalised to deal with such systems in a straightforward manner. Our future interest is in the application of our approach to a broad range of complex physical and robotic systems having interactions with environments.

References

1. C.G. Atkeson, A.W. Moore, S. Schaal, Locally weighted learning for control. Artif. Intell. Rev. **11**(1–5), 75–113 (1997)
2. G. Bätz, U. Mettin, A. Schmidts, M. Scheint, D. Wollherr, A. S. Shiriaev, Ball dribbling with an underactuated continuous-time control phase: theory and experiments, in *IEEE/RSJ International Conference on Intelligent Robots and Systems* (2010), pp. 2890–2895
3. D. Braun, M. Howard, S. Vijayakumar, Optimal variable stiffness control: formulation and application to explosive movement tasks. Auton. Robot. **33**(3), 237–253 (2012)
4. D.J. Braun, F. Petit, F. Huber, S. Haddadin, P. van der Smagt, A. Albu-Schäffer, S. Vijayakumar, Robots driven by compliant actuators: optimal control under actuation constraints. IEEE Trans. Robot. **29**(5), 1085–1101 (2013)
5. A.E. Bryson, Y.-C. Ho, *Applied Optimal Control* (Taylor and Francis, United Kingdom, 1975)
6. M. Buehler, D.E. Koditschek, P.J. Kindlmann, Planning and control of robotic juggling and catching tasks. Int. J. Robot. Res. **13**(2), 101–118 (1994)
7. M. Buss, M. Glocker, M. Hardt, O. von Stryk, R. Bulirsch, G. Schmidt, Nonlinear hybrid dynamical systems: modeling, optimal control, and applications, in *Lecture Notes in Control and Information Science* (Springer, Heidelberg, 2002), pp. 311–335
8. T.M. Caldwell, T.D. Murphey, Switching mode generation and optimal estimation with application to skid-steering. Automatica **47**(1), 50–64 (2011)
9. M.G. Catalano, G. Grioli, M. Garabini, F. Bonomo, M. Mancini, N. Tsagarakis, A. Bicchi. VSA-CubeBot: A modular variable stiffness platform for multiple degrees of freedom robots, in *IEEE International Conference on Robotics and Automation* (2011), pp. 5090–5095
10. M. Gomes, A. Ruina, A five-link 2D brachiating ape model with life-like zero-energy-cost motions. J. Theor. Biol. **237**(3), 265–278 (2005)
11. K. Goris, J. Saldien, B. Vanderborght, D. Lefeber, Mechanical design of the huggable robot probo. Int. J. Humanoid Robot. **8**(3), 481–511 (2011)
12. S. S. Groothuis, G. Rusticelli, A. Zucchelli, S. Stramigioli, R. Carloni, The vsaUT-II: A novel rotational variable stiffness actuator, in *IEEE International Conference on Robotics and Automation* (2012), pp. 3355–3360
13. W. Li, E. Todorov, Iterative linearization methods for approximately optimal control and estimation of non-linear stochastic system. Int. J. Control **80**(9), 1439–1453 (2007)
14. A.W. Long, T.D. Murphey, K.M. Lynch, Optimal motion planning for a class of hybrid dynamical systems with impacts, in *IEEE International Conference on Robotics and Automation* (2011), pp. 4220–4226
15. D. Mitrovic, S. Klanke, M. Howard, S. Vijayakumar, Exploiting sensorimotor stochasticity for learning control of variable impedance actuators, in *IEEE-RAS International Conference on Humanoid Robots* (2010), pp. 536–541
16. D. Mitrovic, S. Klanke, S. Vijayakumar, Optimal control with adaptive internal dynamics models, in *Fifth International Conference on Informatics in Control, Automation and Robotics* (2008)
17. D. Mitrovic, S. Klanke, S. Vijayakumar, Adaptive optimal feedback control with learned internal dynamics models, in *From Motor Learning to Interaction Learning in Robots* (2010), pp. 65–84
18. K. Mombaur, Using optimization to create self-stable human-like running. Robotica **27**(3):321330 (2009)
19. J. Nakanishi, J.A. Farrell, S. Schaal, Composite adaptive control with locally weighted statistical learning. Neural Netw. **18**(1), 71–90 (2005)
20. J. Nakanishi, T. Fukuda, D. Koditschek, A brachiating robot controller. IEEE Trans. Robot. Autom. **16**(2), 109–123 (2000)
21. J. Nakanishi, A. Radulescu, D. J. Braun, S. Vijayakumar, Spatio-temporal stiffness optimization with switching dynamics. Auton. Robot. 1–19 (2016)

22. J. Nakanishi, K. Rawlik, S. Vijayakumar, Stiffness and temporal optimization in periodic movements: an optimal control approach, in *IEEE/RSJ International Conference on Intelligent Robots and Systems* (2011), pp. 718–724
23. D. Nguyen-Tuong, J. Peters, Model learning for robot control: a survey. Cogn. Porocess. **12**(4), 319–340 (2011)
24. F. Petit, M. Chalon, W. Friedl, M. Grebenstein, A. Albu-Schäffer, G. Hirzinger, Bidirectional antagonistic variable stiffness actuation: analysis, design and implementation, in *IEEE International Conference on Robotics and Automation* (2010), pp. 4189–4196
25. B. Piccoli, Hybrid systems and optimal control, in *IEEE Conference on Decision and Control* (1998), pp. 13–18
26. M. Posa, C. Cantu, R. Tedrake, A direct method for trajectory optimization of rigid bodies through contact. Int. J. Robot. Res. **33**(1), 69–81 (2014)
27. M. Posa, S. Kuindersma, R. Tedrake, Optimization and stabilization of trajectories for constrained dynamical systems, in *IEEE International Conference on Robotics and Automation* (2016), pp. 1366–1373
28. A. Radulescu, M. Howard, D. J. Braun, S. Vijayakumar, Exploiting variable physical damping in rapid movement tasks, in *IEEE/ASME International Conference on Advanced Intelligent Mechatronics* (2012), pp. 141–148
29. A. Radulescu, J. Nakanishi, S. Vijayakumar, Optimal control of multi-phase movements with learned dynamics, in *Man–Machine Interactions 4* (Springer, Heidelberg, 2016), pp. 61–76
30. K. Rawlik, M. Toussaint, S. Vijayakumar, An approximate inference approach to temporal optimization in optimal control, in *Advances in Neural Information Processing Systems*, vol. 23 (MIT Press, Cambridge, 2010), pp. 2011–2019
31. N. Rosa Jr., A. Barber, R.D. Gregg, K.M. Lynch, Stable open-loop brachiation on a vertical wall, in *IEEE International Conference on Robotics and Automation* (2012), pp. 1193–1199
32. F. Saito, T. Fukuda, F. Arai, Swing and locomotion control for a two-link brachiation robot. IEEE Control Syst. Mag. **14**(1), 5–12 (1994)
33. S. Schaal, C.G. Atkeson, Constructive incremental learning from only local information. Neural Comput. **10**(8), 2047–2084 (1998)
34. M.S. Shaikh, P.E. Caines, On the hybrid optimal control problem: theory and algorithms. IEEE Trans. Autom. Control **52**(9), 1587–1603 (2007)
35. B. Siciliano, O. Khatib, *Springer Handbook of Robotics* (Springer, Heidelberg, 2008)
36. O. Sigaud, C. Salaün, V. Padois, On-line regression algorithms for learning mechanical models of robots: a survey. Robot. Auton. Syst. **59**, 1115–1129 (2011)
37. Y. Tassa, T. Erez, E. Todorov, Synthesis and stabilization of complex behaviors through online trajectory optimization, in *IEEE/RSJ International Conference on Intelligent Robots and Systems* (2012), pp. 2144–2151
38. M. Van Damme, B. Vanderborght, B. Verrelst, R. Van Ham, F. Daerden, D. Lefeber, Proxy-based sliding mode control of a planar pneumatic manipulator. Int. J. Robot. Res. **28**(2), 266–284 (2009)
39. R. Van Ham, B. Vanderborght, M. Van Damme, B. Verrelst, D. Lefeber, MACCEPA, the mechanically adjustable compliance and controllable equilibrium position actuator: design and implementation in a biped robot. Robot. Auton. Syst. **55**(10), 761–768 (2007)
40. B. Vanderborght, B. Verrelst, R. Van Ham, M. Van Damme, D. Lefeber, B.M.Y. Duran, P. Beyl, Exploiting natural dynamics to reduce energy consumption by controlling the compliance of soft actuators. Int. J. Robot. Res. **25**(4), 343–358 (2006)
41. S. Vijayakumar, S. Schaal, Locally weighted projection regression: An o (n) algorithm for incremental real time learning in high dimensional space, in *International Conference on Machine Learning, Proceedings of the Sixteenth Conference* (2000)
42. L.C. Visser, R. Carloni, S. Stramigioli, Energy-efficient variable stiffness actuators. IEEE Trans. Robot. **27**(5), 865–875 (2011)
43. W. Xi, C.D. Remy, Optimal gaits and motions for legged robots, in *IEEE/RSJ International Conference on Intelligent Robots and Systems* (2014), pp. 3259–3265

44. X. Xu, P.J. Antsaklis, Quadratic optimal control problems for hybrid linear autonomous systems with state jumps, in *American Control Conference* (2003), pp. 3393–3398
45. X. Xu, P.J. Antsaklis, Results and perspectives on computational methods for optimal control of switched systems, in *International Workshop on Hybrid Systems: Computation and Control* (Springer, Heidelberg, 2003), pp. 540–555
46. X. Xu, P.J. Antsaklis, Optimal control of switched systems based on parameterization of the switching instants. IEEE Trans. Autom. Control **49**(1), 2–16 (2004)
47. C. Yang, G. Ganesh, S. Haddadin, S. Parusel, A. Albu-Schäeffer, E. Burdet, Human-like adaptation of force and impedance in stable and unstable interactions. IEEE Trans. Robot. **27**(5), 918–930 (2011)

Printed in the United States
By Bookmasters